Advances in Geometric Programming

MATHEMATICAL CONCEPTS AND METHODS IN SCIENCE AND ENGINEERING

Series Editor: Angelo Miele
Mechanical Engineering and Mathematical Sciences, Rice University

1 **INTRODUCTION TO VECTORS AND TENSORS** • Volume 1: Linear and Multilinear Algebra • *Ray M. Bowen and C.-C. Wang*

2 **INTRODUCTION TO VECTORS AND TENSORS** • Volume 2: Vector and Tensor Analysis • *Ray M. Bowen and C.-C. Wang*

3 **MULTICRITERIA DECISION MAKING AND DIFFERENTIAL GAMES** • *Edited by George Leitmann*

4 **ANALYTICAL DYNAMICS OF DISCRETE SYSTEMS** • *Reinhardt M. Rosenberg*

5 **TOPOLOGY AND MAPS** • *Taqdir Husain*

6 **REAL AND FUNCTIONAL ANALYSIS** • *A. Mukherjea and K. Pothoven*

7 **PRINCIPLES OF OPTIMAL CONTROL THEORY** • *R. V. Gamkrelidze*

8 **INTRODUCTION TO THE LAPLACE TRANSFORM** • *Peter K. F. Kuhfittig*

9 **MATHEMATICAL LOGIC:** An Introduction to Model Theory • *A. H. Lightstone*

11 **INTEGRAL TRANSFORMS IN SCIENCE AND ENGINEERING** • *Kurt Bernardo Wolf*

12 **APPLIED MATHEMATICS:** An Intellectual Orientation • *Francis J. Murray*

14 **PRINCIPLES AND PROCEDURES OF NUMERICAL ANALYSIS** • *Ferenc Szidarovszky and Sidney Yakowitz*

16 **MATHEMATICAL PRINCIPLES IN MECHANICS AND ELECTROMAGNETISM,** Part A: Analytical and Continuum Mechanics • *C.-C. Wang*

17 **MATHEMATICAL PRINCIPLES IN MECHANICS AND ELECTROMAGNETISM,** Part B: Electromagnetism and Gravitation • *C.-C. Wang*

18 **SOLUTION METHODS FOR INTEGRAL EQUATIONS:** Theory and Applications • *Edited by Michael A. Golberg*

19 **DYNAMIC OPTIMIZATION AND MATHEMATICAL ECONOMICS** • *Edited by Pan-Tai Liu*

20 **DYNAMICAL SYSTEMS AND EVOLUTION EQUATIONS:** Theory and Applications • *J. A. Walker*

21 **ADVANCES IN GEOMETRIC PROGRAMMING** • *Edited by Mordecai Avriel*

Advances in Geometric Programming

Edited by
Mordecai Avriel
Technion-Israel Institute of Technology
Haifa, Israel

PLENUM PRESS · NEW YORK AND LONDON

Library of Congress Cataloging in Publication Data

Main entry under title:

Advances in geometric programming.

(Mathematical concepts and methods in science and engineering; v. 21)
Includes index.
1. Geometric programming. I. Avriel, M., 1933-
T57.825.A33 519.7′6 79-20806
ISBN-13: 978-1-4615-8287-8 e-ISBN-13: 978-1-4615-8285-4
DOI: 10.1007/978-1-4615-8285-4

Acknowledgments

Chapter 1 is reprinted from *Lectures in Applied Mathematics,* Volume 11, *"Mathematics of the decision sciences,"* Part I, Pages 401–422, by permission of the American Mathematical Society, 1968.

Chapter 2 is reprinted with permission from *SIAM Review,* Vol. 18. Copyright 1976, Society for Industrial and Applied Mathematics. All rights reserved

Chapters 3–9 and 17 are reprinted from *Journal of Optimization Theory and Applications,* Vol. 26, No. 1, September 1978

Chapter 10 is reprinted from *International Journal for Numerical Methods in Engineering,* Vol. 9, 149–168 (1975)

Chapters 11–16, 18, 20–22 are reprinted from *Journal of Optimization Theory and Applications,* Vol. 26, No. 2, October 1978

Chapter 19 is reprinted from *Journal of Optimization Theory and Applications,* Vol. 28, No. 1, May 1979

Softcover reprint of the hardcover 1st edition 1980

A Division of Plenum Publishing Corporation
227 West 17th Street, New York, N.Y. 10011

Contributors

R. A. Abrams, Graduate School of Business, University of Chicago, Chicago, Illinois

M. Avriel, Faculty of Industrial Engineering and Management, Technion—Israel Institute of Technology, Haifa, Israel

J. D. Barrett, Western Forest Products Laboratory, Vancouver, British Columbia, Canada

R. S. Dembo, School of Organization and Management, Yale University, New Haven, Connecticut

J. G. Ecker, Department of Mathematical Sciences, Rensselaer Polytechnic Institute, Troy, New York

W. Gochet, Department of Applied Economics, Katholieke Universiteit Leuven; and Center for Operations Research and Econometrics, Catholic University of Louvain, Louvain, Belgium

M. Hamala, Comenius University, Matematicky Pavilon PFUK, Mlynska dolina, Bratislava, Czechoslovakia

A. Jain, Stanford Research Institute, Menlo Park, California

T. R. Jefferson, School of Mechanical and Industrial Engineering, University of New South Wales, Kensington, New South Wales, Australia

L. S. Lasdon, Departments of General Business and Mechanical Engineering, University of Texas, Austin, Texas

G. Lidor, Department of Computer Sciences, The City College, New York, New York

L. J. Mancini, Computer Services Department, Standard Oil Company, San Francisco, California

X. M. Martens, Instituut voor Chemie–Ingenieurstechniek, Katholieke Universiteit Leuven, Leuven, Belgium

U. Passy, Faculty of Industrial Engineering and Management, Technion—Israel Institute of Technology, Haifa, Israel

E. L. Peterson, Department of Mathematics and Graduate Program in Operations Research, North Carolina State University, Raleigh, North Carolina

M. Ratner, Computer Science Department, Case Western Reserve University, Cleveland, Ohio

G. V. Reklaitis, School of Chemical Engineering, Purdue University, West Lafayette, Indiana

M. J. Rijckaert, Instituut voor Chemie–Ingenieurstechniek, Katholieke Universiteit Leuven, Leuven, Belgium

P. V. L. N. Sarma, Selas Corporation of America, Dresher, Pennsylvania

C. H. Scott, School of Mechanical and Industrial Engineering, University of New South Wales, Kensington, New South Wales, Australia

Y. Smeers, Department of Engineering, Université Catholique de Louvain; and Center for Operations Research and Econometrics, Catholic University of Louvain, Louvain, Belgium

R. D. Wiebking, Unternehmensberatung Schumann, GmbH, Cologne, West Germany

D. J. Wilde, Department of Mechanical Engineering, Stanford University, Stanford, California

C. T. Wu, School of Business Administration, University of Wisconsin—Milwaukee, Milwaukee, Wisconsin

Preface

In 1961, C. Zener, then Director of Science at Westinghouse Corporation, and a member of the U.S. National Academy of Sciences who has made important contributions to physics and engineering, published a short article in the *Proceedings of the National Academy of Sciences* entitled "A Mathematical Aid in Optimizing Engineering Design." In this article Zener considered the problem of finding an optimal engineering design that can often be expressed as the problem of minimizing a numerical cost function, termed a "generalized polynomial," consisting of a sum of terms, where each term is a product of a positive constant and the design variables, raised to arbitrary powers. He observed that if the number of terms exceeds the number of variables by one, the optimal values of the design variables can be easily found by solving a set of linear equations. Furthermore, certain invariances of the relative contribution of each term to the total cost can be deduced.

The mathematical intricacies in Zener's method soon raised the curiosity of R. J. Duffin, the distinguished mathematician from Carnegie-Mellon University who joined forces with Zener in laying the rigorous mathematical foundations of optimizing generalized polynomials. Interestingly, the investigation of optimality conditions and properties of the optimal solutions in such problems were carried out by Duffin and Zener with the aid of inequalities, rather than the more common approach of the Kuhn–Tucker theory. One of the inequalities that they found useful in studying the optimality properties of generalized polynomials is the classical inequality between the arithmetic and geometric means. Because of this inequality, and some more general ones, called "geometric inequalities," Duffin coined the term "geometric programming" for the problem of optimizing generalized polynomials. In 1963, E. L. Peterson, a student of Duffin, started to work on developing the theory of constrained geometric programming problems. Thus a new branch of optimization was born.

The significance of the theory developed by Duffin, Peterson, and Zener was recognized very early by D. J. Wilde, then Professor of Chemical Engineering at Stanford University, who was equally interested in optimizing engineering design and in the theoretical aspects of optimization. He was fascinated by the simplicity and the potential usefulness of geometric programming and urged his two doctoral students M. Avriel and U. Passy to devote parts of their dissertations to geometric programming. These were completed in 1966 and contained topics that served as a kernel from which many future developments have sprouted. In the book *Geometric Programming*, published in 1967, Duffin, Peterson, and Zener collected their pioneering work. Their book was instrumental in inspiring continued research on theory, computational aspects, and applications.

Since the mid-60's, geometric programming has gradually developed into an important branch of nonlinear optimization. The developments include first of all significant extensions of the type of problems that were considered 10 years ago as geometric programs. Also, two-sided relationships with convex, generalized convex, and nonconvex programming, separable programming, conjugate functions, and Lagrangian duality were established. Numerical solution methods and their convergence properties were studied. Subsequently, computer software that can handle large constrained problems were developed. Applications of geometric programming to more and more problems of engineering optimization and design and problems from many other diverse areas were demonstrated.

In recognition of the important role of geometric programming in optimization, the *Journal of Optimization Theory and Applications* devoted two special issues to this subject. The majority of works appearing in this volume were first published there. In order to make this book as self-contained as possible, three earlier works on geometric programming are also reprinted here. These include (i) the 1968 paper of M. Hamala relating geometric programming duality to conjugate duality, which was originally published as a research report; (ii) the comprehensive survey paper of E. L. Peterson that appeared in 1976 in *SIAM Review*, and on which his other articles in this book are based; and (iii) the 1975 paper of M. Avriel, R. Dembo, and U. Passy that appeared in the *International Journal of Numerical Methods in Engineering*, describing the GGP algorithm that serves as a prototype of several condensation and linearization-based methods. The permission granted to reprint these works is gratefully acknowledged. (For a more detailed description of the articles in this volume, see the Introduction.)

I wish to express my appreciation to Professor Angelo Miele, Editor-in-Chief of the *Journal of Optimization Theory and Applications*, and Series Editor of the *Mathematical Concepts and Methods in Science and Engineering* texts and monographs, for inviting me to edit the two special issues of *JOTA* and also this book. Thanks are also due to Mrs. Batya Maayan for assisting me in these projects.

Haifa Mordecai Avriel

Contents

Introduction

M. Avriel

Geometric programming is probably still considered by many applied mathematicians and operations researchers as a technique for optimizing posynomials (generalized polynomials). The objective of this book is to bring to the attention of interested readers some of the advances made in recent years in geometric programming and related fields which reflect the greatly widened scope of this branch of nonlinear optimization. The advances are in three major categories: analysis, computations, and applications. The papers appearing in the book can be also classified accordingly.

The first two papers are introductory and set the stage for the analysis part of the book. The paper by Hamala presents the reader with the basic notions of convex functions, convex programming, and conjugate duality. Next, it introduces a class of primal–dual program pairs whose optimality properties are derived, and it is shown that primal–dual geometric programs are a special case of this class. Whereas the original development of duality in geometric programming was obtained with the aid of inequalities, Hamala derives analogous results by conjugate duality.

The second introductory paper is an extensive survey by Peterson in which he defines a quite general class of problems termed generalized geometric programs. For such programs he derives a comprehensive theory of optimality and duality, related to the convex analysis of Fenchel and Rockafellar.

An important result is that several well-known classes of optimization problems, possessing certain separability properties, can be reformulated as generalized geometric programs. These are amenable then to a unified treatment by optimality analysis. The next four papers, all written by Peterson, elaborate on the topics introduced in his survey and provide proofs of the results.

The theory of generalized geometric programming is carried further to infinite dimensional spaces in the paper by Jefferson and Scott. They consider convex optimal control problems, derive the appropriate dual problem, and discuss the optimality properties of the primal–dual pair.

In the next paper, Abrams and Wu investigate primal–dual pairs of generalized geometric programs in which one program, say the dual, does not have feasible interior points. In this case the primal problem has an unattained infimum or an unbounded optimal solution set. It is shown that after finitely many restrictions of the dual problem to an affine set and projections of the primal problem onto a subspace, a pair of problems results that has feasible interior points, bounded optimal solution sets, and the same value of the infima as the original pair of programs.

Lidor and Wilde study in the next paper extensions of ordinary (prototype) geometric programs in which some of the primal variables may appear also as exponents or in logarithms. These are called transcendental programs and they resemble in many ways ordinary geometric programs, although, due to lack of convexity, they can have local minima that are nonglobal. Corresponding dual programs are derived that are not "pure" duals, in the sense that they contain also primal variables. Avriel and Passy showed in their doctoral work the mathematical identity between the so-called "chemical equilibrium" problem of reacting species in an ideal system and the dual program of an ordinary geometric program. Lidor and Wilde demonstrate in this paper a relationship between dual transcendental programs and the chemical equilibrium problem in nonideal systems. This paper also concludes the analytic part of this book.

The numerical solution of geometric programs is a subject of great interest, and many specialized algorithms have been developed for this purpose. One of the basic questions concerns solving the primal or the dual program. In case of ordinary geometric programs the dual has a concave objective function and linear constraints, whereas in the primal program the constraints are usually nonlinear. It seems, therefore, logical to solve ordinary geometric programs via the dual. In practice, however, this approach is not always advisable because, if some dual variables must vanish in the optimal solution, the gradient of the objective function becomes unbounded and causes considerable numerical difficulties. Also, since programs more general than ordinary geometric programs either do not have a dual or the dual does not offer apparent computational advantages, several primal-based algorithms have been derived. Most of these use condensation, a technique for combining a sum of positive terms into a single term that enables a nonconvex program to be approximated by a convex one. Further condensation can result in the approximation of a convex program by a linear program. A representative member of such a class of methods is the GGP algorithm described in the paper by Avriel, Dembo, and Passy, which opens the computational part of the book. The main purpose of including this paper in the book is to make this volume more self-contained, as this paper describes many details mentioned in the other computationally oriented works in this book.

The next three papers mainly deal with comparisons of various algorithms for the solution of geometric programming problems. The first paper by Dembo surveys primal-based and dual-based methods for ordinary (prototype) and generalized geometric programs. Special codes for geometric programs are compared among themselves and also with general-purpose algorithms by solving a set of test problems.

Sarma, Martens, Reklaitis, and Rijckaert continue the comparisons for ordinary geometric programs by testing various primal-based and dual-based algorithms. Their main conclusion is that dual-based methods do not offer any significant computational advantages, except in special cases.

Rijckaert and Martens performed a more extensive comparison by using 17 algorithms on 24 test problems. They also report that primal-based methods are generally superior to dual-based ones and that methods using condensation are the fastest and most robust.

The development of procedures for the practical evaluation and comparison of numerical methods and computer software is one of the most interesting problems of optimization awaiting a satisfactory solution. To illustrate this point, in the paper of Rijckaert and Martens it is shown that computer codes specially written for geometric programs clearly perform better than a general-purpose code included in their study. On the other hand, Ratner, Lasdon, and Jain, who developed an excellent and very efficiently programmed computer code that implements the generalized reduced gradient (GRG) method, report that their general-purpose nonlinear programming code can perform just as well as the special-purpose codes when applied to geometric programs. Another interesting observation is that the hitherto generally accepted "standard time" defined by Colville is an inadequate means of compensating for different computing environments. Clearly, much more research is needed in this area.

Although, judging from the above comparisons, the role of duality from a computational standpoint may have been overemphasized, there are several computational aspects of duality that deserve attention. Dembo's paper focuses on the interpretation of Lagrange multipliers that correspond to the constraints of a dual geometric program. He also analyzes the question of subsidiary problems that are needed when the exponents of the primal variables are linearly dependent.

The next two papers offer conceptual algorithms with limited computational experience. In the first paper, Ecker, Gochet, and Smeers present a modified reduced gradient algorithm for solving the dual problem of ordinary geometric programming. The difficulties encountered by other methods that attempt to solve dual geometric programs, such as the Beck and Ecker convex simplex method, are taken into consideration in the development of the new algorithm.

Methods discussed in the previous papers can in the best case find only local minima of nonconvex generalized geometric programs. Such local minima need not be global. Passy proposes here an implicit enumeration method for finding global solutions of nonconvex generalized geometric programs. His method can also be applied to a larger class of nonconvex programs.

The next two papers by Mancini and Wilde explore and demonstrate the possibilities of using interval arithmetic (an extension of ordinary arithmetic in which the basic elements are closed intervals) in geometric programming. The first paper applies interval arithmetic to geometric programs in which either the number of terms appearing in the primal problem exceeds the number of variables by two (one degree of difficulty) or the primal problem is unidimensional. The solution is based on an interval-arithmetic version of Newton's method. The second paper applies interval arithmetic to more general dual geometric programs in order to verify the existence and uniqueness of solutions and to compute error bounds on them.

Next there are two papers in this book that deal with engineering design applications of geometric programming. The usefulness of geometric programming is again demonstrated in this type of application. First, Avriel and Barrett consider a structural design problem of optimizing the geometry of certain wood beams. Second, Ecker and Wiebking formulate and solve the problem of optimally designing a cooling tower used in dissipating the waste heat of steam electrical plants. An interesting aspect of both engineering design papers is that existing theory and computational methods of geometric programming require all constraints to be inequalities. If the design problem has equality constraints, they must be expressed as inequalities in the geometric programming formulation—often an unnecessarily complicating feature. This is yet another unexplored area of geometric programming.

The book is concluded with a classified bibliography of publications in geometric programming, compiled by Rijckaert and Martens.

1

Geometric Programming in Terms of Conjugate Functions[1]

M. HAMALA[2]

Abstract. Our purpose is to show that the duality of geometric programming is a special case of Rockafellar's general theory of duality, and to construct a class of dual programs, which can be considered as a generalization of the usual geometric duality.

1. Introduction

The main purpose of this paper is (i) to show that the geometric duality of Duffin, Peterson, and Zener (Ref. 1) is a special case of Rockafellar's general theory of duality (Ref. 2), and (ii) to construct a general class of dual programs, which can be considered as a generalization of the geometric duality (Ref. 1), and which enable analogous procedures such as geometric programming to be developed.

The generalization proposed here is different from the one given in Ref. 1: the generalized geometric dual program in Ref. 1 is derived from geometric inequalities and in general is not convex. (Provided that the appropriate geometric inequalities are available, the nonconvexity is the main disadvantage of that approach.) The generalized dual program proposed here is derived from the properties of conjugate functions and it is convex. The only disadvantage of this approach is that the dual program has more variables than the primal one. But in many cases the number of variables can be reduced, e.g., in the case of classical geometric programs or linear programs.

It is worth noting that if G^* is the conjugate function of a differentiable convex function of n variables, and $\lambda > 0$, then

$$\sum_{i=1}^{n} x_i y_i \leqq \lambda G(x) + \lambda G^*(y/\lambda),$$

[1] Reprinted from *Lectures in Applied Mathematics,* Volume 11, *Mathematics of the Decision Sciences,* Part I, Pages 401–422 by permission of the American Mathematical Society, 1968.

[2] Author's address: Comenius University, Matematicky pavilon PF UK, Mlynska dolina, 816 31 Bratislava, Czechoslovakia.

and if $\lambda(y) \geq 0$ is a homogeneous function such that $\lambda(\nabla G(x)) = 1$, then

$$\sum x_i y_i \leqq \lambda(y) G(x) + \lambda(y) G^*(y/\lambda(y)) = \lambda(y) G(x) - F(y)$$

is a geometric inequality in the sense defined in Ref. 1.

Now we see that, to apply the approach given in Ref. 1, we need the explicit form of the function λ. In our approach λ is considered simply as an additional new variable.

For the sake of an easier exposition, it is convenient to restate briefly the main ideas of Rockafellar's theory given in Ref. 2. In the first part the necessary prerequisites are developed and in the second part Rockafellar's results are presented. In the third part the special case of Rockafellar's dual programs—the general geometric dual programs—is studied and two theorems are given analogous to those in Ref. 1.

Finally some concrete examples are discussed.

2. Basic Concepts

2.1. Convex Functions

Definition 2.1. The set $C \subset R^n$ is said to be convex if

$$\underset{x_1, x_2 \in C}{\forall} \; \underset{0 \leqq \lambda \leqq 1}{\forall} : \lambda x_1 + (1-\lambda) x_2 \in C.$$

Proposition 2.1. A nonempty convex set $C \subset R^n$ has a nonempty relative interior r int C (see Ref. 3, p. 16).

Definition 2.2. A real-valued function f defined on a nonempty convex set $C \subset R^n$ is said to be *convex on C* if

$$\underset{x_1, x_2 \in C}{\forall} \; \underset{0 \leq \lambda \leq 1}{\forall} : f(\lambda x_1 + (1-\lambda) x_2) \leqq \lambda f(x_1) + (1-\lambda) f(x_2).$$

Proposition 2.2. If f is convex on C and $x_1 \in r$ int C, then

$$\underset{x^* \in R^n}{\exists} \; \underset{x \in C}{\forall} : f(x) - f(x_1) \geqq (x - x_1) x^*.$$

[This result follows from the existence of a nonvertical supporting hyperplane (see Ref. 4, p. 398) and the Hahn–Banach theorem (see Ref. 5, p. 28).]

Definition 2.3. Let f be a function from R^n into $[-\infty, -\infty]$ [i.e., $-\infty \leqq f(x) \leqq +\infty$]. The *epigraph* of f is the subset of R^{n+1} defined by

$$\text{epi } f = \{(x, \mu) \mid f(x) \leqq \mu, x \in R^n, \mu \in R\}.$$

Definition 2.4. A function f from R^n into $[-\infty, +\infty]$ is said to be *convex on* R^n if epi f is a convex set in R^{n+1}.

Proposition 2.3. If $f(x) > -\infty$ [or $f(x) < +\infty$] $\forall x \in R^n$, then Definition 2.4 is equivalent to Definition 2.2. If f is convex on the convex set $C \subset R^n$ and if we define $f(x) = +\infty$ $\forall x \notin C$, then f becomes convex on R^n. This fact allows us to consider all convex functions on C as convex functions on R^n.

Definition 2.5. The effective domain of a convex function f on R^n is a set dom $f = \{x \mid f(x) < +\infty\}$.

Proposition 2.4. Dom f is a convex set in R^n.

Definition 2.6. A convex function f is said to be proper if dom $f \neq \varnothing$ and $f(x) > -\infty$, $\forall x \in R^n$.

Proposition 2.5. If f is a proper convex function, then

$$\underset{\substack{x^* \in R^n,\\ \mu^* \in R}}{\exists} \; \underset{x \in R^n}{\forall} : x^* x - \mu^* \leqq f(x).$$

Proof. Propositions 2.4 and 2.1 imply that $\exists x_1 \in r \text{ int dom } f$ and Proposition 2.5 follows from Proposition 2.2, where $\mu^* = x_1 x^* - f(x_1)$. □

Corollary 2.1. An *improper* convex function f is either constant $+\infty$ or is $-\infty$ on r int dom f.

Proof. (a) Let dom $f = \varnothing$. Then $\forall x \in R^n$ we have $f(x) = +\infty$. (b) Let dom $f \neq \varnothing$ and $x_1 \in r$ int dom f. Suppose $f(x_1)$ is finite. Then by Proposition 2.2 $f(x) > -\infty$ $\forall x \in$ dom f, which contradicts Definition 2.6. So $f(x_1) = -\infty$. □

Definition 2.7. The *closure* of a convex function f is defined by

$$(\text{cl} f)(x) = \sup_{(x^*, \mu^*) \in D} [x^* x - \mu^*],$$

where

$$D = \{(x^*, \mu^*) \mid x^* \in R^n, \mu^* \in R, x^* x - \mu^* \leqq f(x), \forall x \in R^n\}.$$

Corollary 2.2. The following results hold:
(a) If f is improper and $\operatorname{dom} f \neq \emptyset$, then $D = \emptyset$.
(b) If f is improper and $\operatorname{dom} f = \emptyset$, then $D = R^{n+1}$.
(c) If f is a proper convex function, then D is convex in R^{n+1} and $D \neq R^{n+1} \neq \emptyset$.

Proof. (a) and (b): follow from Corollary 2.1. (c): Convexity of D follows from Definition 2.7, and Proposition 2.5 implies that $D \neq \emptyset$. It is obvious that $D \neq R^{n+1}$. □

Proposition 2.6. If f is an improper function, then $\operatorname{cl} f$ is either constant $+\infty$ (if $\operatorname{dom} f = \emptyset$), or constant $-\infty$ (if $\operatorname{dom} f \neq \emptyset$). Proof immediately follows from Corollary 2.2(a) and 2.2(b) and Definition 2.7.

Corollary 2.3. $\operatorname{cl} f$ is a proper convex function if and only if f is proper.

Proof. If f is improper then by Proposition 2.6 $\operatorname{cl} f$ is improper. If f is proper, then $f(x) > -\infty$ $\forall x \in R^n$ and by Corollary 2.2(c) $(\operatorname{cl} f)(x) > -\infty$ $\forall x \in R^n$. Using Proposition 2.3 and the fact that $\sup\,(f+g) \leqq \sup f + \sup g$, we see that $\operatorname{cl} f$ is convex. □

Corollary 2.4. The following relations hold:
(a) $(\operatorname{cl} f)(x) \leqq f(x)$ $\forall x \in R^n$.
(b) $(\operatorname{cl} f)(x) = f(x)$ $\forall x \in r \operatorname{int} \operatorname{dom} f$ and $\forall x \notin \operatorname{cl}(\operatorname{dom} f)$.

Definition 2.8. A convex function f is said to be *closed* if $\operatorname{cl} f = f$.

2.2 Conjugate Functions

Definition 2.9. The conjugate function f^* of a convex function f is defined by

$$f^*(x^*) = \sup_x\,[xx^* - f(x)]$$

Proposition 2.7. f^* is a convex function and $\operatorname{epi} f^* = D$.

Proof.

$$D = \{(x^*, \mu^*) \mid x^*x - f(x) \leq \mu^*, \forall x \in R^n\} = \{(x^*, \mu^*) \mid \sup_x\,[x^*x - f(x)] \leq \mu^*\} = \operatorname{epi} f^*.$$

Using Corollary 2.2 and Definition 2.4 we see that f^* is convex. □

Proposition 2.8. If we define $f^{**}(x)=\sup_{x^*}|[xx^*-f^*(x^*)]$ and $D^*=\{(x,\mu)|x\in R^n, \mu\in R, xx^*-\mu\leqq f^*(x^*), \forall x^*\in R^n\}$, then
(a) $f^{**}=\text{cl}\, f$;
(b) $f^{**}(x)=\sup_{x^*}\inf_y [(x-y)x^*+f(y)]=(\text{cl}\, f)(x)$;
(c) $D^*=\text{epi}\, f^{**}=\text{epi}\,(\text{cl}\, f)$.

Proof.

(a) $$(\text{cl}\, f)(x)=\sup_D [xx^*-\mu^*]=\sup_{f^*(x^*)\leq\mu^*}[x^*x-\mu^*]$$

(from Proposition 2.7)

$$=\sup_{x^*}[xx^*-f^*(x^*)];$$

(b) $$\sup_{x^*}[xx^*-f^*(x^*)]=\sup_{x^*}\{xx^*-\sup_y [yx^*-f(y)]\};$$

(c) Proof is analogous to that for Proposition 2.7. □

Corollary 2.5. The following results hold:
(a) f^* is a closed convex function.
(b) f^* is proper if and only if f is proper.

Proof. (a) By Proposition 2.8(c)

$$(\text{cl}\, f^*)(x^*)=\sup_{D^*}[xx^*-\mu]=\sup_{(\text{cl}\, f)(x)\leq\mu}[x^*x-\mu]=\sup_x [x^*x-(\text{cl}\, f)(x)]$$
$$=(\text{cl}\, f)^*(x^*).$$

By Corollary 2.4 $[xx^*-(\text{cl}\, f)(x)]\geqq[xx^*-f(x)]$, which implies $\text{cl}\, f^*=(\text{cl}\, f)^*\geq f^*$. Using Corollary 2.4 we see that $\text{cl}\, f^*=f^*$.

(b) Obviously f^* is proper if and only if $\text{epi}\, f^*\neq R^{n+1}\neq\emptyset$. Now using Corollary 2.2 and Proposition 2.7 we obtain Corollary 2.5(b). □

Elementary formulas. If f is convex on R^n and $\lambda\in R$, $c\in R^n$, then
(a) $h(x)=f(x)+\lambda$ implies $h^*(x^*)=f^*(x^*)-\lambda$;
(b) $h(x)=f(x+c)$ implies $h^*(x^*)=f^*(x^*)-cx^*$;
(c) $h(x)=\lambda f(x)$, $\lambda>0$, implies $h^*(x^*)=\lambda f^*[(1/\lambda)x^*]$;
(d) $h(x)=f(\lambda x)$, $\lambda\neq 0$, implies $h^*(x^*)=f^*[(1/\lambda)x^*]$.

Example 2.1. Let $f(x)=\log(\sum e^{x_i})$. Then[2]

$$f^*(x^*)=\sup_x [xx^*-\log(\sum e^{x_i})].$$

Obviously $f^*(0)=+\infty(\forall x_i\to-\infty)$, and $f^*(x^*)=+\infty$ if some $x_i^*<0$. The concave function $g(x)=x^*x-\log(\sum e^{x_i})$ attains its maximum if $x_i^*(\sum e^{x_i})=e^{x_i}$ (i.e., $x_i^*>0, \sum x_i^*=1$), which implies $x_i=\log(\lambda x_i^*)$ for arbitrary $\lambda>0$. So we have

$$f^*(x^*)=\sum x_i^* \log(\lambda x_i^*)-\log(\lambda\sum x_i^*)=\sum x_i^*\log x_i^*, \quad \text{if } x^*>0, \sum x_i^*=1.$$

Consider the case when $x_i^*>0$ for $i\in I$, $x_i^*=0$ for $i\notin I$ and $\sum x_i^*=1$. Then

$$g(x)=\sum_{i\in I} x_ix_i^*-\log(\sum e^{x_i})\leqq\sum_{i\in I} x_ix_i^*-\log\left(\sum_{i\in I} e^{x_i}\right)=g_I(x)$$

and $\sup g(x)=\sup g_I(x)$ $(\forall i\notin I, x_i\to-\infty)$. So

$$f^*(x^*)=\sum x_i^*\log x_i^*, \text{ if } x^*\geqq 0, \sum x_i^*=1,$$

where $\delta\log\delta=0$ if $\delta=0$.

Consider the case when $x^*\geqq 0$, $\sum x_i^*=\nu>0$. Denote $a_i=\log(\lambda x_i^*)$, $\lambda>0$ for $i\in I=\{i\,|\,x_i^*>0\}$, $a_i=-\infty$ for $i\notin I$. Then

$$\begin{aligned}\sup_x[\sum x_ix_i^*-\log(\sum e^{x_i})]&\geqq\lim_{x\to a}\{\nu[\sum(x_i^*/\nu)x_i-\log(\sum e^{x_i})]+(\nu-1)\log(\sum e^{x_i})\}\\ &=\nu[\sum(x_i^*/\nu)\log(x_i^*/\nu)]+(\nu-1)\log(\lambda\nu)\\ &\to\infty\begin{cases}\lambda\to\infty & \text{if } \nu>1,\\ \lambda\to 0 & \text{if } \nu<1.\end{cases}\end{aligned}$$

So $f^*(x^*)=+\infty$ if $x^*\notin\{x^*\,|\,x^*\geq 0, \sum x_i^*=1\}$

2.3. Subgradients

Definition 2.10. A subgradient of a convex function f at $x\in R^n$ is a vector $x^*\in R^n$ such that

$$\forall_{z\in R^n}\ f(z)-f(x)\geqq(z-x)x^*.$$

We denote by $\partial f(x)$ the set of all subgradients of f at x.

Proposition 2.9. Some elementary properties of subgradients:

(a) $\partial f(x)$ is a closed convex set (possibly empty).

[2] The convexity of $\log(\sum e^{x_i})$ follows from the Hölder inequality: $\sum a_ib_i\leqq[(\sum a_i^p)^{1/p}\cdot(\sum b_i^q)^{1/q}]$, $a_i>0$, $b_i>0$, $p>1$, $q>0$, $1/p+1/q=1$, when $a_i=e^{\lambda x_i}$, $b_i=e^{(1-\lambda)y_i}$, $1/p=\lambda$, $1/q=1-\lambda$.

(b) If f is proper and $x \in r$ int dom f, then by Proposition 2.2 $\partial f(x) \neq \emptyset$.

(c) $\partial f(x)$ is a one-point set if and only if f is differentiable at x. In this case $\partial f(x) = \{\nabla f(x)\}$.

(d) If f, h are convex functions and $\lambda \geqq 0$, then $\partial(f+h)(x) = \partial f(x) + \partial h(x)$, $\partial(\lambda f)(x) = \lambda \cdot \partial f(x)$.

(e) $0 \in \partial f(x)$ if and only if f attains a minimum at x.

Proposition 2.10. If $\partial f(x)$ is nonempty, then $(\mathrm{cl}\, f)(x) = f(x)$.

Proof. If $x^* \in \partial f(x)$, then by Definition 2.10 $\forall x: f(z) \geqq zx^* - [xx^* - f(x)] = zx^* - \mu^*$ and for $z = x$ we have $f(x) = xx^* - \mu^*$, which implies

$$f(x) = \sup_D [xx^* - \mu^*] = (\mathrm{cl}\, f)(x).$$

Proposition 2.11. $x^* \in \partial f(x)$ if and only if $f^*(x^*) = xx^* - f(x)$.

Proof.

$$x^* \in \partial f(x) \Leftrightarrow \forall z: xx^* - f(x) \geqq zx^* - f(z) \qquad \text{(from Definition 2.10)}$$

$$\Leftrightarrow \sup_z [zx^* - f(z)] = xx^* - f(x) \Leftrightarrow f^*(x^*) = x^*x - f(x). \qquad \square$$

Corollary 2.6. The following results hold:

(a) $x^* \in \partial f(x)$ implies $x \in \partial f^*(x^*)$.

(b) If $f(x) = (\mathrm{cl}\, f)(x)$, then $x \in \partial f^*(x^*)$ implies $x^* \in \partial f(x)$.

Proof. By Proposition 2.11 $x^* \in \partial f(x)$ if and only if $f^*(x^*) = xx^* - f(x)$, and $x \in \partial f^*(x^*)$ if and only if $f^{**}(x) = xx^* - f^*(x^*)$. Using Propositions 2.10 and 2.8(a) we obtain Corollary 2.6. $\square$

2.4. Concave functions

Definition 2.11. The function g is concave on R^n if $(-g)$ is convex.

Proposition 2.12. The basic concepts and properties of concave functions (analogous to those defined above for convex functions) are as follows:

(a) epi $g = \{(x, \nu) \mid g(x) \geqq \nu, x \in R^n, \nu \in R\}$ is a convex set in R^{n+1}.

(b) dom $g = \{x \mid g(x) > -\infty\}$ is a convex set in R^n.

(c) g is proper if dom $g \neq \emptyset$ and $g(x) < +\infty \forall x \in R^n$.

(d) $(\mathrm{cl}\, g)(x) = \inf_E [x^*x - \nu^*]$, $E = \{(x_i^*, \nu^*) \mid x^*x - \nu^* \geq g(x) \forall x \in R^n\}$.

(e) The conjugate of g is $g^*(x^*) = \inf_x [x^*x - g(x)]$.

(f) The subgradient of g at x is a vector x^* such that

$$\forall z: g(z)-g(x) \leqq (z-x)x^*.$$

(g) $0 \in \partial g(x)$ if and only if g attains its maximum at x.

Corollary 2.7. The following results hold:
(a) $\mathrm{cl}\,(-f) = -\mathrm{cl}\, f$.
(b) $(-f)^*(x^*) = -f^*(-x^*)$.
(c) If f is convex and $\lambda \in R$, then $\partial(\lambda f)(x) = \lambda\, \partial f(x)$.

3. Rockafellar Theory of Duality

3.1. Dual Programs. Let ϕ be a function of two vector variables $x \in R^n$, $u \in R^m$, convex on R^{n+m}. Then by Definition 2.9

$$xx^* + uu^* \leqq \phi(x, u) + \phi^*(x^*, u^*) \qquad \forall\, x, x^* \in R^n, \qquad u, u^* \in R^m, \quad (1)$$

which implies[3]

$$0 = x \cdot 0 + 0 \cdot u^* \leqq \phi(x, 0) + \phi^*(0, u^*) \qquad \forall x \in R^n, \qquad u^* \in R^m. \quad (2)$$

The inequality (2) justifies the following definition of dual pair of convex programs.

Definition 3.1. If ϕ is a closed proper convex function on R^{n+m} and $x \in R^n$, $u \in R^m$, then we define

$$\text{primal program:} \min \{\phi(x, 0) \mid (x, 0) \in \mathrm{dom}\, \phi\}, \quad (3)$$

$$\text{dual program:} \quad \max \{-\phi^*(0, u^*) \mid (0, u^*) \in \mathrm{dom}\, \phi^*\}. \quad (4)$$

Note that (3) and (4) can be written, respectively, as $\min_x \phi(x, 0)$ and $\max_{u^*} [-\phi^*(0, u^*)]$.

Remark 3.1. The dual program to (4) is (3).

Proof. By Definition (3.1) the dual to $\min \{\phi^*(0, u^*) \mid (0, u^*) \in \mathrm{dom}\, \phi^*\}$ is $\max \{-\phi^{**}(x, 0) \mid (x, 0) \in \mathrm{dom}\, \phi^{**}\} = -\min \{\phi(x, 0) \mid (x, 0) \in \mathrm{dom}\, \phi\}$. □

[3] The whole theory can be generalized if one replaces R^n and R^m by two arbitrary complementary orthogonal subspaces of R^{n+m}.

Definition 3.2. The primal program (3) is said to be
(a) *consistent* if $\exists x, \phi(x, 0) < +\infty$;
(b) *strictly consistent if*

$$\underset{\epsilon>0}{\exists}\ \underset{\|u\|<\epsilon}{\forall}\ \underset{x}{\exists}\ \phi(x, u) < +\infty;$$

(c) *inconsistent if* $\forall x \in R^n$, $\phi(x, 0) = +\infty$.

From the "weak duality theorem" (2) immediately follows:

Theorem 3.1. $\sup_{u^*} [-\phi^*(0, u^*)] \leqq \inf_x \phi(x, 0)$.

Corollary 3.1. The following results hold:
(a) If $\phi(\hat{x}, 0) = -\phi^*(0, \hat{u}^*)$, then $\hat{x}$, and $\hat{u}^*$ are, respectively, optimal solutions to (3) and (4). Note, that by Proposition 2.11 $\phi(\hat{x}, 0) = -\phi^*(0, \hat{u}^*)$ if and only if $(0, \hat{u}^*) \in \partial\phi(\hat{x}, 0)$.
(b) If $\inf \phi(x, 0) = -\infty$, then (4) is inconsistent and if $\sup [-\phi^*(0, u^*)] = +\infty$, then (3) is inconsistent.

Lemma 3.1. If ϕ is a closed proper convex function on R^{n+m}, then
(a) $\Phi(u) = \inf_x \phi(x, u)$ is convex on R^m;
(b) $\Psi(x^*) = \inf_{u^*} \phi^*(x^*, u^*)$ is convex on R^n.

Proof. (a) We must show that epi $\Phi = \{(u, \nu) \mid \Phi(u) \leqq \nu\}$ is a convex set in R^{m+1}. Obviously epi $\Phi \neq \emptyset$ (ϕ is proper). Let $(u_i, \nu_i) \in$ epi Φ ($i = 1, 2$). Then

$$\underset{\epsilon>0}{\forall}\ \underset{x_i}{\exists}\ \phi(x_i, u_i) < \inf_x \phi(x, u_i) + \epsilon = \Phi(u_i) + \epsilon \leqq \nu_i + \epsilon \qquad (i = 1, 2)$$

If we write $O_3 = \lambda O_1 + (1-\lambda) O_2$ ($O = x, u, \nu$), $0 \leqq \lambda \leqq 1$, then $\phi(x_3, u_3) < \nu_3 + \epsilon$ (epi ϕ is a convex set) and $\Phi(u_3) = \inf_x \phi(x, u_3) \leqq \phi(x_3, u_3) < \nu_3 + \epsilon \quad \forall \epsilon > 0$.
(b) The proof is analogous to the proof of (a). □

Corollary 3.2. The following results hold:
(a) (3) is consistent if and only if $0 \in$ dom Φ.
(b) (3) is strictly consistent if and only if $0 \in$ int dom Φ.
(c) Theorem 3.1 can be written as $-\Psi(0) \leqq \Phi(0)$.

Lemma 3.2. Let ϕ be a closed proper convex function on R^{n+m} and $\Phi(u) = \inf_x \phi(x, u)$, $\Psi(x^*) = \inf_{u^*} \phi^*(x^*, u^*)$. Then
(a) $\phi(x, 0) = \Psi^*(x)$;

(b) $\phi^*(0, u^*) = \Phi^*(u^*)$;
(c) $\mathrm{cl}\,\Psi(0) = -\Phi(0)$;
(d) $\mathrm{cl}\,\Phi(0) = -\Psi(0)$.

Proof.

(a) $$\Psi^*(x) = \sup_{x^*} [xx^* - \Psi(x^*)] = \sup_{x^*} [xx^* - \inf_{u^*} \phi^*(x^*u^*)]$$

$$= \sup_{x^*} \sup_{u^*} [xx^* + 0 \cdot u^* - \phi^*(x^*, u^*)] = \phi^{**}(x, 0) = \phi(x, 0).$$

(b) $$\Phi^*(u^*) = \sup_{u} [uu^* - \Phi(u)] = \sup_{u} [uu^* - \inf_{x} \phi(x, u)]$$

$$= \sup_{u} \sup_{x} [uu^* + x \cdot 0 - \phi(x, u)] = \phi^*(0, u^*).$$

(c) $$\mathrm{cl}\,\Psi(0) = \Psi^{**}(0) = \sup_{x} [x \cdot 0 - \Psi^*(x)]$$

$$= \sup_{x} [x \cdot 0 - \phi(x, 0)] \qquad \text{[from (a)]}$$

$$= -\inf_{x} \phi(x, 0) = -\Phi(0).$$

(d) $$\mathrm{cl}\,[\Phi(0)] = \Phi^{**}(0) = \sup_{u^*} [0 \cdot u^* - \Phi^*(u^*)]$$

$$= \sup_{u^*} [0 \cdot u^* - \phi^*(0, u^*)] \qquad \text{[from (b)]}$$

$$= -\inf_{u^*} \phi^*(0, u^*) = -\Psi(0).$$ □

Theorem 3.2. Let ϕ be a closed proper convex function on R^{n+m}. Then
(a) $\Phi(0) = -\Psi(0)$ if and only if $\mathrm{cl}\,\Phi(0) = \Phi(0)$.
(b) $\mathrm{cl}\,\Phi(0) = \Phi(0)$ if and only if $\mathrm{cl}\,\Psi(0) = \Psi(0)$.

Proof. $$\mathrm{cl}\,\Phi(0) = \Phi(0) \Leftrightarrow -\Psi(0) = \Phi(0) \qquad \text{[from Lemma 3.2(d)]}$$

$$\Leftrightarrow -\Psi(0) = -\mathrm{cl}\,\Psi(0) \qquad \text{[from Lemma 3.2(c)].}$$ □

Corollary 3.3. Let primal program (3) be *strictly* consistent. Then

(a) $\sup[-\phi^*(0, u^*)] = \inf \phi(x, 0)$;

(b) if $\inf \phi(x, 0)$ is finite, there exists an optimal solution $\hat{u}^*$ to (4), and $-\phi^*(0, \hat{u}^*) = \inf \phi(x, 0)$;

(c) if $\hat{x}$ is an optimal solution to (3), there exists an optimal solution $\hat{u}^*$ to (4), and $-\phi^*(0, \hat{u}^*) = \phi(\hat{x}, 0)$.

Proof. (a) By Corollary 3.2(b)

$$\begin{aligned} 0 \in \text{int dom } \Phi &\Rightarrow \text{cl } \Phi(0) = \Phi(0) \qquad &&\text{[from Corollary 2.4(b)]} \\ &\Rightarrow \Phi(0) = -\Psi(0) &&\text{[from Theorem 3.2].} \end{aligned}$$

(b) By Proposition 2.9(b)

$$\begin{aligned} \exists \hat{u}^* \in \partial\Phi(0) &\Rightarrow 0 \in \partial\Phi^*(\hat{u}^*) \qquad &&\text{[from Corollary 2.6]} \\ &\Rightarrow \Phi^* &&\text{attains its minimum at } \hat{u}^* \\ & &&\text{[from Proposition 2.9(e)]} \\ &\Rightarrow \hat{u}^* &&\text{is optimal to (4)} \\ & &&\text{[from Lemma 3.2(b)]} \end{aligned}$$

is optimal to (4).

(c) Follows from (b).

3.2. Saddle Function and Lagrange Multipliers

Definition 3.3. The Kuhn–Tucker function corresponding to the primal program (3) is defined by

$$K(x, u^*) = \inf_u [\phi(x, u) - uu^*] = -\sup_u [uu^* - \phi(x, u)].$$

Lemma 3.3. The Kuhn–Tucker function K is convex on R^n relative to x and concave on R^m relative to u^*.

Proof. K is obviously concave on R^m relative to u^* [because $\inf(f + g) \geqq \inf f + \inf g$]. To prove that K is convex on R^n relative to x we proceed as in Lemma 3.1. □

Note that (a) $\text{cl}_x K(x, u^*) = \sup_{x^*} [xx^* - \phi^*(x^*, u^*)]$, (b) the function $\sup_u [uu^* - \phi(x, u)]$ can be considered as a "partial conjugate" function $\phi_*(x, u^*)$. Analogously if we define ${}_*\phi(x^*, u) = \sup_x [xx^* - \phi(x, u)]$, then

$$({}_*\phi)_*(x^*, u^*) = {}_*(\phi_*)(-x^*, -u^*) = -\phi^*(x^*, -u^*).$$

To define the dual program (4) we can use $({}_*\phi)_*$ instead of $(-\phi^*)$ as is done in Ref. 2, where $({}_*\phi)_*$ is called the adjoint function.

Lemma 3.4. If ϕ is a closed proper convex function on R^{n+m} and $K(x, u^*) = \inf_u [\phi(x, u) - uu^*]$, then
(a) $\phi(x, u) = \sup_{u^*} [K(x, u^*) + uu^*]$.
(b) $-\phi^*(x^*, u^*) = \inf_x [K(x, u^*) - xx^*]$.

Proof.

$$\text{(a)}\ \sup_{u^*} [K(x, u^*) + uu^*] = \sup_{u^*} \inf_{v} [\phi(x, v) - vu^* + uu^*]$$

$$= \text{cl}_u\, \phi(x, u) \qquad \text{[from Proposition 2.8(b)].}$$

$$= \phi(x, u)$$

$$\text{(b)}\ \inf_x [K(x, u^*) - xx^*] = \inf_x \inf_u [\phi(x, u) - uu^* - xx^*] = -\phi^*(x^*, u^*).$$

□

Corollary 3.4. The following relations hold:
(a) $\phi(x, 0) = \sup_{u^*} K(x, u^*)$.
(b) $-\phi^*(0, u^*) = \inf_x K(x, u^*)$.

Theorem 3.3.[4] If $(\hat{x}, \hat{u}^*)$ is a saddle point of K, then $\hat{x}$ is optimal to (3); $\hat{u}^*$ is optimal to (4), and $\phi(\hat{x}, 0) = K(\hat{x}, \hat{u}^*) = -\phi^*(0, \hat{u}^*)$.

Proof. If

$$\forall_{x, u^*}\ K(\hat{x}, u^*) \leqq K(\hat{x}, \hat{u}^*) \leqq K(x, \hat{u}^*),$$

then

$$K(\hat{x}, \hat{u}^*) = \sup_{u^*} K(\hat{x}, u^*) = \phi(\hat{x}, 0) \qquad \text{[from Corollary 3.4(a)]}$$

and

$$K(\hat{x}, \hat{u}^*) = \inf_x K(x, \hat{u}^*) = -\phi^*(0, \hat{u}^*) \qquad \text{[from Corollary 3.4(b)]}$$

which (by Corollary 3.1) implies Theorem 3.3. □

Theorem 3.4. If (3) is strictly consistent and $\hat{x}$ is optimal to (3), then there exists $\hat{u}^*$ such that $(\hat{x}, \hat{u}^*)$ is a saddle point of K.

---

[4] Theorems 3.3 together with 3.4 can be considered as the generalized Kuhn–Tucker theorem with the generalized Slater condition.

Proof. By Corollary 3.3(c) there exists $\hat{u}^*$ optimal to (4) and $\phi(\hat{x}, 0) = -\phi^*(0, \hat{u}^*)$, i.e., (by Corollary 3.4) $\inf_x K(x, \hat{u}^*) = \sup_{u^*} K(\hat{x}, u^*)$. But

$$\inf_x K(x, \hat{u}^*) \leq K(\hat{x}, \hat{u}^*) \leq \sup_{u^*} K(\hat{x}, u^*),$$

which implies

$$K(\hat{x}, u^*) \leq \sup_{u^*} K(\hat{x}, u^*) = K(\hat{x}, \hat{u}^*) = \inf_x K(x, \hat{u}^*) \leq K(x, \hat{u}^*). \qquad \square$$

Corollary 3.5. If $\phi(\hat{x}, 0) = -\phi^*(0, \hat{u}^*)$, then $(\hat{x}, \hat{u}^*)$ is a saddle point of K.

Proof. See the end of the proof of Theorem 3.4.

Remark 3.2. If $(\hat{x}, \hat{u}^*)$ is a saddle point of K, then by Theorem 3.3 and Lemma 3.2 $\phi^{**}(\hat{x}, 0) = \phi(\hat{x}, 0) = \Phi(0) = -\Psi(0) = -\phi^*(0, \hat{u}^*)$, i.e.,

$$\begin{aligned}\sup_{x^*, u^*} [\hat{x}x^* + 0 \cdot u^* - \phi^*(x^*, u^*)] &= -\Psi(0) = \Phi(0)\\ &= -\sup_{x,u} [x0 + u\hat{u}^* - \phi(x, u)],\end{aligned}$$

which implies the next inequalities:

$$\sup_{u^*} [\hat{x}x^* - \phi^*(x^*, u^*)] \leqq -\Psi(0) = \Phi(0) \leq \inf_x [\phi(x, 0) - u\hat{u}^*] \qquad \forall x^*, u$$

or

$$\hat{x}x^* - \Psi(x^*) \leq -\Psi(0) = \Phi(0) \leqq \Phi(u) - uu^* \qquad \forall x^*, u.$$

The last inequalities justify the following definition of Lagrange multipliers.

Definition 3.4. (a) If $\Phi(0)$ is finite, then the Lagrange multiplier for (3) is a vector $u^* \in R^m$ for which

$$\Phi(u) - uu^* \geqq \Phi(0) \qquad \forall_u$$

or, equivalently, by Definition 2.10, $u^* \in \partial\Phi(0)$.

(b) If $\Psi(0)$ is finite, then the Lagrange multiplier for (4) is a vector $x \in R^n$ for which

$$\Psi(x^*) - xx^* \geqq \Psi(0) \qquad \forall x^*$$

or, equivalently, $x \in \partial\Psi(0)$.

Corollary 3.6. If (3) is strictly consistent and $\Phi(0)$ is finite, then there exists a Lagrange multiplier $\hat{u}^*$ for (3).

Proof. See Proposition 2.9(b), Corollary 2.6, and Corollary 3.2(b).

Theorem 3.5. The following results hold:

(a) $\hat{u}^*$ is the Lagrange multiplier for (3) if and only if $\hat{u}^*$ is optimal to (4) and $\Phi(0) = -\phi^*(0, \hat{u}^*)$.

(b) $\hat{x}$ is the Lagrange multiplier for (4) if and only if $\hat{x}$ is optimal to (3) and $\phi(\hat{x}, 0) = -\Psi(0)$.

Proof. (a) I. Let $\hat{u}^* \in \partial\Phi(0)$. Then by Corollary 2.6 $0 \in \partial\Phi^*(\hat{u}^*)$, which by Proposition 2.9(e) and Lemma 3.2(b) implies that $\hat{u}^*$ is optimal to (4). By Proposition 2.10 and Theorem 3.2 $\Phi(0) = -\Psi(0)$. II. Let $\hat{u}^*$ be optimal to (4) and $\Phi(0) = -\phi^*(0, \hat{u}^*)$. Then by Theorem 3.2 $(\text{cl}\,\Phi)(0) = \Phi(0)$, which by Proposition 2.9(e) and Corollary 2.6(b) implies $\hat{u}^* \in \partial\Phi(0)$.

(b) The proof is analogous to the proof of (a).

Corollary 3.7. Let $-\infty < -\Psi(0) = \Phi(0) < +\infty$. Then

(a) $\hat{u}^*$ is the Lagrange multiplier for (3) if and only if $\hat{u}^*$ is optimal to (4);

(b) $\hat{x}$ is the Lagrange multiplier for (4) if and only if $\hat{x}$ is optimal to (3).

Theorem 3.6. The next four statements are equivalent:

(i) $(\hat{x}, \hat{u}^*)$ is a saddle point of K.

(ii) $\hat{x}$ is optimal to (3) and $\hat{u}^*$ is the Lagrange multiplier for (3).

(iii) $\hat{u}^*$ is optimal to (4) and $\hat{x}$ is the Lagrange multiplier for (4).

(iv) $\hat{u}^*$ and $\hat{x}$ are, respectively, the Lagrange multipliers for (3) and (4) and in all cases $\phi(\hat{x}, 0) = -\phi^*(0, \hat{u}^*)$.

Proof. If $(\hat{x}, \hat{u}^*)$ is a saddle point of K, then by Theorems 3.3 and 3.5 $\hat{x}$ is optimal to (3) and at the same time is the Lagrange multiplier for (4) $\hat{u}^*$ is optimal to (4) and at the same time is the Lagrange multiplier for (3) and $\phi(\hat{x}, 0) = -\phi^*(0, \hat{u}^*)$.

If $\hat{u}^*$ is optimal to (4) and $\hat{x}$ is the Lagrange multiplier for (4), then by Theorem 3.5(b) and Corollary 3.5 $\hat{x}$ is optimal to (3), $\phi(\hat{x}, 0) = -\phi^*(0, \hat{u}^*)$, and $(\hat{x}, \hat{u}^*)$ is a saddle point of K.

If $\hat{x}$ is optimal to (3) and $\hat{u}^*$ is the Lagrange multiplier for (3), then by Theorem 3.5(a) $\hat{u}^*$ is optimal to (4) and $\phi(\hat{x}, 0) = -\phi^*(0, \hat{u}^*)$. By Corollary 3.5 $(\hat{x}, \hat{u}^*)$ is a saddle point of K.

If $\hat{u}^* \in \partial\Phi(0)$ and $\hat{x} \in \partial\Psi(0)$, then by Proposition 2.10 and Theorem 3.2 $\Phi(0) = -\Psi(0)$ and by Theorem 3.5 $\hat{x}$ is optimal to (3), $\hat{u}^*$ is optimal to (4). □

Corollary 3.8. The following results hold:

(a) If $\hat{x}$ is optimal to (3), then $\hat{u}^*$ is the Lagrange multiplier for (3) if and only if $(\hat{x}, \hat{u}^*)$ is a saddle point of K.

(b) If $\hat{u}^*$ is optimal to (4), then $\hat{x}$ is the Lagrange multiplier for (4) if and only if $(\hat{x}, \hat{u}^*)$ is the saddle point of K.

4. General Geometric Programming

Proposition 4.1. Let $x_k, y_k \in R^{n_k}$ $(k=0,1,\ldots,m)$, $\lambda_k \in R$ $(k=1,2,\ldots,m)$, and denote $x=(x_0, x_1, \ldots, x_m)$, $y=(y_0, y_1, \ldots, y_m)$, $\lambda=(\lambda_1, \lambda_2, \ldots, \lambda_m)$. Let $x \in R^n$, $u=(y,\lambda) \in R^{n+m}$, and let $\mathscr{P}, \mathscr{D}$ be two complementary orthogonal subspaces of R^n. If G_k $(k=0,1,\ldots,m)$ are proper closed convex functions on R^{n_k}, respectively, then define

$$\phi[x,(y,\lambda)] = \begin{cases} G_0(x_0) & \text{if } G_k(x_k)+\lambda_k \le 0 \ (k=1,2,\ldots,m),\ x-y \in \mathscr{P}, \\ +\infty & \text{otherwise.} \end{cases}$$

Obviously ϕ is a proper closed convex function on R^{2n+m}. The calculation of ϕ^* is as follows (see Elementary Formulas in Section 2.2):

$$\begin{aligned}
&\phi^*[x^*,(y^*,\lambda^*)] \\
&\quad = \sup_{x,y,\lambda} [xx^* + yy^* + \lambda\lambda^* - \phi(x,y,\lambda)] \\
&\quad = \sup_{\substack{\nu \ge 0 \\ x-y \in \mathscr{P}}} \Big[(-y^*)(x-y) + x(y^*+x^*) - \sum_{k=1}^m \lambda_k^*(G_k(x_k)+\nu_k) - G_0(x_0)\Big] \\
&\quad = \begin{cases} \sup\limits_x \Big[x(y^*+x^*) - G_0(x_0) - \sum\limits_{k=1}^m \lambda_k^* G_k(x_k)\Big] & \text{if } \lambda^* \geqq 0,\ y^* \in \mathscr{D}, \\ +\infty & \text{otherwise,} \end{cases} \\
&\quad = \begin{cases} G_0^*(y_0^*+x_0^*) + \sum\limits_{k=1}^m \lambda_k^* G_k^*\left(\dfrac{y_k^*+x_k^*}{\lambda_k^*}\right) & \text{if } \lambda^* \geqq 0,\ y^* \in \mathscr{D}, \\ +\infty & \text{otherwise,} \end{cases}
\end{aligned}$$

where

$$\lambda^* G^*\left(\frac{z^*}{\lambda^*}\right) = \begin{cases} 0 & \text{if } \lambda^*=0,\ z^*=0, \\ +\infty & \text{if } \lambda^*=0,\ z^* \ne 0. \end{cases}$$

Now[5]

$$\phi[x,(0,0)]=\begin{cases} G_0(x_0) & \text{if } G_k(x_k)\leqq 0\ (k=1,2,\ldots,m),\ x\in\mathscr{P},\\ +\infty & \text{otherwise,}\end{cases}$$

$$\phi^*[0,(y,\lambda)]=\begin{cases} G_0^*(y_0)+\sum_{k=1}^{m}\lambda_k G_k^*\left(\dfrac{y_k}{\lambda_k}\right) & \text{if } \lambda\geqq 0,\ y\in\mathscr{D},\\ +\infty & \text{otherwise,}\end{cases}$$

and by Definition 3.1 we have the following pair of dual programs:

$$\min\{G_0(x_0)\mid G_k(x_k)\leqq 0\,(k=1,2,\ldots,m),\ x\in\mathscr{P}\}, \tag{5}$$

$$\max\{V(y,\lambda)\mid(y,\lambda)\in\operatorname{dom} V,\ y\in\mathscr{D}\}, \tag{6}$$

where

$$V(y,\lambda)=-G_0^*(y_0)-\sum_{k=1}^{m}\lambda_k G_k^*\left(\frac{y_k}{\lambda_k}\right)$$

and $\operatorname{dom} V=\{(y,\lambda)\mid\lambda\geqq 0,\ y_0\in\operatorname{dom} G_0^*,\ y_k\in\lambda_k\operatorname{dom} G_k^*\ (k=1,2,\ldots,m)\}$. Note that the substitution $G_k(x_k)\to G_k(x_k+c_k)-\beta_k$, $\beta_0=0$ transforms the dual pair (5), (6) into

$$\min\{G_0(x_0+c_0)\mid G_k(x_k+c_k)\leqq\beta_k\ (k=1,2,\ldots,m),\ x\in\mathscr{P}\}, \tag{7}$$

$$\max\{V(y,\lambda)-\lambda\beta+cy\mid(y,\lambda)\in\operatorname{dom} V,\ y\in\mathscr{D}\}. \tag{8}$$

Remark 4.1. From (5), (6) we see that if G_k are convex functions, such that the effective domains of their conjugates are polyhedral sets (or the whole space), then the dual program (6) is a convex program with linear constraints.

Example 4.1. Let

$$G_k(x^k)=\log\left(\sum_{i\in I_k} e^{x_i}\right)\qquad (k=0,1,2,\ldots,m),\ x^k\in R^{n_k},$$

where

$$I_k=\{i_{k-1}+1,\ldots,i_k\},\ i_{-1}=0,\ i_m=n,\ i_k-i_{k-1}=n_k.$$

Then by Example 2.1

$$G_k^*(y^k)=\begin{cases}\sum_{i\in I_k} y_i\log y_i & \text{if } \sum_{i\in I_k} y_i=1,\ y_i\geq 0,\ i\in I_k,\\ +\infty & \text{otherwise,}\end{cases}$$

[5] For the sake of simplicity of notation we shall omit the conjugate symbol "*" on the variables of the conjugate functions, where it is possible.

i.e.,

$$\text{dom } G_k^*(y^k) = \left\{ y \mid y \in R^{n_k}, \sum_{i \in I_k} y_i = 1, y_i \geq 0, i \in I_k \right\}$$

and

$$\lambda_k G_k^*\left(\frac{y^k}{\lambda_k}\right) = \sum_{I_k} y_i \log y_i - \left(\sum_{I_k} y_i\right) \log \lambda_k \qquad \text{if } \sum_{I_k} y_i = \lambda_k, y_i \geq 0, i \in I_k.$$

Now the dual program (6) can be written as

$$\max\left\{ -\sum_{i=1}^{n} y_i \log y_i + \sum_{k=1}^{m} \lambda_k \log \lambda_k \,\middle|\, \sum_{I_0} y_i = 1, \sum_{I_k} y_i = \lambda_k \ (k = 1, 2, \ldots, m), \right.$$
$$\left. y \geqq 0, y \in \mathscr{D} \right\}$$

and after substitution $x_i \to x_i + \log c_i$, by (7) and (8) we obtain the following dual pair:

$$\min\left\{ \log\left(\sum_{I_0} c_i e^{x_i}\right) \,\middle|\, \log\left(\sum_{I_k} c_i e^{x_i}\right) \leq 0 \qquad (k = 1, 2, \ldots, m) \ x \in \mathscr{P} \right\}, \quad (9)$$

$$\max\left\{ \sum_{i=1}^{n} y_i \log\left(\frac{c_i}{y_i}\right) + \sum_{k=1}^{m} \lambda_k \log \lambda_k \,\middle|\, \sum_{I_0} y_i = 1, \sum_{I_k} y_i = \lambda_k \ (k = 1, 2, \ldots, m), \right.$$
$$\left. y \geq 0, y \in \mathscr{D} \right\}, \quad (10)$$

which is equivalent to the dual geometric programs (see Ref. 1, p. 78). This fact suggests to us to call the dual pair (5), (6) "general geometric" dual programs.

Proposition 4.2 By Definition 3.2 the primal program (5) is said to be

(a) *consistent* if $\exists x \in \mathscr{P}$, $G_k(x_k) \leqq 0$ $(k = 1, \ldots, m)$, $x_0 \in \text{dom } G_0$;

(b) *strictly consistent* if $\exists x \in \mathscr{P}$, $x_k \in \text{int dom } G_k$ $(k = 0, 1, \ldots, m)$, $G_k(x_k) < 0$ $(k = 1, 2, \ldots, m)$;

and the dual program (6) is said to be

(c) *consistent* if $\exists y \in \mathscr{D}$, $\lambda \geqq 0$, $y_0 \in \text{dom } G_0^*$, $y_k \in \lambda_k \text{ dom } G_k^*$;

(d) *strictly consistent* if $\exists y \in \mathscr{D}$, $\lambda > 0$, $y_0 \in \text{int dom } G_0^*$, $y_k \in \text{int } (\lambda_k \text{ dom } G_k^*)$.

Proposition 4.3. By Definition 3.3 the Kuhn–Tucker function corresponding to the primal program (5) is

$$K(x, y^*, \lambda^*) = \inf_{y,\lambda} [\phi(x, y, \lambda) - yy^* - \lambda\lambda^*]$$

$$= \inf_{\substack{\nu \geq 0 \\ x-y \in \mathcal{P}}} \left[G_0(x_0) + \sum_{k=1}^{m} \lambda_k^*(G_k(x_k) + \nu_k) + y^*(x-y) - y^*x\right]$$

$$= \begin{cases} G_0(x_0) + \sum_{k=1}^{m} \lambda_k^* G_k(x_k) - y^*x & \text{if } y^* \in \mathcal{D}, \lambda^* \geq 0, \\ -\infty & \text{otherwise.} \end{cases}$$

Proposition 4.4 If $[\hat{x}, (\hat{y}, \hat{\lambda})]$ is a saddle point of K, then by Theorem 3.3 $\hat{x}$ is optimal to (5), $(\hat{y}, \hat{\lambda})$ is optimal to (6), and

$$\forall_{\substack{y \in \mathcal{D} \\ \lambda \geq 0}} G_0(\hat{x}_0) + \sum_{k=1}^{m} \lambda_k G_k(\hat{x}_k) - y\hat{x} \leqq G_0(\hat{x}_0) + \sum_{k=1}^{m} \hat{\lambda}_k G_k(\hat{x}_k) - \hat{y}\hat{x} = G_0(\hat{x}_0)$$

$$= V(\hat{y}, \hat{\lambda}) \leqq G_0(x_0) + \sum_{k=1}^{m} \hat{\lambda}_k G_k(x_k) - \hat{y}x \qquad \forall_x.$$

From the above inequalities we have

(a) $\hat{\lambda}_k G_k(\hat{x}_k) = 0 \qquad (k = 1, 2, \ldots, m)$

i.e., $\hat{\lambda}_k > 0$ implies $G_k(\hat{x}_k) = 0$ and $G_k(\hat{x}_k) < 0$ implies $\hat{\lambda}_k = 0$;

(b) $\forall_{\substack{x \in \mathcal{P} \\ \lambda \geq 0}} G_0(\hat{x}_0) + \sum_{k=1}^{m} \lambda_k G_k(\hat{x}_k) \leqq G_0(\hat{x}_0) \leqq G_0(x_0) + \sum_{k=1}^{m} \hat{\lambda}_k G_k(x_k)$;

(c) If $\hat{\lambda}_k = 0$, then $\hat{\lambda}_k G_k^*(\hat{y}_k/\hat{\lambda}_k) = 0$, and $\hat{y}_k = 0$.

Now we can prove an analogous theorem as in Ref. 1, p. 201.

Theorem 4.1 If (5) is strictly consistent and $\hat{x}$ is optimal to (5), then

(a) there exists $(\hat{y}, \hat{\lambda})$ optimal to (6);

(b) $G_0(\hat{x}_0) = V(\hat{y}, \hat{\lambda})$;

(c) $(\hat{x}, \hat{\lambda})$ is a saddle point of the Lagrange function $L(x, \lambda) = G_0(x_0) + \sum \lambda_k G_k(x_k)$ relative to $\lambda \geq 0$, $x \in \mathcal{P}$;

(d) $\hat{y}_0 \in \partial G_0(\hat{x}_0)$, $\hat{y}_k \in \hat{\lambda}_k \, \partial G_k(\hat{x}_k)$ $(k = 1, 2, \ldots, m)$ or $\hat{x}_0 \in \partial G_0^*(\hat{y}_0)$, and if $\hat{\lambda}_k > 0$, then $\hat{x}_k \in \partial G_k^*(\hat{y}_k/\hat{\lambda}_k)$.

Proof. (a) and (b) follow from Corollary 3.3(c). (c) follows from Proposition 4.4(b). By Proposition 4.4 $0 \in \partial_x[G_0(\hat{x}_0) + \sum \hat{\lambda}_k G_k(\hat{x}_k) - \hat{y}\hat{x}]$, which by Proposition 2.9(d) and Corollary 2.6 implies Theorem 4.1(d). □

Corollary 4.1. If (5) is strictly consistent, $\hat{x}$ is optimal to (5) and $(\hat{y}, \hat{\lambda})$ is optimal to (6), then $\hat{\lambda}_k = 0$ if and only if $\hat{y}_k = 0$.

Proof. (I) By Proposition 4.4(c) $\hat{\lambda}_k = 0$ implies $\hat{y}_k = 0$. (II) Let $\hat{y}_k = 0$ and suppose $\hat{\lambda}_k > 0$. Then by Theorem 4.1(d) $0 \in \partial G_k(\hat{x}_k)$, i.e., G_k attains its minimum at $\hat{x}_k$ and—because (5) is strictly consistent—$G_k(\hat{x}_k) < 0$. But this contradicts Proposition 4.4(a). □

Definition 4.1. (a) The primal program (5) is said to be subconsistent if

$$\underset{\epsilon>0}{\forall}\ \underset{x\in\mathcal{P}}{\exists} : x_k \in \operatorname{int}\operatorname{dom} G_k\ (k = 0, 1, \ldots, m),\ G_k(x_k) \leqq \epsilon\ (k = 1, 2, \ldots, m).$$

(b) If we denote $\Phi_\epsilon(0) = \inf_x \{G_0(x_0) \mid G_k(x_k) \leqq \epsilon\ (k = 1, 2, \ldots, m), x \in \mathcal{P}\}$ then the subinfimum of (5) is

$$\mathcal{M}_A = \lim_{\epsilon \to 0_+} \Phi_\epsilon(0).$$

Remark 4.2. Let dom G_k $(k = 0, 1, \ldots, m)$ be open sets.
(a) If (5) is consistent, then (5) is subconsistent.
(b) If (5) is strictly consistent, then $\mathcal{M}_A = \Phi(0)$.

Theorem 4.2. If (5) is subconsistent and the subinfimum $\mathcal{M}_A$ is finite, then (6) is consistent and $\sup [V(y, \lambda)] = \mathcal{M}_A$.

Proof. Consider the following pair of dual programs (7), (8):

$$\min \{G_0(x_0) \mid G_k(x_k) \leqq \epsilon\ (k = 1, 2, \ldots, m), x \in \mathcal{P}\}, \tag{11}$$

$$\max \{V(y, \lambda) - \epsilon\lambda \mid (y, \lambda) \in \operatorname{dom} V, y \in \mathcal{D}\}, \tag{12}$$

where

$$\epsilon = (\epsilon, \epsilon, \ldots, \epsilon) \in R^m, \text{ i.e., } \epsilon\lambda = \epsilon\left(\sum_{k=1}^{m} \lambda_k\right).$$

If (5) is subconsistent, then for arbitrary $\epsilon > 0$, (11) is strictly consistent. Obviously $0 < \epsilon_2 < \epsilon_1$ implies $\Phi_{\epsilon_1}(0) \leq \Phi_{\epsilon_2}(0) \leq \mathcal{M}_A$, so

$$\mathcal{M}_A = \lim_{\epsilon \to 0_+} \Phi_\epsilon(0) = \sup_{\epsilon>0} \Phi_\epsilon(0)$$

and we see that $\exists \bar{\epsilon} > 0$ such that $\Phi_\epsilon(0)$ is finite for each $0 < \epsilon \leq \bar{\epsilon}$. Now by Corollary 3.3(b) $\exists y_\epsilon, \lambda_\epsilon$:

$$\underset{0<\epsilon<\bar{\epsilon}}{\forall}\ \Phi_\epsilon(0) = -\Psi_\epsilon(0) = \sup [V(y, \lambda) - \epsilon\lambda]$$
$$= V(y_\epsilon, \lambda_\epsilon) - \epsilon\lambda_\epsilon \leq \sup V(y, \lambda) = M_B,$$

which implies $\mathcal{M}_A \leqq M_B$ and consistency of (6). Suppose that $\mathcal{M}_A < M_B$ and denote

$$\eta = \begin{cases} M_B - \mathcal{M}_A & \text{if } M_B < +\infty, \\ 1 & \text{if } M_B = +\infty. \end{cases}$$

Now

$$\underset{(y_\eta, \lambda_\eta)}{\exists} : V(y_\eta, \lambda_\eta) > M_B - \tfrac{1}{3}\eta,$$

where $(y_n, \lambda_\eta) \in \operatorname{dom} V$, $y_\eta \in \mathcal{D}$. Let $\epsilon_\eta \in (0, \bar{\epsilon})$ such that $\epsilon_\eta \lambda_\eta < \frac{1}{3}\eta$; then

$$V(y_\eta, \lambda_\eta) - \epsilon_\eta \lambda_\eta > M_B - \tfrac{2}{3}\eta > \mathcal{M}_A,$$

and at the same time,

$$V(y_\eta, \lambda_\eta) - \epsilon_\eta \lambda_\eta \leq -\Psi_{\epsilon_\eta}(0) = \Phi_{\epsilon_\eta}(0) \leq \mathcal{M}_A.$$

This contradiction implies that $\mathcal{M}_A = M_B$. □

Example 4.2 Let $x_k \in R$ $(k = 0, 1, \ldots, m)$ i.e., $n_k = 1$, $k = 0, 1, \ldots, m$, and let $G_0(x_0) = x_0$, $G_k(x_k) = b_k - x_k$ $(k = 1, 2, \ldots, m)$, where $b_k \in R$. Then

$$G_0^*(y_0) = \begin{cases} 0 & \text{if } y_0 = 1, \\ +\infty & \text{otherwise,} \end{cases}$$

$$G_k^*(y_k) = \begin{cases} -b_k & \text{if } y_k = -1 \\ +\infty & \text{otherwise} \end{cases} \quad (k = 1, 2, \ldots, m),$$

and by (5), (6) we have the following dual pair:

$$\min \{x_0 \mid x_k \geqq b_k \ (k = 1, 2, \ldots, m), x \in \mathcal{P}\}, \tag{13}$$

$$\max \left\{ \sum_{k=1}^{m} b_k \lambda_k \,\middle|\, \lambda \geq 0,\ y_0 = 1,\ y_k = -\lambda_k \ (k = 1, 2, \ldots, m),\ y \in \mathcal{D} \right\}$$

$$= \max \left\{ -\sum_{k=1}^{m} b_k y_k \,\middle|\, y_0 = 1,\ y_k \leq 0 \ (k = 1, 2, \ldots, m),\ y \in \mathcal{D} \right\}. \tag{14}$$

Let A be $(m \times r)$ matrix, $c \in R^r$, and $B^T = [c^T \mid A^T]$. Now if

$$\mathcal{P} = \{x \mid x \in R^{m+1}, x = Bz, z \in R^r\},$$

then

$$\mathcal{D} = \{y \mid y \in R^{m+1}, yB = 0\},$$

i.e.,

$$x_0 = \sum_{j=1}^{r} c_j x_j, \qquad x_k = \sum_{j=1}^{r} a_{kj} z_j, \quad y_0 c_j + \sum_{k=1}^{m} y_k a_{kj} = 0,$$

and (13), (14) can be written as

$$\min\{cz \mid Az \geqq b\},$$

$$\max\{-yb \mid c + yA = 0,\ y \leq 0\} = \max\{vb \mid vA = c,\ v \geq 0\},$$

i.e., we have obtained the dual pair of linear programs. Note that if $B^T = [c^T \mid A^T \mid E]$ and $b_{m+j} = 0$ ($j = 1, 2, \ldots, r$), then we obtain a symmetric dual pair of linear programs. The Kuhn–Tucker function is

$$K(x, y, \lambda) = x_0 + \sum_{k=1}^{m} \lambda_k (b_k - x_k) - \sum_{k=1}^{m} x_k y_k = cz + \lambda(b - Az) + 0$$

$$= cz - yb + yAz = cz + vb - vAz.$$

Remark 4.3 If primal program (13) is consistent its subinfimum coincides with its minimum, so the duality theory of linear programming follows from Theorem 4.2.

The following formulas could be useful to construct special geometric dual programs:

Proposition 4.5. Let $p > 1$, $1/p + 1/q = 1$ (i.e., $q > 1$). Then

(a) $$f(x) = \begin{cases} (\sum x_i^p)^{1/p} & \text{if } x \geq 0 \\ +\infty & \text{otherwise} \end{cases}$$

$$\Rightarrow f^*(y) = \begin{cases} 0 & \text{if } \sum_I y_i^q \leqq 1,\ I = \{i \mid y_i > 0\}, \\ +\infty & \text{otherwise;} \end{cases}$$

(b) $$f(x) = (\sum |x_i|^p)^{1/p} \Rightarrow f^*(y) = \begin{cases} 0 & \text{if } \sum |y_i|^q \leq 1, \\ +\infty & \text{otherwise.} \end{cases}$$

Proposition 4.6. Let $0 < p < 1$, $1/p + 1/q = 1$ (i.e., $q < 0$). Then

$$g(x) = \begin{cases} (\sum x_i^p)^{1/p} & \text{if } x \geq 0 \\ -\infty & \text{otherwise} \end{cases} \Rightarrow g^*(y) = \begin{cases} 0 & \text{if } y > 0,\ \sum y_i^q \geqq 1, \\ -\infty & \text{otherwise.} \end{cases}$$

Proposition 4.7. Let $p < 0$, $1/p + 1/q = 1$ (i.e., $0 < q < 1$). Then

$$g(x) = \begin{cases} (\sum x_i^p)^{1/p} & \text{if } x > 0, \\ 0 & \text{if } x \geq 0,\ x \not> 0 \\ -\infty & \text{if } x \not\geq 0 \end{cases} \Rightarrow g^*(y) = \begin{cases} 0 & \text{if } y \geq 0,\ \sum y_i^q \geqq 1, \\ -\infty & \text{otherwise.} \end{cases}$$

Proposition 4.8. (The limiting cases of Proposition 4.5 when $p \to +\infty$, $q \to 1$):

(a) $$f(x) = \begin{cases} \max_i (x_i) & \text{if } x \geq 0, \\ +\infty & \text{otherwise}; \end{cases}$$

$$\Rightarrow f^*(y) = \begin{cases} 0 & \text{if } \sum_I y_i \leq 1,\ I = \{i \mid y_i > 0\}, \\ +\infty & \text{otherwise}; \end{cases}$$

(b) $$f(x) = \max_i |x_i| \Rightarrow f^*(y) = \begin{cases} 0 & \text{if } \sum |y_i| \leqq 1, \\ +\infty & \text{otherwise}; \end{cases}$$

(c) $$f(x) = \max_i x_i \Rightarrow f^*(y) = \begin{cases} 0 & \text{if } \sum y_i = 1,\ y \geq 0, \\ +\infty & \text{otherwise}. \end{cases}$$

Proposition 4.9. (The limiting cases of Propositions 4.5 and 4.6 when $p \to 1$, $q \to \pm\infty$):

(a) $$f(x) = \begin{cases} \sum x_i & \text{if } x \geq 0 \\ +\infty & \text{otherwise} \end{cases}$$

$$\Rightarrow f^*(y) = \begin{cases} 0 & \text{if } \forall_i\ y_i \leq 1, \text{[i.e., } \max_i (y_i) \leq 1], \\ +\infty & \text{otherwise}; \end{cases}$$

(b) $$f(x) = \sum |x_i| \Rightarrow f^*(y) = \begin{cases} 0 & \text{if } \forall_i\ |y_i| \leq 1, \text{[i.e., } \max_i |y_i| \leq 1], \\ +\infty & \text{otherwise}; \end{cases}$$

(c) $$g(x) = \begin{cases} \sum x_i & \text{if } x \geq 0, \\ -\infty & \text{otherwise}, \end{cases}$$

$$\Rightarrow g^*(y) = \begin{cases} 0 & \text{if } \forall_i\ y_i \geq 1, \quad \text{[i.e., } \min_i (y_i) \geq 1], \\ -\infty & \text{otherwise}. \end{cases}$$

Proposition 4.10. (The limiting case of Proposition 4.6 when $p \to 0$, $q \to 0$):

$$g(x) = \begin{cases} \left(\prod_{i=1}^n x_i \right)^{1/n} & \text{if } x \geq 0, \\ -\infty & \text{otherwise} \end{cases} \quad \Rightarrow \quad g^*(y) = \begin{cases} 0 & \text{if } \left(\prod_{i=1}^n y_i \right)^{1/n} \geqq \frac{1}{n},\ y > 0, \\ -\infty & \text{otherwise}. \end{cases}$$

Proposition 4.11. (The limiting case of Proposition 4.7 when $p \to -\infty$, $q \to 1$):

(a) $$g(x)=\begin{cases}\min\limits_i (x_i) & \text{if } x \ge 0,\\ -\infty & \text{otherwise}\end{cases} \Rightarrow g^*(y)=\begin{cases}0 & \text{if } y\ge 0, \sum y_i \ge 1,\\ -\infty & \text{otherwise;}\end{cases}$$

(b) $$g(x)=\min_i (x_i) \Rightarrow g^*(y)=\begin{cases}0 & \text{if } y\ge 0, \sum y_i = 1,\\ -\infty & \text{otherwise.}\end{cases}$$

Remark 4.4. [The concave analog of (5), (6)]: Let us use the same notation as in Proposition 4.1. If H_k $(k=0, 1, \ldots, m)$ are proper closed concave functions on R^{n_k}, respectively, then we have the following pair of geometric dual programs:

$$\max\{H_0(x_0) \mid H_k(x_k) \ge 0 \quad (k=1,2,\ldots,m),\ x \in \mathscr{P}\},$$
$$\min\{W(y,\lambda) \mid (y,\lambda) \in \operatorname{dom} W,\ y \in \mathscr{D}\},$$

where

$$W(y,\lambda) = -H_0^*(y_o) - \sum_{k=1}^m \lambda_k H_k^*\left(\frac{y_k}{\lambda_k}\right)$$

and

$$\operatorname{dom} W = \{(y,\lambda) \mid \lambda \ge 0, \quad y_0 \in \operatorname{dom} H_0^*,$$
$$y_k \in \lambda_k \operatorname{dom} H_k^* \ (k=1,2,\ldots,m)\}.$$

Example 4.3. (The linear Chebyshev problem) (for details see Example 4.2): Let

$$G_0(x_1, x_2, \ldots, x_s) = \max_{1\le i\le s} (x_i + d_i), \qquad G_k(x_{s+k}) = b_k - x_{s+k}$$
$$(k=1,2,\ldots,m).$$

Then by Proposition 4.8(c) and Elementary Formula (b) of Section 2,

$$G_0^*(y_1, y_2, \ldots, y_s) = \begin{cases} -\sum\limits_{i=1}^s d_i y_i & \text{if } \sum y_i = 1,\ y \ge 0,\\ +\infty & \text{otherwise,}\end{cases}$$

and we have the following dual pair:

$$\min\left\{\max_{1\le i\le s} (x_i+d_i) \mid x_{s+k} \geqq b_k \quad (k=1,2,\ldots,m), \quad x \in \mathscr{P}\right\} \tag{15}$$

$$\max\left\{\sum_{i=1}^{s} d_i y_i - \sum_{k=1}^{m} b_k y_{r+k} \,\middle|\, \sum_{i=1}^{s} y_i = 1, \quad y_i \geqq 0 \quad (i = 1, 2, \ldots, s), \right.$$
$$\left. y_{s+k} \leq 0 \quad (k = 1, 2, \ldots, m), \quad y \in \mathscr{D}\right\}. \tag{16}$$

If

$$\mathscr{P} = \left\{x \,\middle|\, x_i = \sum_{j=1}^{n} c_{ij} z_j \ (i = 1, 2, \ldots, s), x_{s+k} = \sum_{j=1}^{n} a_{kj} z_j \ (k = 1, 2, \ldots, m)\right\},$$

then (15), (16) can be written as

$$\min\left\{\max_{1 \leq i \leq s}\left(\sum_{j=1}^{n} c_{ij} z_j + d_i\right) \,\middle|\, \sum_{j=1}^{m} a_{kj} z_j \geq b_k \quad (k = 1, 2, \ldots, m)\right\},$$

$$\max\left\{\sum_{i=1}^{s} d_i y_i + \sum_{k=1}^{m} b_k v_k \,\middle|\, \sum_{i=1}^{s} y_i = 1, \sum_{k=1}^{m} v_k a_{kj} = \sum_{i=1}^{s} y_i c_{ij}, \quad y \geq 0, \quad v \geq 0\right\}.$$

Example 4.4. Let

$$G_0(x_1, x_2, \ldots, x_s) = \sum_{i=1}^{s} |x_i + d_i|, G_k(x_{s+k}) = b_k - x_{s+k} \qquad (k = 1, 2, \ldots, m).$$

Then by Proposition 4.9(b) and Elementary Formula (b) in Section 2.2,

$$G_0^*(y_1, y_2, \ldots, y_s) = \begin{cases} -\sum d_i y_i & \text{if } |y_i| \leqq 1 \quad (i = 1, 2, \ldots, s), \\ +\infty & \text{otherwise,} \end{cases}$$

and we have the following dual pair:

$$\min\left\{\sum_{i=1}^{s} |x_i + d_i| \,\middle|\, x_{s+k} \geq b_k \quad (k = 1, 2, \ldots, m), \quad x \in \mathscr{P}\right\} \tag{17}$$

$$\max\left\{\sum_{i=1}^{s} d_i y_i - \sum_{k=1}^{m} b_k y_{s+k} \,\middle|\, |y_i| \leq 1, \ (i = 1, 2, \ldots, s), \right.$$
$$\left. y_{s+k} \leq 0 \ (k = 1, 2, \ldots, m), \ y \in \mathscr{D}\right\}. \tag{18}$$

If

$$\mathscr{P} = \left\{x \,|\, x_i = \sum_{j=1}^{r} c_{ij} z_j \quad (i = 1, 2, \ldots, s), x_{s+k} = \sum_{j=1}^{r} a_{kj} z_j \quad (k = 1, 2, \ldots, m)\right\},$$

then (17), (18) can be written as

$$\min\left\{\sum_{i=1}^{s}\left|\sum_{j=1}^{r} c_{ij} z_j + d_i\right| \,\middle|\, \sum_{j=1}^{r} a_{kj} z_j \geq b_k \ (k = 1, 2, \ldots, m)\right\},$$

$$\max\left\{\sum_{i=1}^{s} d_i y_i + \sum_{k=1}^{m} b_k v_k \,\middle|\, -1 \leq y_i \leq 1 \ (i = 1, 2, \ldots, s), \right.$$
$$\left. \sum_{k=1}^{m} v_k a_{kj} = \sum_{i=1}^{s} y_i c_{ij}, v \geq 0\right\}.$$

Example 4.5. Let

$$H_k(x_{2k}, x_{2k+1}) = [(x_{2k}+d_{2k})(x_{2k+1}+d_{2k+1})]^{1/2} \qquad (k=0,1,\ldots,m), \quad (x+d \geqq 0).$$

Then by Proposition 4.10 and Elementary Formula (b) in Section 2.2,

$$H_k^*(y_{2k}, y_{2k+1}) = \begin{cases} -d_{2k}y_{2k} - d_{2k+1}y_{2k+1} & \text{if } (y_{2k}y_{2k+1})^{1/2} \geq 1/2,\ y>0, \\ -\infty & \text{otherwise,} \end{cases}$$

and we have the following dual pair:

$$\max\{(x_0+d_0)(x_1+d_1) \mid (x_{2k}+d_{2k})(x_{2k+1}+d_{2k+1}) \geq b_k^2 \quad (k=1,2,\ldots,m),\ x\in\mathscr{P}\}, \quad (19)$$

$$\min\left\{\sum_{i=0}^{2m+1} d_i y_i - \sum_{k=1}^{m} b_k\lambda_k \,\middle|\, \lambda \geqq 0,\ (y_0y_1)^{1/2} \geqq \tfrac{1}{2},\ (y_{2k}y_{2k+1})^{1/2} \geqq \tfrac{1}{2}\lambda_k,\ y>0,\ y\in\mathscr{D}\right\}$$

$$= \min\left\{\sum_{i=0}^{2m+1} d_i y_i - 2\sum_{k=1}^{m} b_k (y_{2k}y_{2k+1})^{1/2} \,\middle|\, y_0y_1 \geqq \tfrac{1}{4},\ y>0,\ y\in\mathscr{D}\right\}. \quad (20)$$

If

$$\mathscr{P} = \left\{x \,\middle|\, x_i = \sum_{j=1}^{r} a_{ij}z_j \ (i=0,1,\ldots,2m+1)\right\},$$

then (19), (20) can be written as

$$\max\{(\textstyle\sum a_{0j}z_j + d_0)(\sum a_{1j}z_j+d_1) \mid (\sum a_{2k,j}z_j + d_{2k})(\sum a_{2k+1,j}z_j + d_{2k+1}) \geqq b_k^2\},$$

$$\min\left\{\sum_{i=0}^{2m+1} d_i y_i - 2\sum_{k=1}^{m} b_k (y_{2k}y_{2k+1})^{1/2} \,\middle|\, y_0y_1 \geqq \tfrac{1}{4},\ y>0,\ \sum_{i=0}^{2m+1} y_i a_{ij} = 0\right\}.$$

It is worth noting that the restriction $y_0y_1 \geq \frac{1}{4}$ can be in fact replaced by $y_0y_1 = \frac{1}{4}$.

Example 4.6. Let

$$0<p_k<1, \qquad \frac{1}{p_k}+\frac{1}{q_k}=1 \quad (\text{i.e., } q_k<0),$$

$$H_k(x_k) = \begin{cases} \left[\sum_{I_k} (x_i+d_i)^{p_k}\right]^{1/p_k} & \text{if } x+d \geq 0 \quad (k=0,1,\ldots,m), \\ -\infty & \text{otherwise,} \end{cases}$$

where $I_k = \{i_{k-1}+1,\ldots,i_k\}$, $i_{-1}=0$, $i_m=n$, $i_k - i_{k-1} = n_k$. Then by Proposition 4.6,

$$H_k^*(y_k) = \begin{cases} -\sum_{I_k} d_i y_i & \text{if } y>0,\ \sum_{I_k} y_i^{q_k} \geqq 1, \\ -\infty & \text{otherwise,} \end{cases}$$

and we have the following dual pair:

$$\max\left\{\sum_{I_0}(x_i+d_i)^{p_0}\,\middle|\,\sum_{I_k}(x_i+d_i)^{p_k}\geqq b_k^{p_k}\ (k=1,2,\ldots,m),\ x+d\geq 0,\ x\in\mathcal{P}\right\},$$

$$\min\left\{\sum d_iy_i-\sum b_k\lambda_k\,\middle|\,\lambda\geq 0,\ \sum_{I_0}y_i^{q_0}\geqq 1,\ \sum_{I_k}y_i^{q_k}\geqq\lambda_k^{q_k},\ y\in\mathcal{D}\right\}$$

$$=\min\left\{\sum_{i=1}^{n}d_iy_i-\sum_{k=1}^{m}b_k\left(\sum_{I_k}y_i^{q_k}\right)^{1/q_k}\,\middle|\,\sum_{I_0}y_i^{q_0}\geqq 1,\ y\in\mathcal{D}\right\}.$$

References

1. DUFFIN, R. J., PETERSON, E. L., and ZENER, C., *Geometric Programming*, John Wiley and Sons, New York, 1967.
2. ROCKAFELLAR, R. T., *Nonlinear Programming*, talk given at the American Mathematical Society Summer Seminar on the Mathematics of the Decision Sciences, Stanford University, 1967.
3. EGGLESTON, H. G., *Convexity*, Cambridge University Press, New York, 1958.
4. KARLIN, S., *Mathematical Methods and Theory in Games*, Programming and Economics, Vol. 1., Addison-Wesley Publishing Company, Reading, Massachusetts, 1959.
5. VALENTINE, F. A., *Convex Sets*, McGraw-Hill, New York, 1964.

2

Geometric Programming[1]

E. L. PETERSON[2]

Abstract. Contrary to popular belief, geometric programming is not just a special technique for studying the very important class of posynomial (optimization) problems. It is really a very general mathematical theory that is especially useful for studying a large class of separable problems. Its practical efficacy is due mainly to the fact that many important (seemingly inseparable) problems can actually be formulated as separable geometric programming problems, by fully exploiting their linear algebraic structure. Some examples are: nonlinear network flow problems (both single-commodity and multicommodity), discrete optimal control problems with linear dynamics, optimal location problems of the generalized Fermat type, (l_p constrained) l_p regression problems, chemical equilibrium problems, ordinary programming problems, (quadratically constrained) quadratic programming problems, and general algebraic programming problems. The theory of geometric programming includes (i) very strong existence, uniqueness, and characterization theorems, (ii) useful parametric and economic analyses, (iii) illuminating decomposition principles, and (iv) powerful numerical solution techniques. In addition to indicating the true scope of geometric programming and its potential impact on many fields of application, this paper can serve either as a modern introduction to the subject (including an enlightening review of the original posynomial case) or as an up-to-date handbook of the most fundamental aspects of the subject.

1. Introduction

Since its inception by Zener (Refs. 1 and 2) and Duffin (Refs. 3 and 4), geometric programming has undergone rapid development, especially with

[2] Department of Mathematics and Graduate Program in Operations Research, North Carolina State University, Raleigh 27609.

the appearance of the first book on the subject by Duffin, Peterson, and Zener (Ref. 5). Although its essence and scope have recently been broadened and amplified by Peterson (Refs. 6–9), major advances in theory, computation, and application are still occurring as more workers enter the field. The main purpose of this paper is to summarize the present state of the subject and to indicate some of the directions in which it is developing. To keep the length of this paper within reasonable limits, only the most fundamental aspects are presented, and then only within the context of n-dimensional Euclidean space E_n. (Just to list all relevant papers would itself require several additional pages.) Consequently, some important topics have been omitted. A much more thorough treatment is given in Ref. 10.

With respect to notation, the context alone dictates whether a given vector $\mathbf{v}$ in E_n is to be interpreted as a "column vector," or as a "row vector." In all cases, the symbol $\langle\cdot,\cdot\rangle$ indicates the usual "inner product" function.

2. Problem Formulation and Examples

Geometric programming provides a mechanism for formulating and studying in "separable" form many important (usually inseparable) optimization problems. The key to this mechanism is the exploitation of the linearities that are present in a given problem. Such linearities frequently appear as linear equations or linear inequalities, but they can also appear in much more subtle guises, such as matrices associated with nonlinearities.

We shall begin with unconstrained problems and then proceed to (the more complicated) constrained problems. In each case we consider important examples that arise in various fields of application.

2.1. The Unconstrained Case. Classical optimization theory and ordinary mathematical programming are concerned with the minimization (or maximization) of an arbitrary real-valued function g over some given subset $\mathcal{S}$ of its non-empty domain $\mathcal{C} \subseteq E_n$. In geometric programming, the subset $\mathcal{S}$ is required to be the intersect of the function domain $\mathcal{C}$ with an arbitrary cone $\mathcal{X} \subseteq E_n$ (which is, in fact, a vector space for most examples). For purposes of easy reference and mathematical precision, the resulting *geometric programming Problem* $\mathcal{A}$ is now given the following formal definition in terms of classical terminology and notation.

Problem $\mathcal{A}$. Using the "feasible solution" set

$$\mathcal{S} \triangleq \mathcal{X} \cap \mathcal{C},$$

calculate both the "problem infimum"

$$\varphi \triangleq \inf_{x \in \mathcal{S}} g(x)$$

and the "optimal solution" set

$$\mathcal{S}^* \triangleq (x \in \mathcal{S} \mid g(x) = \varphi\}.$$

Each optimization problem can generally be formulated as Problem $\mathcal{A}$ in more than one way by suitably choosing the function g and the cone $\mathcal{X}$. For example, one can always let g by the "objective function" for the given problem simply by choosing $\mathcal{X}$ to be E_n, but that choice is generally not the best possible choice. The reason is that most problems involve a certain amount of linearity (due to the presence of linear equations, linear inequalities, matrices, etc.), which can be conveniently handled through the introduction of an appropriate *nontrivial* subcone $\mathcal{X} \subset E_n$. The presence of such a subcone $\mathcal{X}$ is one of the distinguishing features of geometric programming.

Example 2.1. Perhaps the most striking example of the utility of geometric programming comes from using it to study the *minimization of signomials*. This was first done by Zener (Refs. 1 and 2) and Duffin (Refs. 3 and 4) and served as the initial development (as well as the main stimulus for subsequent developments) of geometric programming.

A "signomial" (sometimes termed a "generalized polynomial") is any function with the form

$$P(t) = \sum_{i=1}^{n} c_i t_1^{a_{i1}}, t_2^{a_{i2}}, \cdots, t_m^{a_{im}},$$

where the coefficients c_i and the exponents a_{ij} are arbitrary constants, but the independent variables t_j are restricted to be positive. After much experience in the physical sciences, engineering, and operations research, Zener clearly recognized that many optimization problems of practical importance can be accurately modeled with such functions. In many cases they come directly from the laws of nature and/or economics. In other cases, this functional form gives a good fit to empirical data over a wide range of the variables t_j. Actually, the signomials that occur in such cases frequently have positive coefficients, in which event they are termed "posynomials."

The presence of the "exponent matrix" (a_{ij}) (which is, of course, associated with algebraic nonlinearities) is the key to applying geometric programming to signomial optimization. To effectively place the problem of minimizing $P(t)$ in the format of Problem $\mathcal{A}$, simply make the change of variables

$$x_i = \sum_{j=1}^{m} a_{ij} \log t_j, \qquad i = 1, 2, \ldots, n,$$

and then use the laws of exponents to help infer that minimizing $P(t)$ is equivalent to solving Problem $\mathcal{A}$ when

$$\mathcal{C} \triangleq E_n, \qquad g(x) \triangleq \sum_{i=1}^{n} c_i e^{x_i},$$

and

$$\mathcal{H} \triangleq \text{column space of } (a_{ij}).$$

The advantages of studying this Problem $\mathcal{A}$ rather than its signomial predecessor are numerous. For example, unlike the signomial P, the exponential function g is completely separable (in that it is a sum of terms, each of which depends on only a single independent variable x_i). Moreover, if P is actually a posynomial, then g is, of course, strictly convex (even though P itself clearly need not even be convex). Consequently, if t^* minimizes a posynomial P, then the corresponding x^* must be a unique optimal solution to Problem $\mathcal{A}$, in which event the set of all t that minimize P can be obtained from x^* simply by solving the displayed system of equations (a task that is relatively easy because the system is clearly linear in terms of $\log t_j$, $j = 1, 2, \ldots, m$). In Duffin, Peterson, and Zener (Ref. 5), Avriel and Williams (Ref. 11), Duffin and Peterson (Refs. 12–15), as well as Abrams and Bunting (Ref. 16), and some of the references cited therein, these properties and others that are too complicated to describe here have been combined into a very comprehensive existence, uniqueness, and characterization theory for signomial (and especially posynomial) optimization. Moreover, in Falk (Ref. 17) the complete separability induced into signomial optimization forms the basis for a branch-and-bound algorithm that converges to *globally* optimal solutions to (intrinsically nonconvex) signomial optimization problems.

Example 2.2. Our second example comes from the *minimization of quadratic functions*

$$Q(z) = (\tfrac{1}{2})\langle z, Hz\rangle + \langle \mathbf{h}, z\rangle,$$

where H is an arbitrary constant matrix and $\mathbf{h}$ is an arbitrary constant vector.

A factorization of the coefficient matrix H (which is, of course, associated with quadratic nonlinearities) is the key to effectively applying geometric programming to quadratic programming. More specifically, linear algebra is used to compute matrices D and $\mathcal{D}$ such that

$$H = D^{\mathsf{T}}D - \mathcal{D}^{\mathsf{T}}\mathcal{D},$$

where ^{T} indicates the transpose operation. In terms of D and $\mathcal{D}$ the quadratic function

$$Q(\mathfrak{z}) = (\tfrac{1}{2})(\langle D\mathfrak{z}, D\mathfrak{z}\rangle - \langle \mathscr{D}\mathfrak{z}, \mathscr{D}\mathfrak{z}\rangle) + \langle \mathbf{h}, \mathfrak{z}\rangle.$$

Of course, the expression $\langle D\mathfrak{z}, D\mathfrak{z}\rangle$ is not present when $Q(\mathfrak{z})$ is negative semidefinite, and the expression $-\langle \mathscr{D}\mathfrak{z}, \mathscr{D}\mathfrak{z}\rangle$ is not present when $Q(\mathfrak{z})$ is positive semidefinite (i.e., a convex function).

From elementary linear algebra we now infer that minimizing $Q(\mathfrak{z})$ is equivalent to solving Problem $\mathscr{A}$ when

$$\mathscr{C} \triangleq E_{2m+1},$$

$$g(x) \triangleq (\tfrac{1}{2})\left(\sum_{i=1}^{m} x_i^2 - \sum_{i=m+1}^{2m} x_i^2\right) + x_{2m+1},$$

and

$$\mathscr{X} \triangleq \text{column space of} \begin{bmatrix} D \\ \mathscr{D} \\ \mathbf{h} \end{bmatrix}.$$

Notice that, unlike the quadratic function Q, the quadratic function g is completely separable, a fact that can be exploited both theoretically and computationally.

It is useful to introduce some additional parameters into the preceding function g so that a much broader class of optimization problems can be studied. In particular, we redefine g so that

$$g(x) \triangleq \sum_{i=1}^{m} p_i^{-1} |x_i - b_i|^{p_i} - \sum_{i=m+1}^{2m} p_i^{-1} |x_i - b_i|^{p_i} + x_{2m+1} - b_{2m+1},$$

where b_i and p_i are arbitrary constants, and $|\cdot|$ designates the absolute value function. Notice that the function g is still completely separable and can be specialized to the quadratic case by choosing $b_i = 0$ and $p_i = 2$ for each i.

Another interesting specialization is obtained by choosing $p_i = p$ for each i, while choosing $\mathscr{D} = 0$ and $\mathbf{h} = \mathbf{0}$. The resulting problem consists essentially of finding the "best l_p-norm approximation" to the fixed vector $(b_1, \ldots, b_m)$ by vectors in the column space of the matrix D, a fundamental problem in *linear regression analysis*.

A detailed analysis of this rather broad class of optimization problems can be found in Peterson and Ecker (Ref. 18) and the references cited therein.

Example 2.3. Our third example comes from the *optimal location* of a new facility relative to existing facilities. We suppose that there are p existing facilities with fixed locations $\mathbf{b}^1, \mathbf{b}^2, \ldots, \mathbf{b}^p$ in E_m, and we assume that for each facility i there is a cost $d_i(z, \mathbf{b}^i)$ of choosing the new facility location z relative to $\mathbf{b}^i$. In many instances, the functions d_i are just "metrics" that reflect the cost of shipping material between the two locations. (Such metrics are usually determined by the available transportation systems.) The problem then is to choose a new location z that minimizes the total cost $d(z) = \sum_{i=1}^{p} d_i(z, \mathbf{b}^i)$.

In this problem statement there is no matrix that serves as the key to effectively applying geometric programming. However, minimizing $d(z)$ is clearly equivalent to solving Problem $\mathcal{A}$ when

$$\mathcal{C} \triangleq E_{pm}, \qquad g(x) \triangleq \sum_{i=1}^{p} d_i(x^i, \mathbf{b}^i),$$

and

$$\mathcal{X} \triangleq \text{column space of} \begin{bmatrix} \mathcal{I} \\ \mathcal{I} \\ \vdots \\ \mathcal{I} \end{bmatrix},$$

where $x = (x^1, x^2, \ldots, x^p)$ and there are a total of p $(m \times m)$ identity matrices $\mathcal{I}$.

Notice that, unlike the function d, the function g is at least partially separable in that it is a sum of terms, each of which depends on only a single independent vector variable x^i. This separability occurs in even more complicated location problems and has been exploited both theoretically and computationally by Peterson and Wendell (Ref. 19) and the references cited therein.

Example 2.4. Our fourth example comes from *discrete optimal control with linear dynamics* (or dynamic programming with linear transition equations). We suppose that for each "stage" i there is a cost $g_i(s^i, d^i)$ that depends on the ith "state" s^i and the ith "decision" d^i, where the domain of the cost function g_i is the Cartesian product $R_i \times D_i$ of the ith "state set" R_i and the ith "decision set" D_i. We also suppose that the "initial state" s^1 is determined by the "initial decision" d^1 through the equation $s^1 = B_1 d^1$ and that each subsequent state s^i is determined by both the ith decision d^i and the $(i-1)$th state s^{i-1} through the "transition equation" $s^i = A_i s^{i-1} + B_i d^i$, where A_i and B_i are constant matrices. Given that there is a total of p stages, the problem is to make sequential decisions d^i that minimize the total cost $\sum_{i=1}^{p} g_i(s^i, d^i)$.

The presence of the matrices A_i and B_i is the key to applying geometric programming to discrete optimal control. To effectively place the preceding control problem in the format of Problem $\mathscr{A}$, simply let

$$x = (r^1, d^1, r^2, d^2, \ldots, d^{p-1}, r^p, d^p),$$

and then observe that the preceding control problem is equivalent to Problem $\mathscr{A}$ when

$$\mathscr{C} \triangleq \bigtimes_{i=1}^{p} (R_i \times D_i),$$

$$g(x) \triangleq \sum_{i=1}^{p} g_i(r^i, d^i),$$

and

$$\mathscr{X} \triangleq \{x \mid r^1 = B_1 d^1 \text{ and } r^i = A_i r^{i-1} + B_i d^i, i = 2, \ldots, p\}.$$

The partial separability of g and the "sparsity" of the matrix whose columns span $\mathscr{X}$ has been exploited both theoretically and computationally by Dinkel and Peterson (Ref. 20 and the references cited therein).

Example 2.5. Our fifth example comes from an *analysis of multi-commodity transportation networks.* We consider a "graph" whose "arcs" are enumerated from 1 through n in such a way that for $1 \leqq i \leqq \nu$, (return) arc i connects the origin and destination of commodity i, while for $\nu + 1 \leqq i \leqq n$, (roadway) arc i represents a collection of unidirectional lanes over which traffic can flow between two adjacent (intersectional) "nodes." Each return arc i is "directed" from the destination of the corresponding commodity i back to the origin of the same commodity i, and each roadway arc i has the given direction in which traffic is permitted to flow.

Each commodity i can flow from its origin to its destination only over certain (predetermined) feasible (roadway) "paths" $\mathbf{p}^j$, which are enumerated by a finite index set $[i]$. Each path $\mathbf{p}^j$ is, in essence, an n-vector whose kth component p_k^j is to be identified with arc k. In particular, for a given j in $[i]$, component p_k^j is one for (return arc) $k = i$, zero for all other (return arcs) k between 1 and ν, and either one or zero for all (roadway arcs) k between $\nu + 1$ and n, depending respectively on whether roadway arc k is or is not part of path $\mathbf{p}^j$.

For notational convenience in describing feasible flow patterns, all feasible paths $\mathbf{p}^j$ are enumerated sequentially by letting $[i] = \{m_i, m_i + 1, \ldots, n_i\}$, where $1 = m_1 \leqq n_1, n_1 + 1 = m_2 \leqq n_2, \ldots, n_{\nu-1} + 1 = m_\nu \leqq n_\nu$. Thus there is a total of m feasible paths over which traffic can flow, where $m \triangleq n_\nu$. Moreover, a possible commodity flow pattern z is then just a nonnegative m-vector whose jth component z_j is simply the "input flow" on path $\mathbf{p}^j$ of that commodity i for which j is in $[i]$. Of course, each possible commodity flow pattern z produces a possible total flow pattern

$x \triangleq \sum_{j=1}^{m} y_j \mathbf{p}^j$, whose kth component x_k is simply the resulting total flow of all commodities on arc k. The feasible flow patterns are then those possible flow patterns for which $x_i = d_i$ for $1 \leqq i \leqq \nu$, where d_i is a given (nonnegative) total input flow of commodity i.

We now assume that traffic flow on each roadway arc i produces a cost $g_i(x_i)$ that depends only on the total flow x_i. The problem then is to determine those feasible flow patterns that minimize the total cost $\sum_{i=\nu+1}^{n} g_i(x_i)$. Of course, this problem is relevant only if the given transportation network can be centrally controlled—which is usually not the case for highway networks.

For *highway networks*, it is far more realistic to use the same type of separable objective function, but let each function g_i be the (indefinite) integral of a travel cost function c_i (rather than let each g_i be a cost function itself). For many highway networks, the most realistic travel cost function c_i is actually just the travel time $c_i(x_i)$ required to traverse arc i when it is carrying a total traffic flow x_i. In any case, the reason for integrating the cost functions c_i prior to forming the objective function $\sum_{i=\nu+1}^{n} g_i$ is that under certain (relatively weak) conditions, the resulting optimal flow patterns (which are, of course, not usually cost optimal) are in a state of "Wardrop equilibrium"; that is, the origin-to-destination travel cost (e.g., the origin-to-destination travel time) for a given commodity i is identical on all paths used by that commodity and is not greater than what it would be on its unused feasible paths (given the same total flow pattern). Such flow patterns are of interest to highway traffic analysts, because highway traffic scientists contend that the traffic flow patterns of many complicated real-world highway networks are in, or at least tend toward, a state of Wardrop equilibrium.

In any event, the presence of the feasible path vectors $\mathbf{p}^j$ is the key to applying geometric programming to the preceding network problems. To effectively place those problems in the format of *Problem* $\mathscr{A}$, first enlarge the set of functions g_i by letting

$$\mathscr{C}_i \triangleq \begin{cases} \{d_i\}, & 1 \leqq i \leqq \nu, \\ [0, +\infty), & \nu+1 \leqq i \leqq n, \end{cases}$$

and

$$g_i(x_i) \triangleq \begin{cases} 0, & 1 \leqq i \leqq \nu, \\ g_i(x_i), & \nu+1 \leqq i \leqq n. \end{cases}$$

Then, observe that the preceding network problems are equivalent to

Problem $\mathscr{A}$ when

$$\mathscr{C} \triangleq \bigtimes_{i=1}^{n} \mathscr{C}_i,$$

$$g(x) \triangleq \sum_{i=1}^{n} g_i(x_i),$$

and

$$\mathscr{X} \triangleq \left\{ x = \sum_{j=1}^{m} z_j \mathbf{p}^j \,\middle|\, z_j \geqq 0 \text{ for } 1 \leqq j \leqq m \right\}.$$

Note that $\mathscr{X}$ is not a vector space, but is instead a "pointed" polyhedral cone generated by the feasible path vectors $\mathbf{p}^j$. This important class of problems is studied more thoroughly by Hall and Peterson (Ref. 21 and some of the references cited therein).

A closely related (but totally different) class of network flow problems occurs in the context of *electric and hydraulic networks.* Such problems involve only a single commodity (electricity or fluid) and can be effectively placed in the format of Problem $\mathscr{A}$ in the following way: let $c_i(x_i)$ be the voltage drop or pressure drop respectively, across arc i as a function of the commodity flow x_i in arc i [in which event $g_i(x_i)$ is termed the "content" of arc i and is frequently just the power dissipated in arc i], and let $\mathscr{X}$ be the vector space of all (nonunidirectional) feasible flows (i.e., all those flows that satisfy the "Kirchhoff current conservations laws"). Such problems are studied more thoroughly by Duffin (Ref. 22), Minty (Ref. 23), and Rockafellar (Ref. 24) and the references cited therein.

The reader who wishes to avoid the complications inherent with constraints can skip the next section and begin again with Section 2.3.

2.2. The Constrained Case. To generalize geometric programming by incorporating explicit constraints into the preceding problem formulation, we introduce two nonintersecting (possibly empty) positive-integer index sets I and J with finite cardinality $o(I)$ and $o(J)$ respectively. In terms of these index sets I and J we also introduce the following notation and hypotheses:

(1a) For each $k \in \{0\} \cup I \cup J$, there is a function g_k with domain $C_k \subseteq E_{n_k}$, and there is a set $D_j \subseteq E_{n_j}$ for each $j \in J$.

(2a) For each $k \in \{0\} \cup I \cup J$, there is an independent vector variable $\mathbf{x}^k$ in E_{n_k}, and there is an independent vector variable $\boldsymbol{\kappa}$ with components κ_j for each $j \in J$.

(3a) $\mathbf{x}^I$ denotes the Cartesian product of the vector variables $\mathbf{x}^i$, $i \in I$, and $\mathbf{x}^J$ denotes the Cartesian product of the vector variables $\mathbf{x}^j$, $j \in J$. Hence the Cartesian product $(\mathbf{x}^0, \mathbf{x}^I, \mathbf{x}^J) \triangleq \mathbf{x}$ of $\mathbf{x}^0, \mathbf{x}^I$, and $\mathbf{x}^J$ is an independent vector variable in E_n, where

$$n \triangleq n_0 + \sum_I n_i + \sum_J n_j.$$

(4a) There is a cone $X \subseteq E_n$.

For purposes of easy reference and mathematical precision, the resulting *geometric programming Problem A* is now given the following formal definition in terms of classical terminology and notation.

Problem A. Consider the objective function G whose domain

$$C \triangleq \{(\mathbf{x}, \boldsymbol{\kappa}) \mid \mathbf{x}^k \in C_k, \ k \in \{0\} \cup I, \text{ and } (\mathbf{x}^j, \kappa_j) \in C_j^+, \ j \in J\},$$

and whose functional value

$$G(\mathbf{x}, \boldsymbol{\kappa}) \triangleq g_0(\mathbf{x}^0) + \sum_J g_j^+(\mathbf{x}^j, \kappa_j),$$

where

$$C_j^+ \triangleq \{(\mathbf{x}^j, \kappa_j) \mid \text{either } \kappa_j = 0 \text{ and } \sup_{\mathbf{d}^j \in D_j} \langle \mathbf{x}^j, \mathbf{d}^j \rangle < +\infty, \text{ or } \kappa_j > 0 \text{ and } \mathbf{x}^j \in \kappa_j C_j\},$$

and

$$g_j^+(\mathbf{x}^j, \kappa_j) \triangleq \begin{cases} \sup\limits_{\mathbf{d}^j \in D_j} \langle \mathbf{x}^j, \mathbf{d}^j \rangle & \text{if } \kappa_j = 0 \text{ and } \sup\limits_{\mathbf{d}^j \in D_j} \langle \mathbf{x}^j, \mathbf{d}^j \rangle < +\infty, \\ \kappa_j g_j(\mathbf{x}^j / \kappa_j) & \text{if } \kappa_j > 0 \text{ and } \mathbf{x}^j \in \kappa_j C_j. \end{cases}$$

Using the feasible solution set

$$S \triangleq \{(\mathbf{x}, \boldsymbol{\kappa}) \in C \mid \mathbf{x} \in X, \text{ and } g_i(\mathbf{x}^i) \leqq 0, \ i \in I\},$$

calculate both the problem infimum

$$\varphi \triangleq \inf_{(\mathbf{x}, \boldsymbol{\kappa}) \in S} G(\mathbf{x}, \boldsymbol{\kappa})$$

and the optimal solution set

$$S^* \triangleq \{(\mathbf{x}, \boldsymbol{\kappa}) \in S \mid G(\mathbf{x}, \boldsymbol{\kappa}) = \varphi\}.$$

Of course, the unconstrained case occurs when $I = J = \varnothing$, $g_0 : C_0 \triangleq \mathscr{g} : \mathscr{C}$, and $X \triangleq \mathscr{X}$.

When $D_j = E_{n_j}$ (which is frequently the situation), a simplification results from noting that

$$\sup_{\mathbf{d}^j \in D_j} \langle \mathbf{x}^j, \mathbf{d}^j \rangle < +\infty \qquad \text{only if } \mathbf{x}^j = 0, \text{ in which event } \sup_{\mathbf{d}^j \in D_j} \langle \mathbf{x}^j, \mathbf{d}^j \rangle = 0.$$

In particular then, the functional domain

$$C_j^+ = \{(\mathbf{x}^j, \kappa_j) \mid \kappa_j \geqq 0 \text{ and } \mathbf{x}^j \in \kappa_j C_j\},$$

and the functional values

$$g_j^+(\mathbf{x}^j, \kappa_j) = \kappa_j g_j(\mathbf{x}^j/\kappa_j),$$

with the understanding that $0g_j(\mathbf{0}/0) \triangleq 0$.

In defining the feasible solution set S it is important to make a sharp distinction between the *cone condition* $\mathbf{x} \in X$ and the *constraints* $g_i(\mathbf{x}^i) \leqq 0$, $i \in I$, both of which restrict the vector variable $(\mathbf{x}, \boldsymbol{\kappa})$. In many cases the cone X is polyhedral (and hence is finitely generated); and in most examples of practical significance X is actually a vector space (and hence has a finite basis). Consequently, the cone condition $\mathbf{x} \in X$ can frequently be automatically satisfied and therefore explicitly eliminated by a linear transformation of $\mathbf{x}$ that results from the introduction of generating vectors or basis vectors for X (whereas the generally nonlinear constraints $g_i(\mathbf{x}^i) \leqq 0$, $i \in I$, usually can not be explicitly eliminated by even a nonlinear transformation). Nevertheless, even when it is possible to do so, we do not explicitly eliminate the cone condition $\mathbf{x} \in X$, because such a linear transformation would clearly introduce a common vector variable into the arguments of g_0, g_i, and g_j^+. Such a common vector variable only tends to camouflage one of the extremely useful characteristics of geometric programming—its (partial) separability. Such separability is clearly present even when the functions $g_k : C_k$, $k \in \{0\} \cup I \cup J$ are inseparable.

Since each optimization problem can generally be formulated as Problem $\mathscr{A}$ in more than one way by suitably choosing the functions $g_k : C_k$, $k \in \{0\} \cup I \cup J$ and the cone X, a very important aspect of applied geometric programming is the exploitation of this flexibility in such a way that a given inseparable problem is formulated as an equivalent Problem $\mathscr{A}$ with as much function separability as possible. As in the unconstrained case, the key to such a formulation is usually the introduction of an appropriate nontrivial cone X to handle the linearities that are present in a given problem. Such linearities frequently appear as linear equations or linear inequalities, but they can also appear in rather subtle guises, such as matrices associated with nonlinearities.

The function separability induced in the objective function for each of the unconstrained examples given in Section 2.1 can also be induced in any constraint function of the same general type. We now use signomial optimization to illustrate the general procedure for doing so.

Example 2.6. First, we make the following choices:

$$I = \{1, 2, \ldots, p\} \quad \text{and} \quad J = \varnothing,$$

$$C_k \triangleq E_{n_k},$$

and

$$g_k(\mathbf{x}^k) \triangleq \sum_{[k]} c_q e^{x_q} - d_k,$$

where

$$[k] = \{m_k, m_k + 1, \ldots, n_k\}$$

and

$$1 = m_0 \leqq n_0, \quad n_0 + 1 = m_1 \leqq n_1, \quad \ldots, \quad n_{p-1} + 1 = m_p \leqq n_p = n,$$

$$X \triangleq \text{column space of } [a_{qr}],$$

where

$$[a_{qr}] \text{ is any } n \times m \text{ matrix.}$$

Now, note that all functions in Problem $\mathscr{A}$ are completely separable.

To relate problem A to *constrained signomial optimization*, explicitly eliminate the vector space condition $\mathbf{x} \in X$ by the (essentially linear) transformation

$$x_q = \sum_{r=1}^{m} a_{qr} \log t_r, \qquad q = 1, 2, \ldots, n.$$

By virtue of the laws of exponents, Problem A is now clearly equivalent to the following (generally inseparable) signomial optimization problem:

Minimize
$$\sum_{[0]} c_q \prod_{1}^{m} t_r^{a_{qr}}$$

subject to

$$\sum_{[k]} c_q \prod_{1}^{m} t_r^{a_{qr}} \leqq d_k, \qquad k = 1, 2, \ldots, p,$$

and

$$\mathbf{t} > 0.$$

Of course, the preceding procedure is usually reversed in practice; that is, the signomial form of Problem A tends to occur more often than Problem

A in real-world applications, but is transformed into problem A so that the complete separability of the resulting exponential functions can be exploited. Actually, signomial optimization problems (as well as more general "algebraic optimization problems") should usually be reduced to much simpler signomial optimization problems prior to their transformation to an appropriate Problem A. To see how to reduce such problems to signomial optimization problems in which each signomial has at most two terms, both of which have the same sign, consult Duffin (Ref. 25) and Duffin and Peterson (Refs. 13 and 15). If such problems could be further reduced to signomial problems in which each signomial has only a single term, all algebraic optimization problems (and hence essentially all optimization problems involving only continuous functions) could be reduced to (finite-dimensional) linear programming problems. Even though such a reduction will not be accomplished in the future, the reductions given in the preceding references are already starting to be exploited, both theoretically and computationally.

The reader who is interested in the applications of signomial and posynomial optimization should consult the recent book by Zener (Ref. 26) along with recent papers by Wilde (Ref. 27) and Woolsey (Ref. 28) as well as the comprehensive list of papers compiled by Rijckaert (Ref. 29).

The procedure for inducing function separability into constrained versions of each of the other four examples given in Section 2.1 will be left to the imagination of the interested reader, who can also consult the references already cited in Section 2.1.

Linear programming can be viewed as a special case of geometric programming in at least three different ways. We now present the easiest of the three ways.

Example 2.7. First, make the following choices:

$$J = \varnothing,$$

$$g_0: E_1 \to E_1 \quad \text{such that } g_0(x^0) \triangleq x^0,$$

and

$$g_i: E_1 \to E_1 \quad \text{such that } g_i(x^i) \triangleq x^i - b_i, \qquad i \in I,$$

where the b_i, $i \in I$, are arbitrary constants;

$$X \triangleq \{(x^0, \mathbf{x}^I) \in E_n \mid x^0 = \langle \mathbf{a}, \mathbf{z} \rangle \text{ and } \mathbf{x}^I = M\mathbf{z} \text{ for at least one } \mathbf{z} \in E_m \text{ for which } z_j \geqq 0,\ j \in \mathcal{P}\},$$

where $\mathbf{a}$ is an arbitrary vector in E_m, M is an arbitrary $o(I) \times m$ matrix, and $\mathcal{P}$ is any subset of $\{1, 2, \ldots, m\}$. Now, note that the most complicated function g_k in Problem A is just the simplest kind of affine function.

To relate Problem A to *linear programming*, explicitly eliminate the cone condition $\mathbf{x} \in X$ by the linear transformation used in the defining equation for X. Problem A is then clearly equivalent to the following very general linear programming problem:

Minimize $\langle \mathbf{a}, \mathbf{z} \rangle$

subject to $M\mathbf{z} \leqq \mathbf{b}$,

$$z_j \geqq 0, \qquad j \in \mathcal{P}.$$

Two other ways in which to view linear programming from a geometric programming point of view are given by Peterson (Ref. 7), but none of the three ways have yet had other than pedagogical influence on linear programming.

The following example indicates the generality of geometric programming.

Example 2.8. First, make the following choices:

$$J = \varnothing,$$

$$n_k = m \quad \text{and} \quad C_k \triangleq C_0, \qquad k \in \{0\} \cup I,$$

where C_0 is an arbitrary subset of E_m,

$$X \triangleq \text{column space of} \begin{bmatrix} U \\ U \\ \vdots \\ U \end{bmatrix},$$

where there is a total of $1 + o(I)$ identity matrices U that are $m \times m$.

To relate Problem A to *ordinary programming*, explicitly eliminate the vector space condition $\mathbf{x} \in X$ by the linear transformation

$$\begin{pmatrix} \mathbf{x}^0 \\ \mathbf{x}^I \end{pmatrix} = \begin{bmatrix} U \\ U \\ \vdots \\ U \end{bmatrix} \mathbf{z}.$$

Problem A is then clearly equivalent to the following very general "ordinary programming problem":

Minimize $g_0(\mathbf{z})$

subject to $g_i(\mathbf{z}) \leqq 0, \qquad i \in I,$

$$\mathbf{z} \in C_o.$$

Thus ordinary programming can be viewed as a special case of geometric programming.

Yet, in a certain sense, ordinary programming is no more special than geometric programming—as can be seen by the following reverse specialization:

$$C_o \triangleq \{\mathbf{g} = (\mathbf{x}, \boldsymbol{\kappa}) \mid (\mathbf{x}, \boldsymbol{\kappa}) \in C \text{ and } \mathbf{x} \in X\},$$

$$g_0(\mathbf{z}) \triangleq G(\mathbf{x}, \boldsymbol{\kappa}),$$

$$g_i(\mathbf{z}) = g_i(\mathbf{x}^i), \qquad i \in I.$$

Note, though, that the important structural features of geometric programming are obscured in this ordinary prgramming formulation. On the other hand, ordinary programming is actually made partially separable in its geometric programming formulation, namely:

$$\begin{array}{ll} \text{Minimize} & g_0(\mathbf{x}^0) \\ \text{subject to} & g_1(\mathbf{x}^1) \leqq 0,\ g_2(\mathbf{x}^2) \leqq 0, \ldots, g_p(\mathbf{x}^p) \leqq 0, \\ & \mathbf{x}^0 - \mathbf{x}^1 = 0, \\ & \mathbf{x}^1 - \mathbf{x}^2 = 0, \\ & \quad \ddots \qquad \vdots \\ & \mathbf{x}^{p-1} - \mathbf{x}^p = 0, \\ & \mathbf{x}^0,\ \mathbf{x}^1,\ \mathbf{x}^2,\ \ldots\ \mathbf{x}^{p-1},\ \mathbf{x}^p \in C_0, \end{array}$$

where

$\{1, 2, \ldots, p\} = I.$

In all of the examples given so far, the index set J is empty. Possibly the most important example for which J is not empty is the following "generalized chemical equilibrium problem."

Example 2.9. First, make the following choices:

$$I = \varnothing \quad \text{and} \quad J = \{1, 2, \ldots, p\},$$

$$C_k \triangleq \left\{\mathbf{x}^k \in E_{n_k} \mid x_q^k \geq 0 \text{ and } \sum_{[k]} x_q^k = 1\right\},$$

and

$$g_k(\mathbf{x}^k) \triangleq \sum_{[k]} x_q \log\left(\frac{x_q}{c_q}\right) \quad \text{with } c_q > 0$$

where

$$[k]=\{m_k, m_k+1, \ldots, n_k\}$$

and

$$1=m_0 \leqq n_0, \qquad n_0+1=m_1 \leqq n_1, \qquad \ldots, \qquad n_{p-1}+1=m_p \leqq n_p=n,$$

$$D_j = E_{n_j}, \qquad j=1,2,\ldots,p,$$

$$X \triangleq (\text{column space of } [a_{qr}])^{\perp}$$

where

$$[a_{qr}] \text{ is any } n \times m \text{ matrix.}$$

Now, note that Problem A is essentially the following *generalized chemical equilibrium problem*:

Minimize

$$\sum_{[0]} x_q \log\left(\frac{x_q}{c_q}\right) + \sum_{k=1}^{p} \left\{\sum_{[k]} x_q \log\left(\frac{x_q/\kappa_k}{c_q}\right)\right\}$$

subject to

$$x_q \geqq 0, \qquad q=1,2,\ldots,n,$$

$$\sum_{[0]} x_q = 1 \quad \text{and} \quad \sum_{[k]} x_q = \kappa_k, \qquad k=1,2,\ldots,p,$$

$$\sum_{q=1}^{n} a_{qr} x_q = 0, \qquad r=1,2,\ldots,m.$$

If $[0]=\{1\}$ and $c_1=1$, the preceding problem clearly reduces to the following *Gibbs' multiphase chemical equilibrium problem*:

Minimize

$$\sum_{k=1}^{p} \left\{\sum_{[k]} x_q \log\left(\frac{x_q/\kappa_k}{c_q}\right)\right\}$$

subject to

$$x_q \geqq 0, \qquad q=2,\ldots,n,$$

$$\kappa_k = \sum_{[k]} x_q, \qquad k=1,2,\ldots,p,$$

$$\sum_{q=2}^{n} a_{qr} x_q = -a_{1r}, \qquad r=1,2,\ldots,m.$$

According to "Gibbs' variational principle," an optimal solution $(\mathbf{x}^*, \boldsymbol{\kappa}^*)$ to this problem provides the "equilibrium mole fraction" x_q^*/κ_k^* for each "chemical species" $q \in [k]$ that can be "chemically formed" from the m "chemical elements" present in "phase" k of a p-phase "ideal chemical system"—provided that (i) $-\log c_q \triangleq (F_q^0/RT) + \log P$, where F_q^0 is the (empirically determined) "Gibbs' free energy" per "mole" of species q at the given (externally fixed) "reaction temperature" T and 1 "atmosphere of pressure," while R and P are the universal "gas constant" and the given (externally fixed) "reaction pressure," respectively, and (ii) a_{qr}, for $2 \leqq q \leqq n$, is the number of "atoms" of the rth chemical element appearing in a "molecule" of chemical species q, while $-a_{1r}$ is the total number of moles of the rth chemical element in the whole chemical system.

The preceding specialization and many of its consequences were first established by Passy and Wilde (Ref. 30), and were also included as Appendix C in Duffin, Peterson, and Zener (Ref. 5).

In all of the examples given here, either the index set I or the index set J is empty. Possibly the most important example for which neither I nor J is empty is the "generalized min–max location problem" that is described and thoroughly investigated in Section 5 of Peterson and Wendell (Ref. 19).

2.3. A Summary. Experience seems to indicate that any optimization problem involving matrices, linear equations, affine sets, or cones (and even certain other optimization problems, such as optimal location problems) can probably be transformed into a geometric programming problem that is considerably more separable and hence much more amenable to analysis and solution than the original problem. This is especially true for each of the preceding examples, but the complete exploitation of this fact is far beyond the scope of this paper.

In the remaining part of this paper we present only the most basic theory of geometric programming—a theory that is equally applicable to all problem classes and does not, for the most part, actually require separability of the functions $g : \mathscr{C}$ and $g_k : C_k$, $k \in \{0\} \cup I \cup J$. Such function separability does, however, become extremely useful when specific problem classes are to be thoroughly investigated and solved.

3. Basic Theory

The basic theory of geometric programming can be conveniently partitioned into several topics. Certain "optimality conditions" describe important properties possessed by all optimal solutions, and in many cases collectively characterize all optimal solutions. Appropriate "Lagrangians" provide important "saddle-point" characterizations of optimality, and can also be used to introduce the even more significant concepts of "duality." The latter topic provides important "existence and uniqueness theorems" for optimal solutions, as well as useful "algorithmic stopping criteria." Duality is also a key ingredient in "parametric programming" and "post-optimality analysis," which are in turn key ingredients in certain important "decomposition principles."

Since many important problem classes are unconstrained (e.g., the network flow problems given as Example 2.5), and since the theory for the unconstrained case is far simpler than that for the constrained case, we initially limit our attention to the unconstrained case. Actually, in doing so there is no loss of generality, as explained in Section 3.2.

3.1. The Unconstrained Case. Let $\mathcal{Y}$ be the "dual" of the cone $\mathcal{X}$, that is,

$$\mathcal{Y} \triangleq \{y \in E_n \mid 0 \leqq \langle x, y\rangle \text{ for each } x \in \mathcal{X}\}.$$

Rather elementary considerations show that $\mathcal{Y}$ is generally a closed convex cone. Moreover, $\mathcal{Y}$ is "polyhedral" (i.e., "finitely generated") when $\mathcal{X}$ is polyhedral; and $\mathcal{Y}$ is the "orthogonal complement" $\mathcal{X}^{\perp}$ of $\mathcal{X}$ when $\mathcal{X}$ is actually a vector space. In fact, $\mathcal{Y}$ can be computed via elementary linear algebra for each of the examples given in Section 2.1.

Some of the following subsections can be omitted. In particular, either Section 3.1.1 (on optimality conditions) or Section 3.1.3 (on Lagrangian saddle points) can be omitted without serious loss of continuity. Moreover, Section 3.1.2 can be omitted by those readers who are already sufficiently familiar with the "conjugate transformation, subgradients, and convex analysis."

3.1.1. Optimality Conditions. We begin with the following fundamental definition.

Definition 3.1. A *critical solution* (stationary solution, equilibrium solution, P solution) for Problem $\mathcal{A}$ is any vector x^* that satisfies the following *P optimality conditions*:

$$x^* \in \mathcal{X} \cap \mathcal{C}, \qquad \nabla g(x^*) \in \mathcal{Y},$$

and

$$0 = \langle x^*, \nabla g(x^*)\rangle.$$

If the cone $\mathscr{X}$ is actually a vector space (which is the case for each of the examples given in Section 2.1, except Example 2.5), then $\mathscr{Y} = \mathscr{X}^{\perp}$, and hence the last P optimality condition $0 = \langle x^*, \nabla g(x^*)\rangle$ is obviously redundant and can be deleted from the preceding definition. Furthermore, if $\mathscr{X}$ is actually the whole vector space E_n (which is the situation in the unconstrained case of ordinary programming), then $\mathscr{Y} = E_n^{\perp} = \{\mathbf{0}\}$, and hence the remaining P optimality conditions clearly become the (more familiar) optimality conditions

$$x^* \in \mathscr{C} \quad \text{and} \quad \nabla g(x^*) = \mathbf{0}.$$

The following theorem gives two convexity conditions that guarantee the necessity and/or sufficiency of the P optimality conditions for optimality.

Theorem 3.1. Under the hypothesis that g is differentiable at x^*,

(i) given that $\mathscr{X}$ is convex, if x^* is an optimal solution to Problem $\mathscr{A}$, then x^* is a critical solution for Problem $\mathscr{A}$ (but not conversely),

(ii) given that g is convex on $\mathscr{C}$, if x^* is a critical solution for Problem $\mathscr{A}$, then x^* is an optimal solution to Problem $\mathscr{A}$.

The proof of this theorem is not difficult but will be omitted.

It is worth noting that g is differentiable for most of the examples given in Section 2.1. Moreover, $\mathscr{X}$ is polyhedral and hence convex for each of those examples, and g is convex on $\mathscr{C}$ for important special cases of each of those examples. Consequently, the P optimality conditions frequently characterize the optimal solution set $\mathscr{S}^*$ for Problem $\mathscr{A}$.

Characterizations of $\mathscr{S}^*$ that do not require differentiability of g, but do require the concepts described in the following subsection, are given in Section 3.1.3.

3.1.2. The Conjugate Transformation, Subgradients, and Convex Analysis. The conjugate transformation evolved from the classical Legendre transformation but was first studied in great detail only rather recently by Fenchel (Refs. 31 and 32). [For a very thorough and modern treatment of both transformations see the recent book by Rockafellar (Ref. 33).] We now briefly describe only those of their properties that are relevant to geometric programming. All such properties are quite plausible when viewed geometrically in the context of two and three dimensions.

The conjugate transformation maps functions into functions in such a way that the "conjugate transform" $\omega : \Omega$ of a given function $w : W$ has functional values

$$\omega(\boldsymbol{\zeta}) \triangleq \sup_{\mathbf{z} \in W} [\langle \boldsymbol{\zeta}, \mathbf{z}\rangle - w(\mathbf{z})].$$

Of course, the domain Ω of ω is defined to be the set of all those vectors $\boldsymbol{\zeta}$ for which this supremum is finite, and the conjugate transform $\omega : \Omega$ exists only when Ω is not empty.

The conjugate transform of a separable function is clearly the sum of the conjugate transforms of its individual terms—a fact that simplifies the conjugate transform computations for many of the geometric programming examples given in Section 2.1. For purposes of illustration we now perform two of those computations in the convex case—the only case in which the conjugate transformation proves to be extremely effective.

Example 3.1. If

$$w(\mathbf{z}) \triangleq \sum_{i=1}^{\eta} c_i e^{z_i} \quad \text{and} \quad W \triangleq E_\eta, \qquad \text{where } c_i > 0, \quad i = 1, 2, \ldots, \eta,$$

then

$$\omega(\boldsymbol{\zeta}) = \sup_{\mathbf{z} \in E_\eta} \left[\langle \boldsymbol{\zeta}, \mathbf{z} \rangle - \sum_{i=1}^{\eta} c_i e^{z_i} \right] = \sum_{i=1}^{\eta} \sup_{z_i \in E_1} [\zeta_i z_i - c_i e^{z_i}],$$

which is clearly finite if and only if $\zeta_i \geqq 0$, $i = 1, 2, \ldots, \eta$, in which case an application of the differential calculus shows that $\omega(\boldsymbol{\zeta}) = \sum_{i=1}^{\eta} \zeta_i \log(\zeta_i / c_i) - \sum_{i=1}^{\eta} \zeta_i$, with the understanding that $\zeta_i \log \zeta_i \triangleq 0$ when $\zeta_i = 0$. Consequently,

$$\omega(\boldsymbol{\zeta}) = \sum_{i=1}^{\eta} \zeta_i \log(\zeta_i / c_i) - \sum_{i=1}^{\eta} \zeta_i \quad \text{and} \quad \Omega = \{\boldsymbol{\zeta} \in E_\eta \mid \zeta_i \geqq 0, i = 1, 2, \ldots, \eta\}.$$

Example 3.2. If

$$w(\mathbf{z}) \triangleq \sum_{i=1}^{\eta-1} p_i^{-1} |z_i - b_i|^{p_i} + z_\eta - b_\eta \quad \text{and} \quad W \triangleq E_\eta,$$

$$p_i > 1, \quad i = 1, 2, \ldots, \eta - 1,$$

then

$$\omega(\boldsymbol{\zeta}) = \sup_{\mathbf{z} \in E_\eta} \left[\langle \boldsymbol{\zeta}, \mathbf{z} \rangle - \sum_{i=1}^{\eta-1} p_i^{-1} |z_i - b_i|^{p_i} - (z_\eta - b_\eta) \right]$$

$$= \sum_{i=1}^{\eta-1} \sup_{z_i \in E_1} [\zeta_i z_i - p_i^{-1} |z_i - b_i|^{p_i}] + \sup_{z_\eta \in E_1} [\zeta_\eta z_\eta - (z_\eta - b_\eta)],$$

which is clearly finite if and only if $\zeta_\eta = 1$, in which case an application of the differential calculus shows that

$$\omega(\boldsymbol{\zeta}) = \sum_{i=1}^{\eta-1} (q_i^{-1} |\zeta_i|^{q_i} + b_i \zeta_i) + b_\eta,$$

where q_i is determined from p_i by the equation $p_i^{-1} + q_i^{-1} = 1$. Consequently,

$$\omega(\boldsymbol{\zeta}) = \sum_{i=1}^{\eta-1} (q_i^{-1}|\zeta_i|^{q_i} + b_i\zeta_i) + b_\eta \quad \text{and} \quad \boldsymbol{\Omega} = \{\boldsymbol{\zeta} \in E_\eta \mid \zeta_\eta = 1\}.$$

Geometrical insight into the conjugate transformation can be obtained by considering the "subgradient" set for w at $\mathbf{z}$, namely,

$$\partial w(\mathbf{z}) \triangleq \{\boldsymbol{\zeta} \in E_\eta \mid w(\mathbf{z}) + \langle \boldsymbol{\zeta}, \mathbf{z}' - \mathbf{z}\rangle \leqq w(\mathbf{z}') \text{ for each } \mathbf{z}' \in W\}.$$

Subgradients are related to, but considerably different from, the more familiar gradient. The gradient provides a "tangent hyperplane," while a subgradient provides a "supporting hyperplane" (in that the defining inequality obviously states that the hyperplane with equation $w' = w(\mathbf{z}) + \langle \boldsymbol{\zeta}, \mathbf{z}' - \mathbf{z}\rangle$ intersects the "graph" of w at the point $(\mathbf{z}, w(\mathbf{z}))$ and lines entirely "on or below" it). It is, of course, clear that a subgradient may exist and not be unique even when the gradient does not exist. On the other hand, it is also clear that a subgradient may not exist even when the gradient exists. There is, however, an important class of functions whose gradients are also subgradients—the class of convex functions. In fact, the notions of gradient and subgradient coincide for the class of differentiable convex functions defined on open sets, a class that arises in many of the examples given in Section 2.1.

To relate the conjugate transform to subgradients, observe that if $\boldsymbol{\zeta} \in \partial w(\mathbf{z})$, then $\langle \boldsymbol{\zeta}, \mathbf{z}'\rangle - w(\mathbf{z}') \leqq \langle \boldsymbol{\zeta}, \mathbf{z}\rangle - w(\mathbf{z})$ for each $\mathbf{z}' \in W$, which in turn clearly implies that $\boldsymbol{\zeta} \in \boldsymbol{\Omega}$ and that $\omega(\boldsymbol{\zeta}) = -[w(\mathbf{z}) + \langle \boldsymbol{\zeta}, -\mathbf{z}\rangle]$. Hence, $\omega(\boldsymbol{\zeta})$ is simply the negative of the intercept of the corresponding supporting hyperplane with the w' axis. Consequently, the conjugate transform ω exists when w has at least one subgradient $\boldsymbol{\zeta}$, a condition that is known to be fulfilled when w is convex. Actually, the conjugate transform ω restricted (in the set-theoretic sense) to the domain $\bigcup_{\mathbf{z} \in W} \partial w(\mathbf{z})$ is termed the "Legendre transform" of w and has been a major tool in the study of classical mechanics, thermodynamics, and differential equations [as described, for example, by Courant and Hilbert (Ref. 34)]. Usually, the domain $\boldsymbol{\Omega}$ of the conjugate transform ω consists of both $\bigcup_{\mathbf{z} \in W} \partial w(\mathbf{z})$ and some of its limit points.

Each function w and its conjugate transform ω give rise to an important inequality

$$\langle \mathbf{z}, \boldsymbol{\zeta}\rangle \leqq w(\mathbf{z}) + \omega(\boldsymbol{\zeta}),$$

which is termed the "conjugate inequality" (or "Young's inequality") and which is clearly valid for every point $\mathbf{z} \in W$ and every point $\boldsymbol{\zeta} \in \boldsymbol{\Omega}$ [as can be seen from the defining equation for $\omega(\boldsymbol{\zeta})$]. Moreover, we have just shown that equality holds if

$$\boldsymbol{\zeta} \in \partial w(\mathbf{z}),$$

a condition that actually characterizes equality by virtue of another elementary computation.

When it exists, the conjugate transform $\omega : \boldsymbol{\Omega}$ is known to be both convex and "closed," that is, its "epigraph" (which consists of all those points in $E_{\eta+1}$ that are "on or above" its graph) is both convex and (topologically) closed. Moreover, the conjugate transform of $\omega : \boldsymbol{\Omega}$ is the "closed convex hull" $\bar{w} : \bar{W}$ of $w : W$, and thus the conjugate transformation maps the family of all closed convex functions onto itself in one-to-one symmetric fashion. Consequently, the conjugate transformation is its own inverse on the family of all such functions; and given two such "conjugate functions" $w : W$ and $\omega : \boldsymbol{\Omega}$, the relations $\boldsymbol{\zeta} \in \partial w(\mathbf{z})$ and $\mathbf{z} \in \partial \omega(\boldsymbol{\zeta})$ are equivalent and hence "solve" one another.

In geometric programming we must deal with both an arbitrary cone Z and its "dual"

$$\mathrm{Z} \triangleq \{\boldsymbol{\zeta} \in E_\eta \mid 0 \leqq \langle \mathbf{z}, \boldsymbol{\zeta} \rangle \text{ for each } \mathbf{z} \in \mathrm{Z}\},$$

which is clearly itself a cone. Now, it is obvious that the conjugate transform of the zero function with domain Z is just the zero function with domain −Z. Consequently, the theory of the conjugate transformation implies that Z is convex and closed (a fact that can be established by more elementary considerations). Furthermore, if the cone Z is also convex and closed, the symmetry of the conjugate transformation readily implies that the dual of Z is just Z. If, in particular, Z is a vector space in E_η, this symmetry readily implies the better-known symmetry between orthogonal complementary subspaces Z and Z.

In convex analysis the "relative interior" (ri W) of a convex set $W \subseteq E_\eta$ is defined to be the "interior" of W "relative to" the "Euclidean topology" for the "affine hull" of W [i.e., the "smallest affine set (or linear manifold)" containing W]. The reason is that the (ri W) defined in this way is not empty, even when the "interior" of W is empty.

This completes our prerequisites for the remaining subsections of this section.

3.1.3. Lagrangian Saddle Points. Let $h : \mathcal{D}$ be the conjugate transform of the function $g : \mathcal{C}$, that is,

$$\mathcal{D} \triangleq \{y \in E_n \mid \sup_{x \in \mathcal{C}} [\langle y, x \rangle - g(x)] < +\infty\}$$

and

$$h(y) \triangleq \sup_{x \in \mathscr{C}} [\langle y, x \rangle - g(x)].$$

If $h : \mathscr{D}$ exists (which is the case, for example, when $g : \mathscr{C}$ is convex), then $h : \mathscr{D}$ is a closed convex function that inherits any separability present in $g : \mathscr{C}$. In fact, $h : \mathscr{D}$ can be computed via the calculus in the convex case for each example given in Section 2.1. Actually, the required computations for Examples 2.1 and 2.2 have already been given in (the preceding) Section 3.1.2, and the required computations for Examples 2.3–2.5 can be found in the cited references.

The following definition is of fundamental importance.

Definition 3.2. For a consistent Problem $\mathscr{A}$ with a finite infimum φ, a *P vector* is any vector y^* with the two properties

$$y^* \in \mathscr{D}$$

and

$$\varphi = \inf_{x \in \mathscr{X}} L_g(x; y^*),$$

where the (geometric) *Lagrangian*

$$L_g(x; y) \triangleq \langle x, y \rangle - h(y).$$

It should be noted that L_g is generally as easy to compute as $h : \mathscr{D}$.

The following "saddle-point theorem" provides several characterizations of optimality via P vectors.

Theorem 3.2. Given that $g : \mathscr{C}$ is convex and closed, let $x^* \in \mathscr{X}$ and let $y^* \in \mathscr{D}$. Then x^* is optimal for Problem $\mathscr{A}$ and y^* is a P vector for Problem $\mathscr{A}$ if and only if the ordered pair $(x^*; y^*)$ is a "saddle point" for the Lagrangian L_g, that is,

$$\sup_{y \in \mathscr{D}} L_g(x^*; y) = L_g(x^*; y^*) = \inf_{x \in \mathscr{X}} L_g(x; y^*);$$

in which case L_g has the saddle-point value

$$L_g(x^*; y^*) = g(x^*) = \varphi.$$

Moreover,

$$\sup_{y \in \mathscr{D}} L_g(x^*; y) = L_g(x^*; y^*)$$

if and only if x^* and y^* satisfy both the feasibility condition

$$x^* \in \mathscr{C}$$

and the subgradient condition

$$y^* \in \partial g(x^*),$$

in which case

$$L_g(x^*; y^*) = g(x^*).$$

Furthermore,

$$L_g(x^*; y^*) = \inf_{x \in \mathscr{X}} L_g(x; y^*)$$

if and only if x^* and y^* satisfy both the feasibility condition

$$y^* \in \mathscr{Y}$$

and the orthogonality condition

$$0 = \langle x^*, y^* \rangle,$$

in which case

$$L_g(x^*; y^*) = -h(y^*).$$

Since the second assertion of Theorem 3.2 gives certain conditions that are equivalent to the first saddle-point equation, and since the third assertion of Theorem 3.2 gives other conditions that are equivalent to the second saddle-point equation, Theorem 3.2 actually provides four different characterizations of all ordered pairs $(x^*; y^*)$ of optimal solutions x^* and P vectors y^*.

Of course, each of those four characterizations provides a characterization of all optimal solutions x^* in terms of a given P vector y^*, as well as a characterization of all P vectors y^* in terms of a given optimal solution x^*.

The symmetry of the preceding statement suggests that all P vectors may in fact constitute all optimal solutions to a closely related optimization problem. Actually, the appropriate optimization problem can be motivated by the following inequalities

$$\inf_{x \in \mathscr{X}} L_g(x; y^*) \leq \sup_{y \in \mathscr{D}} [\inf_{x \in \mathscr{X}} L_g(x; y)] \leq \inf_{x \in \mathscr{X}} [\sup_{y \in \mathscr{D}} L_g(x; y)] \leq \sup_{y \in \mathscr{D}} L_g(x^*; y),$$

which are valid for each $y^* \in \mathscr{D}$ and each $x^* \in \mathscr{X}$, by virtue of the single fact that x and y reside in independent sets $\mathscr{X}$ and $\mathscr{D}$, respectively [i.e., the vector $(x; y)$ resides in a Cartesian product $\mathscr{X} \times \mathscr{D}$]. In particular, note that each of these inequalities must be an equality when $(x^*; y^*)$ is a saddle point for L_g, in which case x^* is obviously an optimal solution to the minimization problem

$$\inf_{x\in\mathscr{X}} [\sup_{y\in\mathscr{D}} L_g(x; y)],$$

and y^* is obviously an optimal solution to the maximization problem

$$\sup_{y\in\mathscr{D}} [\inf_{x\in\mathscr{X}} L_g(x; y)].$$

Now, when g is convex and closed, the definition of L_g and the symmetry of the conjugate transformation clearly imply that the preceding minimization problem is essentially Problem $\mathscr{A}$. Consequently, it is not unnatural to consider the preceding maximization problem simultaneously with Problem $\mathscr{A}$ and term it the "geometric dual problem" $\mathscr{B}$.

Problem $\mathscr{B}$. Using the feasible solution set

$$\mathscr{T} \triangleq \{y \in \mathscr{D} \mid \inf_{x\subset\mathscr{X}} L_g(x; y) \text{ is finite}\}$$

and the objective function

$$\mathscr{H}(y) \triangleq \inf_{x\in\mathscr{X}} L_g(x; y),$$

calculate both the problem supremum

$$\Psi \triangleq \sup_{y\in\mathscr{T}} \mathscr{H}(y)$$

and the optimal solution set

$$\mathscr{T}^* \triangleq \{y \in \mathscr{T} \mid \mathscr{H}(y) = \Psi\}.$$

Even though Problem $\mathscr{B}$ is essentially a "maximin problem"—a type of problem that tends to be relatively difficult to analyze—the minimization problems that must be solved to obtain the objective function $\mathscr{H}: \mathscr{T}$ have trivial solutions. In particular, the definition of L_g and the hypothesis that $\mathscr{X}$ is a cone clearly imply that $\inf_{x\in\mathscr{X}} L_g(x; y)$ is finite if and only if $y \in \mathscr{Y}$, in which case $\inf_{x\in\mathscr{X}} L_g(x; y) = -h(y)$. Consequently, $\mathscr{T} = \mathscr{Y} \cap \mathscr{D}$ and $\mathscr{X}(y) = -h(y)$, so Problem $\mathscr{B}$ can actually be rephrased in the following more direct way:

Using the feasible solution set

$$\mathscr{T} = \mathscr{Y} \cap \mathscr{D},$$

calculate both the problem infimum

$$\psi \triangleq \inf_{y \in \mathcal{T}} h(y) = -\Psi$$

and the optimal solution set

$$\mathcal{T}^* = \{y \in \mathcal{T} \mid h(y) = \psi\}.$$

When phrased in this way, Problem $\mathcal{B}$ closely resembles Problem $\mathcal{A}$ and is in fact a geometric programming problem. Of course, the geometric dual Problem $\mathcal{B}$ can actually be defined in this way—an approach that is exploited in the following subsection. Nevertheless, the preceding derivation serves as an important link between Lagrangians and duality.

3.1.4. Duality. Let $h : \mathcal{D}$ be the conjugate transform of the function $g : \mathcal{C}$ (as defined at the beginning of Section 3.1.3).

Now, consider the following geometric programming problem $\mathcal{B}$.

Problem $\mathcal{B}$. Using the feasible solution set

$$\mathcal{T} \triangleq \mathcal{Y} \cap \mathcal{D},$$

calculate both the problem infimum

$$\psi \triangleq \inf_{y \in \mathcal{T}} h(y)$$

and the optimal solution set

$$\mathcal{T}^* \triangleq \{y \in \mathcal{T} \mid h(y) = \psi\}.$$

It should be noted that Problem $\mathcal{B}$ is generally as easy to compute as $h : \mathcal{D}$ and $\mathcal{Y}$. Moreover, Problem $\mathcal{B}$ is always a convex programming problem, because both $h : \mathcal{D}$ and $\mathcal{Y}$ are always convex and closed (even when $g : \mathcal{C}$ and $\mathcal{X}$ are not convex and closed).

Problems $\mathcal{A}$ and $\mathcal{B}$ are termed *geometric dual problems*. When both $g : \mathcal{C}$ and $\mathcal{X}$ are convex and closed, this duality is clearly symmetric, in that Problem $\mathcal{A}$ can then be constructed from Problem $\mathcal{B}$ in the same way that Problem $\mathcal{B}$ has just been constructed from Problem $\mathcal{A}$. This symmetry induces a symmetry on the theory that relates $\mathcal{A}$ to $\mathcal{B}$, in that each statement about $\mathcal{A}$ and $\mathcal{B}$ automatically produces an equally valid "dual statement" about $\mathcal{B}$ and $\mathcal{A}$. To be concise, each dual statement will be left to the reader's imagination.

It is worth mentioning that there are cases in which Problems $\mathcal{A}$ and $\mathcal{B}$ have additional interesting symmetries. In particular, Problem $\mathcal{B}$ turns out to be a "reversed-time" discrete optimal control problem when Problem $\mathcal{A}$

is taken to be a (forward-time) discrete optimal control problem (i.e., Example 2.4) whose linear dynamics are such that each matrix B_i is identical to a nonsingular matrix B that commutes with each matrix A_i for $i = 1, 2, \ldots, p$.

Unlike the usual min–max formulations of duality in mathematical programming, both Problem $\mathscr{A}$ and its geometric dual Problem $\mathscr{B}$ are minimization problems. The relative simplicity of this min–min formulation will soon become clear, but the reader who is accustomed to the usual min–max formulation must bear in mind that a given duality theorem will generally have slightly different statements depending on the formulation in use. In particular, a theorem that asserts the equality of the min and max in the usual formulation will assert that the sum of the mins is zero (i.e. $\varphi + \psi = 0$) in the present formulation.

The following definition is almost as important as the definition of the dual problems $\mathscr{A}$ and $\mathscr{B}$.

Definition 3.3. The *extremality conditions* (for unconstrained geometric programming) are:

(I) $$x \in \mathscr{X} \quad \text{and} \quad y \in \mathscr{Y},$$

(II) $$0 = \langle x, y \rangle,$$

(III) $$y \in \partial g(x).$$

Extremality conditions (I) are simply the "cone conditions" for Problems $\mathscr{A}$ and $\mathscr{B}$, respectively. Extremality condition (II) is termed the "orthogonality condition," and extremality condition (III) is termed the "subgradient condition."

If the cone $\mathscr{X}$ is actually a vector space (which is the case for each of the examples given in Section 2.1, except Example 2.5), then $\mathscr{Y} = \mathscr{X}^{\perp}$, and hence the orthogonality condition (II) is redundant and can be deleted from the preceding definition (and from wherever that definition is used).

The following "duality theorem" is the basis for many others to come.

Theorem 3.3. If x and y are feasible solutions to Problems $\mathscr{A}$ and $\mathscr{B}$, respectively [in which case the extremality conditions (I) are satisfied], then

$$0 \leq g(x) + h(y),$$

with equality holding if and only if the extremality conditions (II) and (III) are satisfied, in which case x and y are optimal solutions to Problems $\mathscr{A}$ and $\mathscr{B}$, respectively.

In essence the proof of this key theorem consists only of combining the defining inequality $0 \leqq \langle x, y \rangle$ for $\mathcal{Y}$ with the conjugate inequality $\langle x, y \rangle \leqq g(x) + h(y)$ for $h : \mathcal{D}$.

The following important corollary is an immediate consequence of Theorem 3.3.

Corollary 3.1. If the dual Problems $\mathcal{A}$ and $\mathcal{B}$ are both consistent, then

(i) the infimum φ for Problem $\mathcal{A}$ is finite, and

$$0 \leqq \varphi + h(y) \qquad \text{for each } y \in \mathcal{T},$$

(ii) the infimum ψ for Problem $\mathcal{B}$ is finite, and

$$0 \leqq \varphi + \psi.$$

The strictness of the inequality in conclusion (ii) plays a crucial role in almost all that follows.

Definition 3.4. Consistent dual Problems $\mathcal{A}$ and $\mathcal{B}$ for which

$$0 < \varphi + \psi$$

have a *duality gap* of $\varphi + \psi$.

It is well known that duality gaps do not occur in finite linear programming, but they do occasionally occur in infinite linear programming, where this phenomenon was first encountered by Duffin (Ref. 35) and Kretchmer (Ref. 36). Although duality gaps occur very frequently in the present (generally nonconvex) formulation of geometric programming, we shall eventually see that they can occur only very rarely in the convex case, in that they can then be excluded by very weak conditions on the geometric dual problems $\mathcal{A}$ and $\mathcal{B}$. Yet, they do occur in the convex case, and examples (due originally to J. J. Stoer) can be found in Appendix C of Peterson (Ref. 6).

Geometric programming problems $\mathcal{A}$ that are convex are usually much more amenable to study than those that are nonconvex, mainly because of the relative lack of duality gaps in the convex case. Duality gaps are undesirable from a theoretical point of view because we shall see that relatively little can be said about the corresponding geometric dual problems. They are also undesirable from a computational point of view because they usually destroy the possibility of using the inequality $0 \leqq g(x) + h(y)$ to provide an *algorithmic stopping criterion.*

Such a criterion results from specifying a positive tolerance ϵ so that the numerical algorithms being used to minimize both $g(x)$ and $h(y)$ are terminated when they produce a pair of feasible solutions $x^{\dagger}$ and $y^{\dagger}$ for which

$$g(x^{\dagger})+h(y^{\dagger})\leqq 2\epsilon.$$

Because conclusion (i) to Corollary 3.1 along with the definition of φ shows that $-h(y^{\dagger})\leqq\varphi\leqq g(x^{\dagger})$, we conclude from the preceding tolerance inequality that

$$\left|\varphi-\frac{g(x^{\dagger})-h(y^{\dagger})}{2}\right|\leqq\epsilon.$$

Hence φ can be approximated by $[g(x^{\dagger})-h(y^{\dagger})]/2$ with an error no greater than $\pm\epsilon$. Moreover, duality (i.e., symmetry) implies that ψ can be approximated by $[h(y^{\dagger})-g(x^{\dagger})]/2$ with an error no greater than $\pm\epsilon$.

Note though that Problems $\mathcal{A}$ and $\mathcal{B}$ have a duality gap if and only if there is a positive tolerance ϵ so small that $2\epsilon<\varphi+\psi$. Because the definitions for φ and ψ imply that $\varphi+\psi\leqq g(x)+h(y)$ for feasible solutions x and y, we infer that when ϵ satisfies the preceding inequality there are no feasible solutions $x^{\dagger}$ and $y^{\dagger}$ for which $g(x^{\dagger})+h(y^{\dagger})\leqq 2\epsilon$, in which event the algorithms being used are never terminated.

The following corollary provides a useful characterization of dual optimal solutions x^* and y^* in terms of the extremality conditions.

Corollary 3.2 If the extremality conditions have a solution x' and y', then

(i) $$x'\in\mathcal{S}^* \quad\text{and}\quad y'\in\mathcal{T}^*,$$

(ii) $$\mathcal{T}^*=\{y\in\mathcal{Y}\cap\partial g(x')\,|\,0=\langle x',y\rangle\},$$

(iii) $$0=\varphi+\psi.$$

On the other hand, if the dual Problems $\mathcal{A}$ and $\mathcal{B}$ are both consistent, and if $0=\varphi+\psi$, then $x\in\mathcal{S}^*$ and $y\in\mathcal{T}^*$ if and only if x and y satisfy the extremality conditions.

The proof of this corollary is an immediate consequence of Theorem 3.3 and the conjugate transform relation $\partial g(x)\subseteq\mathcal{D}$.

The first part of Corollary 3.2 shows that Problems $\mathcal{A}$ and $\mathcal{B}$ can be viewed as "variational principles" for finding solutions to the extremality conditions. Actually, in many contexts, the extremality conditions are, in one form or another, the natural objects of study (rather than either Problem $\mathcal{A}$ or Problem $\mathcal{B}$). For the highway network problem described in Example 2.5, the extremality conditions (I)–(III) are equivalent to the conditions that define "Wardrop equilibrium." For the electric and hydraulic network problems alluded to in the same example, the cone conditions (I) are simply the "Kirchhoff current and potential conservation laws," respectively; the orthogonality condition (II) is redundant because $\mathscr{X}$ and $\mathscr{Y}$ are orthogonal complementary vector spaces; and the subgradient condition (III) is just "Ohm's law." Moreover, in the context of electric and hydraulic networks, Problems $\mathscr{A}$ and $\mathscr{B}$ are frequently termed the "Maxwell–Duffin complementary variational principles." Needless to say, Corollary 3.2 shows that the lack of a duality gap is fundamental in all such contexts.

It is worth noting that if $g : \mathscr{C}$ is convex and closed, then the symmetry of the conjugate transformation implies that the subgradient condition (III) can be replaced by the equivalent subgradient condition

(IIIa) $$x \in \partial h(y)$$

without changing the validity of Corollary 3.2.

In that case Corollary 3.2 and its (unstated) dual are of direct use when $0 = \varphi + \psi$ and *both* $\mathscr{S}^*$ and $\mathscr{T}^*$ are known to be nonempty, because they then provide a method for calculating all optimal solutions from the knowledge of only a single optimal solution. For example, if x^* in $\mathscr{S}^*$ is a known optimal solution to Problem $\mathscr{A}$, then

$$\mathscr{T}^* = \{y \in \mathscr{Y} \cap \partial g(x^*) \mid 0 = \langle x^*, y\rangle\},$$

and for each $y^* \in \mathscr{T}^*$, the set

$$\mathscr{S}^* = \{x \in \mathscr{X} \cap \partial h(y^*) \mid 0 = \langle x, y^*\rangle\}.$$

The definition of the Lagrangian L_g (in Section 3.1.3) and the fact that $\mathscr{X}$ is a cone readily imply that the cone condition $y \in \mathscr{Y}$ and the orthogonality condition $0 = \langle x, y\rangle$ can both be replaced by the single equivalent condition $L_g(x; y) = \inf_{x' \in \mathscr{X}} L_g(x' : y)$. Moreover, if $g : \mathscr{C}$ is convex and closed, conjugate transform theory readily implies that the subgradient condition $y \in \partial g(x)$ can be replaced by the equivalent condition $\sup_{y' \in \mathscr{D}} L_g(x; y') = L_g(x; y)$. Consequently, the saddle-point condition discussed in Section 3.1.3 is equivalent to the extremality conditions when $g : \mathscr{C}$ is convex and closed. Nevertheless, it seems that the extremality conditions given in the definition are the most convenient to work with.

The following theorem provides an important tie between dual Problem $\mathscr{B}$ and the P vectors defined in Section 3.1.3.

Theorem 3.4. Given that Problem $\mathscr{A}$ is consistent with a finite infimum φ,

(i) If Problem $\mathscr{A}$ has a P vector, then Problem $\mathscr{B}$ is consistent and $0 = \varphi + \psi$,

(ii) If Problem $\mathscr{B}$ is consistent and $0 = \varphi + \psi$, then

$$\{y^* \mid y^* \text{ is a } P \text{ vector for Problem } \mathscr{A}\} = \mathscr{T}^*.$$

The proof of this theorem is not difficult but will be omitted.

An important consequence of Theorem 3.4 is that, when they exist, all P vectors for Problem $\mathscr{A}$ can be obtained simply by computing the dual optimal solution set $\mathscr{T}^*$. However, there are cases in which the vectors in $\mathscr{T}^*$ are not P vectors for Problem $\mathscr{A}$, though such cases can occur only when $0 < \varphi + \psi$, in which event Theorem 3.4 implies that there can be no P vectors for Problem $\mathscr{A}$.

The absence of a duality gap (i.e., the assumption that $0 = \varphi + \psi$) is crucial to the preceding computational and theoretical techniques, as well as others to come. Although there are numerous conditions that guarantee the absence of a duality gap, the most useful and widely used ones can be viewed as special manifestations of the hypotheses in the following theorem. In addition to guaranteeing the absence of a duality gap, this (geometric programming) version of "Fenchel's theorem" also serves as a very important existence theorem.

Theorem 3.5. Suppose that both $g : \mathscr{C}$ and $\mathscr{X}$ are convex and closed. If the dual Problem $\mathscr{B}$ has a feasible solution $y^0 \in (\text{ri } \mathscr{Y}) \cap (\text{ri } \mathscr{D})$, and if Problem $\mathscr{B}$ has a finite infimum ψ, then $0 = \varphi + \psi$ and $\mathscr{S}^* \neq \emptyset$.

A proof of this theorem that is quite different from Fenchel's original proof [as given by Fenchel (Ref. 32) or Rockafellar (Ref. 33)] can be found in Peterson (Ref. 6).

There are several facts about relative interiors that help in the application of Theorem 3.5. For example, if the cone $\mathscr{Y}$ is actually a vector space, then $(\text{ri } \mathscr{Y}) = \mathscr{Y}$, and hence the hypothesis $y^0 \in (\text{ri } \mathscr{Y}) \cap (\text{ri } \mathscr{D})$ is implied by the hypothesis $y^0 \in \mathscr{Y} \cap (\text{ri } \mathscr{D})$. Also, if the set $\mathscr{D}$ turns out to be a vector space, then $(\text{ri } \mathscr{D}) = \mathscr{D}$, and hence the hypothesis $y^0 \in (\text{ri } \mathscr{Y}) \cap (\text{ri } \mathscr{D})$ is implied by the hypothesis $y^0 \in (\text{ri } \mathscr{Y}) \cap \mathscr{D}$, which in turn is always satisfiable when $\mathscr{D} = E_n$.

Theorem 3.5 and the preceding facts are the key ingredients needed to show that there are no duality gaps for many of the examples given in Section 2.1, such as:

1. Posynomial programming problems (Example 2.1) that are "canonical" [as defined by Duffin and Peterson (Ref. 45) or Duffin, Peterson, and Zener (Ref. 5)].

2. Convex l_p programming problems (Example 2.2) that are "canonical" [as defined by Peterson and Ecker (Ref. 18)].

3. Optimal location problems (Example 2.3) whose costs $d_i(\mathbf{z}, b^i) = \|\mathbf{z} - \mathbf{b}^i\|_i$ for some "norm" $\|\cdot\|_i$ [as first shown by Peterson and Wendell (Ref. 19)].

4. Discrete optimal control problems (Example 2.4) that are "canonical" [as defined by Dinkel and Peterson (Ref. 20)].

5. Highway network equilibrium problems (Example 2.5) whose roadway arc travel times are monotone nondecreasing and unbounded from above as functions of the corresponding arc total traffic flows [as first shown by Hall and Peterson (Ref. 21)].

6. Electric and hydraulic network equilibrium problems (also Example 2.5) that contain only "current sources," "potential sources," and linear or nonlinear "monotone resistors" [as first shown essentially by Duffin (Ref. 22), Minty (Ref. 23) and Rockafellar (Ref. 24)].

The classification "canonical" is actually synonymous with the existence of a dual feasible solution $y^0 \in (\text{ri } \mathcal{Y}) \cap (\text{ri } \mathcal{D})$, termed *Fenchel's hypothesis for the dual Problem $\mathcal{B}$*. Although Theorem 3.5 obviously cannot be applied directly to noncanonical problems, it frequently can be appplied indirectly to such problems. In particular, given that both $g : \mathcal{C}$ and $\mathcal{X}$ are convex and closed, Peterson (Ref. 9) has shown that when Fenchel's hypothesis for the dual Problem $\mathcal{B}$ is not satisfied there is a *recession direction for Problem $\mathcal{A}$*, namely, a nonzero (direction) vector $\delta \in \mathcal{X}$ such that for each $x \in \mathcal{C}$ the function $g(x + \cdot\delta)$ is defined and monotone nonincreasing on the set of all nonnegative real numbers. Although the existence of a recession direction δ for Problem $\mathcal{A}$ does not require that $\mathcal{A}$ be consistent, if $\mathcal{A}$ happens to be consistent, the existence of δ clearly does imply that $\mathcal{S}^*$ is either empty or a union of half-lines. In any event, the dimension n of Problem $\mathcal{A}$ can be reduced via projection procedures given by Abrams (Ref. 37). In fact, such a "reduction" frequently can be carried to such an extent that Fenchel's hypothesis is satisfied for the geometric dual $\mathcal{B}'$ of the resulting "canonical form" $\mathcal{A}'$ for Problem $\mathcal{A}$. Moreover, in such situations, Peterson (Ref. 9) has also shown that this reduction can always be carried to such an extent that the canonical form $\mathcal{A}'$ has a (nonempty) bounded optimal solution set $\mathcal{S}'^*$. Actually, detailed reductions of this type had previously been given for the preceding examples 1 through 5 in the cited references.

It is worth mentioning here that, when both $g : \mathcal{C}$ and $\mathcal{X}$ are convex and closed, the absence of a duality gap can actually be characterized in terms of the changes induced in the problem infimum φ by small changes in certain problem input parameters—a characterization that is given in the following subsection.

3.1.5. Parametric Programming and Post-Optimality Analysis. For both practical and theoretical reasons, Problem $\mathcal{A}$ should not be studied entirely in isolation. It should also be embedded in a parametrized family $\mathcal{F}$ of closely related geometric programming problems $\mathcal{A}(u)$ that are generated by simply translating (the domain $\mathcal{C}$ of) g through all possible displacements $-u \in E_n$, while keeping $\mathcal{X}$ fixed. (For gaining insight, we recommend making a sketch of a typical case in which n is 2 and $\mathscr{X}$ is a one-dimensional vector space.) Problem $\mathscr{A}$ then appears in the parametrized family $\mathscr{F}$ as Problem $\mathscr{A}(\mathbf{0})$ and is studied in relation to all other geometric programming Problems $\mathscr{A}(u)$, with special attention given to those Problems $\mathscr{A}(u)$ in $\mathscr{F}$ that are close to $\mathscr{A}(\mathbf{0})$ in the sense that (the "norm" of) u is small.

The parametrized family $\mathscr{F}$ of all Problems $\mathscr{A}(u)$ (for fixed $g : \mathscr{C}$ and $\mathscr{X}$) is termed a *geometric programming family*. For purposes of easy reference and mathematical precision, Problem $\mathscr{A}(u)$ is now given the following formal definition, which should be compared with the formal definition of Problem $\mathscr{A}$ at the beginning of Section 2.1.

Problem $\mathscr{A}(u)$. Using the feasible solution set

$$\mathscr{S}(u) \triangleq \mathscr{X} \cap (\mathscr{C} - u),$$

calculate both the problem infimum

$$\varphi(u) \triangleq \inf_{x \in \mathscr{S}(u)} g(x+u)$$

and the optimal solution set

$$\mathscr{S}^*(u) \triangleq \{x \in \mathscr{S}(u) \mid g(x+u) = \varphi(u)\}.$$

Note that (in a rather general set-theoretic sense) the symbols $\mathscr{A}$, $\mathscr{S}$, φ, and $\mathscr{S}^*$ now represent functions of u, though they originally represented only the particular functional values $\mathscr{A}(\mathbf{0})$, $\mathscr{S}(\mathbf{0})$, $\varphi(\mathbf{0})$, and $\mathscr{S}^*(\mathbf{0})$, respectively. Needless to say, the reader must keep this notational discrepancy in mind when comparing subsequent developments with previous developments.

For a given u, Problem $\mathscr{A}(u)$ is either consistent or inconsistent, depending on whether the feasible solution set $\mathscr{S}(u)$ is nonempty or empty. It is, of course, obvious that the parametrized family $\mathscr{F}$ contains infinitely many consistent Problems $\mathscr{A}(u)$. The domain of the infimum function φ is taken to be the corresponding nonempty set $\mathscr{U}$ of all those vectors u for which $\mathscr{A}(u)$ is consistent. Thus, the range of φ may contain the point $-\infty$, but if $\varphi(u) = -\infty$ then the optimal solution set $\mathscr{S}^*(u)$ is clearly empty.

Due to the preeminence of Problem $\mathscr{A}(\mathbf{0})$, we shall find it useful to interpret Problem $\mathscr{A}(u)$ as a perturbed version of $\mathscr{A}(\mathbf{0})$, so we term the set

$$\mathscr{U} \triangleq \{u \in E_n \mid \mathscr{S}(u) \text{ is not empty}\}$$

the *feasible perturbation set* for Problem $\mathscr{A}(\mathbf{0})$ (relative to the family $\mathscr{F}$).

The functions φ and $\mathscr{S}^*$ usually show the dependence of optimality on actual external influences and hence are of prime interest in "cost-benefit analysis" and other such subjects. In fact, for the examples given in Section 2.1, it is easy to see that the perturbation vector u:

(a) in Example 2.1 alters the (log of the absolute value of the) signomial coefficients c_i (which are generally determined by such external influences as design requirements, performance requirements, material costs, and so forth);

(b) in the linear regression analysis case of Example 2.2 alters the vector $(b_1, \ldots, b_m)$ being optimally approximated;

(c) in Example 2.3 alters the fixed facility locations $\mathbf{b}^i$, provided that each cost $d_i(\mathbf{z}, \mathbf{b}^i) = \|\mathbf{z} - \mathbf{b}^i\|_i$ for some "norm" $\|\cdot\|_i$,

(d) in Example 2.4 translates the "state sets" R_i and the "decision sets" D_i,

(e) in Example 2.5 alters the total input flows d_i, as well as other network parameters.

The properties of the parametrized family $\mathcal{F}$ brought out in the following theorem are of fundamental theoretical significance and also have rather obvious applications in parameteric programming.

Theorem 3.6. The feasible perturbation set $\mathcal{U}$ is given by the formula

$$\mathcal{U} = \mathcal{C} - \mathcal{X}.$$

Moreover, if both $\mathcal{C}$ and $\mathcal{X}$ are convex, then so is $\mathcal{U}$, and the point-to-set function $\mathcal{S}$ is "concave" on $\mathcal{U}$, in that

$$\delta_1 \mathcal{S}(u^1) + \delta_2 \mathcal{S}(u^2) \subseteq \mathcal{S}(\delta_1 u^1 + \delta_2 u^2)$$

for each "convex combination" $\delta_1 u^1 + \delta_2 u^2$ of arbitrary points u^1, $u^2 \in \mathcal{U}$. Furthermore, if both $g : \mathcal{C}$ and $\mathcal{X}$ are convex, then so is Problem $\mathcal{A}(u)$ for each $u \in \mathcal{U}$, and the infimum function φ is either finite and convex on $\mathcal{U}$ or $\varphi(u) \equiv -\infty$ for each $u \in (\mathrm{ri}\ \mathcal{U})$.

A proof for a slightly more limited version of this theorem is given in Peterson (Ref. 6).

It should be mentioned that there are instances in which both $g : \mathcal{C}$ and $\mathcal{X}$ are convex (as well as closed) and for which $\varphi(u)$ is finite for at least one $u \in (\mathrm{rb}\ \mathcal{U})$ (the "relative boundary" of $\mathcal{U}$) even though $\varphi(u) = -\infty$ for each $u \in (\mathrm{ri}\ \mathcal{U})$. An example (due originally to J. J. Stoer) can be found in Appendix C of Peterson (Ref. 6).

In cases for which $\mathcal{X}$ is actually a vector space, the following theorem reduces a study of the parametrized family $\mathcal{F}$ to a study of only those Problems $\mathcal{A}(u)$ in $\mathcal{F}$ for which $u \in \mathcal{Y} = \mathcal{X}^\perp$. In such cases, it is convenient to adopt the notation $u_{\mathcal{X}}$ and $u_{\mathcal{Y}}$ for the "orthogonal projection" of an arbitrary vector $u \in E_n$ onto the orthogonal complementary subspaces $\mathscr{X}$ and $\mathscr{Y}$ respectively.

Theorem 3.7. Suppose that $\mathscr{X}$ and $\mathscr{Y}$ are orthogonal complementary subspaces of E_n. Then, for each vector $u \in E_n$, either the feasible solution sets $\mathscr{S}(u)$ and $\mathscr{S}(u_{\mathscr{Y}})$ are both empty, or both are nonempty, with the latter being the case if and only if $u \in \mathscr{U}$, in which case

$$\mathscr{S}(u) = \mathscr{S}(u_{\mathscr{Y}}) - u_{\mathscr{X}}$$

and

$$\varphi(u) = \varphi(u_{\mathscr{Y}}).$$

Furthermore, if $u \in \mathscr{U}$, then either the optimal solution sets $\mathscr{S}^*(u)$ and $\mathscr{S}^*(u_{\mathscr{Y}})$ are both empty, or both are nonempty and

$$\mathscr{S}^*(u) = \mathscr{S}^*(u_{\mathscr{Y}}) - u_{\mathscr{X}}.$$

The proof of this theorem is not difficult and can be found in Peterson (Ref. 6).

In addition to its rather obvious applications in parametric programming the preceding reduction theorem and its (unstated) dual can be used to relate the present geometric programming formulation of duality to both the original Fenchel formulation of duality [Fenchel (Ref. 32)] and the more recent Rockafellar formulations of duality [Rockafellar (Refs. 33, 38, and 39)]. All such relations can be found in Peterson (Ref. 6).

The following (geometric programming) version of a theorem due originally to Rockafellar (Refs. 33, 38 and 39) provides a direct link between the infimum function φ and the dual problem $\mathscr{B}$. This link serves as the key to deriving very important properties of φ via conjugate transform theory.

Theorem 3.8. The infimum function φ is finite everywhere on its domain $\mathscr{U}$ and possesses a conjugate transform if and only if the dual Problem $\mathscr{B}$ is consistent, in which case the dual objective function $h : \mathscr{T}$ is the conjugate transform of $\varphi : \mathscr{U}$.

Although this theorem has a relatively simple and direct proof (due originally to Rockafellar), it can also be viewed as an immediate corollary to other theorems that produce microeconomic interpretations of the dual Problem $\mathscr{B}$. In fact, a detailed analysis of the situation [given in Peterson (Ref. 6)] shows that a rather interesting class of microeconomic problems can be solved explicitly in terms of Problem $\mathscr{B}$.

The following corollary plays a crucial role in most applications of Theorem 3.8.

Corollary 3.3. If the dual Problem $\mathcal{B}$ is consistent, then the conjugate transform of its objective function $h : \mathcal{T}$ is the closed convex hull $\bar{\varphi} : \bar{\mathcal{U}}$ of the infimum function $\varphi : \mathcal{U}$ (which is, of course, identical to $\varphi : \mathcal{U}$ when $\varphi : \mathcal{U}$ happens to be both convex and closed).

This corollary is an immediate consequence of Theorem 3.8 and conjugate transform theory.

The preceding theorem and its corollary provide a *method for constructing, without the use of numerical optimization techniques, the closed convex hull* $\bar{\varphi} : \bar{\mathcal{U}}$ of $\varphi : \mathcal{U}$ (which, by virtue of Theorem 3.6, is essentially the desired infimum function $\varphi : \mathcal{U}$ when both $g : \mathcal{C}$ and $\mathcal{X}$ are convex). In particular, if the dual feasible solution set $\mathcal{T}$ is not empty (which, as indicated by Theorem 3.8, is the only really interesting nontrivial convex case), it can of course be covered with a "mesh"

$$\mathcal{M} \triangleq \{y^1, y^2, \ldots, y^s\} \subseteq \mathcal{T} \triangleq \mathcal{Y} \cap \mathcal{D}.$$

From Corollary 3.3 and conjugate transform theory, $\bar{\varphi}$ is clearly bounded from below on $\bar{\mathcal{U}}$ by the (polyhedral) approximating function $\bar{\varphi}_\ell$ whose functional values

$$\bar{\varphi}_\ell(u) \triangleq \max_{i=1,2,\ldots,s} [\langle y^i, u\rangle - h(y^i)] \qquad \text{for each } u \in \bar{\mathcal{U}}.$$

Moreover, the conjugate inequality (for $\bar{\varphi} : \bar{\mathcal{U}}$ and $h : \mathcal{T}$) can be used to show that

$$\bar{\varphi}(u) = \langle y^i, u\rangle - h(y^i) \qquad \text{for each } u \in \partial h(y^i) - \{x \in \mathcal{X} \mid \langle x, y^i\rangle = 0\},$$
$$i = 1, 2, \ldots, s,$$

and the convexity of $\bar{\varphi}$ then implies that $\bar{\varphi}$ is bounded from above by "affine interpolations" between such functional values. Furthermore, it is a consequence of conjugate transform theory that these lower and upper approximations can be made with arbitrary accuracy simply by choosing the mesh $\mathcal{M}$ to be sufficiently "dense" in $\mathcal{T}$. To be practical, though, this method requires an explicit construction of the dual objective function $h : \mathcal{T}$, a construction that is rather easy for virtually all of the important (convex) examples given in Section 2.1.

The preceding corollary also helps to motivate the following definition.

Definition 3.5. Problem $\mathcal{A}(u)$ is said to be *quasi-consistent* if the closed convex hull $\bar{\varphi} : \bar{\mathcal{U}}$ of the infimum function $\varphi : \mathcal{U}$ exists, and if $u \in \bar{\mathcal{U}}$, in which case $\bar{\varphi}(u)$ is termed the *quasi-infimum* for Problem $\mathcal{A}(u)$.

Since $\mathcal{U} \subseteq \bar{\mathcal{U}}$, each consistent Problem $\mathcal{A}(u)$ is quasi-consistent, but not conversely.

The following theorem leads to a complete explanation of duality gaps, and is also at the heart of many post-optimal "sensitivity analyses."

Theorem 3.9. Suppose that the dual Problem $\mathscr{B}$ is consistent. Then its infimum ψ is finite if and only if Problem $\mathscr{A}(\mathbf{0})$ is quasi-consistent, in which case

$$0 = \bar{\varphi}(\mathbf{0}) + \psi \quad \text{and} \quad \partial\bar{\varphi}(\mathbf{0}) = \mathscr{T}^*.$$

The proof of this theorem is a rather direct consequence of Corollary 3.3 and conjugate transform theory. The details for somewhat limited versions of this theorem can be found in either Peterson (Ref. 6) or Rockafellar (Ref. 33).

The following corollary identifies duality gaps as simply the difference between $\varphi(\mathbf{0})$ and $\bar{\varphi}(\mathbf{0})$.

Corollary 3.4. Suppose that the dual Problems $\mathscr{A}(\mathbf{0})$ and $\mathscr{B}$ are both consistent. Then, Problem $\mathscr{A}(\mathbf{0})$ is quasi-consistent and

$$\varphi(\mathbf{0}) - \bar{\varphi}(\mathbf{0}) = \varphi(\mathbf{0}) + \psi.$$

The proof of this corollary comes from Corollary 3.1 and the addition of $\varphi(\mathbf{0}) - \bar{\varphi}(\mathbf{0})$ to both sides of the equation $0 = \bar{\varphi}(\mathbf{0}) + \psi$ given in Theorem 3.9.

The preceding corollary helps to motivate the following definition.

Definition 3.6. A consistent Problem $\mathscr{A}(u)$ for which $\varphi(u) = \bar{\varphi}(u)$ is said to be *normal.*

The following corollary formalizes the equivalence of normality and the lack of a duality gap, while showing that either condition guarantees the validity of the equation on which many sensitivity analyses are based.

Corollary 3.5. Suppose that the dual Problems $\mathscr{A}(\mathbf{0})$ and $\mathscr{B}$ are both consistent. Then Problem $\mathscr{A}(\mathbf{0})$ is normal if and only if Problems $\mathscr{A}(\mathbf{0})$ and $\mathscr{B}$ have no duality gap, in which case

$$\partial\varphi(\mathbf{0}) = \mathscr{T}^*.$$

The proof of this corollary uses both Theorem 3.9 and Corollary 3.4 along with the (easily derived) fact that $\partial\varphi(\mathbf{0}) = \partial\bar{\varphi}(\mathbf{0})$ when $\varphi(\mathbf{0}) = \bar{\varphi}(\mathbf{0})$.

An additional prerequisite for sensitivity analyses is provided by the following definition.

Definition 3.7. A consistent Problem $\mathscr{A}(u)$ with a finite infimum $\varphi(u)$ is termed *stably set* (relative to the family $\mathscr{F}$) when the (one-sided) "directional derivative"

$$D_d\varphi(u) \triangleq \lim_{s \to 0^+} \frac{\varphi(u+sd)+\varphi(u)}{s}$$

exists and is finite for each feasible direction d (i.e., each direction d such that $u+sd \in \mathscr{U}$ for sufficiently small $s>0$).

The following theorem ties stability directly to the existence of certain subgradients, and shows how each guarantees normality.

Theorem 3.10. Let both $g:\mathscr{C}$ and $\mathscr{K}$ be convex, and suppose that Problem $\mathscr{A}(\mathbf{0})$ is consistent and has a finite infimum $\varphi(\mathbf{0})$. Then Problem $\mathscr{A}(\mathbf{0})$ is stably set if and only if $\partial\varphi(\mathbf{0})$ is not empty, in which case Problem $\mathscr{A}(\mathbf{0})$ is normal.

This theorem is a simple consequence of Theorem 3.6 and the differentiability properties of convex functions [as described, for example, by Rockafellar (Ref. 33)].

Given a consistent Problem $\mathscr{A}(\mathbf{0})$ with a known finite infimum $\varphi(\mathbf{0})$, *sensitivity analysis* consists of estimating other infima $\varphi(u)$ for very small u. First-order estimates can, of course, be based on the directional derivatives $D_u\varphi(\mathbf{0})$ when Problem $\mathscr{A}(\mathbf{0})$ is stably set. In that case, the defining equation for $D_u\varphi(\mathbf{0})$ provides the usual estimation formula

$$\varphi(u) \approx \varphi(\mathbf{0}) + D_u\varphi(\mathbf{0}),$$

whose use requires a computation of $D_u\varphi(\mathbf{0})$. Toward that end, Fenchel (Ref. 32) and, more recently, Rockafellar (Ref. 33) have appropriately extended the well-known formula $D_u\varphi(\mathbf{0}) = \langle u, \nabla\varphi(\mathbf{0})\rangle$ by showing that

$$D_u\varphi(\mathbf{0}) = \max_{y \in \partial\varphi(\mathbf{0})} \langle u, y\rangle$$

when φ is convex on $\mathscr{U}$—which is indeed the case when both $g:\mathscr{C}$ and $\mathscr{K}$ are convex, by virtue of Theorem 3.6. Moreover, Corollary 3.5 shows that the preceding formula can be rewritten as

$$D_u\varphi(\mathbf{0}) = \max_{y \in \mathscr{T}^*} \langle u, y\rangle$$

when the dual Problem $\mathscr{B}$ is consistent and there is no duality gap—which is almost always the case, as indicated by Theorem 3.6, Theorem 3.8, and (Fenchel's) Theorem 3.5. Consequently, it is of interest to know $\mathscr{T}^*$ in addition to $\varphi(\mathbf{0})$ and $\mathscr{S}^*(\mathbf{0})$, so that the preceding displayed formulas can be used to estimate $\varphi(u)$ for a very small u. But $\mathcal{T}^*$ can usually be calculated from an arbitrary $x^* \in \mathcal{S}^*(\mathbf{0})$ by employing the extremality conditions, as explained after Corollary 3.2.

To further characterize duality gaps, we also embed the dual Problem $\mathcal{B}$ in a parametrized family $\mathcal{G}$ of closely related geometric programming problems $\mathcal{B}(v)$ that are generated by simply translating (the domain $\mathcal{D}$ of) h through all possible displacements $-v \in E_n$, while keeping $\mathcal{Y}$ fixed. Problem $\mathcal{B}$ then appears in the parametrized family $\mathcal{G}$ as Problem $\mathcal{B}(\mathbf{0})$ and is studied in relation to all other geometric programming problems $\mathcal{B}(v)$, with special attention given to those Problems $\mathcal{B}(v)$ in $\mathcal{G}$ that are close to $\mathcal{B}(\mathbf{0})$ in the sense that (the "norm" of) v is small.

Naturally, the symbols $\mathcal{B}$, $\mathcal{T}$, ψ, and $\mathcal{T}^*$ now represent functions of v, though they originally represented only the particular functional values $\mathcal{B}(\mathbf{0})$, $\mathcal{T}(\mathbf{0})$, $\psi(\mathbf{0})$, and $\mathcal{T}^*(\mathbf{0})$, respectively. Needless to say, the reader must keep this notational discrepancy in mind when comparing subsequent developments with previous developments.

Since the main results having to do with the *dual geometric programming family* $\mathcal{G}$ are essentially dual to those already given for the family $\mathcal{F}$, we shall feel free to use them without further discussion.

The following important theorem involves the dual families $\mathcal{F}$ and $\mathcal{G}$ in a reflexive (i.e., self-dual) way.

Theorem 3.11. Let both $g:\mathcal{C}$ and $\mathcal{X}$ be convex and closed, and suppose that the dual problems $\mathcal{A}(\mathbf{0})$ and $\mathcal{B}(\mathbf{0})$ are both consistent. Then the following three conditions are equivalent:

(i) Problem $\mathcal{A}(\mathbf{0})$ is normal,
(ii) Problem $\mathcal{B}(\mathbf{0})$ is normal,
(iii) Problems $\mathcal{A}(\mathbf{0})$ and $\mathcal{B}(\mathbf{0})$ do not have a duality gap.

Moreover, if any of these three conditions are satisfied, then

$$\partial\varphi(\mathbf{0}) = \mathcal{T}^* \quad \text{and} \quad \partial\psi(\mathbf{0}) = \mathcal{S}^*.$$

This theorem can be proved simply by a repeated application of Corollary 3.5 and its (unstated) dual.

There are various degrees of consistency, ranging all the way from (the very weak) quasi-consistency through (the intermediate) consistency to "strong consistency."

Definition 3.8. Problem $\mathcal{B}(v)$ is said to be *strongly consistent if* $v \in (\text{ri}\ \mathcal{V})$.

Since $(\text{ri}\ \mathcal{V}) \subseteq \mathcal{V}$, each strongly consistent Problem $\mathcal{B}(v)$ is consistent, but not conversely.

The following theorem indicates the importance of strong consistency.

Theorem 3.12. Each strongly consistent dual Problem $\mathcal{B}(\mathbf{0})$ with a finite infimum $\psi(\mathbf{0})$ is stably set (and hence normal).

This theorem is an immediate consequence of the convexity of $h:\mathcal{D}$, the (unstated) dual of Theorem 3.6, and the differentiability properties of convex functions [which are given by both Fenchel (Ref. 32) and Rockafellar (Ref. 33)].

Since (ri $\mathcal{V}$) is "almost all" of $\mathcal{V}$, Theorem 3.12 implies that almost all consistent Problems $\mathcal{B}(\mathbf{0})$ with a finite infimum $\psi(\mathbf{0})$ are normal, so Theorem 3.11 shows that duality gaps are rather rare phenomena when both $g:\mathcal{C}$ and $\mathcal{X}$ are convex and closed. Nevertheless, duality gaps can occur when both $g:\mathcal{C}$ and $\mathcal{X}$ are convex and closed, and examples (due originally to J. J. Stoer) can be found in Appendix C of Peterson (Ref. 6).

3.1.6. Decomposition Principles. In all Problems $\mathcal{A}(\mathbf{0})$ known to the author to be of practical significance (including all examples given in Section 2.1) the cone $\mathcal{X}$ is polyhedral and hence "finitely generated." Suppose, then, without any known loss of practical significance, that there is at least one $n\times m$ matrix $\mathcal{M}$ with a corresponding index set $\mathcal{P}\subseteq\{1,2,\ldots,m\}$ for which

$$\mathcal{X}=\{x\in E_n \mid x=\mathcal{M}z \text{ for at least one } z\in E_m \text{ for which } z_j\geqq 0,\ j\in\mathcal{P}\}.$$

The index set $\mathcal{P}$ can of course be taken to be the empty set when $\mathcal{X}$ is in fact a vector space.

The main prerequisites for decomposing Problem $\mathcal{A}(\mathbf{0})$ into smaller (more manageable) subproblems are "sparsity" of the matrix $\mathcal{M}$ and separability of the function $g:\mathcal{C}$. Sparsity of $\mathcal{M}$ is frequently a natural occurrence with the modeling of large systems, while separability of $g:\mathcal{C}$ comes from making the appropriate problem transformations (as illustrated by the examples given in Section 2.1).

Three different decomposition princples for three different types of sparsity are described here. The three different types of sparsity are indicated by the three different types of "block diagonal structure" illustrated in Fig. 1. The enumerated submatrices $\mathcal{M}_k$ are of course the only submatrices of $\mathcal{M}$ that need not be zero matrices. Assume in general that s such submatrices $\mathcal{M}_1,\mathcal{M}_2,\ldots,\mathcal{M}_s$, with $s\geqq 2$, are arranged diagonally. (In particular, s is 4 for each of the three illustrated examples.) Matrices $\mathcal{M}$ of Type 1 are then those that have no additional nonzero submatrices; matrices $\mathcal{M}$ of Type 2 are those that have a single additional nonzero submatrix $\mathcal{M}_0$ consisting of entire columns of $\mathcal{M}$; and matrices $\mathcal{M}$ of Type 3 are those that have a single additional nonzero submatrix $\mathfrak{M}_{s+1}$ consisting of entire rows of

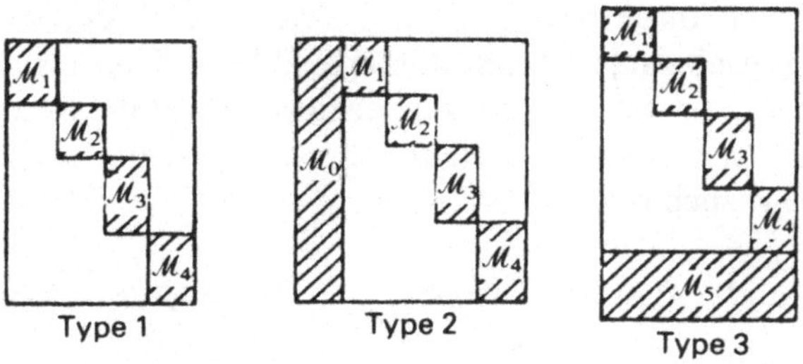

Fig. 1. Sparsity types.

$\mathcal{M}$. Actually, some matrices of neither type can be effectively transformed into one of the three types simply by row and/or column permutations.

Type 1 sparsity is the easiest to exploit, and its exploitation is at the heart of exploiting both Type 2 sparsity and, to some extent, Type 3 sparsity. Types 1 and 2 are exploited directly, but Type 3 requires the use of geometric Lagrangians and duality.

Each of the three sparsity types induces a partitioning of the rows of $\mathcal{M}$ and hence the components of x in such a way that

$$x = (x^1, x^2, \ldots, x^s) \qquad \text{for problems of Types 1 and 2,}$$

while

$$x = (x^1, x^2, \ldots, x^s, x^{s+1}) \qquad \text{for problems of Type 3,}$$

where the components of the vector variable x^k are enumerated exactly the same as those rows of $\mathcal{M}$ that contain rows of the submatrix $\mathcal{M}_k$. Of course, the rows of $\mathcal{M}_{s+1}$ for a given problem of Type 3 are the "coupling rows" that must be contended with in reducing such a problem to an equivalent problem of Type 1.

Similarly, each of the three sparsity types induces a partitioning of the columns of $\mathcal{M}$ and hence the components of z in such a way that

$$z = (z^1, z^2, \ldots, z^s) \qquad \text{for problems of Types 1 and 3,}$$

while

$$z = (z^0, z^1, z^2, \ldots, z^s) \qquad \text{for problems of Type 2,}$$

where the components of the vector variable z^k are enumerated exactly the same as those columns of $\mathcal{M}$ that contain columns of the submatrix $\mathcal{M}_k$. Of course, the columns of $\mathcal{M}_0$ for a given problem of Type 2 are the "coupling columns" that must be contended with in reducing such a problem to an equivalent problem of Type 1.

To render all three problem types amenable to decomposition, the function $g : \mathscr{C}$ must be at least partially separable, and its partial separability must be compatible with the preceding partitioning of the components of x. In particular, then, assume that there are functions $g_k : \mathscr{C}_k$, $k = 1, 2, \ldots, r, r+1$, such that

$$\mathscr{C} = \mathop{\times}_{k=1}^{r} \mathscr{C}_k \quad \text{and} \quad g(x) = \sum_{k=1}^{r} g_k(x^k) \qquad \text{for problems of Types 1 and 2,}$$

while

$$\mathscr{C} = \mathop{\times}_{k=1}^{r+1} \mathscr{C}_k \quad \text{and} \quad g(x) = \sum_{k=1}^{r+1} g_k(x^k) \qquad \text{for problems of Type 3.}$$

This assumption is, of course, automatically satisfied when $g : \mathscr{C}$ is completely separable, a condition that holds for many of the examples given in Section 2.1.

Decomposition principles for Type 1 and Type 2 problems utilize the cones

$$\mathscr{X}_k \triangleq \{x^k \in E_{n_k} \mid x^k = \mathscr{M}_k z^k \text{ for at least one } z^k \in E_{m_k} \text{ for which } z_j^k \geqq 0, j \in \mathscr{P}\}.$$

There are, of course, r such cones $\mathscr{X}_k$, $k = 1, 2, \ldots, r$, for a problem of Type 1, and $(r+1)$ such cones $\mathscr{X}_k$, $k = 0, 1, 2, \ldots, r$, for a problem of Type 2. Only the extra (coupling) cone $\mathscr{X}_0$ for a problem of Type 2 is a subcone of $\mathscr{X}$.

Problems of Type 1. Observe that the cone $\mathscr{X}$ is separable in that

$$\mathscr{X} = \mathop{\times}_{k=1}^{r} \mathscr{X}_k.$$

This separation of the cone $\mathscr{X}$ into the "direct sum" of the cones $\mathscr{X}_k$, $k = 1, 2, \ldots, r$, and the separation of the function $g : \mathscr{C}$ into a sum of the functions $g_k : \mathscr{C}_k$, $k = 1, 2, \ldots, r$, immediately imply that Problem $\mathscr{A}(\mathbf{0})$, now designated Problem $\mathscr{A}^1(\mathbf{0})$, can be solved by solving the smaller geometric programming problems $\mathscr{A}_k(\mathbf{0})$ that are constructed from the respective functions $g_k : \mathscr{C}_k$ and the respective cones $\mathscr{X}_k$, $k = 1, 2, \ldots, r$. In particular, the (desired) infimum $\varphi^1(\mathbf{0})$ for Problem $\mathscr{A}^1(\mathbf{0})$ can clearly be determined from the infima $\varphi_k(\mathbf{0})$ for the respective Problems $\mathscr{A}_k(\mathbf{0})$, $k = 1, 2, \ldots, r$, by the formula

$$\varphi^1(\mathbf{0}) = \sum_{k=1}^{r} \varphi_k(\mathbf{0}).$$

Moreover, the (desired) optimal solution set $\mathscr{S}^{1*}(\mathbf{0})$ for Problem $\mathscr{A}^1(\mathbf{0})$ can obviously be determined from the optimal solution sets $\mathscr{S}_k^*(\mathbf{0})$ for the respective Problems $\mathscr{A}_k(\mathbf{0})$, $k = 1, 2, \ldots, \imath$, by the formula

$$\mathscr{S}^{1*}(\mathbf{0}) = \bigtimes_{k=1}^{\imath} \mathscr{S}_k^*(\mathbf{0}).$$

This direct decomposition of Problem $\mathscr{A}^1(\mathbf{0})$ into $\imath$ smaller Problems $\mathscr{A}_k(\mathbf{0})$, $k = 1, 2, \ldots, \imath$, generally increases computational efficiency and can in fact be a necessity when $\mathscr{A}^1(\mathbf{0})$ is itself too large for computer storage.

It is important to note that the previous assumptions about Problem $\mathscr{A}^1(\mathbf{0})$ are inherited by all Problems $\mathscr{A}^1(u)$ in the geometric programming family $\mathscr{F}^1$. In particular, the cone $\mathscr{X}$ and hence the block diagonal structure of its matrix representation $\mathscr{M}$ remain invariant of u, and the separability of the function $g : \mathscr{C}$ is clearly inherited by all functions $g(\cdot + u) : (\mathscr{C} - u)$. Consequently, the decomposition principle just described in the context of Problem $\mathscr{A}^1(\mathbf{0})$ is just as applicable to all other Problems $\mathscr{A}^1(u)$ in the family $\mathscr{F}^1$.

In treating such Problems $\mathscr{A}^1(u)$, the components of u must be partitioned in the same way that the components of x have been partitioned, namely,

$$u = (u^1, u^2, \ldots, u^{\imath}),$$

where the components of the vector variable u^k are enumerated exactly the same as those rows of $\mathscr{M}$ that contain rows of the submatrix $\mathscr{M}_k$.

Now, an application of the decomposition principle just described in the context of Problem $\mathscr{A}^1(\mathbf{0})$ to each Problem $\mathscr{A}^1(u)$ in a given family $\mathscr{F}^1$ shows that the (desired) functions $\varphi^1 : \mathscr{U}^1$ and $\mathscr{S}^{1*} : \mathscr{U}^1$ are determined by the corresponding functions $\varphi_k : \mathscr{U}_k$ and $\mathscr{S}_k^* : \mathscr{U}_k$ that are associated with the geometric programming families $\mathscr{F}_k$ that are of course constructed from the respective functions $g_k : \mathscr{C}_k$ and the respective cones $\mathscr{X}_k$, $k = 1, 2, \ldots, \imath$. This direct decomposition of the family $\mathscr{F}^1$ into $\imath$ smaller families $\mathscr{F}_k$, $k = 1, 2, \ldots, \imath$, can be concisely described by the formulas

$$\mathscr{U}^1 = \bigtimes_{k=1}^{\imath} \mathscr{U}_k,$$

$$\varphi^1(u) = \sum_{k=1}^{\imath} \varphi_k(u^k),$$

and

$$\mathscr{S}^{1*}(u) = \bigtimes_{k=1}^{\imath} \mathscr{S}_k^*(u^k),$$

where

$$\mathscr{U}_k = \mathscr{C}_k - \mathscr{X}_k, \qquad k = 1, 2, \ldots, \imath.$$

Consequently, the functions $\varphi^1 : \mathscr{U}^1$ and $\mathscr{S}^{1*} : \mathscr{U}^1$ can be studied by studying the functions $\varphi_k : \mathscr{U}_k$ and $\mathscr{S}_k^* : \mathscr{U}_k$, $k = 1, 2, \ldots, r$.

In particular, given that the (one-sided) directional derivative $D_{d^k}\varphi_k(u^k)$ of the function $\varphi_k : \mathscr{U}_k$ (at a point $u^k \in \mathscr{U}_k$ in the direction $d^k \in E_{n_k}$) exists for $k = 1, 2, \ldots, r$, and given that both $u = (u^1, u^2, \ldots, u^r)$ and $d = (d^1, d^2, \ldots, d^r)$, the (one-sided) directional derivative $D_d\varphi^1(u)$ of the function $\varphi^1 : \mathscr{U}^1$ (at the point $u \in \mathscr{U}^1$ in the direction $d \in E_n$) clearly exists and is given by the formula

$$D_d\varphi^1(u) = \sum_{k=1}^{r} D_{d^k}\varphi_k(u^k).$$

It is only on rare occasions that the functions $\varphi_k : \mathscr{U}_k$, $k = 1, 2, \ldots, r$, can be obtained in terms of elementary formulas. Consequently, the directional derivatives $D_{d^k}\varphi_k(u^k)$, $k = 1, 2, \ldots, r$, usually have to be determined by numerical differentiation or other numerical methods.

If a given function $\varphi_k : \mathscr{U}_k$ is convex (which is the case when both $g_k : \mathscr{C}_k$ and $\mathscr{X}_k$ are convex), then other relatively mild conditions (as discussed in Section 3.1.5) guarantee that

$$D_{d^k}\varphi_k(u^k) = \max_{y^k \in \mathscr{T}_k^*(\mathbf{0}; u^k)} \langle d^k, y^k \rangle,$$

where $\mathscr{T}_k^*(\mathbf{0}, u^k)$ is the optimal solution set for the geometric dual $\mathscr{B}_k(\mathbf{0}; u^k)$ of Problem $\mathscr{A}_k(u^k)$. Since $\mathscr{T}_k^*(\mathbf{0}; u^k)$ can usually be calculated from a single known optimal solution $x^{k^*} \in \mathscr{S}_k^*(u^k)$ by employing the appropriate extremality conditions (as explained after Corollary 3.2), and since $\mathscr{T}_k^*(\mathbf{0}; u^k)$ is frequently polyhedral (and sometimes a singleton), the directional derivatives $D_d\varphi^1(u)$ can often be determined by the preceding two displayed formulas by little more than linear programming (and sometimes much less). This means of course that first order methods can often be used in conjunction with the preceding decomposition principle to minimize $\varphi^1(u)$ over a given subset of $\mathscr{U}^1$—a fundamental technique to be used in decomposing problems of Type 2.

Problems of Type 2. Observe that a given Problem $\mathscr{A}(\mathbf{0})$ of Type 2, now designated Problem $\mathscr{A}^2(\mathbf{0})$, reduces to a Problem $\mathscr{A}^1(u)$ of Type 1 when the coupling vector variable z^0 is (temporarily) fixed and u is chosen to be $\mathscr{M}_0 z^0$. Now, very elementary arguments show that the (desired) infimum $\varphi^2(\mathbf{0})$ for Problem $\mathscr{A}^2(\mathbf{0})$ can in fact be determined by the formula

$$\varphi^2(\mathbf{0}) = \inf_{u \in \mathscr{X}_0 \cap \mathscr{U}^1} \varphi^1(u).$$

The minimization problem that appears in this formula is obviously a geometric programming problem and is termed the *master problem.* Its objective function $\varphi^1 : \mathcal{U}^1$ has of course already been (partially) separated into a sum of *subproblem* infima functions $\varphi_k : \mathcal{U}_k$, $k = 1, 2, \ldots, \imath$, by the decomposition principle just described for problems of Type 1. That decomposition principle is, of course, to be used here not only to calculate the functional values $\varphi^1(u)$ but also to calculate any directional derivatives $D_d\varphi^1(u)$ that are needed to implement an appropriate algorithm for solving the master problem. Once the master problem's optimal solution set

$$\mathcal{U}^* \triangleq \{u \in \mathcal{X}_0 \cap \mathcal{U}^1 \mid \varphi^1(u) = \varphi^2(\mathbf{0})\}$$

has been obtained, the (desired) optimal solution set $\mathcal{S}^{2*}(\mathbf{0})$ for Problem $\mathcal{A}^2(\mathbf{0})$ can clearly be determined by the formula

$$\mathcal{S}^{2*}(\mathbf{0}) = \bigcup_{u^* \in \mathcal{U}^*} \left\{u^* + \bigtimes_{k=1}^{\imath} \mathcal{S}_k^*(u^{*k})\right\}.$$

In the process of determining a $u^* \in \mathcal{U}^*$ with the aid of this "tearing procedure," one can of course expect to determine an $x^{k*} \in \mathcal{S}_k^*(u^{*k})$, $k = 1, 2, \ldots, \imath$, in which event $u^* + (x^{1*}, x^{2*}, \ldots, x^{\imath*})$ is one of the desired optimal solutions to Problem $\mathcal{A}^2(\mathbf{0})$. This assumes of course that such optimal solutions exist.

In summary, Problem $\mathcal{A}^2(\mathbf{0})$ can be torn into $\imath$ smaller Problems $\mathcal{A}_k(u^k)$, $k = 1, 2, \ldots, \imath$, that are judiciously selected by the master problem (with the possible help of geometric programming duality in the convex case). This tearing *may* increase computational efficiency but can in fact be a necessity when $\mathcal{A}^2(\mathbf{0})$ is itself too large for computer storage.

It is worth noting that the preceding decomposition principle can easily be applied to each Problem $\mathcal{A}^2(u)$ in a given family $\mathcal{F}^2$ whose (unperturbed) Problem $\mathcal{A}^2(\mathbf{0})$ is of Type 2. However, the result of such an application is notationally cumbersome to describe and is left to the imagination of the interested reader.

Decomposition principles for Type 3 problems implicitly utilize the dual Problem $\mathcal{B}(\mathbf{0})$. In doing so, they require $g : \mathcal{C}$ to be convex and closed, and they require the absence of a duality gap.

Problems of Type 3. Recall from the end of Section 3.1.3 that Problem $\mathcal{A}(\mathbf{0})$, now designated Problem $\mathcal{A}^3(\mathbf{0})$, can be described in terms of the geometric Lagrangian L_g when $g : \mathcal{C}$ is convex and closed. In particular, the (desired) infimum

$$\varphi^3(\mathbf{0}) = \inf_{x \in \mathcal{X}} [\sup_{y \in \mathcal{D}} L_g(x; y)].$$

Now, the separability inherited by L_g from $g:\mathscr{C}$ (via $h:\mathscr{D}$) clearly implies that

$$\inf_{x\in\mathscr{X}}[\sup_{y\in\mathscr{D}} L_g(x;y)] = \inf_{x\in\mathscr{X}}[\sup_{y^{r+1}\in\mathscr{D}_{r+1}} \mathscr{L}_g(x;y^{r+1})],$$

where the "contracted geometric Lagrangian"

$$\mathscr{L}_g(x;y^{r+1}) \triangleq \sum_{k=1}^{r} g_k(x^k) + \langle y^{r+1}, x^{r+1}\rangle - h_{r+1}(y^{r+1}).$$

Moreover, the absence of a duality gap implies that

$$\inf_{x\in\mathscr{X}}[\sup_{y^{r+1}\in\mathscr{D}_{r+1}} \mathscr{L}_g(x;y^{r+1})] = \sup_{y^{r+1}\in\mathscr{D}_{r+1}}[\inf_{x\in\mathscr{X}} \mathscr{L}_g(x;y^{r+1})]$$

and hence that

$$\varphi^3(\mathbf{0}) = \sup_{y^{r+1}\in\mathscr{D}_{r+1}}[\inf_{x\in\mathscr{X}} \mathscr{L}_g(x;y^{r+1})].$$

Consequently, the sparsity of $\mathscr{M}$ produces a *master problem*

$$\varphi^3(\mathbf{0}) = \sup_{y^{r+1}\in\mathscr{D}_{r+1}} \Psi(y^{r+1})$$

whose objective function value

$$\Psi(y^{r+1}) \triangleq \sum_{k=1}^{r} \Psi_k(y^{r+1}) - h_{r+1}(y^{r+1}),$$

where the *subproblem* infima

$$\Psi_k(y^{r+1}) \triangleq \inf_{z^k}[g_k(\mathscr{M}_k z^k) + \langle y^{r+1}, \mathscr{M}^k_{r+1} z^k\rangle], \qquad k = 1, 2, \ldots, r,$$

with $\mathscr{M}^k_{r+1}$ representing that submatrix of $\mathscr{M}_{r+1}$ whose columns correspond to the components of z^k.

It is worth noting that the kth subproblem is in fact a geometric programming problem with an objective function value $g_k(x^k) + \langle y^{r+1}, x^{r+1}\rangle$ and a cone generated by the columns of the matrix

$$\begin{bmatrix} \mathscr{M}_k \\ \mathscr{M}^k_{r+1} \end{bmatrix}$$

relative to $\mathscr{P}$. Moreover, the master problem turns out to be a "suboptimized" version of the geometric dual problem $\mathscr{B}(\mathbf{0})$; that is, $\Psi(y^{r+1})$ is finite if and only if there exists a vector $(y^1, y^2, \ldots, y^r)$ such that the augmented vector $y = (y^1, y^2, \ldots, y^r, y^{r+1}) \in \mathscr{T}(\mathbf{0})$, in which case $\Psi(y^{r+1})$ is just the negative of the (sub)infimum of $h(y)$ over all such $(y^1, y^2, \ldots, y^r)$.

The preceding fact implies that the master problem consists of *maximizing a concave* function Ψ over a convex set (which makes it a convex programming problem). Given that $g_k : \mathscr{C}_k$, $k = 1, 2, \ldots, \imath, \imath+1$, is convex and closed, any directional derivatives $D_d\Psi(y^{\imath+1})$ that are needed to implement an appropriate algorithm for solving the master problem can be obtained from the subgradient representation

$$-D_d\Psi(y^{\imath+1}) = \sup_{u^{\imath+1} \in -\partial\Psi(y^{\imath+1})} \langle d, u^{\imath+1} \rangle,$$

where the subgradient set

$$-\partial\Psi(y^{\imath+1}) = \Big\{ u^{\imath+1} \,\Big|\, \text{there exist } x^k \in \mathscr{X}_k,\ k = 1, 2, \ldots, \imath, \imath+1,$$

$$\text{for which } 0 = \sum_1^{\imath+1} \langle x^k, y^k \rangle,\ x^k \in \partial h_k(y^k),$$

$$k = 1, 2, \ldots, \imath, \text{ and } x^{\imath+1} + u^{\imath+1} \in \partial h_{\imath+1}(y^{\imath+1}) \Big\}.$$

Of course, $y^{\imath+1}$ is actually an optimal solution to the master problem if and only if $\mathbf{0} \in -\partial\Psi(y^{\imath+1})$, in which event the (desired) optimal solution set

$$\mathscr{S}^{3*}(\mathbf{0}) = \Big\{ (x^1, x^2, \ldots, x^\imath, x^{\imath+1}) \,\Big|\, x^k \in \mathscr{X}_k,\ k = 1, 2, \ldots, \imath, \imath+1,$$

$$0 = \sum_1^{\imath+1} \langle x^k, y^k \rangle, \text{ and } x^k \in \partial h_k(y^k),\ k = 1, 2, \ldots, \imath, \imath+1 \Big\}.$$

It is worth noting that the preceding decomposition principle can easily be applied to each Problem $\mathscr{A}^3(u)$ in a given family $\mathscr{F}^3$ whose (unperturbed) Problem $\mathscr{A}^3(\mathbf{0})$ is of Type 3. However, the result of such an application is notationally cumbersome to describe and is left to the imagination of the interested reader.

All of the preceding decomposition principles can be combined to treat Problems $\mathscr{A}(\mathbf{0})$ for which $\mathscr{M}$ has the type of block diagonal structure shown in Fig. 2. As indicated by the partitioning of the matrix $\mathscr{M}$ between the submatrices $\mathscr{M}_0$ and $\mathscr{M}_5$ (in Fig. 2), a series of Type 3 decompositions is to be guided by a Type 2 decomposition.

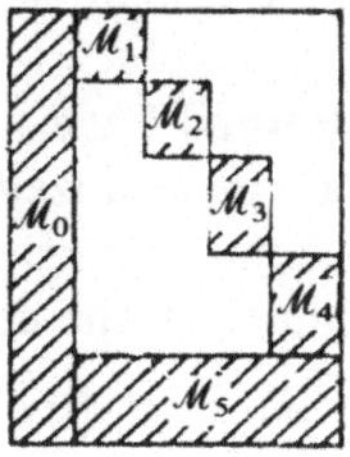

Fig. 2. Sparsity type 4.

More detailed descriptions of the preceding decomposition principles can be found in Peterson (Refs. 40 and 41).

3.2. Relations between the Constrained and Unconstrained Cases. The constrained case can, of course, be specialized to the unconstrained case, simply by letting both index sets I and J be empty while choosing $g_0 : C_0$ to be $\mathscr{g} : \mathscr{C}$ and X to be $\mathscr{X}$. A somewhat surprising fact is that this specialization can be reversed, that is, the unconstrained case can actually be specialized to the constrained case.

To do so, let the functional domain

$$\mathscr{C} \triangleq \{(\mathbf{x}^0, \mathbf{x}^I, \boldsymbol{\alpha}, \mathbf{x}^J, \boldsymbol{\kappa}) \in E_{\mathscr{n}} \mid \mathbf{x}^0 \in C_0; \mathbf{x}^i \in C_i, \alpha_i \in E_1,$$

and

$$g_i(\mathbf{x}^i) + \alpha_i \leqq 0, \, i \in I; \, (\mathbf{x}^j, \boldsymbol{\kappa}_j) \in C_j^+, \, j \in J\};$$

and let the functional value

$$g(\mathbf{x}^0, \mathbf{x}^I, \boldsymbol{\alpha}, \mathbf{x}^J, \boldsymbol{\kappa}) \triangleq g_0(\mathbf{x}^0) + \sum_J g_j^+(\mathbf{x}^j, \boldsymbol{\kappa}_j) \triangleq G(\mathbf{x}, \boldsymbol{\kappa}),$$

while letting the cone

$$\mathscr{X} \triangleq \{(\mathbf{x}^0, \mathbf{x}^I, \boldsymbol{\alpha}, x^J, \boldsymbol{\kappa}) \in E_{\mathscr{n}} \mid (\mathbf{x}^0, \mathbf{x}^I, \mathbf{x}^J) \in X; \boldsymbol{\alpha} = \mathbf{0}; \boldsymbol{\kappa} \in E_{o(J)}\}.$$

Then (the unconstrained) Problem $\mathscr{A}$ is clearly identical to (the constrained) Problem A. The additional independent vector variable $\boldsymbol{\alpha}$ with components α_i, $i \in I$, may seem superfluous, but is included so that the induced family $\mathscr{F}$ is essentially identical to a family F that includes certain important constraint parameters μ_i (as described in Section 3.3.5).

Of course, the preceding choice of $\mathscr{g} : \mathscr{C}$ and $\mathscr{X}$ also induces a choice of the dual problem $\mathscr{B}$ and dual family $\mathscr{G}$ corresponding to Problem $\mathscr{A}$ (i.e., Problem A). In fact, computations of the resulting conjugate transform $\mathscr{h} : \mathscr{D}$ and dual cone $\mathscr{Y}$ [given by Peterson (Refs. 8 and 9)] show that the induced dual problem $\mathscr{B}$ and dual family $\mathscr{G}$ are essentially identical to the dual Problem B and dual family G described in Sections 3.3.4 and 3.3.5, respectively. This identification turns out to be highly significant, because it provides an efficient mechanism for extending to the constrained case most of the important theorems already described for the unconstrained case.

3.3. The Constrained Case. Only the key ideas and results from the unconstrained case are generalized here, but hopefully the reader can easily fill in the remaining details by using Section 3.1 as a guide.

Let Y be the "dual" of the cone X, that is,

$$Y \triangleq \{\mathbf{y} \in E_n \mid 0 \leqq \langle \mathbf{x}, \mathbf{y} \rangle \text{ for each } \mathbf{x} \in X\}.$$

Of course, the fact that $\mathbf{x} \triangleq (\mathbf{x}^0, \mathbf{x}^I, \mathbf{x}^J)$ means that $\mathbf{y} \triangleq (\mathbf{y}^0, \mathbf{y}^I, \mathbf{y}^J)$, with $\mathbf{y}^I$ and $\mathbf{y}^J$ constructed in the same manner as $\mathbf{x}^I$ and $\mathbf{x}^J$.

3.3.1. Optimality Conditions. We begin with the following fundamental definition.

Definition 3.9. A *critical solution* (stationary solution, equilibrium solution, P solution) for Problem A is any vector $(\mathbf{x}^*, \boldsymbol{\kappa}^*)$ for which there is a vector $\boldsymbol{\lambda}^*$ in $E_{o(I)}$ such that $(\mathbf{x}^*, \boldsymbol{\kappa}^*)$ and $\boldsymbol{\lambda}^*$ jointly satisfying the following P *optimality conditions*:

$$\mathbf{x}^* \in X,$$

$$g_i(\mathbf{x}^{*i}) \leqq 0, \qquad i \in I,$$

$$\lambda_i^* \geqq 0, \qquad i \in I,$$

$$\lambda_i^* g_i(\mathbf{x}^{*i}) = 0, \qquad i \in I,$$

$$\mathbf{y}^* \in Y,$$

$$0 = \langle \mathbf{x}^*, \mathbf{y}^* \rangle,$$

and

$$\langle \mathbf{x}^{*j}, \mathbf{y}^{*j} \rangle = g_j^+(\mathbf{x}^{*j}, \kappa_j^*), \qquad j \in J,$$

where

$$\mathbf{y}^{*0} \triangleq \nabla g_0(\mathbf{x}^{*0}), \qquad \mathbf{y}^{*i} \triangleq \lambda_i^* \nabla g_i(\mathbf{x}^{*i}), \qquad i \in I,$$

and

$$\mathbf{y}^{*j} \triangleq \nabla g_j(\mathbf{x}^{*j}/\kappa_j^*), \qquad j \in J.$$

For ordinary programming (i.e., Example 2.8 in Section 2.2), it is worth noting that the P optimality conditions are essentially the (more familiar) "Kuhn–Tucker optimality conditions":

$$g_i(\mathbf{z}^*) \leqq 0, \qquad i \in I,$$

$$\lambda_i^* \geqq 0, \qquad i \in I,$$

$$\lambda_i^* g_i(\mathbf{z}^*) = 0, \qquad i \in I,$$

and

$$\nabla g_0(\mathbf{z}^*) + \sum_I \lambda_i^* \nabla g_i(\mathbf{z}^*) = 0.$$

On the other hand, the following important concept from ordinary programming plays a crucial role in the constrained case of geometric programming.

Definition 3.10. For a consistent Problem A with a finite infimum φ, a *Kuhn–Tucker vector* is any vector $\boldsymbol{\lambda}^*$ in $E_{o(I)}$ with the two properties

$$\lambda_i^* \geqq 0 \qquad i \in I,$$

and

$$\varphi = \inf_{\substack{(\mathbf{x},\boldsymbol{\kappa}) \in C \\ \mathbf{x} \in X}} L_o(\mathbf{x}, \boldsymbol{\kappa}; \boldsymbol{\lambda}^*),$$

where the (ordinary) *Lagrangian*

$$L_o(\mathbf{x}, \boldsymbol{\kappa}; \boldsymbol{\lambda}) \triangleq G(\mathbf{x}, \boldsymbol{\kappa}) + \sum_I \lambda_i g_i(\mathbf{x}^i).$$

The following theorem gives two convexity conditions that guarantee the necessity and/or sufficiency of the P optimality conditions for optimality.

Theorem 3.13. Under the hypotheses that g_k is differentiable at $\mathbf{x}^{*k}$, $k \in \{0\} \cup I \cup J$ and that g_j^+ is differentiable at $(\mathbf{x}^{*j}, \kappa_j^*)$, $j \in J$,

(i) given that X is convex, if $(\mathbf{x}^*, \boldsymbol{\kappa}^*)$ is an optimal solution to Problem A and if $\boldsymbol{\lambda}^*$ is a Kuhn–Tucker vector for Problem A, then $(\mathbf{x}^*, \boldsymbol{\kappa}^*)$ is a critical solution for Problem A relative to $\boldsymbol{\lambda}^*$ (but not conversely),

(ii) given that g_k is convex on C_k, $k \in \{0\} \cup I \cup J$, if $(\mathbf{x}^*, \boldsymbol{\kappa}^*)$ is a critical solution for Problem A relative to $\boldsymbol{\lambda}^*$, then $(\mathbf{x}^*, \boldsymbol{\kappa}^*)$ is an optimal solution to Problem A and $\boldsymbol{\lambda}^*$ is a Kuhn–Tucker vector for Problem A.

3.3.2. Lagrangian Saddle Points. Let $h_k : D_k$ be the conjugate transform of the function $g_k : C_k$, $k \in \{0\} \cup I \cup J$, and let these sets D_j determine the sets D_j, $j \in J$ postulated at the beginning of Section 2.2.

The following definition is of fundamental importance.

Definition 3.11. Consider the function H whose domain

$$D \triangleq \{(\mathbf{y}, \boldsymbol{\lambda}) \mid \mathbf{y}^k \in D_k,\ k \in \{0\} \cup J, \text{ and } (\mathbf{y}^i, \boldsymbol{\lambda}_i) \in D_i^+,\ i \in I\},$$

and whose functional value

$$H(\mathbf{y}, \boldsymbol{\lambda}) \triangleq h_0(\mathbf{y}^0) + \sum_I h_i^+(\mathbf{y}^i, \lambda_i),$$

where

$$D_i^+ \triangleq \{(\mathbf{y}^i, \lambda_i) \mid \text{either } \lambda_i = 0 \text{ and } \sup_{\mathbf{c}^i \in C_i} \langle \mathbf{y}^i, \mathbf{c}^i \rangle < +\infty, \text{ or } \lambda_i > 0 \text{ and } \mathbf{y}^i \in \lambda_i D_i\},$$

and

$$h_i^+(\mathbf{y}^i, \lambda_i) \triangleq \begin{cases} \sup\limits_{\mathbf{c}^i \in C_i} \langle \mathbf{y}^i, \mathbf{c}^i \rangle & \text{if } \lambda_i = 0 \text{ and } \sup\limits_{\mathbf{c}^i \in C_i} \langle \mathbf{y}^i, \mathbf{c}^i \rangle < +\infty, \\ \lambda_i h_i(\mathbf{y}^i/\lambda_i) & \text{if } \lambda_i > 0 \text{ and } \mathbf{y}^i \in \lambda_i D_i. \end{cases}$$

For a consistent Problem A with a finite infimum φ, a *P vector* is any vector $(\mathbf{y}^*, \boldsymbol{\lambda}^*)$ with the two properties

$$(\mathbf{y}^*, \boldsymbol{\lambda}^*) \in D$$

and

$$\varphi = \inf_{\substack{\mathbf{x} \in X \\ \boldsymbol{\kappa} \geqq 0}} L_g(\mathbf{x}, \boldsymbol{\kappa}; \mathbf{y}^*, \boldsymbol{\lambda}^*),$$

where the (geometric) *Lagrangian*

$$L_g(\mathbf{x}, \boldsymbol{\kappa}; \mathbf{y}, \boldsymbol{\lambda}) \triangleq \langle \mathbf{x}, \mathbf{y} \rangle - H(\mathbf{y}, \boldsymbol{\lambda}) - \mathop{E}_{J} \kappa_j h_j(\mathbf{y}^j).$$

For a given Problem A note that the geometric Lagrangian L_g is entirely different from the ordinary Lagrangian L_o. Unlike the geometric Lagrangian L_g, the ordinary Lagrangian L_o simply reduces to the objective function G when $I = \varnothing$, but L_o exists even when the conjugate transforms $h_k : D_k$, $k \in \{0\} \cup I \cup J$ do not exist. However, Corollary 3.7 shows that Kuhn–Tucker vectors and P vectors are closely related.

The following saddle-point theorem provides several characterizations of optimality via P vectors.

Theorem 3.14. Given that $g_k : C_k$, $k \in \{0\} \cup I \cup J$ is convex and closed, let $(\mathbf{x}^*, \boldsymbol{\kappa}^*)$ be such that $\mathbf{x}^* \in X$ and $\boldsymbol{\kappa}^* \geqq 0$, and let $(\mathbf{y}^*, \boldsymbol{\lambda}^*) \in D$. Then $(\mathbf{x}^*, \boldsymbol{\kappa}^*)$ is optimal for Problem A and $(\mathbf{y}^*, \boldsymbol{\lambda}^*)$ is a P vector for Problem A if and only if the ordered pair $(\mathbf{x}^*, \boldsymbol{\kappa}^*; \mathbf{y}^*, \boldsymbol{\lambda}^*)$ is a "saddle point" for the Lagrangian L_g, that is,

$$\sup_{(\mathbf{y}, \boldsymbol{\lambda}) \in D} L_g(\mathbf{x}^*, \boldsymbol{\kappa}^*; \mathbf{y}, \boldsymbol{\lambda}) = L_g(\mathbf{x}^*, \boldsymbol{\kappa}^*; \mathbf{y}^*, \boldsymbol{\lambda}^*) = \inf_{\substack{\mathbf{x} \in X \\ \boldsymbol{\kappa} \geqq 0}} L_g(\mathbf{x}, \boldsymbol{\kappa}; \mathbf{y}^*, \boldsymbol{\lambda}^*),$$

in which case L_g has the saddle-point value

$$L_g(\mathbf{x}^*, \boldsymbol{\kappa}^*; \mathbf{y}^*, \boldsymbol{\lambda}^*) = G(\mathbf{x}^*, \boldsymbol{\kappa}^*) = \varphi.$$

Moreover,

$$\sup_{(\mathbf{y}, \boldsymbol{\lambda}) \in D} L_g(\mathbf{x}^*, \boldsymbol{\kappa}^*; \mathbf{y}, \boldsymbol{\lambda}) = L_g(\mathbf{x}^*, \boldsymbol{\kappa}^*; \mathbf{y}^*, \boldsymbol{\lambda}^*)$$

if and only if $(\mathbf{x}^*, \boldsymbol{\kappa}^*)$ and $(\mathbf{y}^*, \boldsymbol{\lambda}^*)$ satisfy both the feasibility conditions

$$(\mathbf{x}^*, \boldsymbol{\kappa}^*) \in C$$

$$g_i(\mathbf{x}^{*i}) \leqq 0, \qquad i \in I,$$

and the subgradient and "complementary slackness" conditions

$$\mathbf{y}^{*0} \in \partial g_0(\mathbf{x}^{*0}),$$

either $\lambda_i^* = 0$ and $\langle \mathbf{x}^{*i}, \mathbf{y}^{*i} \rangle = \sup_{\mathbf{c}^i \in C_i} \langle \mathbf{c}^i, \mathbf{y}^{*i} \rangle$, or $\lambda_i^* > 0$ and $\mathbf{y}^{*i} \in \lambda_i^* \partial g_i(\mathbf{x}^{*i})$, $i \in I$,

$$\lambda_i^* g_i(\mathbf{x}^{*i}) = 0, \qquad i \in I,$$

either $\kappa_j^* = 0$ and $\langle \mathbf{x}^{*j}, \mathbf{y}^{*j} \rangle = \sup_{\mathbf{d}^j \in D_j} \langle \mathbf{x}^{*j}, \mathbf{d}^j \rangle$, or $\kappa_j^* > 0$ and $\mathbf{y}^{*j} \in \partial g_j(\mathbf{x}^{*j}/\kappa_j^*)$, $j \in J$,

in which case

$$L_g(\mathbf{x}^*, \boldsymbol{\kappa}^*; \mathbf{y}^*, \boldsymbol{\lambda}^*) = G(\mathbf{x}^*, \boldsymbol{\kappa}^*).$$

Furthermore,

$$L_g(\mathbf{x}^*, \boldsymbol{\kappa}^*; \mathbf{y}^*, \boldsymbol{\lambda}^*) = \inf_{\substack{\mathbf{x} \in X \\ \boldsymbol{\kappa} \geqq 0}} L_g(\mathbf{x}, \boldsymbol{\kappa}; \mathbf{y}^*, \boldsymbol{\lambda}^*)$$

if and only if $(\mathbf{x}^*, \boldsymbol{\kappa}^*)$ and $(\mathbf{y}^*, \boldsymbol{\lambda}^*)$ satisfy both the feasibility conditions

$$\mathbf{y}^* \in Y,$$

$$h_j(\mathbf{y}^{*j}) \leqq 0, \qquad j \in J,$$

and the orthogonality and complementary slackness conditions

$$0 = \langle \mathbf{x}^*, \mathbf{y}^* \rangle,$$

$$\kappa_j^* h_j(\mathbf{y}^{*j}) = 0, \qquad j \in J,$$

in which case

$$L_g(\mathbf{x}^*, \boldsymbol{\kappa}^*; \mathbf{y}^*, \boldsymbol{\lambda}^*) = -H(\mathbf{y}^*, \boldsymbol{\lambda}^*).$$

3.3.3. The Geometric Inequality. To introduce duality into the constrained case, the conjugate inequality (given in Section 3.1.2) must be extended in a very special way.

The resulting "geometric inequality" can actually be derived directly from the conjugate inequality by introducing a scalar variable $\tau \geqq 0$. First, suppose that $\tau > 0$ and that $\boldsymbol{\zeta}/\tau$ is in $\boldsymbol{\Omega}$, so that $\boldsymbol{\zeta}/\tau$ can be substituted for $\boldsymbol{\zeta}$ in the conjugate inequality. Then multiply the resulting inequality by τ to establish the nontrivial part of the geometric inequality

$$\langle \mathbf{z}, \boldsymbol{\zeta} \rangle \leqq \tau w(\mathbf{z}) + \omega^+(\boldsymbol{\zeta}, \tau) \qquad \text{for } \mathbf{z} \in W \text{ and } (\boldsymbol{\zeta}, \tau) \in \boldsymbol{\Omega}^+,$$

where

$$\boldsymbol{\Omega}^{+} \triangleq \{(\boldsymbol{\zeta}, \tau) \in E_{\eta+1} \,|\, \text{either } \tau = 0 \text{ and } \sup_{\mathbf{z}' \in W} \langle \mathbf{z}', \boldsymbol{\zeta} \rangle < +\infty, \text{ or } \tau > 0 \text{ and } \boldsymbol{\zeta} \in \tau \boldsymbol{\Omega}\},$$

and

$$\omega^{+}(\boldsymbol{\zeta}, \tau) \triangleq \begin{cases} \sup_{\mathbf{z}' \in W} \langle \mathbf{z}', \boldsymbol{\zeta} \rangle & \text{if } \tau = 0 \text{ and } \sup_{\mathbf{z}' \in W} \langle \mathbf{z}', \boldsymbol{\zeta} \rangle < +\infty, \\ \tau \omega(\boldsymbol{\zeta}/\tau) & \text{if } \tau > 0 \text{ and } \boldsymbol{\zeta} \in \tau \boldsymbol{\Omega}. \end{cases}$$

Of course, the trivial part of this geometric inequality is an immediate consequence of the definition of $\omega^{+}(\boldsymbol{\zeta}, \tau)$ for $\tau = 0$. Moreover, it is clear from the equality characterization of the conjugate inequality that equality holds if and only if

$$\text{either} \quad \tau = 0 \quad \text{and} \quad \langle \mathbf{z}, \boldsymbol{\zeta} \rangle = \sup_{\mathbf{z}' \in W} \langle \mathbf{z}', \boldsymbol{\zeta} \rangle, \quad \text{or} \quad \tau > 0 \quad \text{and} \quad \boldsymbol{\zeta} \in \tau \partial w(\mathbf{z}).$$

Of course, another geometric inequality can be derived from the same conjugate inequality simply by introducing another scalar variable $t \geqq 0$ and substituting $\mathbf{z}/t$ for $\mathbf{z}$ in the conjugate inequality. The details of that inequality are left to the imagination of the reader.

If w is convex and closed, the symmetry of the conjugate transformation clearly implies that the condition $\boldsymbol{\zeta} \in \tau \partial w(\mathbf{z})$ can be replaced by the condition $\mathbf{z} \in \partial \omega(\boldsymbol{\zeta}/\tau)$ in the characterization of equality for the geometric inequality, in which case the relations $\mathbf{z} \in \partial \omega(\boldsymbol{\zeta}/\tau)$ and $\boldsymbol{\zeta} \in \tau \partial w(\mathbf{z})$ are equivalent and hence "solve" one another when $\tau > 0$.

This completes our prerequisites for all the remaining subsections.

3.3.4. Duality. Let $h_k : D_k$ be the conjugate transform of the function $g_k : C_k$, $k \in \{0\} \cup I \cup J$, and let these sets D_j determine the sets D_j, $j \in J$, postulated at the beginning of Section 2.2.

Now consider the following geometric programming Problem B.

Problem *B*. Consider the objective function H whose domain

$$D \triangleq \{(\mathbf{y}, \boldsymbol{\lambda}) \,|\, \mathbf{y}^k \in D_k, \, k \in \{0\} \cup J, \text{ and } (\mathbf{y}^i, \boldsymbol{\lambda}_i) \in D_i^{+}, \, i \in I\},$$

and whose functional value

$$H(\mathbf{y}, \boldsymbol{\lambda}) \triangleq h_0(\mathbf{y}^0) + \sum_I h_i^{+}(\mathbf{y}^i, \boldsymbol{\lambda}_i),$$

where

$$D_i^{+} \triangleq \{(\mathbf{y}^i, \boldsymbol{\lambda}_i) \,|\, \text{either } \lambda_i = 0 \text{ and } \sup_{\mathbf{c}^i \in C_i} \langle \mathbf{y}^i, \mathbf{c}^i \rangle < +\infty, \text{ or } \lambda_i > 0 \text{ and } \mathbf{y}^i \in \lambda_i D_i\},$$

and

$$h_i^+(\mathbf{y}^i, \lambda_i) \triangleq \begin{cases} \sup\limits_{\mathbf{c}^i \in C_i} \langle \mathbf{y}^i, \mathbf{c}^i \rangle & \text{if } \lambda_i = 0 \text{ and } \sup\limits_{\mathbf{c}^i \in C_i} \langle \mathbf{y}^i, \mathbf{c}^i \rangle < +\infty, \\ \lambda_i h_i(\mathbf{y}^i/\lambda_i) & \text{if } \lambda_i > 0 \text{ and } \mathbf{y}^i \in \lambda_i D_i. \end{cases}$$

Using the feasible solution set

$$T \triangleq \{(\mathbf{y}, \boldsymbol{\lambda}) \in D \mid \mathbf{y} \in Y, \quad \text{and} \quad h_j(\mathbf{y}^j) \leqq 0, j \in J\},$$

calculate both the problem infimum

$$\psi \triangleq \inf_{(\mathbf{y}, \boldsymbol{\lambda}) \in T} H(\mathbf{y}, \boldsymbol{\lambda})$$

and the optimal solution set

$$T^* \triangleq \{(\mathbf{y}, \boldsymbol{\lambda}) \in T \mid H(\mathbf{y}, \boldsymbol{\lambda}) = \psi\}.$$

Problems A and B are, of course, termed *geometric dual problems.* When $g_k : C_k$, $k \in \{0\} \cup I \cup J$ and X are convex and closed, this duality is clearly symmetric, in that Problem A can then be constructed from Problem B in the same way that Problem B has just been constructed from Problem A. Actually, this duality is the only completely symmetric duality that is presently known for general (closed) convex programming with explicit constraints.

In linear programming (Example 2.7), elementary computations show that dual problem B is, in essence, just the usual linear dual problem. However, in ordinary programming (Example 2.8), elementary considerations show that dual Problem B is not just the (Wolfe) "ordinary dual problem" [as properly defined for the first time by Falk (Ref. 42)]. Actually, the ordinary dual problem results from a (sub)optimization of the geometric dual Problem B over y—a fact that indicates why the ordinary dual problem can almost never be computed as easily as the geometric dual problem. Unlike the ordinary dual problem, the geometric dual problem can frequently be computed in terms of elementary functions, particularly for the convex examples described or alluded to in Section 2.2.

Somewhat surprisingly, the dual Problem B corresponding to the generalized chemical equilibrium problem (Example 2.9) is, in essence, just the posynomial optimization problem that is a special case of Example 2.6. In particular, rather elementary computations show that the corresponding dual Problem B is simply Example 2.6 with each $d_k = 0$, and each $(\sum_{[k]} c_q e^{x_q})$ replaced by $\log(\sum_{[k]} c_q e^{x_q})$. Furthermore, the given transformation from $\mathbf{x}$ to $\mathbf{t}$ followed by an exponentiation of the resulting objective function and constraints obviously produces the displayed posynomial optimization problem with each d_k replaced by $e^0 = 1$. Since a posynomial constraint can clearly be consistent only when its upper bound $d_k > 0$, division of each of the original posynomial constraints by the corresponding d_k shows that there is actually no loss of generality in letting each $d_k = 1$. This connection between chemical equilibrium and posynomial optimization has been exploited computationally by Passy and Wilde (Ref. 30), and has also been intimately related to the "Darwin–Fowler method in statistical mechanics by Duffin and Zener (Ref. 43). In addition, more general chemical systems have been studied within the context of geometric programming by Duffin and Zener (Ref. 44).

The following definition is almost as important as the definition of the dual Problems A and B.

Definition 3.12. The *extremality conditions* (for constrained geometric programming) are:

(I) $$\mathbf{x} \in X \quad \text{and} \quad \mathbf{y} \in Y,$$

(II) $$g_i(\mathbf{x}^i) \leqq 0, i \in I, \quad \text{and} \quad h_j(\mathbf{y}^j) \leqq 0, j \in J,$$

(III) $$0 = \langle \mathbf{x}, \mathbf{y} \rangle,$$

(IV) $$\mathbf{y}^0 \in \partial g_0(\mathbf{x}^0),$$

(V) either $\lambda_i = 0$ and $\langle \mathbf{x}^i, \mathbf{y}^i \rangle = \sup_{\mathbf{c}^i \in C_i} \langle \mathbf{c}^i, \mathbf{y}^i \rangle$ or $\lambda_i > 0$ and $\mathbf{y}^i \in \lambda_i \, \partial g_i(\mathbf{x}^i)$, $i \in I$,

(VI) either $\kappa_j = 0$ and $\langle \mathbf{x}^j, \mathbf{y}^j \rangle = \sup_{\mathbf{d}^j \in D_j} \langle \mathbf{x}^j, \mathbf{d}^j \rangle$ or $\kappa_j > 0$ and $\mathbf{y}^j \in \partial g_j(\mathbf{x}^j/\kappa_j)$, $j \in J$,

(VII) $$\lambda_i g_i(\mathbf{x}^i) = 0, \quad i \in I, \quad \text{and} \quad \kappa_j h_j(\mathbf{y}^j) = 0, \qquad j \in J.$$

When $C_i = E_{n_i}$ (which is frequently the situation), a simplification results from noting that $\langle \mathbf{x}^i, \mathbf{y}^i \rangle = \sup_{\mathbf{c}^i \in C_i} \langle \mathbf{c}^i, \mathbf{y}^i \rangle$ if and only if $\mathbf{y}^i = \mathbf{0}$. In particular, then, the corresponding extremality condition (V) can be replaced by the equivalent extremality condition

$$\lambda_i \geqq 0 \quad \text{and} \quad \mathbf{y}^i \in \lambda_i \, \partial g_i(\mathbf{x}^i), \qquad i \in I.$$

Extremality conditions (I) are simply the "cone conditions" for Problems A and B, respectively, and extremality conditions (II) are simply the "constraints" for Problems A and B, respectively. Extremality condition (III) is termed the "orthogonality condition," extremality conditions (IV)–(VI) are termed the "subgradient conditions," and extremality conditions (VII) are (of course) termed the "complementary slackness conditions."

The following duality theorem is the basis for many others.

Theorem 3.15. If $(\mathbf{x}, \boldsymbol{\kappa})$ and $\mathbf{y}, \boldsymbol{\lambda})$ are feasible solutions to Problems A and B, respectively [in which case the extremality conditions (I)–(II) are satisfied], then

$$0 \leqq G(\mathbf{x}, \boldsymbol{\kappa}) + H(\mathbf{y}, \boldsymbol{\lambda}),$$

with equality holding if and only if the extremality conditions (III)–(VII) are satisfied, in which case $(\mathbf{x}, \boldsymbol{\kappa})$ and $(\mathbf{y}, \boldsymbol{\lambda})$ are optimal solutions to Problems A and B, respectively.

In essence, the proof of this key theorem consists of little more than combining the defining inequality $0 \leqq \langle \mathbf{x}, \mathbf{y} \rangle$ with an inequality that comes from summing the conjugate inequality $\langle \mathbf{x}^0, \mathbf{y}^0 \rangle \leqq g_0(\mathbf{x}^0) + h_0(\mathbf{y}^0)$, the geometric inequalities $\langle \mathbf{x}^i, \mathbf{y}^i \rangle \leqq \lambda_i g_i(\mathbf{x}^i) + h_i^+(\mathbf{y}^i, \boldsymbol{\lambda}_i)$, $i \in I$, and the geometric inequalities $\langle \mathbf{x}^j, \mathbf{y}^j \rangle \leqq g_j^+(\mathbf{x}^j, \kappa_j) + \kappa_j h_j(\mathbf{y}^j)$, $j \in J$.

The preceding theorem has two important corollaries that are left to the imagination of the reader because they differ only slightly from the two corollaries to (the analogous) Theorem 3.3. The first corollary is fundamental to the definition of "duality gap" and is helpful in studying the constrained version of the algorithmic stopping criterion presented in Section 3.1.4. The second corollary provides a useful characterization of dual optimal solutions $(\mathbf{x}^*, \boldsymbol{\kappa}^*)$ and $(\mathbf{y}^*, \boldsymbol{\lambda}^*)$ in terms of the extremality conditions.

It is worth noting that if $g_0 : C_0$ is convex and closed, then the symmetry of the conjugate transformation implies that the subgradient condition (IV) can be replaced by the equivalent subgradient condition

$$\text{(IVa)} \qquad \mathbf{x}^0 \in \partial h_0(\mathbf{y}^0).$$

Likewise, if $g_i : C_i$, $i \in I$, is convex and closed, then the subgradient condition (V) can be replaced by the equivalent subgradient condition:

$$\text{(Va) either } \lambda_i = 0 \text{ and } \langle \mathbf{x}^i, \mathbf{y}^i \rangle = \sup_{\mathbf{c}^i \in C_i} \langle \mathbf{c}^i, \mathbf{y}^i \rangle, \text{ or } \lambda_i > 0 \text{ and } \mathbf{x}^i \in \partial h_i(\mathbf{y}^i/\lambda_i),\ i \in I,$$

and if $g_j : C_j$, $j \in J$, is convex and closed, then the subgradient condition (VI) can be replaced by the equivalent subgradient condition

$$\text{(VIa) either } \kappa_j = 0 \text{ and } \langle \mathbf{x}^j, \mathbf{y}^j \rangle = \sup_{\mathbf{d}^j \in D_j} \langle \mathbf{y}^j, \mathbf{d}^j \rangle, \text{ or } \kappa_j > 0 \text{ and } \mathbf{x}^j \in \kappa_j\, \partial h_j(\mathbf{y}^j),\ j \in J.$$

These equivalent subgradient conditions (IVa)–(VIa) are, of course, especially helpful when using the extremality conditions to compute all primal optimal solutions $(\mathbf{x}^*, \boldsymbol{\kappa}^*)$ in S^* from the knowledge of only a single dual optimal solution $(\mathbf{y}^*, \boldsymbol{\lambda}^*)$ in T^*.

It should be emphasized that Problem A need not always be solved directly. Under appropriate conditions it can actually be solved indirectly by solving either the extremality conditions (I)–(VII) or Problem B. In some cases it may be advantageous to solve the extremality conditions (I)–(VII), especially when they turn out to be (essentially) linear (e.g., linearly constrained quadratic programming). In other cases it may be advantageous to solve Problem B, especially when the index set J is empty (e.g., quadratically constrained quadratic programming and posynomial constrained posynomial programming), in which event Problem B has no constraints (even when Problem A does). Of course, in all such cases the absence of a duality gap is crucial.

The definition of the Lagrangian L_g (in Section 3.3.2) and the fact that X is a cone readily imply that the cone condition $\mathbf{y} \in Y$, the constraints $h_j(\mathbf{y}^j) \leqq 0$, $j \in J$, the orthogonality condition $0 = \langle \mathbf{x}, \mathbf{y} \rangle$, and the complementary slackness conditions $\kappa_j h_j(\mathbf{y}^j) = 0$, $j \in J$, can all be replaced by the single equivalent condition

$$L_g(\mathbf{x}, \boldsymbol{\kappa}; \mathbf{y}, \boldsymbol{\lambda}) = \inf_{\substack{\mathbf{x}' \in X \\ \boldsymbol{\kappa}' \geqq \mathbf{0}}} L_g(\mathbf{x}', \boldsymbol{\kappa}'; \mathbf{y}, \boldsymbol{\lambda}).$$

Moreover, if $g_k : C_k$, $k \in \{0\} \cup I \cup J$ is convex and closed, conjugate transform theory implies that the constraints $g_i(\mathbf{x}^i) \leqq 0$, $i \in I$, the subgradient conditions (IV)–(VI), and the complementary slackness conditions $\lambda_i g_i(\mathbf{x}^i) = 0$, $i \in I$, can all be replaced by the two equivalent conditions $\boldsymbol{\kappa} \geqq \mathbf{0}$ and $\sup_{(\mathbf{y}', \boldsymbol{\lambda}') \in D} L_g(\mathbf{x}, \boldsymbol{\kappa}; \mathbf{y}', \boldsymbol{\lambda}') = L_g(\mathbf{x}, \boldsymbol{\kappa}; \mathbf{y}, \boldsymbol{\lambda})$. Consequently, the saddle-point condition discussed in Section 3.3.2 is equivalent to the extremality conditions when $g_k : C_k$, $k \in \{0\} \cup I \cup J$, is convex and closed. Nevertheless, it seems that the extremality conditions given in the definition are the most convenient to work with.

The following theorem provides an important tie between dual Problem B and the P vectors defined in Section 3.3.2.

Theorem 3.16. Given that Problem A is consistent with a finite infimum φ,

(i) if Problem A has a P vector, then Problem B is consistent and $0 = \varphi + \psi$,

(ii) if Problem B is consistent and $0 = \varphi + \psi$, then

$$\{(\mathbf{y}^*, \boldsymbol{\lambda}^*) \mid (\mathbf{y}^*, \boldsymbol{\lambda}^*) \text{ is a } P \text{ vector for Problem } A\} = T^*.$$

The following theorem provides an important tie between dual Problem B and the Kuhn–Tucker vectors defined in Section 3.3.1.

Theorem 3.17. Given that Problems A and B are both consistent and that $0 = \varphi + \psi$, if there is a "minimizing sequence" $\{(\mathbf{y}^q, \boldsymbol{\lambda}^q)\}_1^\infty$ for Problem B (i.e., $(\mathbf{y}^q, \boldsymbol{\lambda}^q) \in T$ and $\lim_{q\to+\infty} H(\mathbf{y}^q, \boldsymbol{\lambda}^q) = \psi$) such that $\lim_{q\to+\infty} \boldsymbol{\lambda}^q$ exists and is finite, then $\boldsymbol{\lambda}^* \triangleq \lim_{q\to+\infty} \boldsymbol{\lambda}^q$ is a Kuhn–Tucker vector for Problem A.

The following corollary ties the dual optimal solution set T^* directly to Kuhn–Tucker vectors.

Corollary 3.6. Given that Problems A and B are both consistent and that $0 = \varphi + \psi$, each dual optimal solution $(\mathbf{y}^*, \boldsymbol{\lambda}^*) \in T^*$ provides a Kuhn–Tucker vector $\boldsymbol{\lambda}^*$ for Problem A.

The following corollary ties the set of all P vectors $(\mathbf{y}^*, \boldsymbol{\lambda}^*)$ directly to the set of all Kuhn–Tucker vectors $\boldsymbol{\lambda}^*$.

Corollary 3.7. Given that Problem A is consistent with a finite infimum φ, each P vector $(\mathbf{y}^*, \boldsymbol{\lambda}^*)$ for Problem A provides a Kuhn–Tucker vector $\boldsymbol{\lambda}^*$ for Problem A.

The proof of this corollary requires Theorem 3.16 as well as Corollary 3.6.

As the preceding theory indicates, the absence of a duality gap is a highly desired situation. The following (geometric programming) version of "Fenchel's theorem" provides useful conditions that guarantee both the absence of a duality gap and the existence of primal optimal solutions $(\mathbf{x}^*, \boldsymbol{\kappa}^*) \in S^*$.

Theorem 3.18. Suppose that both $g_k : C_k$, $k \in \{0\} \cup I \cup J$ and X are convex and closed. If

(i) Problem B has a feasible solution $(\mathbf{y}', \boldsymbol{\lambda}')$ such that

$$h_j(\mathbf{y}^j) < 0, \qquad j \in J,$$

(ii) Problem B has a finite infimum ψ,

(iii) there exists a vector $(\mathbf{y}^+, \boldsymbol{\lambda}^+)$ such that

$$\begin{aligned} \mathbf{y}^+ &\in (\operatorname{ri} Y), \\ \mathbf{y}^{+k} &\in (\operatorname{ri} D_k), \qquad k \in \{0\} \cup J, \\ (\mathbf{y}^{+i}, \lambda_i^+) &\in (\operatorname{ri} D_i^+), \qquad i \in I, \end{aligned}$$

then $0 = \varphi + \psi$ and $S^* \neq \varnothing$.

This theorem is perhaps the deepest theorem in geometric programming. Its most direct proof utilizes (the corresponding unconstrained)

Theorem 3.5 along with rather intricate arguments based on convex analysis.

Note that the vector $(\mathbf{y}^+, \boldsymbol{\lambda}^+)$ in hypothesis (iii) need not be a feasible solution to Problem B. However, $(\mathbf{y}^+, \boldsymbol{\lambda}^+)$ is obviously such a solution when J is empty (which is the case for all examples described in Section 2.2, except Example 2.9). Consequently, hypothesis (i) can clearly be replaced by the hypothesis

(i′) J is empty,

without disturbing the validity of Theorem 3.18. Moreover, when the cone Y is in fact a vector space (which is the case for all examples described or alluded to in Section 2.2, except Example 2.7), the condition $\mathbf{y}^+ \in (\text{ri } Y)$ is implied by the weaker condition $\mathbf{y}^+ \in Y$.

Theorem 3.18 can be used to show that there are no duality gaps for the constrained versions of problem classes 1–3 listed after Theorem 3.5. Furthermore, its (unstated) dual can be used to strengthen the "Kuhn–Tucker–Slater theorem" in ordinary programming.

3.3.5. Parametric Programming and Postoptimality Analysis. As mentioned in Section 3.2, the parametrized family F into which Problem A is to be embedded can be obtained by translating the prescribed domain $\mathscr{C}$ (given at the beginning of Section 3.2) through all possible displacements $-\boldsymbol{\mathscr{u}}$, while keeping the prescribed cone $\mathscr{X}$ fixed.

Now, the structure of $\mathscr{C}$ obviously induces a partitioning of the components of $\boldsymbol{\mathscr{u}}$ into vectors $\mathbf{u}^0, \mathbf{u}^I, \boldsymbol{\mu}, \mathbf{u}^J$, and $\mathbf{u}_J$ that correspond respectively to vectors $\mathbf{x}^0, \mathbf{x}^I, \boldsymbol{\alpha}, \mathbf{x}^J$, and $\boldsymbol{\kappa}$. Clearly, each component u_j of $\mathbf{u}_J$ does not influence the problem infimum, but simply translates through $-u_j$ only the optimal value of κ_j (if such a value exists). Hence setting u_j equal to zero deletes from the resulting family $\mathscr{F}$ only Problems $\mathscr{A}(\boldsymbol{\mathscr{u}})$ that are essentially superfluous. Consequently, the family F is actually taken to be the resulting family $\mathscr{F}$ with all such superfluous problems deleted. Of course, the vectors $\mathbf{u}^0, \mathbf{u}^I, \mathbf{u}^J$, and $\boldsymbol{\mu}$ still needed to parametrize F constitute a single vector parameter $(\mathbf{u}, \boldsymbol{\mu})$ where $(\mathbf{u}^0, \mathbf{u}^I, \mathbf{u}^J) \triangleq \mathbf{u}$.

The parametrized family F of all Problems $A(\mathbf{u}, \boldsymbol{\mu})$ (for fixed $g_k : C_k$, $k \in \{0\} \cup I \cup J$ and X) is termed a *geometric programming family*. For purposes of easy reference and mathematical precision, Problem $A(\mathbf{u}, \boldsymbol{\mu})$ is now given the following formal definition, which should be compared with the formal definition of Problem A at the beginning of Section 2.2.

Problem $A(\mathbf{u}, \boldsymbol{\mu})$. Consider the objective function $G(\cdot + \mathbf{u}, \kappa)$ whose domain

$$C(\mathbf{u}) \triangleq \{(\mathbf{x}, \boldsymbol{\kappa}) \mid \mathbf{x}^k + \mathbf{u}^k \in C_k,\ k \in \{0\} \cup I, \text{ and } (\mathbf{x}^j + \mathbf{u}^j, \kappa^j) \in C_j^+,\ j \in J\},$$

and whose functional value

$$G(\mathbf{x}+\mathbf{u}, \boldsymbol{\kappa}) \triangleq g_0(\mathbf{x}^0+\mathbf{u}^0)+\sum_J g_j^+(\mathbf{x}^j+\mathbf{u}^j, \kappa_j),$$

where

$$C_j^+ \triangleq \{(\mathbf{c}^j, \kappa_j) \mid \text{either } \kappa_j=0 \text{ and } \sup_{\mathbf{d}^j\in D_j} \langle \mathbf{c}^j, \mathbf{d}^j\rangle<+\infty, \text{ or } \kappa_j>0 \text{ and } \mathbf{c}^j \in \kappa_j C_j\}$$

and

$$g_j^+(\mathbf{c}^j, \kappa_j) \triangleq \begin{cases} \sup_{\mathbf{d}^j\in D_j} \langle \mathbf{c}^j, \mathbf{d}^j\rangle & \text{if } \kappa_j=0 \text{ and } \sup_{\mathbf{d}^j\in D_j} \langle \mathbf{c}^j, \mathbf{d}^j\rangle<+\infty, \\ \kappa_j g_j(\mathbf{c}^j/\kappa_j) & \text{if } \kappa_j>0 \text{ and } \mathbf{c}^j \in \kappa_j C_j. \end{cases}$$

Using the feasible solution set

$$S(\mathbf{u}, \boldsymbol{\mu}) \triangleq \{(\mathbf{x}, \boldsymbol{\kappa}) \in C(\mathbf{u}) \mid \mathbf{x}\in X, \text{ and } g_i(\mathbf{x}^i+\mathbf{u}^i)+\mu_i \leqq 0, i \in I\},$$

calculate both the problem infimum

$$\varphi(\mathbf{u}, \boldsymbol{\mu}) \triangleq \inf_{(\mathbf{x},\boldsymbol{\kappa})\in S(\mathbf{u},\boldsymbol{\mu})} G(\mathbf{x}+\mathbf{u}, \boldsymbol{\kappa})$$

and the optimal solution set

$$S^*(\mathbf{u}, \boldsymbol{\mu}) \triangleq \{(\mathbf{x}, \boldsymbol{\kappa}) \in S(\mathbf{u}, \boldsymbol{\mu}) \mid G(\mathbf{x}+\mathbf{u}, \boldsymbol{\kappa}) = \varphi(\mathbf{u}, \boldsymbol{\mu})\}.$$

Of course, Problem A appears in the parametrized family F as Problem $A(\mathbf{0}, \mathbf{0})$, and the symbols A, S, φ, and S^* now represent functions of $(\mathbf{u}, \boldsymbol{\mu})$, though they originally represented only the particular functional values $A(\mathbf{0}, \mathbf{0})$, $S(\mathbf{0}, \mathbf{0})$, $\varphi(\mathbf{0}, \mathbf{0})$, and $S^*(\mathbf{0}, \mathbf{0})$, respectively—a notational discrepancy that must be kept in mind when comparing subsequent developments with previous developments.

For the examples given in Section 2.2, it is easy to see that the perturbation vector $(\mathbf{u}, \boldsymbol{\mu})$:

(a) in Example 2.6 alters the (log of the absolute value of the) signomial coefficients c_q and the constraint upper bounds d_k.

(b) in Example 2.7 alters the affine objective function constant 0 and the linear constraint upper bounds b_i.

(c) in (the ordinary programming) Example 2.8 translates the common function domain C_0 in several simultaneous directions while altering the constraint upper bounds 0.

(d) in Example 2.9 alters the total number of moles of each element.

To extend the unconstrained theorems given in Section 3.1.5 into corresponding constrained theorems:

(a) each hypothesis that the set $\mathscr{C}$ be convex should be replaced by the hypothesis that the sets C_0 and C_j^+, $j \in J$, along with the functions $g_i : C_i$, $i \in I$, be convex,

(b) each hypothesis that the function $\mathcal{g} : \mathcal{C}$ be convex (closed) should be replaced by the hypothesis that the functions $g_k : C_k$, $k \in \{0\} \cup I$, and the functions $g_j^+ : C_j^+$, $j \in J$, be convex (closed),

(c) each hypothesis that the cone $\mathcal{X}$ be convex (closed) should be replaced by the hypothesis that the cone X be convex (closed),

(d) make the obvious notational alterations in both the hypotheses and the conclusions, as well as the definitions.

In carrying out (d), the symbol $\mathcal{G}$ should be replaced by the symbol G, the symbol $\boldsymbol{v}$ should be replaced by the symbol $(\mathbf{v}, \boldsymbol{\nu})$, where $(\mathbf{v}^0, \mathbf{v}^I, \mathbf{v}^J) \triangleq \mathbf{v}$ translates sets and $\boldsymbol{\nu}$ influences constraint upper bounds.

Needless to say, the remaining details are left to the interested reader.

3.3.6. Decomposition Principles. To generalize the decomposition principles given in Section 3.1.6, simply view the constrained case in the context of the unconstrained case via the prescribed choices of $\mathcal{g} : \mathcal{C}$ and $\mathcal{X}$ given in Section 3.2.

In doing so it is important to realize that the components of $x = (\mathbf{x}^0, \mathbf{x}^I, \boldsymbol{\alpha}, \mathbf{x}^J, \boldsymbol{\kappa})$ can be placed in any order to achieve a block diagonal structure for some matrix representation $\mathcal{M}$ of $\mathcal{X}$. Moreover, it is clear from the formula for $\mathcal{X}$ that the possibility of achieving such a block diagonal structure for some $\mathcal{M}$ depends entirely on the possibility of ordering the components of $\mathbf{x} = (\mathbf{x}^0, \mathbf{x}^I, \mathbf{x}^J)$ in such a way that a block diagonal structure is achieved for some matrix representation M of X.

It is equally important to realize that the function $\mathcal{g} : \mathcal{C}$ inherits any separability that is present in the functions $g_0 : C_0$ and $g_j^+ : C_j^+$, $j \in J$. Although $\mathcal{g} : \mathcal{C}$ clearly does not generally inherit any of the separability that is present in a given function $g_i : C_i$ (unless a corresponding Kuhn–Tucker multiplier λ_i^* is known, in which case the constraint $g_i(\mathbf{x}^i) + \alpha_i \leqq 0$ can be deleted from the defining equation for $\mathcal{C}$ while the expression $\lambda_i^*[g_i(\mathbf{x}^i) + \alpha_i]$ is added to the defining equation for $\mathcal{g}(\mathbf{x}^0, \mathbf{x}^I, \boldsymbol{\alpha}, \mathbf{x}^J, \boldsymbol{\kappa})$), $\mathcal{g} : \mathcal{C}$ does inherit sufficient partial separability when the components of $\mathbf{x}^i$ belong to a single vector x^k.

References

1. ZENER, C., *A Mathematical Aid in Optimizing Engineering Designs*, Proceedings of the National Academy of Sciences U.S.A., Vol. 47, pp. 537–539, 1961.
2. ZENER, C., *A Further Mathematical Aid in Optimizing Engineering Designs*, Proceedings of the National Academy of Sciences U.S.A., Vol. 48, pp. 518–522, 1962.
3. DUFFIN, R. J., *Dual Programs and Minimum Cost*, SIAM Journal on Applied Mathematics, Vol. 10, p. 119, 1962.
4. DUFFIN, R. J., *Cost Minimization Problems Treated by Geometric Means*, Operations Research, Vol. 10, p. 668, 1962.

5. DUFFIN, R. J., PETERSON, E. L., and ZENER, C., *Geometric Programming—Theory and Applications,* John Wiley and Sons, New York, New York, 1967.
6. PETERSON, E. L., *Symmetric Duality for Generalized Unconstrained Geometric Programming,* SIAM Journal on Applied Mathematics, Vol. 19, pp. 487–526, 1970.
7. PETERSON, E. L., *Mathematical Foundations of Geometric Programming: Appendix to Geometric Programming and Some of Its Extensions,* Optimization and Design, Edited by M. Avriel, M. J. Rijckaert, and D. J. Wilde, Prentice Hall, Englewood Cliffs, New Jersey, pp. 244–289, 1973.
8. PETERSON, E. L., *Generalization and Symmetrization of Duality in Geometric Programming,* to appear, 1975.
9. PETERSON, E. L., *Fenchel's Hypothesis and the Existence of Recession Directions in Convex Programming,* to appear, 1975.
10. PETERSON, E. L., *The Mathematical Foundations of Convex and Nonconvex Programming,* to appear, 1982.
11. AVRIEL, M., and WILLIAMS, A. C., *Complementary Geometric Programming,* SIAM Journal on Applied Mathematics, Vol. 19, pp. 125–141, 1970.
12. DUFFIN, R. J., and PETERSON, E. L., *Geometric Programs Treated with Slack Variables,* Applicable Analysis, Vol. 2, pp. 255–267, 1972.
13. DUFFIN, R. J., and PETERSON, E. L., *The Proximity of Algebraic Geometric Programming to Linear Programming,* Mathematical Programming, Vol. 3, pp. 250–253, 1972.
14. DUFFIN, R. J., and PETERSON, E. L., *Reversed Geometric Programs Treated by Harmonic Means,* Indiana University Mathematics Journal, Vol. 22, 531–550, 1972.
15. DUFFIN, R. J., and PETERSON, E. L., *Geometric Programming with Signomials,* Journal of Optimization Theory and Applications, Vol. 11, pp. 3–35, 1973.
16. ABRAMS, R. A., and BUNTING, M. L., *Reducing Reversed Posynomial Programs,* SIAM Journal on Applied Mathematics, Vol. 27, pp. 629–640, 1974.
17. FALK, J. E., *Global Solutions of Signomial Programs,* George Washington University, Program in Logistics, Technical Paper No. T-274, Washington, D.C., 1973.
18. PETERSON, E. L., and ECKER, J. G., *Geometric Programming: Duality in Quadratic Programming and l_p-approximation,* I, Proceedings of the International Mathematical Programming Symposium, Edited by H. W. Kuhn, Princeton University Press, Princeton, New Jersey, pp. 445–480, 1970.
19. PETERSON, E. L., and WENDELL, R. E., *Optimal Location by Geometric Programming,* to appear, 1975.
20. DINKEL, J. J. and PETERSON, E. L., *Discrete Optimal Control (with Linear Dynamics) via Geometric Programming,* Journal of Mathematical Analysis and Applications, to appear, 1975.
21. HALL, M. A., and PETERSON, E. L., *Traffic Equilibria Analysed via Geometric Programming,* Proceedings of the International Symposium on Traffic Equilibrium Methods, Edited by M. Florian, Springer-Verlag, Berlin, Germany, 1976.
22. DUFFIN, R. J., *Nonlinear Networks,* II*a*, Bulletin of the American Mathematical Society, Vol. 53, pp. 963–971, 1947.

23. MINTY, G. J., *Monotone Networks*, Proceedings of the Royal Society of London Ser. A, Vol. 257, pp. 194–212, 1960.
24. ROCKAFELLAR, R. T., *Convex Programming and Systems of Elementary Monotonic Relations*, Journal of Mathematical Analysis and Applications, Vol. 19, pp. 543–564, 1967.
25. DUFFIN, R. J., *Linearizing Geometric Programs, SIAM Review*, Vol. 12, pp. 211–227, 1970.
26. ZENER, C., *Engineering Design by Geometric Programming*, John Wiley and Sons, New York, New York, 1971.
27. WILDE, D. J., Selected preprints and reprints available through the Department of Mechanical Engineering, Stanford University, Stanford, California, 1975.
28. WOOLSEY, R. E. D., Selected preprints and reprints available through the Department of Mineral Resources, Colorado School of Mines, Golden, Colorado, 1975.
29. RIJCKAERT, M. J., *Engineering Applications of Geometric Programming*, Optimization and Design, Edited by M. Avriel, M. J. Rijckaert and D. J. Wilde, Prentice Hall, Englewood Cliffs, New Jersey, pp. 196–220, 1973.
30. PASSY, U., and WILDE, D. J., *A Geometric Programming Algorithm for Solving Chemical Equilibrium Problems*, SIAM Journal on Applied Mathematics, Vol. 16, pp. 363–373, 1968.
31. FENCHEL, W., *On Conjugate Convex Functions*, Canadian Journal of Mathematics, Vol. 1, pp. 73–77, 1949.
32. FENCHEL, W., "Convex Cones, Sets and Functions," Mathematics Department Mimeographed Lecture Notes, Princeton University, Princeton, N.J., 1951.
33. ROCKAFELLAR, R. T., *Convex Analysis*, Princeton University Press, Princeton, New Jersey, 1970.
34. COURANT, R., and HILBERT, D., *Methods of Mathematical Physics*, Vol. 1, Interscience, New York, 1953.
35. DUFFIN, R. J., *Infinite Programs, Linear Inequalities and Related Systems*, Edited by H. W. Kuhn and A. W. Tucker, Princeton University Press, Princeton, New Jersey, pp. 157–170, 1956.
36. KRETSCHMER, K. S., *Programmes in Paired Spaces*, Canadian Journal of Mathematics, Vol. 13, p. 221, 1961.
37. ABRAMS, R. A., *Projections of Convex Programs with Unattained Infima*, SIAM Journal on Control, Vol. 13, pp. 706–718, 1975.
38. ROCKAFELLAR, R. T., *Duality and Stability in Extremum Problems Involving Convex Functions*, Pacific Journal of Mathematics, Vol. 21, pp. 167–187, 1967.
39. ROCKAFELLAR, R. T., *Duality in Nonlinear Programming*, American Mathematical Society Lecture on Applied Mathematics, Vol. 11, Edited by G. B. Dantzig and A. F. Veinott, American Mathematical Society, Providence, R.I., pp. 401–422, 1968.
40. PETERSON, E. L., *The Decomposition of Large Generalized Geometric Programming Problems by Tearing*, Decomposition of Large-Scale Systems, Edited by D. M. Himmelblau, North Holland Publishing Company, Amsterdam, Holland, p. 525, 1973.

41. PETERSON, E. L., *The Decomposition of Large Generalized Geometric Programming Problems by Geometric Lagrangians*, to appear, 1976.
42. FALK, J. E., *Lagrange Multipliers and Nonlinear Programming*, Journal of Mathematical Analysis, Vol. 19, pp. 141–159, 1967.
43. DUFFIN, R. J., and ZENER, C., *Geometric Programming, Chemical Equilibrium, and the Anti-Entropy Function*, Proceedings of the National Academy of Sciences, Vol. 63, pp. 629–639, 1969.
44. DUFFIN, R. J., and ZENER, C., *Geometric Programming and the Darwin–Fowler Method in Statistical Mechanics*, Journal of Physical Chemistry, Vol. 74, pp. 2419–2423, 1970.
45. DUFFIN, R. J., and PETERSON, E. L., *Duality Theory for Geometric Programming*, SIAM Journal on Applied Mathematics, Vol. 14, pp. 1307–1349, 1966.

3

Optimality Conditions in Generalized Geometric Programming[1]

E. L. PETERSON[2]

Abstract. Generalizations of the Kuhn–Tucker optimality conditions are given, as are the fundamental theorems having to do with their necessity and sufficiency.

1. Introduction

Optimization problems from the real world usually possess a linear-algebraic component, either directly in the form of problem linearities (e.g., those involving the node-arc incidence matrices in network optimization) or indirectly in the more subtle form of certain problem nonlinearities (e.g., those involving the coefficient matrices in quadratic programming or, perhaps more appropriately, those involving the exponent matrices in signomial programming). As demonstrated in the author's recent survey paper (Ref. 1) and some of the references cited therein, such a component can frequently be exploited by taking a (generalized) geometric programming approach. In fact, geometric programming is primarily a body of techniques and theorems for inducing and exploiting as much linearity as possible.

Unlike the vast literature (including Refs. 2–8) on optimality conditions in general nonlinear programming, this paper presents optimality conditions that are tailored to geometric programming. Although the fundamental theorems having to do with their necessity and sufficiency have already been described in Ref. 1, proofs are given here for the first time.

[1] This research was sponsored by the Air Force Office of Scientific Research, Air Force Systems Command, USAF, under Grant No. AFOSR-73-2516.

[2] Professor, Department of Engineering Sciences and Applied Mathematics, Department of Industrial Engineering and Management Sciences, and Department of Mathematics, Northwestern University, Evanston, Illinois.

In geometric programming, problems with only linear constraints are treated in essentially the same way as problems without constraints. Only problems with nonlinear constraints require additional attention, and hence are classified as constrained problems.

Since many important problems are unconstrained (e.g., most network optimization problems), and since the theory for the unconstrained case is much simpler than its counterpart for the constrained case, the unconstrained case is treated separately (even though its theory is actually embedded in the theory for the constrained case).

Mathematically, this paper is essentially self-contained.

2. Unconstrained Case

Given a nonempty cone $\mathscr{X} \subseteq E_n$ (n-dimensional Euclidean space), and given a function g with a nonempty domain $\mathscr{C} \subseteq E_n$, the resulting *geometric programming problem*, Problem 2.1, is defined in the following way.

Problem 2.1. Using the feasible solution set

$$\mathscr{S} \triangleq \mathscr{X} \cap \mathscr{C},$$

calculate both the problem infimum

$$\varphi \triangleq \inf_{x \in \mathscr{S}} g(x)$$

and the optimal solution set

$$\mathscr{S}^* \triangleq \{x \in \mathscr{S} \mid g(x) = \varphi\}.$$

Needless to say, the *ordinary programming case* occurs when $\mathscr{X}$ is actually the entire vector space E_n.

Our optimality conditions for the preceding Problem 2.1 utilize the *dual cone*

$$\mathscr{Y} \triangleq \{y \in E_n \mid 0 \leq \langle x, y \rangle \text{ for each } x \in \mathscr{X}\}.$$

They are stated as part of the following definition.

Definition 2.1. A *critical solution* (stationary solution, equilibrium solution, P-solution) for Problem 2.1 is any vector x^* that satisfies the following P-*optimality conditions*:

$$x^* \in \mathscr{X} \cap \mathscr{C}, \qquad \nabla g(x^*) \in \mathscr{Y},$$

and

$$0 = \langle x^*, \nabla g(x^*) \rangle.$$

If the cone $\mathscr{X}$ is actually a vector space, then $\mathscr{Y} = \mathscr{X}^{\perp}$; hence, the P-optimality condition $0 = \langle x^*, \nabla g(x^*)\rangle$ is redundant and can be deleted. Furthermore, in the ordinary programming case, the vector space $\mathscr{Y} = E_n^{\perp} = \{0\}$, so the remaining P-optimality conditions become the (more familiar) *ordinary optimality conditions*

$$x^* \in \mathscr{C} \qquad \text{and} \qquad \nabla g(x^*) = 0.$$

The following theorem gives two convexity conditions that guarantee the necessity and sufficiency of the P-optimality conditions.

Theorem 2.1. Under the hypothesis that g is differentiable at x^*, (i) given that $\mathscr{X}$ is convex, if x^* is an optimal solution to Problem 2.1, then x^* is a critical solution for Problem 2.1 (but not conversely); (ii) given that g is convex on $\mathscr{C}$, if x^* is a critical solution for Problem 2.1, then x^* is an optimal solution to Problem 2.1.

Proof. To prove part (i), first recall that the optimality of x^* implies that $x^* \in \mathscr{X} \cap \mathscr{C}$. Then, notice that the optimality of x^*, the differentiability of g at x^*, and the convexity of $\mathscr{X}$ imply that

$$\langle \nabla g(x^*), x\rangle \geq 0 \qquad \text{for each } x \in \mathscr{X}.$$

Likewise, the optimality of x^*, the differentiability of g at x^*, and the observation that $x^* + s(\ -x^*) \in \mathscr{X}$ for $s \leq 1$ imply that

$$\langle \nabla g(x^*), -x^*\rangle \geq 0.$$

Consequently,

$$\nabla g(x^*) \in \mathscr{Y} \qquad \text{and} \qquad 0 = \langle x^*, \nabla g(x^*)\rangle.$$

Counterexamples to the converse of part (i) are numerous and easy to construct. In fact, the reader is probably already familiar with counterexamples from the ordinary programming case.

To prove part (ii), first recall that the convexity of g and the differentiability of g at x^* imply that

$$g(x) - g(x^*) \geq \langle \nabla g(x^*), x - x^*\rangle \qquad \text{for each } x \in \mathscr{C}.$$

Then, notice that the assumptions

$$0 = \langle x^*, \nabla g(x^*)\rangle \qquad \text{and} \qquad \nabla g(x^*) \in \mathscr{Y}$$

imply that

$$\langle \nabla g(x^*), x - x^*\rangle = \langle \nabla g(x^*), x\rangle \geq 0 \qquad \text{for each } x \in \mathscr{X}.$$

From the preceding displayed relations, we see that

$$g(x) - g(x^*) \geq 0 \qquad \text{for each } x \in \mathcal{X} \cap \mathcal{C}.$$

Consequently, the assumption $x^* \in \mathcal{X} \cap \mathcal{C}$ shows that x^* is optimal for Problem 2.1. □

It is worth noting that g is differentiable everywhere for most of the examples given in Section 2.1 of Ref. 1. Moreover, $\mathcal{X}$ is polyhedral and hence convex for each of those examples, and g is convex for important special cases of each of those examples. Consequently, the P-optimality conditions frequently characterize the optimal solution set $\mathcal{S}^*$ for Problem 2.1.

3. Constrained Case

To extend geometric programming by the explicit inclusion of (generally nonlinear) constraint functions, we introduce two nonintersecting (possibly empty) positive-integer index sets I and J with finite cardinality $o(I)$ and $o(J)$, respectively. In terms of these index sets I and J, we also introduce the following notation and hypotheses:

(i) For each $k \in \{0\} \cup I \cup J$, there is a function g_k with a nonempty domain $C_k \subseteq E_{n_k}$, and there is a nonempty set $D_j \subseteq E_{n_j}$ for each $j \in J$.

(ii) For each $k \in \{0\} \cup I \cup J$, there is an independent vector variable x^k in E_{n_k}, and there is an independent vector variable κ with components κ_j for each $j \in J$.

(iii) x^I denotes the Cartesian product of the vector variables x^i, $i \in I$, and x^J denotes the Cartesian product of the vector variables x^j, $j \in J$. Hence, the Cartesian product $(x^0, x^I, x^J) \triangleq x$ is an independent vector variable in E_n, where

$$n \triangleq n_0 + \sum_I n_i + \sum_J n_j.$$

(iv) There is a nonempty cone $X \subseteq E_n$.

The resulting *geometric programming problem*, Problem 3.1, is defined in the following way.

Problem 3.1. Consider the objective function G whose domain is

$$C \triangleq \{(x, \kappa) \mid x^k \in C_k,\ k \in \{0\} \cup I, \text{ and } (x^j, \kappa_j) \in C_j^+,\ j \in J\}$$

and whose functional value is

$$G(x, \kappa) \triangleq g_0(x^0) + \sum_J g_j^+(x^j, \kappa_j),$$

where

$$C_j^+ \triangleq \{(x^j, \kappa_j) \mid \text{either } \kappa_j = 0 \text{ and } \sup_{d^j \in D_j} \langle x^j, d^j \rangle < +\infty, \text{ or } \kappa_j > 0 \text{ and } x^j \in \kappa_j C_j\}$$

and

$$g_j^+(x^j, \kappa_j) \triangleq \begin{cases} \sup_{d^j \in D_j} \langle x^j, d^j \rangle & \text{if } \kappa_j = 0 \text{ and } \sup_{d^j \in D_j} \langle x^j, d^j \rangle < +\infty, \\ \kappa_j g_j(x^j/\kappa_j) & \text{if } \kappa_j > 0 \text{ and } x^j \in \kappa_j C_j. \end{cases}$$

Using the feasible solution set

$$S \triangleq \{(x, \kappa) \in C \mid x \in X, \text{ and } g_i(x^i) \le 0, i \in I\},$$

calculate both the problem infimum

$$\varphi \triangleq \inf_{(x,\kappa) \in S} G(x, \kappa)$$

and the optimal solution set

$$S^* \triangleq \{(x, \kappa) \in S \mid G(x, \kappa) = \varphi\}.$$

Needless to say, the unconstrained case occurs when

$$I = J = \varnothing, \qquad g_0 : C_0 \triangleq g : \mathscr{C}, \qquad \text{and} \qquad X = \mathscr{X}.$$

On the other hand, the *ordinary programming case* occurs when

$$J = \varnothing,$$

$$n_k = m \qquad \text{and} \qquad C_k \triangleq C_o \text{ for some set } C_o \subseteq E_m, \qquad k \in \{0\} \cup I,$$

and

$$X \triangleq \text{column space of} \begin{bmatrix} U \\ U \\ \vdots \\ U \end{bmatrix},$$

where there is a total of $1 + o(I)$ identity matrices U that are $m \times m$. In particular, an explicit elimination of the vector space condition $x \in X$ by the linear transformation

$$\begin{bmatrix} x^0 \\ x^I \end{bmatrix} = \begin{bmatrix} U \\ U \\ \vdots \\ U \end{bmatrix} z$$

shows that Problem 3.1 is then equivalent to the very general ordinary programming problem:

$$\text{minimize } g_0(z)$$
$$\text{subject to } g_i(z) \le 0, \qquad i \in I,$$
$$z \in C_0.$$

Our optimality conditions for the preceding Problem 3.1 utilize the dual cone

$$Y \triangleq \{y \in E_n \mid 0 \le \langle x, y\rangle \text{ for each } x \in X\},$$

whose vector variable y has the same Cartesian-product structure as the vector variable x. They are stated as part of the following definition.

Definition 3.1. A *critical solution* (stationary solution, equilibrium solution, P-solution) for Problem 3.1 is any vector (x^*, κ^*) for which there is a vector λ^* in $E_{o(I)}$ such that (x^*, κ^*) and λ^* jointly satisfy the following P-*optimality conditions*:

$$x^* \in X,$$
$$g_i(x^{*i}) \le 0, \qquad i \in I,$$
$$\lambda_i^* \ge 0, \qquad i \in I,$$
$$\lambda_i^* g_i(x^{*i}) = 0, \qquad i \in I,$$
$$y^* \in Y,$$
$$0 = \langle x^*, y^*\rangle,$$

and

$$\langle x^{*j}, y^{*j}\rangle = g_j^+(x^{*j}, \kappa_j^*), \qquad j \in J,$$

where

$$y^{*0} \triangleq \nabla g_0(x^{*0}),$$
$$y^{*i} \triangleq \lambda_i^* \nabla g_i(x^{*i}), \qquad i \in I,$$
$$y^{*j} \triangleq \nabla g_j(x^{*j}/\kappa_j^*), \qquad j \in J.$$

Needless to say, if the cone X is actually a vector space, then $Y = X^\perp$; hence, the P-optimality condition $0 = \langle x^*, y^*\rangle$ is redundant and can be deleted. Furthermore, in the ordinary programming case, the vector space

$$Y = \{y \in E_n \mid y^0 + \sum_I y^i = 0\},$$

so the remaining P-optimality conditions are essentially the (more familiar) *Kuhn–Tucker optimality conditions*

$$g_i(z^*)\le 0, \qquad i\in I,$$
$$\lambda_i^*\ge 0, \qquad i\in I,$$
$$\lambda_i^* g_i(z^*)=0, \qquad i\in I,$$
$$\nabla g_0(z^*)+\sum_I \lambda_i^*\nabla g_i(z^*)=0.$$

On the other hand, the following important concept from ordinary programming plays a crucial role in the theory to come.

Definition 3.2. For a consistent Problem 3.1 with a finite infimum φ, a *Kuhn–Tucker vector* is any vector λ^* in $E_{o(I)}$ with the two properties

$$\lambda_i^*\ge 0, \qquad i\in I,$$

and

$$\varphi=\inf_{\substack{(x,\kappa)\in C\\ x\in X}} L_0(x,\kappa;\lambda^*),$$

where the (ordinary) *Lagrangian* is given by

$$L_0(x,\kappa;\lambda)\triangleq G(x,\kappa)+\sum_I \lambda_i g_i(x^i).$$

It is important to realize that the preceding definition of Kuhn–Tucker vectors differs considerably from the widely used definition involving the Kuhn–Tucker optimality conditions. Even in the ordinary convex programming case, the two definitions are not equivalent, though it is well-known that the preceding definition simply admits a somewhat larger set of vectors in that case.

The following theorem gives two convexity conditions that guarantee the necessity and sufficiency of the P-optimality conditions.

Theorem 3.1. Under the hypotheses that g_k is differentiable at x^{*k}, $k\in\{0\}\cup I$, and that g_j^+ is differentiable at (x^{*j},κ_j^*), $j\in J$, (i) given that X is convex, if (x^*,κ^*) is an optimal solution to Problem 3.1 and if λ^* is a Kuhn–Tucker vector for Problem 3.1, then (x^*,κ^*) is a critical solution for Problem 3.1 relative to λ^* (but not conversely); (ii) given that g_k is convex on C_k, $k\in\{0\}\cup I\cup J$, if (x^*,κ^*) is a critical solution for Problem 3.1 relative to λ^*, then (x^*,κ^*) is an optimal solution to Problem 3.1 and λ^* is a Kuhn–Tucker vector for Problem 3.1.

Proof. To prove part (i), first recall that the optimality of (x^*, κ^*) implies that $x^* \in X$ and that

$$g_i(x^{*i}) \le 0, \qquad i \in I.$$

Then, note that the defining properties for a Kuhn–Tucker vector λ^* assert that $\lambda_i^* \ge 0$, $i \in I$, and that

$$\varphi \le G(x^*, \kappa^*) + \sum_I \lambda_i^* g_i(x^{*i}).$$

Since

$$\varphi = G(x^*, \kappa^*),$$

the preceding inequalities collectively imply that

$$\lambda_i^* g_i(x^{*i}) = 0, \qquad i \in I,$$

from which we infer that

$$L_o(x^*, \kappa^*; \lambda^*) = \inf_{\substack{(x,\kappa)\in C \\ x \in X}} L_o(x, \kappa; \lambda^*).$$

Since our hypotheses clearly imply that $L_o(\cdot, \times; \lambda^*)$ is differentiable at (x^*, κ^*), the preceding equation and the convexity of X imply that

$$\langle \nabla_x L_o(x^*, \kappa^*; \lambda^*), x\rangle \ge 0 \qquad \text{for each } x \in X.$$

Likewise, the differentiability of $L_o(\cdot, \times; \lambda^*)$ at (x^*, κ^*), the preceding equation, and the observation that $x^* + s(-x^*) \in X$ for $s \le 1$ imply that

$$\langle \nabla_x L_o(x^*, \kappa^*; \lambda^*), -x^*\rangle \ge 0.$$

Consequently,

$$\nabla_x L_o(x^*, \kappa^*; \lambda^*) \in Y \qquad \text{and} \qquad 0 = \langle x^*, \nabla_x L_o(x^*, \kappa^*; \lambda^*)\rangle,$$

which means that

$$y^* \in Y \qquad \text{and} \qquad 0 = \langle x^*, y^*\rangle.$$

Finally, since our hypothesis that g_j^+ is differentiable at (x^{*j}, κ_j^*), $j \in J$, clearly implies that $\kappa^* > 0$, the differentiability of $L_o(\cdot, \times; \lambda^*)$ at (x^*, κ^*) and the preceding displayed equation imply that

$$\nabla_\kappa L_o(x^*, \kappa^*; \lambda^*) = 0,$$

which means that

$$\langle x^{*j}, y^{*j}\rangle = g_j^+(x^{*j}, \kappa_j^*), \qquad j \in J.$$

Counterexamples to the converse of part (i) are numerous and easy to construct. In fact, the reader is probably already familiar with counterexamples from the ordinary programming case.

To prove part (ii), first observe that $L_0(\cdot, \times; \lambda^*)$ is convex on C and that $L_0(\cdot, \times; \lambda^*)$ is differentiable at (x^*, κ^*). These two observations together imply that

$$L_0(x, \kappa; \lambda^*) - L_0(x^*, \kappa^*; \lambda^*) \geq \langle \nabla_{(x,\kappa)} L_0(x^*, \kappa^*; \lambda^*), (x, \kappa) - (x^*, \kappa^*) \rangle \qquad \text{for each } (x, \kappa) \in C.$$

Since the assumption that

$$\langle x^{*j}, y^{*j} \rangle = g_j^+(x^{*j}, \kappa_j^*), \qquad j \in J,$$

simply means that

$$\nabla_\kappa L_0(x^*, \kappa^*; \lambda^*) = 0,$$

elementary linear algebra shows that

$$\langle \nabla_{(x,\kappa)} L_0(x^*, \kappa^*; \lambda^*), (x, \kappa) - (x^*, \kappa^*) \rangle = \langle \nabla_x L_0(x^*, \kappa^*; \lambda^*), x - x^* \rangle \qquad \text{for each } (x, \kappa).$$

Since it is clear that

$$\nabla_x L_0(x^*, \kappa^*; \lambda^*) = y^*,$$

the assumptions that

$$0 = \langle x^*, y^* \rangle \qquad \text{and} \qquad y^* \in Y$$

imply that

$$\langle \nabla_x L_0(x^*, \kappa^*; \lambda^*), x - x^* \rangle = \langle \nabla_x L_0(x^*, \kappa^*; \lambda^*), x \rangle \geq 0 \qquad \text{for each } (x, \kappa) \text{ for which } x \in X.$$

From the preceding displayed relations, we infer that

$$L_0(x, \kappa; \lambda^*) - L_0(x^*, \kappa^*; \lambda^*) \geq 0 \qquad \text{for each } (x, \kappa) \in C \text{ for which } x \in X.$$

Using this inequality and the assumption that

$$\lambda_i^* g_i(x^{*i}) = 0, \qquad i \in I,$$

we see that

$$G(x^*, \kappa^*) \leq G(x, \kappa) + \sum_I \lambda_i^* g_i(x^i) \qquad \text{for each } (x, \kappa) \in C \text{ for which } x \in X.$$

On the other hand, the assumption that

$$\lambda_i^* \geq 0, \qquad i \in I,$$

guarantees that

$$G(x, \kappa)+\sum_I \lambda_i^* g_i(x^i)\le G(x, \kappa) \qquad \text{for each } (x, \kappa)\in C \text{ for which } g_i(x^i)\le 0,\ i\in I.$$

From the preceding two displayed inequalities involving G, we infer that

$$G(x^*, \kappa^*)\le G(x, \kappa) \qquad \text{for each } (x, \kappa)\in S.$$

Consequently, the assumptions that

$$g_i(x^{*i})\le 0, \qquad i\in I,$$

and that $x^*\in X$ imply that (x^*, κ^*) is optimal for Problem 3.1, which means of course that

$$\varphi = G(x^*, \kappa^*).$$

Using these facts and the assumption that

$$\lambda_i^* g_i(x^{*i})=0, \qquad i\in I,$$

we infer from the last displayed inequality involving L_0 that λ^* is a Kuhn–Tucker vector for Problem 3.1. □

It is worth noting that, for most of the examples given or alluded to in Section 2.2 of Ref. 1, the g_k, $k\in\{0\}\cup I$, are differentiable everywhere while either J is empty or the g_j^+, $j\in J$, are differentiable everywhere, except at the origin. Moreover, X is polyhedral, and hence convex for each of those examples, and the g_k, $k\in\{0\}\cup I\cup J$, are convex for important special cases of each of those examples. Consequently, the P-optimality conditions frequently characterize the optimal solution set S^* for Problem 3.1.

Characterizations of S^* that do not require differentiability of the g_k, $k\in\{0\}\cup I$, and the g_j^+, $j\in J$, but do require conjugate transform theory are given in Ref. 9.

References

1. PETERSON, E. L., *Geometric Programming*, SIAM Review, Vol. 18, pp. 1–51, 1976.
2. HALKIN, H., and NEUSTADT, L. W., *General Necessary Conditions for Optimization Problems*, Proceedings of the National Academy of Sciences, Vol. 29, pp. 422–435, 1966.
3. GUIGNARD, M., *Generalized Kuhn–Tucker Conditions for Mathematical Programming Problems in Banach Space*, SIAM Journal on Control, Vol. 7, pp. 232–241, 1969.
4. LUENBERGER, D. L., *Optimization by Vector Space Methods*, John Wiley and Sons, New York, New York, 1969.

5. NEUSTADT, L. W., *A General Theory of Extremals*, Journal of Computer Systems and Science, Vol. 3, pp. 12–65, 1969.
6. HOLMES, R. B., *A Course on Optimization and Best Approximation*, Springer-Verlag, Heidelberg, Germany, 1972.
7. BAZARRA, M., and GOODE, W., *Necessary Optimality Criteria in Mathematical Programming in the Presence of Differentiability*, Journal of Mathematical Analysis and Applications, Vol. 40, pp. 16–23, 1972.
8. LUENBERGER, D. L., *Introduction to Linear and Nonlinear Programming*, Addison-Wesley Publishing Company, Reading, Massachusetts, 1973.
9. PETERSON, E. L., *Lagrangian Saddle Points and Duality in Generalized Geometric Programming*, Journal of Optimization Theory and Applications, Vol. 26, No. 1, 1978.

4

Saddle Points and Duality in Generalized Geometric Programming[1]

E. L. PETERSON[2]

Abstract. Extensions of the ordinary Lagrangian are used both in saddle-point characterizations of optimality and in a development of duality theory.

1. Introduction

Optimization problems from the real world usually possess a linear–algebraic component, either directly in the form of problem linearities (e.g., those involving the node-arc incidence matrices in network optimization) or indirectly in the more subtle form of certain problem nonlinearities (e.g., those involving the coefficient matrices in quadratic programming or, perhaps more appropriately, those involving the exponent matrices in signomial programming). As demonstrated in the author's recent survey paper (Ref. 1) and some of the references cited therein, such a component can frequently be exploited by taking a (generalized) geometric programming approach. In fact, geometric programming is primarily a body of techniques and theorems for inducing and exploiting as much linearity as possible.

The duality that exploits such linearity differs considerably from ordinary Lagrangian duality. Since Section 33 of Rockafellar's book (Ref. 2) indicates that any such duality can be viewed as originating from an appropriate Lagrangian (or saddle function), it is natural to seek such a

[1] This research was sponsored by the Air Force Office of Scientific Research, Air Force Systems Command, USAF, under Grant No. AFOSR-73-2516.
[2] Professor, Department of Engineering Sciences and Applied Mathematics, Department of Industrial Engineering and Management Sciences, and Department of Mathematics, Northwestern University, Evanston, Illinois.

Lagrangian for geometric programming. Although such a Lagrangian and the corresponding saddle-point characterizations of optimality have already been described in Ref. 1, proofs are given here for the first time.

In geometric programming, problems with only linear constraints are treated in essentially the same way as problems without constraints. Only problems with nonlinear constraints require additional attention, and hence are classified as constrained problems.

Since many important problems are unconstrained (e.g., most network optimization problems), and since the theory for the unconstrained case is much simpler than its counterpart for the constrained case, the unconstrained case is treated separately (even though its theory is actually embedded in the theory for the constrained case).

The only prerequisites for Section 2 are the basic facts about the conjugate transformation described in Section 3.1.2 of Ref. 1 (and established in Ref. 2). An additional prerequisite for Section 3 is the geometric inequality established in Section 3.3.3 of Ref. 1.

2. Unconstrained Case

In harmony with Sections 2.1 and 3.1 of Ref. 1, suppose that $g:\mathscr{C}$ is a (proper) function g with a nonempty (effective) domain $\mathscr{C} \subseteq E_n$ (n-dimensional Euclidean space), and assume that $\mathscr{X}$ is a nonempty cone in E_n. For purposes of easy reference and mathematical precision, the resulting *geometric programming problem* is now given the following formal definition in terms of classical terminology and notation.

Problem 2.1. Using the feasible solution set

$$\mathscr{S} \triangleq \mathscr{X} \cap \mathscr{C},$$

calculate both the problem infimum

$$\varphi \triangleq \inf_{x \in \mathscr{S}} g(x)$$

and the optimal solution set

$$\mathscr{S}^* \triangleq \{x \in \mathscr{S} \mid g(x) = \varphi\}.$$

Needless to say, the *ordinary programming case* occurs when $\mathscr{X}$ is actually the entire vector space E_n.

2.1. Saddle Points. Our Lagrangian for Problem 2.1 utilizes the *conjugate transform* $h:\mathscr{D}$ of $g:\mathscr{C}$, whose domain is

$$\mathscr{D} \triangleq \{y \in E_n \mid \sup_{x \in \mathscr{C}} [\langle y, x\rangle - g(x)] \text{ is finite}\}$$

and whose functional value is

$$h(y) \triangleq \sup_{x \in \mathscr{C}} [\langle y, x\rangle - g(x)].$$

The corresponding saddle-point characterization of optimality also utilizes the *dual cone*

$$\mathscr{Y} \triangleq \{y \in E_n \mid 0 \leq \langle x, y\rangle \text{ for each } x \in \mathscr{X}\}.$$

The following definition lays a foundation for that characterization.

Definition 2.1. For a consistent Problem 2.1 with a finite infimum φ a *P-vector* is any vector y^* with the two properties

$$y^* \in \mathscr{D},$$

$$\varphi = \inf_{x \in \mathscr{X}} L_g(x; y^*),$$

where the (geometric) *Lagrangian* is given by

$$L_g(x; y) \triangleq \langle x, y\rangle - h(y).$$

It is worth noting that, even in the ordinary programming case, the geometric Lagrangian L_g is unlike the ordinary Lagrangian $L_0 = g$. In fact, L_g exists only if g has a conjugate transform h.

The following theorem gives a saddle-point characterization of optimal solutions x^* and P-vectors y^*. It also provides a basis for other important characterizations of such vectors.

Theorem 2.1. Given that $g: \mathscr{C}$ is convex and closed, let $x^* \in \mathscr{X}$ and let $y^* \in \mathscr{D}$. Then, x^* is optimal for Problem 2.1 and y^* is a P-vector for Problem 2.1 iff the ordered pair $(x^*; y^*)$ is a *saddle point* for the Lagrangian L_g, that is,

$$\sup_{y \in \mathscr{D}} L_g(x^*; y) = L_g(x^*; y^*) = \inf_{x \in \mathscr{X}} L_g(x; y^*),$$

in which case L_g has the saddle-point value

$$L_g(x^*; y^*) = g(x^*) = \varphi.$$

Moreover,

$$\sup_{y \in \mathscr{D}} L_g(x^*; y) = L_g(x^*; y^*)$$

iff x^* and y^* satisfy both the feasibility condition

$$x^* \in \mathscr{C}$$

and the subgradient condition

$$y^* \in \partial g(x^*),$$

in which case

$$L_g(x^*; y^*) = g(x^*).$$

Furthermore,

$$L_g(x^*; y^*) = \inf_{x \in \mathscr{X}} L_g(x; y^*)$$

iff x^* and y^* satisfy both the feasibility condition

$$y^* \in \mathscr{Y}$$

and the orthogonality condition

$$0 = \langle x^*, y^* \rangle,$$

in which case

$$L_g(x^*; y^*) = -h(y^*).$$

Proof. The following two lemmas must be used repetitively.

Lemma 2.1. Given that $g: \mathscr{C}$ is convex and closed, a vector x satisfies the restraint $x \in \mathscr{C}$ iff $\sup_{y \in \mathscr{D}} L_g(x; y)$ is finite, in which case

$$\sup_{y \in \mathscr{D}} L_g(x; y) = g(x),$$

$$\{y \in \mathscr{D} \mid L_g(x; y) = g(x)\} = \partial g(x).$$

Proof. It is immediate from the definition of $L_g(x; y)$ and conjugate transform theory.

Lemma 2.2. A vector y in $\mathscr{D}$ satisfies the cone condition $y \in \mathscr{Y}$ iff $\inf_{x \in \mathscr{X}} L_g(x; y)$ is finite, in which case

$$\inf_{x \in \mathscr{X}} L_g(x; y) = -h(y),$$

$$\{x \in \mathscr{X} \mid L_g(x; y) = -h(y)\} = \{x \in \mathscr{X} \mid 0 = \langle x, y \rangle\}.$$

Proof. It is immediate from the assumption that $\mathscr{X}$ is a cone; because, in that case, it is clear that $y \in \mathscr{Y}$ iff $\inf_{x \in \mathscr{X}} \langle x, y \rangle$ is finite, in which case $\inf_{x \in \mathscr{X}} \langle x, y \rangle = 0$.

Now, assuming that x^* is optimal for Problem 2.1 and that y^* is a P-vector for Problem 2.1, we deduce from Lemma 2.1 that

$$\sup_{y \in \mathcal{D}} L_g(x^*; y) = g(x^*) = \varphi = \inf_{x \in \mathcal{X}} L_g(x; y^*),$$

by virtue of the defining properties for optimal solutions and P-vectors. Notice that the first and second equations show that

$$L_g(x^*; y^*) \le g(x^*) = \varphi,$$

and observe that the second and third equations show that

$$g(x^*) = \varphi \le L_g(x^*; y^*);$$

so we infer that

$$L_g(x^*; y^*) = g(x^*) = \varphi.$$

Consequently, $(x^*; y^*)$ is a saddle point for L_g.

Conversely, assuming that $(x^*; y^*)$ is a saddle point for L_g, we deduce from Lemma 2.1 that $x^* \in \mathcal{C}$; so x^* is feasible by virtue of the hypothesis $x^* \in \mathcal{X}$. From Lemma 2.1, we also infer that

$$\sup_{y \in \mathcal{D}} L_g(x^*; y) = g(x^*);$$

so the saddle-point equations imply that

$$g(x^*) = \inf_{x \in \mathcal{X}} L_g(x; y^*),$$

which in turn means that

$$g(x^*) \le L_g(x; y^*) \qquad \text{for each } x \in \mathcal{X}.$$

Moreover, Lemma 2.1 also implies that

$$L_g(x; y^*) \le g(x) \qquad \text{for each } x \in \mathcal{S},$$

by virtue of the hypothesis $y^* \in \mathcal{D}$. It then follows from these two displayed inequalities that

$$g(x^*) \le g(x) \qquad \text{for each } x \in \mathcal{S}.$$

Thus, x^* is optimal for Problem 2.1, and hence $\varphi = g(x^*)$. This equation and the preceding displayed equation show that y^* is a P-vector for Problem 2.1, by virtue of the hypothesis $y^* \in \mathcal{D}$.

Now, assuming that

$$\sup_{y \in \mathcal{D}} L_g(x^*; y) = L_g(x^*; y^*),$$

we infer from Lemma 2.1 that $x^* \in \mathscr{C}$ and that

$$L_g(x^*; y^*) = g(x^*),$$

which in turn imply that $y^* \in \partial g(x^*)$ by virtue of the hypothesis $y^* \in \mathscr{D}$ and Lemma 2.1. Conversely, assuming that $x^* \in \mathscr{C}$ and that $y^* \in \partial g(x^*)$, we infer from Lemma 2.1 that

$$\sup_{y \in \mathscr{D}} L_g(x^*; y) = g(x^*) = L_g(x^*; y^*).$$

Finally, assuming that

$$\inf_{x \in \mathscr{X}} L_g(x; y^*) = L_g(x^*; y^*),$$

we infer from Lemma 2.2 that $y^* \in \mathscr{Y}$ and that

$$L_g(x^*; y^*) = -h(y^*),$$

which in turn implies that $0 = \langle x^*, y^* \rangle$ by virtue of the definition of $L_g(x^*; y^*)$. Conversely, assuming that $y^* \in \mathscr{Y}$ and that $0 = \langle x^*, y^* \rangle$, we infer from Lemma 2.2 that

$$\inf_{x \in \mathscr{X}} L_g(x; y^*) = -h(y^*) = L_g(x^*; y^*). \qquad \square$$

Since the second assertion of Theorem 2.1 gives certain conditions that are equivalent to the first saddle-point equation, and since the third assertion of Theorem 2.1 gives certain other conditions that are equivalent to the second saddle-point equation, Theorem 2.1 actually provides four different characterizations of all ordered pairs $(x^*; y^*)$ of optimal solutions x^* and P-vectors y^*.

Of course, each of those four characterizations provides a characterization of all optimal solutions x^* in terms of a given P-vector y^*, as well as a characterization of all P-vectors y^* in terms of a given optimal solution x^*.

Still another characterization of all optimal solutions x^* to certain Problems 2.1 has been given by the author (Ref. 3).

2.2. Duality. Corresponding to Problem 2.1 is the following *geometric dual problem*.

Problem 2.2. Using the feasible solution set

$$\mathscr{T} \triangleq \{y \in \mathscr{D} \mid \inf_{x \in \mathscr{X}} L_g(x; y) \text{ is finite}\}$$

and the objective function

$$\mathscr{H}(y) \triangleq \inf_{x \in \mathscr{X}} L_g(x; y),$$

calculate both the problem supremum

$$\Psi \triangleq \sup_{y \in \mathcal{T}} \mathcal{H}(y)$$

and the optimal solution set

$$\mathcal{T}^* \triangleq \{y \in \mathcal{T} \mid \mathcal{H}(y) = \Psi\}.$$

Even though Problem 2.2 is essentially a *maximin problem*, a type of problem that tends to be relatively difficult to analyze, the minimization problems that must be solved to obtain the objective function $\mathcal{H}:\mathcal{T}$ have trivial solutions. In particular, Lemma 2.2 clearly implies that

$$\mathcal{T} = \mathcal{Y} \cap \mathcal{D} \qquad \text{and} \qquad \mathcal{H}(y) = -h(y),$$

so Problem 2.2 can actually be rephrased in the following more direct way.

Problem 2.2. Using the feasible solution set

$$\mathcal{T} \triangleq \mathcal{Y} \cap \mathcal{D},$$

calculate both the problem infimum

$$\psi \triangleq \inf_{y \in \mathcal{T}} h(y) = -\Psi$$

and the optimal solution set

$$\mathcal{T}^* \triangleq \{y \in \mathcal{T} \mid h(y) = \psi\}.$$

When phrased in this way, Problem 2.2 closely resembles Problem 2.1 and is in fact a geometric programming problem. Of course, the geometric dual Problem 2.2 can actually be defined in this way, but the preceding derivation serves as an important link between geometric Lagrangians and geometric duality (analogous to the link between ordinary Lagrangians and ordinary duality).

To further strengthen that link, we first need to develop the most basic duality theory, a theory in which the following definition is almost as important as the definition of the dual Problems 2.1 and 2.2.

Definition 2.2. The *extremality conditions* (for unconstrained geometric programming) are:

(I) $x \in \mathcal{X}$ and $y \in \mathcal{Y}$,

(II) $0 = \langle x, y\rangle$,

(III) $y \in \partial g(x)$.

The following *duality theorem* is the basis for many important theorems.

Theorem 2.2. If x and y are feasible solutions to Problems 2.1 and 2.2, respectively [in which case the extremality conditions (I) are satisfied], then

$$0 \le g(x) + h(y),$$

with equality holding iff the extremality conditions (II) and (III) are satisfied, in which case x and y are optimal solutions to Problems 2.1 and 2.2, respectively.

Proof. The following fact is formalized as a lemma in order to facilitate a comparison between the constrained and unconstrained cases.

Lemma 2.3. If $x \in \mathscr{C}$ and $y \in \mathscr{D}$, then

$$\langle x, y \rangle \le g(x) + h(y),$$

with equality holding iff the extremality condition (III) is satisfied.

Proof. Simply invoke the conjugate inequality presented in Section 3.1.2 of Ref. 1.

Now, the fact that x and y are in the cone $\mathscr{X}$ and its dual $\mathscr{Y}$, respectively, combined with Lemma 2.3, shows that

$$0 \le \langle x, y \rangle \le g(x) + h(y),$$

with equality holding in both of these inequalities iff the equality conditions stated in the theorem are satisfied. □

The following important corollary is an immediate consequence of Theorem 2.2.

Corollary 2.1. If the dual Problems 2.1 and 2.2 are both consistent, then

(i) the infimum φ for Problem 2.1 is finite, and

$$0 \le \varphi + h(y) \qquad \text{for each } y \in \mathscr{T},$$

(ii) the infimum ψ for Problem 2.2 is finite, and

$$0 \le \varphi + \psi.$$

The strictness of the inequality in conclusion (ii) plays a crucial role in almost all duality theorems.

Definition 2.3. Consistent dual Problems 2.1 and 2.2, for which

$$0 < \varphi + \psi,$$

have a *duality gap* of $\varphi + \psi$.

A much more thorough discussion of duality theory and the role played by duality gaps is given in Ref. 1 and some of the references cited therein.

The link between geometric Lagrangians and geometric duality can now be further strengthened by the following tie between dual Problem 2.2 and the P-vectors for Problem 2.1 defined in Section 2.1.

Theorem 2.3. Given that Problem 2.1 is consistent with a finite infimum φ,

(i) if Problem 2.1 has a P-vector, then Problem 2.2 is consistent and

$$0 = \varphi + \psi,$$

(ii) if Problem 2.2 is consistent and $0 = \varphi + \psi$, then

$$\{\mathscr{y}^* \mid \mathscr{y}^* \text{ is a } P\text{-vector for problem 2.1}\} = \mathscr{T}^*.$$

Proof. If $\mathscr{y}^*$ is a P-vector for Problem 2.1, then Lemma 2.2 implies that $\mathscr{y}^*$ is feasible for Problem 2.2 and that $\varphi = -\mathscr{h}(\mathscr{y}^*)$, which in turn implies that $\psi = \mathscr{h}(\mathscr{y}^*)$ by virtue of conclusion (i) to Corollary 2.1. Consequently, $0 = \varphi + \psi$ and $\mathscr{y}^* \in \mathscr{T}^*$.

On the other hand, if Problem 2.2 is consistent and $0 = \varphi + \psi$, then each vector $\mathscr{y}^*$ in $\mathscr{T}^*$ has the property that $\varphi = -\mathscr{h}(\mathscr{y}^*)$, and hence each such vector $\mathscr{y}^*$ is a P-vector for Problem 2.1 by virtue of the first two paragraphs of this subsection. □

An important consequence of Theorem 2.3 is that, when they exist, all P-vectors for Problem 2.1 can be obtained simply by computing the dual optimal solution set $\mathscr{T}^*$. However, there are cases in which the vectors in $\mathscr{T}^*$ are not P-vectors for Problem 2.1, though such cases can occur only when $0 < \varphi + \psi$, in which event Theorem 2.3 implies that there can be no P-vectors for Problem 2.1.

3. Constrained Case

In harmony with Sections 2.2 and 3.3 of Ref. 1, we introduce two nonintersecting (possibly empty) positive-integer index sets I and J, with finite cardinality $o(I)$ and $o(J)$, respectively. In terms of these index sets I and J, we also introduce the following notation and hypotheses.

(i) For each $k \in \{0\} \cup I \cup J$, suppose that $g_k : C_k$ is a (proper) function g_k with a nonempty (effective) domain $C_k \subseteq E_{n_k}$ (n_k-dimensional Euclidean space); and, for each $j \in J$, let D_j be the (effective) domain of the *conjugate transform* $h_j : D_j$ of $g_j : C_j$.

(ii) For each $k \in \{0\} \cup I \cup J$, let x^k be an independent vector variable in E_{n_k}, and let κ be an independent vector variable with components κ_j for each $j \in J$.

(iii) Denote the Cartesian product of the vector variables x^i, $i \in I$, by the symbol x^I, and denote the Cartesian product of the vector variables x^j, $j \in J$, by the symbol x^J. Then, the Cartesian product $(x^0, x^I, x^J) \triangleq x$ is a vector variable in E_n, where

$$n \triangleq n_0 + \sum_I n_i + \sum_J n_j.$$

(iv) Assume that X is a nonempty cone in E_n.

For purposes of easy reference and mathematical precision, the resulting *geometric programming problem* is now given the following formal definition in terms of classical terminology and notation.

Problem 3.1. Consider the objective function G whose domain is

$$C \triangleq \{(x, \kappa) \mid x^k \in C_k,\ k \in \{0\} \cup I, \text{ and } (x^j, \kappa_j) \in C_j^+,\ j \in J\}$$

and whose functional value is

$$G(x, \kappa) \triangleq g_0(x^0) + \sum_J g_j^+(x^j, \kappa_j),$$

where

$$C_j^+ \triangleq \{(x^j, \kappa_j) \mid \text{either } \kappa_j = 0 \text{ and } \sup_{d^j \in D_j} \langle x^j, d^j\rangle < +\infty, \text{ or } \kappa_j > 0 \text{ and } x^j \in \kappa_j C_j\}$$

and

$$g_j^+(x^j, \kappa_j) \triangleq \begin{cases} \sup_{d^j \in D_j} \langle x^j, d^j\rangle & \text{if } \kappa_j = 0 \text{ and } \sup_{d^j \in D_j} \langle x^j, d^j\rangle < +\infty, \\ \kappa_j g_j(x^j/\kappa_j) & \text{if } \kappa_j > 0 \text{ and } x^j \in \kappa_j C_j. \end{cases}$$

Using the feasible solution set

$$S \triangleq \{(x, \kappa) \in C \mid x \in X, \text{ and } g_i(x^i) \le 0,\ i \in I\},$$

calculate both the problem infimum

$$\varphi \triangleq \inf_{(x,\kappa) \in S} G(x, \kappa)$$

and the optimal solution set

$$S^* \triangleq \{(x, \kappa) \in S \mid G(x, \kappa) = \varphi\}.$$

Needless to say, the unconstrained case occurs when

$$I = J = \varnothing, \qquad g_0 : C_0 \triangleq \mathscr{g} : \mathscr{C}, \qquad \text{and} \qquad X \triangleq \mathscr{X}.$$

On the other hand, the *ordinary programming case* occurs when

$$J=\varnothing,$$

$$n_k = m \text{ and } C_k \triangleq C_0 \text{ for some set } C_0 \subseteq E_m,\ k\in\{0\}\cup I,$$

$$X \triangleq \text{column space of } \begin{bmatrix} U \\ U \\ \vdots \\ U \end{bmatrix},$$

where there is a total of $1+o(I)$ identity matrices U that are $m\times m$. In particular, an explicit elimination of the vector space condition $x\in X$ by the linear transformation

$$\begin{pmatrix} x^0 \\ x^I \end{pmatrix} = \begin{bmatrix} U \\ U \\ \vdots \\ U \end{bmatrix} z$$

shows that Problem 3.1 is then equivalent to the very general ordinary programming problem:

$$\text{minimize } g_0(z),$$

$$\text{subject to } g_i(z)\le 0, \qquad i\in I,$$

$$z\in C_0.$$

3.1. Saddle Points. Our Lagrangian for Problem 3.1 utilizes the *conjugate transform* $h_k: D_k$ of $g_k: C_k$, whose domain is

$$D_k \triangleq \{y^k\in E_{n_k} \mid \sup_{x^k\in C_k} [\langle y^k, x^k\rangle - g_k(x^k)] \text{ is finite}\}$$

and whose functional value is

$$h_k(y^k) \triangleq \sup_{x^k\in C_k} [\langle y^k, x^k\rangle - g_k(x^k)].$$

Notationally, it involves both the vector variable (x,κ) and an analogous vector variable (y,λ), where the vector variable y has the same Cartesian-product structure as the vector variable x and where the vector variable λ has components λ_i for each $i\in I$. The corresponding saddle-point characterization of optimality also utilizes the *dual cone*

$$Y \triangleq \{y\in E_n \mid 0\le \langle x, y\rangle \text{ for each } x\in X\}.$$

The following definition lays a foundation for that characterization.

Definition 3.1. Consider the function H whose domain is

$$D \triangleq \{(y, \lambda) \mid y^k \in D_k,\ k \in \{0\} \cup J, \text{ and } (y^i, \lambda_i) \in D_i^+,\ i \in I\}$$

and whose functional value is

$$H(y, \lambda) \triangleq h_0(y^0) + \sum_I h_i^+(y^i, \lambda_i),$$

where

$$D_i^+ \triangleq \{(y^i, \lambda_i) \mid \text{either } \lambda_i = 0 \text{ and } \sup_{c^i \in C_i} \langle c^i, y^i \rangle < +\infty, \text{ or } \lambda_i > 0 \text{ and } y^i \in \lambda_i D_i\}$$

and

$$h_i^+(y^i, \lambda_i) \triangleq \begin{cases} \sup_{c^i \in C_i} \langle c^i, y^i \rangle & \text{if } \lambda_i = 0 \text{ and } \sup_{c^i \in C_i} \langle c^i, y^i \rangle < +\infty, \\ \lambda_i h_i(y^i/\lambda_i) & \text{if } \lambda_i > 0 \text{ and } y^i \in \lambda_i D_i. \end{cases}$$

For a consistent Problem 3.1 with a finite infimum φ, a *P-vector* is any vector (y^*, λ^*) with the two properties

$$(y^*, \lambda^*) \in D,$$

$$\varphi = \inf_{\substack{x \in X \\ \kappa \geq 0}} L_g(x, \kappa; y^*, \lambda^*),$$

where the (geometric) *Lagrangian* is given by

$$L_g(x, \kappa; y, \lambda) \triangleq \langle x, y \rangle - H(y, \lambda) - \sum_J \kappa_j h_j(y^j).$$

Needless to say, even in the ordinary programming case, the geometric Lagrangian L_g is unlike the ordinary Lagrangian

$$L_o \triangleq g_0 + \sum_I \lambda_i g_i.$$

In fact, L_g exists only if g_k has a conjugate transform h_k for $k \in \{0\} \cup I \cup J$.

On the other hand, the following important concept from ordinary programming plays a crucial role in the theory to come.

Definition 3.2. For a consistent Problem 3.1 with a finite infimum φ, a *Kuhn–Tucker vector* is any vector λ^* in $E_{o(I)}$ with the two properties

$$\lambda_i^* \geq 0, \qquad i \in I,$$

$$\varphi = \inf_{\substack{(x, \kappa) \in C \\ x \in X}} L_o(x, \kappa; \lambda^*),$$

where the (ordinary) *Lagrangian* is given by

$$L_o(x, \kappa; \lambda) \triangleq G(x, \kappa) + \sum_I \lambda_i g_i(x^i).$$

It is important to realize that the preceding definition of Kuhn–Tucker vectors differs considerably from the widely used definition involving the Kuhn–Tucker optimality conditions. Even in the ordinary convex programming case, the two definitions are not equivalent, though it is well-known that the preceding definition simply admits a somewhat larger set of vectors in that case.

The following theorem gives a saddle-point characterization of optimal solutions (x^*, κ^*) and P-vectors (y^*, λ^*). It also provides a basis for other important characterizations of such vectors.

Theorem 3.1. Given that $g_k: C_k$, $k \in \{0\} \cup I \cup J$, is convex and closed, let (x^*, κ^*) be such that $x^* \in X$ and $\kappa^* \geq 0$, and let $(y^*, \lambda^*) \in D$. Then, (x^*, κ^*) is optimal for Problem 3.1 and (y^*, λ^*) is a P-vector for Problem 3.1 iff the ordered pair $(x^*, \kappa^*; y^*, \lambda^*)$ is a *saddle point* for the Lagrangian L_g, that is,

$$\sup_{(y,\lambda)\in D} L_g(x^*, \kappa^*; y, \lambda) = L_g(x^*, \kappa^*; y^*, \lambda^*) = \inf_{\substack{x\in X \\ \kappa\geq 0}} L_g(x, \kappa; y^*, \lambda^*),$$

in which case L_g has the saddle-point value

$$L_g(x^*, \kappa^*; y^*, \lambda^*) = G(x^*, \kappa^*) = \varphi.$$

Moreover,

$$\sup_{(y,\lambda)\in D} L_g(x^*, \kappa^*; y, \lambda) = L_g(x^*, \kappa^*; y^*, \lambda^*)$$

iff (x^*, κ^*) and (y^*, λ^*) satisfy both the feasibility conditions

$$(x^*, \kappa^*) \in C,$$

$$g_i(x^{*i}) \leq 0, \qquad i \in I,$$

and the *subgradient and complementary slackness conditions*

$$y^{*0} \in \partial g_0(x^{*0}),$$

$$\text{either } \lambda_i^* = 0 \text{ and } \langle x^{*i}, y^{*i}\rangle = \sup_{c^i \in C_i} \langle c^i, y^{*i}\rangle,$$

$$\text{or } \lambda_i^* > 0 \text{ and } y^{*i} \in \lambda_i^* \, \partial g_i(x^{*i}),\ i \in I,$$

$$\lambda_i^* g_i(x^{*i}) = 0, \qquad i \in I,$$

$$\text{either } \kappa_j^* = 0 \text{ and } \langle x^{*j}, y^{*j}\rangle = \sup_{d^j \in D_j} \langle x^{*j}, d^j\rangle,$$

$$\text{or } \kappa_j^* > 0 \text{ and } y^{*j} \in \partial g_j(x^{*j}/\kappa_j^*),\ j \in J,$$

in which case

$$L_g(x^*, \kappa^*; y^*, \lambda^*) = G(x^*, \kappa^*).$$

Furthermore,

$$L_g(x^*, \kappa^*; y^*, \lambda^*) = \inf_{\substack{x \in X \\ \kappa \geq 0}} L_g(x, \kappa; y^*, \lambda^*)$$

iff (x^*, κ^*) and (y^*, λ^*) satisfy both the feasibility conditions

$$y^* \in Y,$$
$$h_j(y^{*j}) \leq 0, \qquad j \in J,$$

and the *orthogonality and complementary slackness conditions*

$$0 = \langle x^*, y^* \rangle,$$
$$\kappa_j^* h_j(y^{*j}) = 0, \qquad j \in J,$$

in which case

$$L_g(x^*, \kappa^*; y^*, \lambda^*) = -H(y^*, \lambda^*).$$

Proof. The following lemmas must be used repetitively.

Lemma 3.1. Given that $g_k: C_k$, $k \in \{0\} \cup I \cup J$, is convex and closed, a vector (x, κ) with $\kappa \geq 0$ satisfies the restraint $(x, \kappa) \in C$ and the constraints $g_i(x^i) \leq 0$, $i \in I$, iff $\sup_{(y,\lambda) \in D} L_g(x, \kappa; y, \lambda)$ is finite, in which case

$$\sup_{(y,\lambda) \in D} L_g(x, \kappa; y, \lambda) = G(x, \kappa)$$

and

$$\{(y, \lambda) \in D \mid L_g(x, \kappa; y, \lambda) = G(x, \kappa)\} = \{(y, \lambda) \mid y^0 \in \partial g_0(x^0);$$
$$\text{either } \lambda_i = 0 \text{ and } \langle x^i, y^i \rangle = \sup_{c^i \in C_i} \langle c^i, y^i \rangle,$$
$$\text{or } \lambda_i > 0 \text{ and } y^i \in \lambda_i\, \partial g_i(x^i),\ i \in I;\ \lambda_i g_i(x^i) = 0,\ i \in I;$$
$$\text{either } \kappa_j = 0 \text{ and } \langle x^j, y^j \rangle = \sup_{d^j \in D_j} \langle x^j, d^j \rangle,$$
$$\text{or } \kappa_j > 0 \text{ and } y^j \in \partial g_j(x^j/\kappa_j),\ j \in J\}.$$

Proof. First, observe that

$$\sup_{(y,\lambda)\in D} L_g(x,\kappa;y,\lambda)=\sup_{(y,\lambda)\in D}[\langle x,y\rangle-H(y,\lambda)-\sum_J \kappa_j h_j(y^j)]$$

$$=\sup_{(y,\lambda)\in D}[\langle x^0,y^0\rangle+\sum_I\langle x^i,y^i\rangle+\sum_J\langle x^j,y^j\rangle$$

$$-h_0(y^0)-\sum_I h_i^+(y^i,\lambda_i)-\sum_J \kappa_j h_j(y^j)]$$

$$=\sup_{y^0\in D_0}[\langle x^0,y^0\rangle-h_0(y^0)]+\sum_I\sup_{(y^i,\lambda_i)\in D_i^+}[\langle x^i,y^i\rangle$$

$$+0\lambda_i-h_i^+(y^i,\lambda_i)]+\sum_J\sup_{y^j\in D_j}[\langle x^j,y^j\rangle-\kappa_j h_j(y^j)].$$

From conjugate transform theory, we know that the preceding supremum with index 0 is finite iff $x^0\in C_0$, in which case

$$\sup_{y^0\in D_0}[\langle x^0,y^0\rangle-h_0(y^0)]=g_0(x^0),$$

$$\{y^0\in D_0\,|\,\langle x^0,y^0\rangle-h_0(y^0)=g_0(x^0)\}=\partial g_0(x^0).$$

Analogous information about the preceding suprema with indices i and j is provided by the following two sublemmas, which collectively complete the proof of Lemma 3.1.

Sublemma 3.1. Given that $g_i: C_i$ is convex and closed, the $\sup_{(y^i,\lambda_i)\in D_i^+}[\langle x^i,y^i\rangle+\alpha_i\lambda_i-h_i^+(y^i,\lambda_i)]$ is finite iff both

$$x^i\in C_i \qquad \text{and} \qquad g_i(x^i)+\alpha_i\le 0,$$

in which case

$$\sup_{(y^i,\lambda_i)\in D_i^+}[\langle x^i,y^i\rangle+\alpha_i\lambda_i-h_i^+(y^i,\lambda_i)]=0$$

and

$$\{(y^i,\lambda_i)\in D_i^+\,|\,\langle x^i,y^i\rangle+\alpha_i\lambda_i-h_i^+(y^i,\lambda_i)=0\}$$

$$=\{(y^i,\lambda_i)\in D_1^+\,|\text{ either }\lambda_i=0\text{ and }\langle x^i,y^i\rangle=\sup_{c^i\in C_i}\langle c^i,y^i\rangle,$$

$$\text{or }\lambda_i>0\text{ and }y^i\in\lambda_i\,\partial g_i(x^i);\ \lambda_i[g_i(x^i)+\alpha_i]=0\}.$$

Proof. First, observe that

$$\sup_{(y^i,\lambda_i)\in D_i^+} [\langle x^i, y^i\rangle + \alpha_i\lambda_i - h_i^+(y^i, \lambda_i)]$$

$$= \sup_{\lambda_i\geq 0} [\sup_{y^i} \{\langle x^i, y^i\rangle + \alpha_i\lambda_i - h_i^+(y^i, \lambda_i) \mid (y^i, \lambda_i)\in D_i^+\}]$$

$$= \sup_{\lambda_i\geq 0} [\alpha_i\lambda_i + \sup_{y^i} \{\langle x^i, y^i\rangle - h_i^+(y^i, \lambda_i) \mid (y^i, \lambda_i)\in D_i^+\}]$$

$$= \sup_{\lambda_i\geq 0} \left\{ \alpha_i\lambda_i + \begin{bmatrix} \sup_{y^i} \{\langle x^i, y^i\rangle - \sup_{c^i\in C_i} \langle c^i, y^i\rangle \mid \sup_{c^i\in C_i} \langle c^i, y^i\rangle < +\infty\} & \\ \text{if } \lambda_i = 0 & \\ \sup_{y^i} \{\langle x^i, y^i\rangle - \lambda_i h_i(y^i/\lambda_i) \mid y^i/\lambda_i \in D_i\} & \text{if } \lambda_i > 0 \end{bmatrix} \right\}$$

$$= \sup_{\lambda_i\geq 0} \left\{ \alpha_i\lambda_i + \begin{bmatrix} 0 & \text{if } \lambda_i = 0 \text{ and } x^i \in \bar{C}_i \\ +\infty & \text{if } \lambda_i = 0 \text{ and } x^i \notin \bar{C}_i \\ +\infty & \text{if } \lambda_i > 0 \text{ and } x^i \notin C_i \\ \lambda_i g_i(x^i) & \text{if } \lambda_i > 0 \text{ and } x^i \in C_i \end{bmatrix} \right\},$$

where the final step makes use of the fact that the zero function with domain $\bar{C}_i$ (the topological closure of C_i) is the conjugate transform of the conjugate transform of the zero function with domain C_i. Now, note that the last expression is finite only if $x^i \in C_i$, in which case the last expression clearly equals

$$\sup_{\lambda_i\geq 0} [\alpha_i\lambda_i + \lambda_i g_i(x^i)].$$

But this expression is obviously finite iff

$$g_i(x^i) + \alpha_i \leq 0,$$

in which case this expression is clearly zero.

Finally, given that $x^i \in C_i$, if $(y^i, \lambda_i) \in D_i^+$, then the geometric inequality established in Section 3.3.3 of Ref. 1 implies that

$$\langle x^i, y^i\rangle + \alpha_i\lambda_i - h_i^+(y^i, \lambda_i) \leq \lambda_i[g_i(x^i) + \alpha_i],$$

with equality holding iff

$$\text{either } \lambda_i = 0 \text{ and } \langle x^i, y^i\rangle = \sup_{c^i\in C_i} \langle c^i, y^i\rangle, \qquad \text{or } \lambda_i > 0 \text{ and } y^i \in \lambda_i\, \partial g_i(x^i).$$

Moreover, given that $g_i(x^i)+\alpha_i \le 0$, if $(y^i, \lambda_i) \in D_i^+$, then the fact that $\lambda_i \ge 0$ implies that

$$\lambda_i[g_i(x^i)+\alpha_i] \le 0,$$

with equality holding iff

$$\lambda_i[g_i(x^i)+\alpha_i] = 0.$$

Taken together, these two inequalities and the corresponding characterizations of equality clearly imply the last equation of Sublemma 3.1.

Sublemma 3.2. Given that $g_j: C_j$ is convex and closed and given that $\kappa_j \ge 0$, the $\sup_{y^j \in D_j} [\langle x^j, y^j\rangle - \kappa_j h_j(y^j)]$ is finite iff $(x^j, \kappa_j) \in C_j^+$, in which case

$$\sup_{y^j \in D_j} [\langle x^j, y^j\rangle - \kappa_j h_j(y^j)] = g_j^+(x^j, \kappa_j)$$

and

$$\begin{aligned}\{y^j \in D_j \mid \langle x^j, y^j\rangle - \kappa_j h_j(y^j) = g_j^+(x^j, \kappa_j)\} \\ = \{y^j \in D_j \mid \text{either } \kappa_j = 0 \text{ and } \langle x^j, y^j\rangle = \sup_{d^j \in D_j} \langle x^j, d^j\rangle, \\ \text{or } \kappa_j > 0 \text{ and } y^j \in \partial g_j(x^j/\kappa_j)\}.\end{aligned}$$

Proof. First, observe that

$$\sup_{y^j \in D_j} [\langle x^j, y^j\rangle - \kappa_j h_j(y^j)] = \begin{cases} \sup_{y^j \in D_j} \langle x^j, y^j\rangle & \text{if } \kappa_j = 0, \\ \kappa_j g_j(x^j/\kappa_j) & \text{if } \kappa_j > 0 \text{ and } x^j \in \kappa_j C_j, \\ +\infty & \text{if } \kappa_j > 0 \text{ and } x^j \notin \kappa_j C_j. \end{cases}$$

Now, note that this expression is finite only if $(x^j, \kappa_j) \in C_j^+$, in which case this expression is clearly $g_j^+(x^j, \kappa_j)$.

Finally, given that $(x^j, \kappa_j) \in C_j^+$, if $y^j \in D_j$, then the geometric inequality established in Section 3.3.3 of Ref. 1 asserts that

$$\langle x^j, y^j\rangle - \kappa_j h_j(y^j) \le g_j^+(x^j, \kappa_j),$$

with equality holding iff

$$\text{either } \kappa_j = 0 \text{ and } \langle x^j, y^j\rangle = \sup_{d^j \in D_j} \langle x^j, d^j\rangle, \qquad \text{or } \kappa_j > 0 \text{ and } x^j \in \kappa_j\, \partial h_j(y^j).$$

This inequality and the corresponding characterization of equality imply the last equation of Sublemma 3.2, because the relation $x^j \in \kappa_j\, \partial h_j(y^j)$ is equivalent to the relation $y^j \in \partial g_j(x^j/\kappa_j)$ when $\kappa_j > 0$ (by virtue of conjugate transform theory).

The proof of Lemma 3.1 is now complete.

Lemma 3.2. A vector (y, λ) in D satisfies the cone condition $y \in Y$ and the constraints $h_j(y^j) \leq 0$, $j \in J$, iff $\inf_{\substack{x \in X \\ \kappa \geq 0}} L_g(x, \kappa; y, \lambda)$ is finite, in which case

$$\inf_{\substack{x \in X \\ \kappa \geq 0}} L_g(x, \kappa; y, \lambda) = -H(y, \lambda)$$

and

$$\{(x, \kappa) | x \in X, \kappa \geq 0; \text{ and } L_g(x, \kappa; y, \lambda) = -H(y, \lambda)\}$$
$$= \{(x, \kappa) | x \in X, \kappa \geq 0; 0 = \langle x, y \rangle; \text{ and } \kappa_j h_j(y^j) = 0, j \in J\}.$$

Proof. First, observe that

$$\inf_{\substack{x \in X \\ \kappa \geq 0}} L_g(x, \kappa; y, \lambda) = \inf_{\substack{x \in X \\ \kappa \geq 0}} [\langle x, y \rangle - H(y, \lambda) - \sum_J \kappa_j h_j(y^j)]$$

$$= \inf_{x \in X} \langle x, y \rangle + \inf_{\kappa \geq 0} \sum_J \kappa_j [-h_j(y^j)] - H(y, \lambda).$$

The proof of Lemma 3.2 is now immediate from the assumption that X is a cone and elementary considerations.

Now, assuming that (x^*, κ^*) is optimal for Problem 3.1 and that (y^*, λ^*) is a P-vector for Problem 3.1, we deduce from Lemma 3.1 that

$$\sup_{(y, \lambda) \in D} L_g(x^*, \kappa^*; y, \lambda) = G(x^*, \kappa^*) = \varphi = \inf_{\substack{x \in X \\ \kappa \geq 0}} L_g(x, \kappa; y^*, \lambda^*),$$

by virtue of the defining properties for optimal solutions and P-vectors. Notice that the first and second equations show that

$$L_g(x^*, \kappa^*; y^*, \lambda^*) \leq G(x^*, \kappa^*) = \varphi,$$

and observe that the second and third equations show that

$$G(x^*, \kappa^*) = \varphi \leq L_g(x^*, \kappa^*; y^*, \lambda^*);$$

so we infer that

$$L_g(x^*, \kappa^*; y^*, \lambda^*) = G(x^*, \kappa^*) = \varphi.$$

Consequently, $(x^*, \kappa^*; y^*, \lambda^*)$ is a saddle point for L_g.

Conversely, assuming that $(x^*, \kappa^*; y^*, \lambda^*)$ is a saddle point for L_g, we deduce from Lemma 3.1 that

$$(x^*, \kappa^*) \in C \qquad \text{and} \qquad g_i(x^{*i}) \leq 0, \qquad i \in I;$$

so (x^*, κ^*) is feasible by virtue of the hypotheses $x^* \in X$ and $\kappa^* \geq 0$. From

Lemma 3.1, we also infer that

$$\sup_{(y,\lambda)\in D} L_g(x^*, \kappa^*; y, \lambda) = G(x^*, \kappa^*);$$

so the saddle-point equations imply that

$$G(x^*, \kappa^*) = \inf_{\substack{x\in X\\ \kappa\geq 0}} L_g(x, \kappa; y^*, \lambda^*),$$

which in turn means that

$$G(x^*, \kappa^*) \leq L_g(x, \kappa; y^*, \lambda^*) \qquad \text{for each } x\in X \text{ and each } \kappa \geq 0.$$

Moreover, Lemma 3.1 also implies that

$$L_g(x, \kappa; y^*, \lambda^*) \leq G(x, \kappa) \qquad \text{for each } (x, \kappa)\in S,$$

by virtue of the hypothesis $(y^*, \lambda^*)\in D$. It then follows from these two displayed inequalities that

$$G(x^*, \kappa^*) \leq G(x, \kappa) \qquad \text{for each } (x, \kappa)\in S.$$

Thus, (x^*, κ^*) is optimal for Problem 3.1, and hence

$$\varphi = G(x^*, \kappa^*).$$

This equation and the preceding displayed equation show that (y^*, λ^*) is a P-vector for Problem 3.1, by virtue of the hypothesis $(y^*, \lambda^*)\in D$.

Now, assuming that

$$\sup_{(y,\lambda)\in D} L_g(x^*, \kappa^*; y, \lambda) = L_g(x^*, \kappa^*; y^*, \lambda^*),$$

we infer from Lemma 3.1 that

$$(x^*, \kappa^*)\in C \qquad \text{and} \qquad g_i(x^{*i}) \leq 0, \qquad i\in I,$$

and that

$$L_g(x^*, \kappa^*; y^*, \lambda^*) = G(x^*, \kappa^*).$$

This in turn implies that

$$y^{*0} \in \partial g_0(x^{*0}),$$

$$\text{either } \lambda_i^* = 0 \text{ and } \langle x^{*i}, y^{*i}\rangle = \sup_{c^i\in C_i} \langle c^i, y^{*i}\rangle,$$

$$\text{or } \lambda_i^* > 0 \text{ and } y^{*i} \in \lambda_i^*\, \partial g_i(x^{*i}),\ i\in I,$$

$$\lambda_i^* g_i(x^{*i}) = 0, \qquad i\in I,$$

$$\text{either } \kappa_j^* = 0 \text{ and } \langle x^{*j}, y^{*j}\rangle = \sup_{d^j\in D_j} \langle x^{*j}, d^j\rangle$$

$$\text{or } \kappa_j^* > 0 \text{ and } y^{*j} \in \partial g_j(x^{*j}/\kappa_j^*),\ j\in J,$$

by virtue of the hypothesis $(y^*, \lambda^*) \in D$ and Lemma 3.1. Conversely, assuming that

$$(x^*, \kappa^*) \in C \qquad \text{and } g_i(x^{*i}) \leq 0, \qquad i \in I,$$

and that (y^*, λ^*) satisfies the preceding displayed relations, we infer from Lemma 3.1 that

$$\sup_{(y,\lambda)\in D} L_g(x^*, \kappa^*; y, \lambda) = G(x^*, \kappa^*) = L_g(x^*, \kappa^*; y^*, \lambda^*).$$

Finally, assuming that

$$L_g(x^*, \kappa^*; y^*, \lambda^*) = \inf_{\substack{x\in X\\ \kappa\geq 0}} L_g(x, \kappa; y^*, \lambda^*),$$

we infer from Lemma 3.2 that

$$y^* \in Y \qquad \text{and} \qquad h_j(y^{*j}) \leq 0, \qquad j \in J,$$

and that

$$L_g(x^*, \kappa^*; y^*, \lambda^*) = -H(y^*, \lambda^*).$$

This in turn implies that

$$0 = \langle x^*, y^* \rangle,$$

$$\kappa_j^* h_j(y^{*j}) = 0, \qquad j \in J,$$

by virtue of the hypotheses $x^* \in X$ and $\kappa^* \geq 0$. Conversely, assuming that

$$y^* \in Y \qquad \text{and} \qquad h_j(y^{*j}) \leq 0, \qquad j \in J,$$

and that (x^*, κ^*) satisfies the preceding displayed relations, we infer from Lemma 3.2 that

$$\inf_{\substack{x\in X\\ \kappa\geq 0}} L_g(x, \kappa; y^*, \lambda^*) = -H(y^*, \lambda^*) = L_g(x^*, \kappa^*; y^*, \lambda^*). \qquad \square$$

Since the second assertion of Theorem 3.1 gives certain conditions that are equivalent to the first saddle-point equation, and since the third assertion of Theorem 3.1 gives certain other conditions that are equivalent to the second saddle-point equation, Theorem 3.1 actually provides four different characterizations of all ordered pairs $(x^*, \kappa^*; y^*, \lambda^*)$ of optimal solutions (x^*, κ^*) and P-vectors (y^*, λ^*).

Of course, each of those four characterizations provides a characterization of all optimal solutions (x^*, κ^*) in terms of a given P-vector (y^*, λ^*), as well as a charcterization of all P-vectors (y^*, λ^*) in terms of a given optimal solution (x^*, κ^*).

Still another characterization of all optimal solutions (x^*, κ^*) to certain Problems 3.1 has been given by the author (Ref. 3).

3.2. Duality. Corresponding to Problem 3.1 is the following *geometric dual problem.*

Problem 3.2. Using the feasible solution set

$$T \triangleq \{(y, \lambda) \in D \mid \inf_{\substack{x \in X \\ \kappa \geq 0}} L_g(x, \kappa; y, \lambda) \text{ is finite}\}$$

and the objective function

$$\mathcal{H}(y, \lambda) \triangleq \inf_{\substack{x \in X \\ \kappa \geq 0}} L_g(x, \kappa; y, \lambda),$$

calculate both the problem supremum

$$\Psi \triangleq \sup_{(y,\lambda) \in T} \mathcal{H}(y, \lambda)$$

and the optimal solution set

$$T^* \triangleq \{(y, \lambda) \in T \mid \mathcal{H}(y, \lambda) = \Psi\}.$$

Even though Problem 3.2 is essentially a *maximin problem*, a type of problem that tends to be relatively difficult to analyze, the minimization problems that must be solved to obtain the objective function $\mathcal{H}: T$ have trivial solutions. In particular, Lemma 3.2 clearly implies that

$$T = \{(y, \lambda) \in D \mid y \in Y, \text{ and } h_j(y^j) \leq 0, j \in J\} \qquad \text{and} \qquad \mathcal{H}(y, \lambda) = -H(y, \lambda),$$

so Problem 3.2 can actually be rephrased in the following more direct way.

Problem 3.2. Consider the objective function H whose domain is

$$D \triangleq \{(y, \lambda) \mid y^k \in D_k, k \in \{0\} \cup J, \text{ and } (y^i, \lambda_i) \in D_i^+, i \in I\}$$

and whose functional value is

$$H(y, \lambda) \triangleq h_0(y^0) + \sum_I h_i^+(y^i, \lambda_i),$$

where

$$D_i^+ \triangleq \{(y^i, \lambda_i) \mid \text{either } \lambda_i = 0 \text{ and } \sup_{c^i \in C_i} \langle c^i, y^i \rangle < +\infty, \text{ or } \lambda_i > 0 \text{ and } y^i \in \lambda_i D_i\}$$

and

$$h_i^+(y^i, \lambda_i) \triangleq \begin{cases} \sup_{c^i \in C_i} \langle c^i, y^i \rangle & \text{if } \lambda_i = 0 \text{ and } \sup_{c^i \in C_i} \langle c^i, y^i \rangle < +\infty, \\ \lambda_i h_i(y^i/\lambda_i) & \text{if } \lambda_i > 0 \text{ and } y^i \in \lambda_i D_i. \end{cases}$$

Using the feasible solution set

$$T \triangleq \{(y, \lambda) \in D \mid y \in Y, \text{ and } h_j(y^j) \le 0, j \in J\},$$

calculate both the problem infimum

$$\psi \triangleq \inf_{(y,\lambda)\in T} H(y,\lambda) = -\Psi$$

and the optimal solution set

$$T^* \triangleq \{(y, \lambda) \in T \mid H(y, \lambda) = \psi\}.$$

When phrased in this way, Problem 3.2 closely resembles Problem 3.1 and is in fact a geometric programming problem. Of course, the geometric dual Problem 3.2 can actually be defined in this way, but the preceding derivation serves as an important link between geometric Lagrangians and geometric duality (analogous to the link between ordinary Lagrangians and ordinary duality).

To further strengthen that link, we first need to develop the most basic duality theory, a theory in which the following definition is almost as important as the definition of the dual Problems 3.1 and 3.2.

Definition 3.3. The *extremality conditions* (for constrained geometric programming) are:

(I) $x \in X$ and $y \in Y$,

(II) $g_i(x^i) \le 0, i \in I$ and $h_j(y^j) \le 0, j \in J$,

(III) $0 = \langle x, y \rangle$,

(IV) $y^0 \in \partial g_0(x^0)$,

(V) either $\lambda_i = 0$ and $\langle x^i, y^i \rangle = \sup_{c^i \in C_i} \langle c^i, y^i \rangle$, or $\lambda_i > 0$ and $y^i \in \lambda_i \partial g_i(x^i), i \in I$,

(VI) either $\kappa_j = 0$ and $\langle x^j, y^j \rangle = \sup_{d^j \in D_j} \langle x^j, d^j \rangle$, or $\kappa_j > 0$ and $y^j \in \partial g_j(x^j/\kappa_j), j \in J$,

(VII) $\lambda_i g_i(x^i) = 0, i \in I$, and $\kappa_j h_j(y^j) = 0, j \in J$.

The following *duality theorem* is the basis for many important theorems.

Theorem 3.2. If (x, κ) and (y, λ) are feasible solutions to Problems 3.1 and 3.2, respectively [in which case the extremality conditions (I) and (II) are satisfied], then

$$0 \le G(x, \kappa) + H(y, \lambda),$$

with equality holding iff the extremality conditions (III) through (VII) are satisfied, in which case (x, κ) and (y, λ) are optimal solutions to Problems 3.1 and 3.2, respectively.

Proof. The following lemma will also be used in the proof of other theorems.

Lemma 3.3. If $(x, \kappa) \in C$ and $(y, \lambda) \in D$, then

$$\langle x, y\rangle \le G(x, \kappa) + \sum_I \lambda_i g_i(x^i) + H(y, \lambda) + \sum_J \kappa_j h_j(y^j),$$

with equality holding iff the extremality conditions (IV) through (VI) are satisfied. Moreover, if $h_j(y^j) \le 0$, $j \in J$ [i.e., the second part of extremality condition (II) is satisfied], then

$$G(x, \kappa) + \sum_I \lambda_i g_i(x^i) + H(y, \lambda) + \sum_J \kappa_j h_j(y^j) \le G(x, \kappa) + \sum_I \lambda_i g_i(x^i) + H(y, \lambda),$$

with equality holding iff the second part of extremality condition (VII) is satisfied. Furthermore, if $g_i(x^i) \le 0$, $i \in I$ [i.e., the first part of extremality condition (II) is satisfied], then

$$G(x, \kappa) + \sum_I \lambda_i g_i(x^i) + H(y, \lambda) \le G(x, \kappa) + H(y, \lambda),$$

with equality holding iff the first part of extremality condition (VII) is satisfied.

Proof. From the conjugate inequality presented in Section 3.1.2 of Ref. 1, we know that

$$\langle x^0, y^0\rangle \le g_0(x^0) + h_0(y^0),$$

with equality holding iff the extremality condition (IV) is satisfied. From the geometric inequality established in Section 3.3.3 of Ref. 1, we know that

$$\langle x^i, y^i\rangle \le \lambda_i g_i(x^i) + h_i^+(y^i, \lambda_i),$$

with equality holding iff the extremality condition (V) is satisfied. Likewise, we know that

$$\langle x^j, y^j\rangle \le g_j^+(x^j, \kappa_j) + \kappa_j h_j(y^j),$$

with equality holding iff the extremality condition (VI) is satisfied. Adding all $1 + o(I) + o(J)$ of these inequalities and taking account of the defining equations for x, y, G, H proves the first assertion. The second assertion is an immediate consequence of the fact that $\kappa_j \geq 0$ when $(x^j, \kappa_j) \in C_j^+$, $j \in J$; and the third assertion is an immediate consequence of the fact that $\lambda_i \geq 0$ when $(y^i, \lambda_i) \in D_i^+$, $i \in I$. This concludes the proof of Lemma 3.3.

Now, the fact that x and y are in the cone X and its dual Y, respectively, combined with a sequential application of all three assertions of Lemma 3.3 shows that

$$
\begin{aligned}
0 \leq \langle x, y \rangle &\leq G(x, \kappa) + \sum_I \lambda_i g_i(x^i) + H(y, \lambda) + \sum_J \kappa_j h_j(y^j) \\
&\leq G(x, \kappa) + \sum_I \lambda_i g_i(x^i) + H(y, \lambda) \\
&\leq G(x, \kappa) + H(y, \lambda),
\end{aligned}
$$

with equality holding in all four inequalities iff the equality conditions stated in the theorem are satisfied. □

The following important corollary is an immediate consequence of Theorem 3.2.

Corollary 3.1. If the dual Problems 3.1 and 3.2 are both consistent, then

(i) the infimum φ for Problem 3.1 is finite, and

$$0 \leq \varphi + H(y, \lambda) \qquad \text{for each } (y, \lambda) \in T,$$

(ii) the infimum ψ for Problem 3.2 is finite, and

$$0 \leq \varphi + \psi.$$

The strictness of the inequality in conclusion (ii) plays a crucial role in almost all duality theorems.

Definition 3.4. Consistent dual Problems 3.1 and 3.2, for which

$$0 < \varphi + \psi,$$

have a *duality gap* of $\varphi + \psi$.

A much more thorough discussion of duality theory and the role played by duality gaps is given in Ref. 1 and some of the references cited therein.

The link between geometric Lagrangians and geometric duality can now be further strengthened by the following tie between dual Problem 3.2 and the P-vectors for Problem 3.1 defined in Section 3.1.

Theorem 3.3. Given that Problem 3.1 is consistent with a finite infimum φ,

(i) if Problem 3.1 has a P-vector, then Problem 3.2 is consistent and

$$0=\varphi+\psi,$$

(ii) if Problem 3.2 is consistent and $0=\varphi+\psi$, then

$$\{(y^*,\lambda^*)|(y^*,\lambda^*) \text{ is a } P\text{-vector for Problem 3.1}\}=T^*.$$

Proof. If (y^*,λ^*) is a P-vector for Problem 3.1, then Lemma 3.2 implies that (y^*,λ^*) is feasible for Problem 3.2 and that

$$\varphi=-H(y^*,\lambda^*),$$

which in turn implies that

$$\psi=H(y^*,\lambda^*)$$

by virtue of conclusion (i) to Corollary 3.1. Consequently,

$$0=\varphi+\psi \qquad \text{and} \qquad (y^*,\lambda^*)\in T^*.$$

On the other hand, if Problem 3.2 is consistent and $0=\varphi+\psi$, then each vector (y^*,λ^*) in T^* has the property that

$$\varphi=-H(y^*,\lambda^*),$$

and hence each such vector (y^*,λ^*) is a P-vector for Problem 3.1 by virtue of the first two paragraphs of this subsection. □

An important consequence of Theorem 3.3 is that, when they exist, all P-vectors for Problem 3.1 can be obtained simply by computing the dual optimal solution set T^*. However, there are cases in which the vectors in T^* are not P-vectors for Problem 3.1, though such cases can occur only when $0<\varphi+\psi$, in which event Theorem 3.3 implies that there can be no P-vectors for Problem 3.1.

The following theorem provides an important tie between dual Problem 3.2 and the Kuhn–Tucker vectors for Problem 3.1 defined in Section 3.1.

Theorem 3.4. Given that Problems 3.1 and 3.2 are both consistent and that $0=\varphi+\psi$, if there is a minimizing sequence $\{(y^q,\lambda^q)\}_1^\infty$ for Problem 3.2 [i.e., $(y^q,\lambda^q)\in T$ and $\lim_{q\to+\infty} H(y^q,\lambda^q)=\psi$] such that $\lim_{q\to+\infty}\lambda^q$ exists and is finite, then $\lambda^*\triangleq\lim_{q\to+\infty}\lambda^q$ is a Kuhn–Tucker vector for Problem 3.1.

Proof. The feasibility of (y^q,λ^q) implies that $\lambda^q\geq 0$ for each q, so

$$\lambda_i^*\geq 0, \qquad i\in I.$$

Now, the feasibility of (y^q, λ^q) and a sequential application of the first two assertions of Lemma 3.3 show that

$$\langle x, y^q\rangle \le G(x, \kappa)+\sum_I \lambda_i^q g_i(x^i)+H(y^q, \lambda^q)+\sum_J \kappa_j h_j(y^{qj})$$

$$\le G(x, \kappa)+\sum_I \lambda_i^q g_i(x^i)+H(y^q, \lambda^q)$$

for each $(x, \kappa)\in C$ and for each q. Consequently, for each $(x, \kappa)\in C$ such that $x\in X$, we deduce that

$$0\le G(x, \kappa)+\sum_I \lambda_i^q g_i(x^i)+H(y^q, \lambda^q)$$

for each q, because $y^q \in Y$. This inequality and the hypotheses

$$\lim_{q\to+\infty} H(y^q, \lambda^q)=\psi \qquad \text{and} \qquad \lim_{q\to+\infty} \lambda^q=\lambda^*$$

clearly imply that

$$0\le G(x, \kappa)+\sum_I \lambda_i^* g_i(x^i)+\psi$$

for each $(x, \kappa)\in C$ such that $x\in X$. Using the fact that

$$L_0(x, \kappa; \lambda^*)\triangleq G(x, \kappa)+\sum_I \lambda_i^* g_i(x^i)$$

and the hypothesis $0=\varphi+\psi$, we infer from the preceding inequality that

$$\varphi\le \inf_{\substack{(x,\kappa)\in C\\ x\in X}} L_0(x, \kappa; \lambda^*).$$

Now, choose a minimizing sequence $\{(x^q, \kappa^q)\}_1^\infty$ for Problem 3.1, and then observe for each q that

$$L_0(x^q, \kappa^q; \lambda^*)\le G(x^q, \kappa^q),$$

because $\lambda_i^*\ge 0$ and $g_i(x^{qi})\le 0$, $i\in I$. From the construction of $\{(x^q, \kappa^q)\}_1^\infty$, we know that $(x^q, \kappa^q)\in C$ and $x^q\in X$ for each q and that

$$\varphi=\lim_{q\to+\infty} G(x^q, \kappa^q);$$

so we conclude from the preceding two displayed inequalities that

$$\varphi=\inf_{\substack{(x,\kappa)\in C\\ x\in X}} L_0(x, \kappa; \lambda^*). \qquad \square$$

The following corollary ties the dual optimal solution set T^* directly to Kuhn–Tucker vectors for Problem 3.1.

Corollary 3.2. Given that Problems 3.1 and 3.2 are both consistent and that $0 = \varphi + \psi$, each $(y^*, \lambda^*) \in T^*$ provides a Kuhn–Tucker vector λ^* for Problem 3.1.

The following corollary ties the set of all P-vectors (y^*, λ^*) for Problem 3.1 directly to the set of all Kuhn–Tucker vectors λ^* for Problem 3.1.

Corollary 3.3. Given that Problem 3.1 is consistent with a finite infimum φ, each P-vector (y^*, λ^*) for Problem 3.1 provides a Kuhn–Tucker vector λ^* for Problem 3.1.

Proof. Use Theorem 3.3 along with Corollary 3.2.

Finally, it is worth noting that Theorem 3.4 and its two corollaries also provide certain connections between dual Problem 3.2 and the *ordinary dual problem* corresponding to Problem 3.1. Those connections are left to the reader's imagination while more subtle connections are given in Ref. 4.

References

1. PETERSON, E. L., *Geometric Programming*, SIAM Review, Vol. 18, pp. 1–51, 1976.
2. ROCKAFELLAR, R. T., *Convex Analysis*, Princeton University Press, Princeton, New Jersey, 1970.
3. PETERSON, E. L., *Optimality Conditions in Generalized Geometric Programming*, Journal of Optimization Theory and Applications, Vol. 26, No. 1, 1978.
4. PETERSON, E. L., *Geometric Duality vis-a-vis Ordinary Duality* (to appear).

5

Constrained Duality via Unconstrained Duality in Generalized Geometric Programming[1]

E. L. PETERSON[2]

Abstract. A specialization of unconstrained duality (involving problems without explicit constraints) to constrained duality (involving problems with explicit constraints) provides an efficient mechanism for extending to the latter many important theorems that were previously established for the former.

1. Introduction

Although the implications of our specialization have already been discussed in the author's recent survey paper (Ref. 1), a proof of it is being given here for the first time. This proof depends only on rather elementary properties of the conjugate transformation reviewed in Section 3.1.2 of Ref. 1.

2. Constrained and Unconstrained Duality

In harmony with Ref. 1, we introduce two nonintersecting (possibly empty) positive-integer index sets I and J, with finite cardinality $o(I)$ and $o(J)$, respectively. In terms of these index sets I and J, we also introduce the following notation and hypotheses.

[1] This research was sponsored by the Air Force Office of Scientific Research, Air Force Systems Command, USAF, under Grant No. AFOSR-73-2516.

[2] Professor, Department of Engineering Sciences and Applied Mathematics, Department of Industrial Engineering and Management Sciences, and Department of Mathematics, Northwestern University, Evanston, Illinois.

(i) For each $k \in \{0\} \cup I \cup J$, suppose that $g_k : C_k$ is a (proper) function with a nonempty (effective) domain $C_k \subseteq E_{n_k}$; and, for each $j \in J$, let D_j be the (effective) domain of the conjugate transform $h_j : D_j$ of $g_j : C_j$.

(ii) For each $k \in \{0\} \cup I \cup J$, let x^k be an independent vector variable in E_{n_k}, and let κ be an independent vector variable with components κ_j for each $j \in J$.

(iii) Denote the Cartesian product of the vector variables x^i, $i \in I$, by the symbol x^I, and denote the Cartesian product of the vector variables x^j, $j \in J$, by the symbol x^J. Then, the Cartesian product $(x^0, x^I, x^J) \triangleq x$ is a vector variable in E_n, where

$$n \triangleq n_0 + \sum_I n_i + \sum_J n_j.$$

(iv) Assume that X is a nonempty cone in E_n.

For purposes of easy reference and mathematical precision, the resulting *geometric programming problem* is now given the following formal definition in terms of classical terminology and notation.

Problem 2.1. Consider the objective function G whose domain is

$$C \triangleq \{(x, \kappa) \mid x^k \in C_k,\ k \in \{0\} \cup I, \text{ and } (x^j, \kappa_j) \in C_j^+,\ j \in J\}$$

and whose functional value is

$$G(x, \kappa) \triangleq g_0(x^0) + \sum_J g_j^+(x^j, \kappa_j),$$

where

$$C_j^+ \triangleq \{(x^j, \kappa_j) \mid \text{either } \kappa_j = 0 \text{ and } \sup_{d^j \in D_j} \langle x^j, d^j \rangle < +\infty, \text{ or } \kappa_j > 0 \text{ and } x^j \in \kappa_j C_j\}$$

and

$$g_j^+(x^j, \kappa_j) \triangleq \begin{cases} \sup\limits_{d^j \in D_j} \langle x^j, d^j \rangle & \text{if } \kappa_j = 0 \text{ and } \sup\limits_{d^j \in D_j} \langle x^j, d^j \rangle < +\infty, \\ \kappa_j g_j(x^j/\kappa_j) & \text{if } \kappa_j > 0 \text{ and } x^j \in \kappa_j C_j. \end{cases}$$

Using the feasible solution set

$$S \triangleq \{(x, \kappa) \in C \mid x \in X, \text{ and } g_i(x^i) \leq 0,\ i \in I\},$$

calculate both the problem infimum

$$\varphi \triangleq \inf_{(x, \kappa) \in S} G(x, \kappa)$$

and the optimal solution set

$$S^* \triangleq \{(x, \kappa) \in S \mid G(x, \kappa) = \varphi\}.$$

The practical significance of this problem is illustrated in Ref. 1 and some of the references cited therein.

Closely related to Problem 2.1 is its geometric dual, Problem 2.2. To obtain Problem 2.2 from Problem 2.1, we need the following additional notation and hypotheses.

(v) For each $k \in \{0\} \cup I \cup J$, suppose that the function $g_k : C_k$ has a conjugate transform $h_k : D_k$ with a nonempty (effective) domain $D_k \subseteq E_{n_k}$ (which is always the case when $g_k : C_k$ is convex).

(vi) For each $k \in \{0\} \cup I \cup J$, let y^k be an independent vector variable in E_{n_k}, and let λ be an independent vector variable with components λ_i for each $i \in I$.

(vii) Denote the Cartesian product of the vector variables y^i, $i \in I$, by the symbol y^I, and denote the Cartesian product of the vector variables y^j, $j \in J$, by the symbol y^J. Then, the Cartesian product $(y^0, y^I, y^J) \triangleq y$ is a vector variable in E_n, where

$$n = n_0 + \sum_I n_i + \sum_J n_j.$$

(viii) Assume that the cone X has a nonempty dual Y in E_n (which is always the case).

For purposes of easy reference and mathematical precision, the resulting *geometric dual problem* is now given the following formal definition in terms of classical terminology and notation.

Problem 2.2. Consider the objective function H whose domain is

$$D \triangleq \{(y, \lambda) \mid y^k \in D_k,\ k \in \{0\} \cup J, \text{ and } (y^i, \lambda_i) \in D_i^+,\ i \in I\}$$

and whose functional value is

$$H(y, \lambda) \triangleq h_0(y^0) + \sum_I h_i^+(y^i, \lambda_i),$$

where

$$D_i^+ \triangleq \{(y^i, \lambda_i) \mid \text{either } \lambda_i = 0 \text{ and } \sup_{c^i \in C_i} \langle y^i, c^i \rangle < +\infty, \text{ or } \lambda_i > 0 \text{ and } y^i \in \lambda_i D_i\}$$

and

$$h_i^+(y^i, \lambda_i) \triangleq \begin{cases} \sup_{c^i \in C_i} \langle y^i, c^i \rangle & \text{if } \lambda_i = 0 \text{ and } \sup_{c^i \in C_i} \langle y^i, c^i \rangle < +\infty, \\ \lambda_i h_i(y^i/\lambda_i) & \text{if } \lambda_i > 0 \text{ and } y^i \in \lambda_i D_i. \end{cases}$$

Using the feasible solution set

$$T \triangleq \{(y, \lambda) \in D \mid y \in Y, \text{ and } h_j(y^j) \le 0,\ j \in J\},$$

calculate both the problem infimum

$$\psi \triangleq \inf_{(y,\lambda)\in T} H(y,\lambda)$$

and the optimal solution set

$$T^* \triangleq \{(y,\lambda)\in T \mid H(y,\lambda)=\psi\}.$$

Problems 2.1 and 2.2 are, of course, termed *geometric dual problems.* When $g_k : C_k$, $k \in \{0\}\cup I \cup J$ and X are convex and closed, this duality is clearly symmetric, in that Problem 2.1 can then be constructed from Problem 2.2 in the same way that Problem 2.2 has just been constructed from Problem 2.1. Actually, this duality is the only completely symmetric duality that is presently known for general (closed) convex programming with explicit constraints. Moreover, Ref. 1 and some of the references cited therein show that all other duality in mathematical programming can be viewed as a special case.

The preceding constrained duality can of course be specialized to unconstrained duality, simply by letting both index sets I and J be empty. In doing so, we drop the (now unnecessary) subscript 0 from the symbols $g_0 : C_0$ and $h_0 : D_0$; and we also replace all remaining symbols with their script counterparts in order to avoid ambiguous notation when reversing this specialization.

We suppose then that $\mathscr{g} : \mathscr{C}$ is a (proper) function $\mathscr{g}$ with a nonempty (effective) domain $\mathscr{C} \subseteq E_n$, and we assume that $\mathscr{X}$ is a nonempty cone in E_n. The geometric programming problem being considered can now be given the following concise definition.

Problem 2.3. Using the feasible solution set

$$\mathscr{S} \triangleq \mathscr{X} \cap \mathscr{C},$$

calculate both the problem infimum

$$\varphi \triangleq \inf_{x\in\mathscr{S}} \mathscr{g}(x)$$

and the optimal solution set

$$\mathscr{S}^* \triangleq \{x \in \mathscr{S} \mid \mathscr{g}(x) = \varphi\}.$$

If the function $\mathscr{g} : \mathscr{C}$ has a conjugate transform $\mathscr{h} : \mathscr{D}$, then the geometric dual problem clearly exists and is of course defined in terms of $\mathscr{h} : \mathscr{D}$ and the dual $\mathscr{Y}$ of the cone $\mathscr{X}$. In particular then, the geometric dual problem can now be given the following concise definition.

Problem 2.4. Using the feasible solution set

$$\mathcal{T} \triangleq \mathcal{Y} \cap \mathcal{D},$$

calculate both the problem infimum

$$\psi \triangleq \inf_{y \in \mathcal{T}} h(y)$$

and the optimal solution set

$$\mathcal{T}^* \triangleq \{y \in \mathcal{T} \mid h(y) = \psi\}.$$

In contrast with constrained duality, notice the innate simplicity of unconstrained duality. That simplicity was a great aid in uncovering most of the theorems described in Section 3.1 of Ref. 1. The somewhat surprising fact is that such theorems can actually be applied to (the seemingly more general) constrained duality, and hence can be exploited in establishing most of the theorems described in Section 3.3 of Ref. 1. The mechanism for doing so is the following specialization of unconstrained duality.

3. Key Specialization

Introducing an additional independent vector variable α with components α_i for each $i \in I$, we let the functional domain be

$$\mathcal{C} \triangleq \{(x^0, x^I, \alpha, x^J, \kappa) \in E_n \mid x^0 \in C_0;\ x^i \in C_i,\ \alpha_i \in E_1, \text{ and } g_i(x^i) + \alpha_i \le 0,\ i \in I;\ (x^j, \kappa_j) \in C_j^+,\ j \in J\},$$

and we let the functional value be

$$g(x^0, x^I, \alpha, x^J, \kappa) \triangleq g_0(x^0) + \sum_J g_j^+(x^j, \kappa_j) \triangleq G(x, \kappa).$$

We also define the cone

$$\mathcal{X} \triangleq \{(x^0, x^I, \alpha, x^J, \kappa) \in E_n \mid (x^0, x^I, x^J) \in X;\ \alpha = 0;\ \kappa \in E_{o(J)}\}.$$

Then, Problem 2.3 is clearly identical to Problem 2.1, and the crucial question now is whether Problem 2.4 is identical to Problem 2.2 (a question whose answer turns out to be negative if the seemingly inessential vector variable α is omitted from the definitions of the given function $g : \mathcal{C}$ and the given cone $\mathcal{X}$). To obtain the answer, we need to compute both the conjugate transform $h : \mathcal{D}$ of the given function $g : \mathcal{C}$ and the dual $\mathcal{Y}$ of the given cone $\mathcal{X}$.

To compute $\mathscr{k}:\mathscr{D}$, first note that

$$\mathscr{k}(y^0, y^I, \lambda, y^J, \beta) = \sup_{(x^0,x^I,\alpha,x^J,\kappa)\in\mathscr{C}} \{\langle y^0, x^0\rangle + \sum_I \langle y^i, x^i\rangle + \sum_J \langle y^j, x^j\rangle$$

$$+\sum_I \lambda_i\alpha_i + \sum_J \beta_j\kappa_j - g_0(x^0) - \sum_J g_j^+(x^j, \kappa_j)\},$$

which is clearly finite only if $\lambda_i \geq 0$, $i \in I$, in which case we readily see that this expression becomes

$$\mathscr{k}(y^0, y^I, \lambda, y^J, \beta) = \sup_{x^0\in C_0} [\langle y^0, x^0\rangle - g_0(x^0)] + \sum_I \sup_{x^i\in C_i} [\langle y^i, x^i\rangle - \lambda_i g_i(x^i)]$$

$$+\sum_J \sup_{(x^j,\kappa_j)\in C_j^+} [\langle y^j, x^j\rangle + \beta_j\kappa_j - g_j^+(x^j, \kappa_j)].$$

Consequently, $(y^0, y^I, \lambda, y^J, \beta)\in\mathscr{D}$ iff both $\lambda_i \geq 0$, $i \in I$, and each term on the right-hand side of the preceding equation is finite. Of course, the first term is finite iff $y^0 \in D_0$, in which case the first term is equal to $h_0(y^0)$. The finiteness of the remaining terms can be conveniently characterized with two lemmas.

The following lemma characterizes the finiteness of the terms involving the index set I.

Lemma 3.1. Given that $\lambda_i \geq 0$, $\sup_{x^i\in C_i} [\langle y^i, x^i\rangle - \lambda_i g_i(x^i)]$ is finite iff $(y^i, \lambda_i) \in D_i^+$, in which case

$$\sup_{x^i\in C_i} [\langle y^i, x^i\rangle - \lambda_i g_i(x^i)] = h_i^+(y^i, \lambda_i).$$

Proof. Simply observe that

$$\sup_{x^i\in C_i} [\langle y^i, x^i\rangle - \lambda_i g_i(x^i)] = \begin{cases} \sup_{x^i\in C_i} \langle y^i, x^i\rangle & \text{if } \lambda_i = 0, \\ \lambda_i h_i(y^i/\lambda_i) & \text{if } \lambda_i > 0 \text{ and } y^i \in \lambda_i D_i, \\ +\infty & \text{if } \lambda_i > 0 \text{ and } y^i \notin \lambda_i D_i, \end{cases}$$

and then use the defining formula for $h_i^+ : D_i^+$. □

The next lemma characterizes the finiteness of the terms involving the index set J.

Lemma 3.2. $\sup_{(x^j,\kappa_j)\in C_j^+} [\langle y^j, x^j\rangle + \beta_j\kappa_j - g_j^+(x^j, \kappa_j)]$ is finite iff both $y^j \in D_j$ and $h_j(y^j) + \beta_j \leq 0$, in which case

$$\sup_{(x^j,\kappa_j)\in C_j^+} [\langle y^j, x^j\rangle + \beta_j\kappa_j - g_j^+(x^j, \kappa_j)] = 0.$$

Proof. First, observe that

$$\sup_{(x^j, \kappa_j) \in C_j^+} [\langle y^j, x^j \rangle + \beta_j \kappa_j - g_j^+(x^j, \kappa_j)]$$

$$= \sup_{\kappa_j \geq 0} [\sup_{x^j} \{\langle y^j, x^j \rangle + \beta_j \kappa_j - g_j^+(x^j, \kappa_j) \mid (x^j, \kappa_j) \in C_j^+\}]$$

$$= \sup_{\kappa_j \geq 0} [\beta_j \kappa_j + \sup_{x^j} \{\langle y^j, x^j \rangle - g_j^+(x^j, \kappa_j) \mid (x^j, \kappa_j) \in C_j^+\}]$$

$$= \sup_{\kappa_j \geq 0} \left\{ \beta_j \kappa_j + \begin{bmatrix} \sup_{x^j} \{\langle y^j, x^j \rangle - \sup_{d^j \in D_j} \langle x^j, d^j \rangle \mid \sup_{d^j \in D_j} \langle x^j, d^j \rangle < +\infty\} & \text{if } \kappa_j = 0 \\ \sup_{x^j} \{\langle y^j, x^j \rangle - \kappa_j g_j(x^j/\kappa_j) \mid x^j/\kappa_j \in C_j\} & \text{if } \kappa_j > 0 \end{bmatrix} \right\}$$

$$= \sup_{\kappa_j \geq 0} \left\{ \beta_j \kappa_j + \begin{bmatrix} 0 & \text{if } \kappa_j = 0 \text{ and } y^j \in \bar{D}_j \\ +\infty & \text{if } \kappa_j = 0 \text{ and } y^j \notin \bar{D}_j \\ +\infty & \text{if } \kappa_j > 0 \text{ and } y^j \notin D_j \\ \kappa_j h_j(y^j) & \text{if } \kappa_j > 0 \text{ and } y^j \in D_j \end{bmatrix} \right\},$$

where the final step makes use of the fact that the zero function with domain $\bar{D}_j$ (the topological closure of D_j) is the conjugate transform of the conjugate transform of the zero function with domain D_j. Now, note that the last expression is finite only if $y^j \in D_j$, in which case the last expression clearly equals

$$\sup_{\kappa_j \geq 0} [\beta_j \kappa_j + \kappa_j h_j(y^j)].$$

But this expression is obviously finite iff

$$h_j(y^j) + \beta_j \leq 0,$$

in which case this expression is clearly zero. □

We have now shown that the functional domain is

$$\mathscr{D} = \{(y^0, y^I, \lambda, y^J, \beta) \in E_n \mid y^0 \in D_0;\ (y^i, \lambda_i) \in D_i^+, i \in I;\ y^j \in D_j,$$
$$\beta_j \in E_1, \text{ and } h_j(y^j) + \beta_j \leq 0, j \in J\},$$

and we have also shown that the functional value is

$$\mathscr{h}(y^0, y^I, \lambda, y^J, \beta) = h_0(y^0) + \sum_I h_i^+(y^i, \lambda_i) \triangleq H(y, \lambda).$$

Moreover, elementary considerations show that

$$\mathscr{Y} = \{(y^0, y^I, \lambda, y^J, \beta) \in E_n \mid (y^0, y^I, y^J) \in Y;\ \lambda \in E_{o(I)};\ \beta = 0\}.$$

Therefore, Problem 2.4 is clearly identical to Problem 2.2, and hence constrained duality can be viewed as a special case of unconstrained duality.

The mechanism for obtaining theorems concerning constrained duality from comparable theorems concerning unconstrained duality is now twofold: (i) using the definition of the function $\mathscr{g}:\mathscr{C}$ in terms of the functions $g_0:C_0$, $g_i:C_i$, and $g_j^+:C_j^+$, find hypotheses about $g_0:C_0$, $g_i:C_i$, and $g_j^+:C_j^+$ that imply the required hypotheses about $\mathscr{g}:\mathscr{C}$; and (ii) using the definition of the cone $\mathscr{X}$ in terms of the cone X, find hypotheses about X that imply the required hypotheses about $\mathscr{X}$.

An inspection of the definitions of $\mathscr{g}:\mathscr{C}$ and $\mathscr{X}$ indicates that a knowledge of the algebraic and topological properties of Cartesian products is one major prerequisite for exploiting this mechanism. The only other major prerequisite for doing so is of course a knowledge of the theorems concerning unconstrained duality.

As previously mentioned, this mechanism has already been exploited in obtaining most of the theorems given in Section 3.3 of Ref. 1 from comparable theorems given in Section 3.1 of Ref. 1.

Reference

1. PETERSON, E. L., *Geometric Programming*, SIAM Review, Vol. 18, pp. 1–51, 1976.

6

Fenchel's Duality Theorem in Generalized Geometric Programming[1]

E. L. PETERSON[2]

Abstract. Fenchel's duality theorem is extended to generalized geometric programming with explicit constraints—an extension that also generalizes and strengthens Slater's version of the Kuhn–Tucker theorem.

1. Introduction

Although many implications of this extension have already been discussed in the author's recent survey paper (Ref. 1), a proof of it is given here for the first time.

This proof utilizes the unconstrained version that has already been established by independent and somewhat different arguments in Refs. 2 and 3. In doing so, it exploits the main result from Ref. 4 and also requires some of the convexity theory in Ref. 3—especially the theory having to do with the *relative interior* (ri S) of an arbitrary convex set $S \subseteq E_N$ (N-dimensional Euclidean space).

2. Main Result

In harmony with Ref. 1, we use the notation reviewed in Section 2 of Ref. 4 (the preceding paper).

---

[1] This research was sponsored by the Air Force Office of Scientific Research, Air Force Systems Command, USAF, under Grant No. AFOSR-73-2516.

[2] Professor, Department of Engineering Sciences and Applied Mathematics, Department of Industrial Engineering and Management Sciences, and Department of Mathematics, Northwestern University, Evanston, Illinois.

Our main result can be stated in the context of (geometric) dual Problems 2.1 and 2.2 as follows.

Theorem 2.1. If (i) Problem 2.2 has a feasible solution (y', λ') such that

$$h_j(y'^j) < 0, \qquad j \in J,$$

(ii) Problem 2.2 has a finite infimum ψ, and (iii) there exists a vector (y^+, λ^+) such that

$$y^+ \in (\text{ri } Y),$$
$$y^{+k} \in (\text{ri } D_k), \qquad k \in \{0\} \cup J,$$
$$(y^{+i}, \lambda_i^+) \in (\text{ri } D_i^+), \qquad i \in I,$$

then (I) Problem 2.1 has both a nonempty feasible solution set S and a finite infimum φ, and

$$0 = \varphi + \psi,$$

(II) Problem 2.1 has a nonempty optimal solution set S^*.

The unconstrained version of this theorem (i.e., the version in which $I = J = \emptyset$) is essentially Fenchel's duality theorem—a theorem that is stated and proved as Theorem 31.4 on page 335 of Ref. 3. Stated in the context of (geometric) dual Problems 2.1 and 2.2, the unconstrained version is as follows.

Theorem 2.2. If Problem 2.4 has both a feasible solution $\mathscr{y}^0 \in (\text{ri } \mathscr{Y}) \cap (\text{ri } \mathscr{D})$ and a finite infimum ψ, then (I) Problem 2.3 has both a nonempty feasible solution set $\mathscr{S}$ and a finite infimum φ, and

$$0 = \varphi + \psi,$$

(II) Problem 2.3 has a nonempty optimal solution set $\mathscr{S}^*$.

Our proof of Theorem 2.1 utilizes Theorem 2.2 and the main result from Ref. 4 (the preceding paper). The main result from Ref. 4 is that dual Problems 2.1 and 2.2 can be viewed as a special case of dual Problems 2.3 and 2.4.

In particular, let

$$\mathscr{X} \triangleq \{(x^0, x^I, \alpha, x^J, \kappa) \in E_n \mid (x^0, x^I, x^J) \in X;\ \alpha = 0;\ \kappa \in E_{o(J)}\}.$$

Also, let

$$\mathscr{C} \triangleq \{(x^0, x^I, \alpha, x^J, \kappa) \in E_n \mid x^0 \in C_0;\ x^i \in C_i,\ \alpha_i \in E_1, \text{ and } g_i(x^i) + \alpha_i \leq 0,$$
$$i \in I;\ (x^j, \kappa_j) \in C_j^+,\ j \in J\},$$

and let

$$\mathscr{g}(x^0, x^I, \alpha, x^J, \kappa) \triangleq g_0(x^0) + \sum_J g_j^+(x^j, \kappa_j).$$

Then, Section 3 of Ref. 4 shows that

$$\mathscr{Y} = \{(y^0, y^I, \lambda, y^J, \beta) \in E_n \mid (y^0, y^I, y^J) \in Y; \beta = 0; \lambda \in E_{o(I)}\}.$$

Section 3 of Ref. 4 also shows that

$$\mathscr{D} = \{(y^0, y^I, \lambda, y^J, \beta) \in E_n \mid y^0 \in D_0; (y^i, \lambda_i) \in D_i^+, i \in I; y^j \in D_j,$$
$$\beta_j \in E_1, \text{ and } h_j(y^j) + \beta_j \leq 0, j \in J\},$$

and that

$$\mathscr{h}(y^0, y^I, \lambda, y^J, \beta) = h_0(y^0) + \sum_I h_i^+(y^i, \lambda_i).$$

To prove Theorem 2.1, we obviously need only show that the Fenchel hypothesis in Theorem 2.2 [i.e., the hypothesis that there exists a vector $\mathscr{y}^0 \in (\text{ri } \mathscr{Y}) \cap (\text{ri } \mathscr{D})$] is equivalent to hypotheses (i) and (iii) in Theorem 2.1.

Toward that end, we first use the formulas for $\mathscr{Y}$ and $\mathscr{D}$ to derive comparable formulas for (ri $\mathscr{Y}$) and (ri $\mathscr{D}$)—two derivations that make crucial use of the following basic facts:

(A) (ri U) = U when U is a vector space,

(B) $(\text{ri } V) = \bigtimes_1^\eta (\text{ri } V_k)$ when $V = \bigtimes_1^\eta V_k$ and the sets V_k are convex,

(C) (ri W) = (int W), the *interior* of W, when W is a convex set with the same *dimension* as the space in which it is embedded.

Fact (A) is established on page 44 of Ref. 3; Fact (B) can be obtained inductively from the formula at the top of page 49 of Ref. 3; and Fact (C) is explained on page 44 of Ref. 3.

Now, the formula for $\mathscr{Y}$ along with Facts (A) and (B) implies that

$$(\text{ri } \mathscr{Y}) = \{(y^0, y^I, \lambda, y^J, \beta) \in E_n \mid (y^0, y^I, y^J) \in (\text{ri } Y); \lambda \in E_{o(I)}; \beta = 0\}.$$

Moreover, the formula for $\mathscr{D}$ along with Facts (A) and (B) implies that

$$(\text{ri } \mathscr{D}) = \{(y^0, y^I, \lambda, y^J, \beta) \in E_n \mid y^0 \in (\text{ri } D_0); \lambda_i > 0 \text{ and } y^i \in \lambda_i(\text{ri } D_i), i \in I;$$
$$y^j \in (\text{ri } D_j), \beta_j \in E_1, \text{ and } h_j(y^j) + \beta_j < 0, j \in J\},$$

by virtue of both the equation

$$(\text{ri } D_i^+) = \{(y^i, \lambda_i) \mid \lambda_i > 0 \text{ and } y^i \in \lambda_i(\text{ri } D_i)\}$$

and the equation

$$(\text{ri } \{(y^j, \beta_j) \mid y^j \in D_j \text{ and } h_j(y^j) + \beta_j \leq 0\})$$
$$= \{(y^j, \beta_j) \mid \beta_j \in E_1, y^j \in (\text{ri } D_j), \text{ and } h_j(y^j) + \beta_j < 0\}.$$

To derive the latter equation, simply use Theorem 6.8 on page 49 of Ref. 3 along with Fact (C). To derive the former equation, first consider the point-to-set mapping $Y_i^+ : \Lambda_i^+$, where

$$Y_i^+[\lambda_i] \triangleq \{y^i \mid (y^i, \lambda_i) \in D_i^+\},$$
$$\Lambda_i^+ \triangleq \{\lambda_i \mid Y_i^+[\lambda_i] \text{ is not empty}\}.$$

Now, Corollary 6.8.1 on page 50 of Ref. 3 implies that

$$(\text{ri } D_i^+) = \{(y^i, \lambda_i) \mid \lambda_i \in (\text{ri } \Lambda_i^+) \text{ and } y^i \in (\text{ri } Y_i^+[\lambda_i])\}.$$

Moreover, the definition of D_i^+ clearly shows that

$$\Lambda_i^+ = \{\lambda_i \geq 0\},$$

which means of course that

$$(\text{ri } \Lambda_i^+) = \{\lambda_i > 0\}.$$

Furthermore, for $\lambda_i > 0$, the definition of D_i^+ clearly shows that

$$Y_i^+[\lambda_i] = \lambda_i D_i,$$

which means that

$$(\text{ri } Y_i^+[\lambda_i]) \equiv \lambda_i(\text{ri } D_i) \qquad \text{for } \lambda_i \in (\text{ri } \Lambda_i^+),$$

by virtue of Corollary 6.6.1 on page 48 of Ref. 3. Consequently, our derivation of the preceding formula for (ri $\mathcal{D}$) is complete.

In particular then, the Fenchel hypothesis in Theorem 2.2 simply asserts that

there exists a vector $(y^0, y^I, \lambda, y^J, 0) = \mathscr{y}^0$ such that

$$(y^0, y^I, y^J) \in (\text{ri } Y); \qquad y^0 \in (\text{ri} D_0);$$
$$\lambda_i > 0 \text{ and } y^i \in \lambda_i(\text{ri } D_i), \qquad i \in I;$$
$$y^j \in (\text{ri } D_j) \text{ and } h_j(y^j) < 0, \qquad j \in J.$$

To complete our proof, we now show that this hypothesis is in fact equivalent to the hypothesis that

there exists a vector $(y'^0, y'^I, \lambda', y'^J)$ such that

$$(y'^0, y'^I, y'^J) \in Y; \qquad y'^0 \in D_0;$$
$$(y'^i, \lambda_i') \in D_i^+, \qquad i \in I;$$

$$y'^j \in D_j \text{ and } h_j(y'^j) < 0, \qquad j \in J,$$

and that there exists a vector $(y^{+0}, y^{+I}, \lambda^+, y^{+J})$ such that

$$(y^{+0}, y^{+I}, y^{+J}) \in (\text{ri } Y); \qquad y^{+0} \in (\text{ri } D_0);$$
$$\lambda_i^+ > 0 \text{ and } y^{+i} \in \lambda_i(\text{ri } D_i), \qquad i \in I;$$
$$y^{+j} \in (\text{ri } D_j), \qquad j \in J.$$

Obviously, a vector (y^0, y^I, λ, y^J) that satisfies the former hypothesis satisfies both parts of the latter hypothesis. On the other hand, Theorem 6.1 on page 45 of Ref. 3 and Theorem 7.1 on page 51 of Ref. 3 imply that a convex combination $\alpha(y'^0, y'^I, \lambda', y'^J) + \beta(y^{+0}, y^{+I}, \lambda^+, y^{+J})$ of vectors $(y'^0, y'^I, \lambda', y'^J)$ and $(y^{+0}, y^{+I}, \lambda^+, y^{+J})$ that satisfy the latter hypothesis will satisfy the former hypothesis for sufficiently small $\beta > 0$. □

3. Slater's Theorem

Although the condition $h_j(y'^j) < 0$, $j \in J$, in hypothesis (i) of Theorem 2.1 resembles the well-known *Slater constraint qualification*, it is of course to be deleted when J is empty—which is the situation in most applications. However, the analogous condition $g_i(x'^i) < 0$, $i \in I$, in hypothesis (i) of the (unstated) dual of Theorem 2.1 (obtained from Theorem 2.1 by interchanging the symbols 2.1 and 2.2, the symbols x and y, the symbols κ and λ, the symbols g and h, the symbols i and j, the symbols I and J, the symbols φ and ψ, the symbols X and Y, the symbols C and D, the symbols S and T, and the symbols S^* and T^*) is essentially the Slater constraint qualification. In fact, we shall now see that the *ordinary programming* case of the dual of Theorem 2.1 actually strengthens Slater's version of the *Kuhn–Tucker theorem.*

The ordinary programming case occurs when

$$J = \emptyset,$$

$$n_k = m \text{ and } C_k \triangleq C_o \text{ for some set } C_o \subseteq E_m, \qquad k \in \{0\} \cup I,$$

$$X \triangleq \text{column space of } \begin{bmatrix} U \\ U \\ \vdots \\ U \end{bmatrix},$$

where there is a total of $1 + o(I)$ identity matrices U that are $m \times m$.

In particular, an explicit elimination of the vector space condition $x \in X$ by the linear transformation

$$\begin{pmatrix} x^0 \\ x^I \end{pmatrix} = \begin{bmatrix} U \\ U \\ \vdots \\ U \end{bmatrix} z$$

shows that the resulting Problem 2.1 is equivalent to the very general ordinary programming problem:

$$\text{minimize } g_0(z),$$
$$\text{subject to } g_i(z) \leq 0, \qquad i \in I,$$
$$z \in C_o.$$

Now, the Slater constraint qualification for the preceding problem simply requires the existence of a feasible solution z' such that

$$g_i(z') < 0, \qquad i \in I.$$

Moreover, Slater's version of the Kuhn–Tucker theorem asserts that the existence of such a *Slater solution* z' and the existence of a finite infimum φ are sufficient to guarantee the existence of a Kuhn–Tucker (Lagrange) multiplier vector λ^*.

To strengthen the preceding theorem with the aid of the dual of Theorem 2.1, first note that the image $x' = (z', z', \ldots, z')$ of a Slater solution z' under the given linear transformation satisfies hypothesis (i) of the dual of Theorem 2.1. Then, note that the existence of a finite infimum φ is simply hypothesis (ii) of the dual of Theorem 2.1. Now, the convexity of C_o implies the existence of a vector $z^+ \in (\text{ri } C_o)$, by virtue of Theorem 6.2 on page 45 of Ref. 3. Moreover, its image $x^+ = (z^+, z^+, \ldots, z^+)$ under the given linear transformation clearly satisfies hypothesis (iii) of the dual of Theorem 2.1—because $(\text{ri } X) = X$ by virtue of Fact (A), and because $J = \varnothing$. Consequently, the dual of Theorem 2.1 implies that both T and T^* are nonempty and that $0 = \varphi + \psi$. In view of Corollary 3.2 of Ref. 5, we conclude from the nonemptyness of T^* that a Kuhn–Tucker (Lagrange) vector λ^* exists. Finally, note that we have also shown the existence of another vector y^*; so the Slater version of the Kuhn–Tucker theorem has actually been strengthened.

More significant implications of Theorem 2.1 are given on page 47 of Ref. 1.

References

1. PETERSON, E. L., *Geometric Programming*, SIAM Review, Vol. 18, pp. 1–51, 1976.
2. PETERSON, E. L., *Symmetric Duality for Generalized Unconstrained Geometric Programming*, SIAM Journal of Applied Mathematics, Vol. 19, pp. 487–538, 1970.
3. ROCKAFELLAR, R. T., *Convex Analysis*, Princeton University Press, Princeton, New Jersey, 1970.
4. PETERSON, E. L., *Constrained Duality via Unconstrained Duality in Generalized Geometric Programming*, Journal of Optimization Theory and Applications, Vol. 26, No. 1, 1978.
5. PETERSON, E. L., *Saddle Points and Duality in Generalized Geometric Programming*, Journal of Optimization Theory and Applications, Vol. 26, No. 1, 1978.
6. PETERSON, E. L., *Geometric Duality vis-a-vis Ordinary Duality* (to appear).

7

Generalized Geometric Programming Applied to Problems of Optimal Control: I. Theory[1]

T. R. JEFFERSON[2] AND C. H. SCOTT[2]

Abstract. The interest in convexity in optimal control and the calculus of variations has gone through a revival in the past decade. In this paper, we extend the theory of generalized geometric programming to infinite dimensions in order to derive a dual problem for the convex optimal control problem. This approach transfers explicit constraints in the primal problem to the dual objective functional.

1. Introduction

The fruitful exploitation of convexity in the area of nonlinear mathematical programming by Fenchel (Ref. 1), Rockafellar (Ref. 2), Peterson (Ref. 3) and others has been followed by the realization that such properties are useful in the analysis of problems in the calculus of variations and optimal control theory. Rockafellar (Ref. 4), Vinter (Ref. 5), Wets and Van Syke (Ref. 6), and Mossino (Ref. 7) have gained valuable insights into variational methods through convexity arguments.

An important technique that has hitherto not been applied to infinite-dimensional problems is the recently developed theory of generalized geometric programming of Peterson (Ref. 3). This has proved very effective in the formulation and solution of convex programs in finite-dimensional space. Generalized geometric programming takes the structure of a

[1] The authors are indebted to the referees for suggestions leading to improvement of the paper.

[2] Lecturer, School of Mechanical and Industrial Engineering, University of New South Wales, Kensington, New South Wales, Australia.

mathematical program into account by identifying and utilizing the following properties: (i) convexity, (ii) linearity, (iii) separability, and (iv) duality.

Convexity and linearity are two very important functional forms in mathematical programming and they provide strong results regarding optima. Separability, even if only partial, provides insight into the solution of the mathematical program. The dual program, taken in the geometric programming sense, often provides a much simpler problem to work with for the following two reasons: (i) the dual program is optimized over the orthogonal complementary linear subspace to that of the primal; that means that, if the primal linear subspace is of high dimension, the dual linear subspace is of low dimension; and (ii) the constraints in the primal program are absorbed into the dual objective function; in this way, while the primal program may have nonlinear constraints, the dual may be simply an optimization over a polyhedral set.

As Rockafellar (Ref. 4) treats all constraints implicitly, his approach requires the calculation of very difficult conjugate transforms when the problem has state constraints or mixed state–control constraints. Other approaches to duality in convex optimal control problems either ignore pure state constraints and mixed state–control constraints (Ref. 5) or treat them implicitly (Refs. 5–7). On the other hand, our approach recognizes the practical difficulty of calculating conjugate transforms by treating mixed state–control constraints explicitly. Although we have not done so, pure state constraints should also be treated in this way if they complicate calculation of the conjugate transforms. This would often be the case. The net result of our approach is to transfer these complicating constraints to the objective of the dual problem where they are more readily handled from a computational point of view.

A primary ingredient in the generalized geometric programming approach is the identification of a suitable subspace in the problem. For the optimal control problem with linear dynamics, we recognize the appropriate subspace to be the solution of the system of dynamical equations.

In this paper, we will develop an extension of geometric programming to convex problems of optimal control from a theoretical viewpoint. A further paper will examine the computational aspects of the technique (Ref. 8).

We choose to work in $L^2([0, T])$, the quotient space of the space formed by the set of all functions which are Lebesgue integrable with respect to the square norm. $L^2([0, T])$ is a Banach space which is its own dual space. In addition, we will not present the full general and symmetric form of the problem as this would obscure the computational advantages of the geometric programming formulation.

2. Statement of the Problem

We consider optimal control problems of the following form:

Program A

$$\text{minimize} \int_0^T f_0(t, u_0(t), x_0(t))\, dt, \tag{1}$$

subject to

$$f_i(t, u_i(t), x_i(t)) \leq 0, \qquad \text{a.e. on } [0, T], \qquad i = 1, \ldots, n, \tag{2}$$

where

$$u_i \in U_i, \qquad x_i \in L_p^2([0, T]) \qquad \text{for } i = 0, \ldots, n, \tag{3}$$

f_i is convex in u_i and x_i for $i = 0, \ldots, n$; u_i is a vector in the convex region U_i. U_i is a subset of $L_m^2([0, T])$.

Define

$$x \triangleq (x_0, x_1, x_2, \ldots, x_n)^T, \qquad u \triangleq (u_0, u_1, u_2, \ldots, u_n)^T, \tag{4}$$

where T denotes transpose. There is a further constraint on the vector variables x and u:

$$\begin{bmatrix} x \\ u \end{bmatrix} \in \chi, \tag{5}$$

where χ is a subspace of $L^2([0, T])$. We show in Section 7 that the standard optimal control theory formulation is contained in the above.

The geometric dual of Program A is given by:

Program B

$$\text{minimize} \left[\int_0^T g_0(t, v_0(t), y_0(t))\, dt + \int_0^T \sum_{i=1}^n g_i^+(t, v_i(t), y_i(t); \lambda_i(t))\, dt \right], \tag{6}$$

subject to

$$v_0 \in V_0, \qquad y_0 \in D_0, \tag{7}$$

and

$$v_i/\lambda_i \in D_i, \qquad i = 1, \ldots, n.$$

Define

$$y = (y_0, y_1, \ldots, y_n)^T, \qquad v = (v_0, v_1, \ldots, v_n)^T. \tag{8}$$

Also,

$$\begin{bmatrix} y \\ v \end{bmatrix} \in \chi^\perp, \tag{9}$$

where $\chi^\perp$ is the orthogonal complementary subspace to χ and

$$\int_0^T g_i^+(t, v_i(t), y_i(t); \lambda_i(t))\, dt$$

is defined by Eq. (19). $\langle x, y\rangle$ denotes an inner product of the vectors x and y and is defined by the integral

$$\int_0^T y(t)x(T-t)\, dt. \tag{10}$$

The terms $\int g_i$, $i = 0, 1, \ldots, n$, are conjugate transforms of the terms $\int f_i$, $i = 0, 1, \ldots, n$, respectively. See Section 5 for definition of a conjugate transform.

The dual Program B, due to its structure, may be easier to solve than the Primal Program A. Sometimes it is finite-dimensional, and hence easily solved by standard calculus.

Let

$$H_i(T-t) = \lambda_i(t) g_i(t, v_i(t)/\lambda_i(t), y_i(t)/\lambda_i(t)),$$
$$\text{for } \lambda_i(t) > 0, \qquad i = 1, \ldots, n,$$
$$H_0(T-t) = g_0(t, v_0(t), y_0(t)).$$

We find that the primal and dual solutions are related by

$$u_i(t) = (\partial H_i/\partial v_i)(t), \qquad x_i(t) = (\partial H_i/\partial y_i)(t) \qquad \text{for } i = 0, \ldots, n, \tag{11}$$

almost everywhere for fixed $t \in [0, T]$ and provided g_i, $i = 0, \ldots, n$, are differentiable with respect to v_i belonging to the relative interior of V_i and y_i belonging to the relative interior of D_i.

Alternatively, we let

$$h_i(T-t) = \lambda_i(T-t) f_i(t, u_i(t), x_i(t)) \qquad \text{for } \lambda_i(t) > 0, \qquad i = 1, \ldots, n,$$
$$h_0(T-t) = f_0(t, u_0(t), x_0(t)),$$

so that

$$v_i(t) = (\partial h_i/\partial u_i)(t), \qquad y_i(t) = (\partial h_i/\partial x_i)(t), \qquad i = 0, \ldots, n, \tag{12}$$

for all fixed $t \in [0, T]$ and provided f_i, $i = 0, \ldots, n$, are differentiable with respect to u_i belonging to the relative interior of U_i.

At the boundary and where the appropriate function is not differentiable, we have the dual point (function) belonging to a set called the subgradient set defined in Section 3. In addition, we shall show in Section 6 below that the sum of the primal and dual objective functions is zero provided the functions f_i belong to $L^2[0, T]$, are closed convex functions with respect to u_i, x_i jointly, and form a consistent problem.

These optimality conditions combined with the feasibility conditions on the primal variables provide us with sufficient information to determine a primal optimal point. First, let us review the major properties of convexity in a Banach space.

3. Convexity

Definition 3.1. A subset C of a vector space X is said to be convex if

$$x_1, x_2 \in C \Rightarrow x = \lambda x_1 + (1-\lambda)x_2 \in C \tag{13}$$

for all scalars $0 \leq \lambda \leq 1$.

Definition 3.2. A function f defined on a convex set C is said to be convex if, for $x_1, x_2 \in C$, we have

$$\lambda f(x_1) + (1-\lambda)f(x_2) \geq f(\lambda x_1 + (1-\lambda)x_2), \tag{14}$$

for all scalars λ, $0 \leq \lambda \leq 1$.

The definitions are clear when $X = L^2([0, T])$.

We deal with convex functionals

$$\int f(x) = \int_0^T f(t, x(t))\, dt, \qquad x \in L^2([0, T]).$$

It is clear that $\int f(x)$ is convex in x if f is convex, since integration is a linear monotone operator and as such, when composed with a convex function, preserves convexity.

Definition 3.3. The pair $[\int f(x), C]$ is called closed if the set

$$\left\{(x, \alpha) \,\middle|\, \int f(x) \leq \alpha, x \in C\right\} \tag{15}$$

is closed. We shall assume that, for every functional $\int f(x)$ defined on C, the pair $[\int f(x), C]$ is closed.

We can associate a set of dual points (functions) with each pair $x_0 \in C$ and convex functional $\int f(x)$. This set, called the subgradient, is defined below.

Definition 3.4. The subgradient set of $\int f$ at x_0 is given by

$$\left\{y \,\middle|\, \int_0^T f(x_0(T-t))\,dt - \int_0^T y(t)(x_0(T-t)-x(T-t))\,dt \leq \int_0^T f(x(T-t))\,dt, \forall x \in C\right\}. \quad (16)$$

Remark 3.1. With the use of the inner product $\langle x, y\rangle$ introduced earlier, one may observe that the above definition of subgradient is the usual one:

$$\partial I(x_0) = \{y \mid I(x_0) - I(x) \leq \langle y, x_0 - x\rangle \text{ for all } x \in C\},$$

where

$$I(x) = \int f(x) = \int_0^T f(t, x(t))\,dt = \int_0^T f(T-t, x(T-t))\,dt.$$

4. Measurability

This section follows the work of Rockafellar (Ref. 9). Rockafellar reduces the class of convex functionals considered to be those which have normal integrands. These have the following properties:

(i) They are of the form $\int f(t, x(t))\,dt$, where f is a nontrivial closed convex function of $x(t)$ bounded from below for each t for each $x \in C$, i.e., $f(t, x(t)) \geq m$ for all $t \in [0, T]$ and $x \in C$.

(ii) There exists a countable subset $U \subset C$ such that (a) $f(t, x(t))$ is measurable in t for any $x \in U$ and (b) defining $U_t = \{x(t) \mid x \in U\}$ and $C_t = \{x(t) \mid x \in C\}$, then, for each t, U_t is dense in C_t.

With these conditions, Rockafellar is able to show that the conjugate transform of such a convex functional is measurable; hence, Program B is well defined. Indeed, without property (ii)-(b), one can construct integrands such that the integrand conjugate transform is not measurable. As we cannot think of a real optimal control problem which would not satisfy these conditions, they need not concern us further.

5. Conjugacy

Definition 5.1. (*Ref. 10*). The conjugate transform of a functional $\int f(x)$ defined on $x \in C$, $[\int f(x), C]$, is $[\int g(y), D]$, where

$$\int_0^T g(t, y)\,dt = \sup_{x \in C}\left\{\int_0^T y(t)x(T-t)\,dt - \int_0^T f(T-t, x(T-t))\,dt\right\}, \quad (17)$$

$$D = \left\{ y \middle| \sup_{x \in C} \langle x, y \rangle - \int f(x) < \infty \right\}. \tag{18}$$

Definition 5.2. The positive homogeneous extension of a conjugate transform is $[\int g^+(y;\lambda), D^+]$, where

$$\int g^+(t, y(t); \lambda(t)) = \begin{cases} \sup_{x \in C} \langle x, y \rangle & \text{if } \lambda(t) = 0 \text{ and } \sup_{x \in C} \langle x, y \rangle < +\infty, \\ \int \lambda(t) g(t, y(t)/\lambda(t)) & \text{if } \lambda(t) > 0 \text{ for } t \in [0, T], \end{cases} \tag{19}$$

$$D^+ = \mathop{\mathsf{X}}_{t \in [0,T]} D^+(t), \tag{20}$$

$$D^+(t) = \{(y(t), \lambda(t)) | \sup_{x \in C} \langle x, y \rangle < \infty, \lambda(t) = 0\}$$
$$\cup \{(y(t), \lambda(t)) | y(t) \in D; \lambda(t) > 0\}. \tag{21}$$

The conjugate transform and its positive homogeneous extension are both closed convex functionals by construction. If $[\int g(y), D]$ and $[\int f(x), C]$ are closed conjugate functionals, we have by construction that

$$\int_0^T g(t, y(t))\, dt + \int_0^T f(T-t, x(T-t))\, dt \geq \int_0^T y(t) x(T-t)\, dt$$
$$\text{for } y \in D,\ x \in C. \tag{22}$$

Following this and the work of Peterson (Ref. 3) in finite-dimensional spaces, we will now relate the set

$$\{x \,|\, x \in C \text{ and } f(T-t, x(T-t)) \leq 0 \text{ a.e. on } [0, T]\}$$

and the function

$$\int_0^T g^+(t, y(t); \lambda(t))\, dt, \qquad (y, \lambda) \in D^+,$$

where g^+ and D^+ are defined by Eqs. (19) and (21).

For particular t, the theorem of Peterson may be applied to show that

$$[0, \{x(T-t) | x \in C \text{ and } f(T-t, x(T-t)) \leq 0\}],$$
$$[g^+(t, y(t); \lambda(t)), \{(y(t), \lambda(t)) | (y, \lambda) \in D^+\}]$$

are conjugate transforms of each other. Integrating over t, we get that

$$[0, \{x \mid x \in C \text{ and } f(T-t, x(T-t)) \leq 0 \text{ a.e. on } [0, T]\}],$$
$$\left[\int_0^T g^+(t, y(t); \lambda(t))\, dt, D^+\right]$$

are conjugate transforms of each other.

6. Optimality

By construction, we have from Eq. (17) that

$$\int f_0 + \int g_0 \geq \langle u_0, v_0 \rangle + \langle x_0, y_0 \rangle \geq 0 \tag{23}$$

for feasible (u_0, x_0) and (v_0, y_0). Similarly, from Eq. (17),

$$0 + \int g_i^+ \geq \langle u_i, v_i \rangle + \langle x_i, y_i \rangle \geq 0 \tag{24}$$

for feasible (u_i, x_i) and (v_i, y_i, λ_i) and $i = 1, \ldots, n$. Summing Eqs. (23) and (24), we obtain that

$$\int f_0 + \int g_0 + \sum_{i=1}^{n} \int g_i^+ \geq \sum_i \langle u_i, v_i \rangle + \sum_i \langle x_i, y_i \rangle, \tag{25}$$

using notation from Eqs. (4) and (8).

Theorem 6.1. Provided the functionals $\int f_0$, $\int g_0$, and $\int g_i^+$, $i = 1, \ldots, n$, are closed and convex, and Programs A and B are consistent, at optimality we have the following relationship:

$$\int f_0 + \int g_0 + \sum_{i=1}^{n} g_i^+ = 0. \tag{26}$$

Proof. Since the functionals $\int g_0$ and $\int g_i^+$, $i = 1, \ldots, n$, are closed and convex, the sum functional $\int g_0 + \sum_i \int g_i^+$ is closed and convex. Suppose that Program B attains its minimum at a point (y^*, v^*). Consider the subgradient set of $\int g_0 + \sum \int g_i^+$ at (y^*, v^*), and call it $\partial G(y^*, v^*)$. These points (functions) belong to the set

$$\mathop{\times}_i L_p^2([0, T]) \mathop{\times}_i U_i.$$

Since the functionals are closed and convex, we have that

$$\int f_0 + \int g_0 + \sum_i \int g_i^+ = \langle u, v^* \rangle + \langle x, y^* \rangle \qquad \text{for } (x, u) \in \partial G(y^*, v^*).$$

If

$$\partial G(y^*, v^*) \cap \chi \neq \phi,$$

then we choose any point in the intersection and

$$\langle u, v^* \rangle + \langle x, y^* \rangle = 0.$$

Since

$$\begin{bmatrix} u \\ x \end{bmatrix} \in \chi \qquad \text{and} \qquad \begin{bmatrix} y^* \\ v^* \end{bmatrix} \in \chi^{\perp},$$

the result then follows immediately.

Hence, we suppose that

$$\partial G(y^*, v^*) \cap \chi = \phi.$$

Let $R(y^*, v^*)$ be the closed convex cone of feasible directions at (y^*, v^*). Now, $\partial G(y^*, v^*) \cap R(y^*, v^*)$ is a compact convex set and χ is a subspace. By the Hahn–Banach theorem, there exists a function (y', n') such that

$$\langle v', u \rangle + \langle y', x \rangle = 0 \qquad \text{for } (x, u) \in \chi, \tag{27}$$

$$\langle v', u \rangle + \langle y', x \rangle < 0 \qquad \text{for } (x, u) \in \partial G(y^*, v^*) \cap R(y^*, v^*). \tag{28}$$

Equation (27) implies that

$$(y^* + \alpha y', v^* + \alpha v') \in \chi^{\perp}.$$

Equation (28) and continuity tell us that there exists a small positive α such that

$$\int g_0 + \sum_i \int g_i^+ \qquad \text{at } (y^* + \alpha y', v^* + \alpha v')$$

is less than at (y^*, v^*). This contradicts the optimality of (y^*, v^*); hence,

$$\partial G(y^*, v^*) \cap \chi \neq \phi,$$

and the dual and primal objectives sum to zero.

7. Convex Optimal Control Problem

This is to find a vector $u_0(t) \in L_m^2([0, T])$ which minimizes the functional

$$J_0(u_0(t)) = \int_0^T f_0(t, x_0(t), u_0(t))\, dt, \tag{29}$$

subject to the system dynamics

$$dx_0/dt = A(t)x_0 + B(t)u_0, \qquad \text{a.e. on } [0, T], \tag{30}$$

with

$$x_0(0) = s \text{ given}, \qquad x_0 \in L_p^2([0, T]),$$

and $A(t)$ and $B(t)$ are $p \times p$ and $p \times m$ real-valued matrices, and subject to control constraints

$$u_0 \in U_0, \tag{31}$$

to state constraints

$$x_0 \in L_p^2([0, T]) \tag{32}$$

where U_0 is a closed convex set, and to mixed constraints of the form

$$f_i(t, x_0(t), u_0(t)) \leq 0, \qquad \text{a.e. on}[0, T], \tag{33}$$

where $i = 1, \ldots, n$; $f_i(t, x_0, u_0)$ are closed and convex in x and u, for all t and for $i = 0, \ldots, n$. We take

$$x_0(T) = 0.$$

We now transform the above standard formulation into the geometric programming form (Program A). The first step is to solve the system dynamics given by Eq. (30). This yields

$$x_0(t) = \Phi(t, 0)x_0(0) + \int_0^t \Phi(t, \tau)B(\tau)u_0(\tau)\, d\tau, \tag{34}$$

where Φ satisfies the equation

$$d\Phi/dt = A(t)\Phi(t, \tau),$$

with

$$\Phi(\tau, \tau) = I.$$

We also introduce the prescription

$$\begin{aligned} x_0 &= x_i, \qquad \text{a.e. on } [0, T], \qquad i = 1, \ldots, n, \\ u_0 &= u_i, \qquad \text{a.e. on } [0, T], \qquad i = 1, \ldots, n. \end{aligned} \tag{35}$$

We note that Eqs. (34) and (35) determine a subspace. Hence, the convex optimal control problem may be formulated as the geometric programming problem:

$$\text{minimize } J(x, u, d), \tag{36}$$

$$\text{subject to } (x, u, d) \in \chi^{\perp}, \qquad u \in U, \qquad d \in D,$$

$$f_i(t, u_i(t), x_i(t)) \leq 0, \qquad \text{a.e. on } [0, T], \qquad i = 1, \ldots, n,$$

where

$$U = \mathop{\mathbf{X}}_i U_i, \qquad u_i \in U_i,$$

$$J(x, u, d) = \int_0^T f_0(x_0, u_0, t)\, dt,$$

$$D = \{d \mid d(0) = s\},$$

$$\chi = \{(x, u, d)^T \mid x(t) = \Phi(t, 0)d(0)$$

$$+ \int_0^T \Phi(t, \tau)B(\tau)\, d\tau;\; x_0 = x_i,\; i = 0, \ldots, n;$$

$$u_0 = u_i,\; i = 0, \ldots, n;\ \text{a.e. on } [0, T]\}.$$

The results of Sections 5 and 6 then give the dual problem, Program B, to be: minimize

$$K(y, v, e), \tag{37}$$

subject to

$$(y, v, e) \in \chi^\perp, \qquad v_0 \in V_0, \qquad y_0 \in D_0,$$

$$v_i/\lambda_i \in V_i, \qquad y_i/\lambda_i \in D_i, \qquad i = 1, \ldots, n,$$

where

$$K(y, v, e) = \int_0^T g_0(t, e, v_0, y_0)\, dt + \int_0^T \sum_{i=1}^n g_i^+(t, v_i, y_i; \lambda_i(t))\, dt.$$

Here, $\int g_i^+$ is defined as in Section 5, $g_0(t, e, v_0, y_0)$ is the conjugate transform of $f_0(t, x, u, d)$, e is the dual variable corresponding to the primal variable d, and the orthogonal complementary subspace $\chi^\perp$ is given by (see Appendix A)

$$\Big\{(y, v, e) \Big| \int_0^T \sum_i y_i(t)\Phi(T - t, 0)\, dt = -e(T),\ e(\tau) = 0 \text{ a.e. on } [0, T], \text{ and}$$

$$\int_0^T \sum_i y_i(t)\Phi(T - t, T - \tau)B(T - \tau)\, dt = -\sum_i v_i(\tau) \text{ a.e. on } [0, T]\Big\}. \tag{38}$$

Furthermore, Eq. (26) implies that

$$J(x, u, d) + K(y, v, e) = 0 \tag{39}$$

at optimality. Such results are very useful as an algorithm-stopping criterion.

8. Conclusions

We note that the dual problem, Program B as given in (37), is still an optimal control problem by virtue of the orthogonal complementary subspace $\chi^{\perp}$, but it now unconstrained as far as control–state constraints go. Note that the roles of the dual variables v and y are interchanged with respect to those of the primal variables u and x. In some circumstances, the dual problem reduces to a simple finite-dimensional optimization problem (Ref. 8).

9. Appendix: Evaluation of $\chi^{\perp}$

Define

$$d(\tau)=\begin{cases} s, & \tau=0,\\ \text{unspecified elsewhere.} \end{cases}$$

Hence, Eq. (34) for $x_0(t)$ may be written as

$$x_0(t)=\int_0^t \Phi(t,\tau)[d(\tau)\delta(\tau)+B(\tau)u_0(\tau)]\,d\tau, \tag{40}$$

where $\delta(\tau)$ is the Dirac delta function. By *expression* (37), the dual variables (y, v, e) must satisfy

$$\langle y, x\rangle+\langle v, u\rangle+\langle e, d\rangle=0,$$

that is,

$$\sum_{i=0}^{n}\langle y_i, x_i\rangle+\sum_{i=0}^{n}\langle v_i, u_i\rangle+\langle e, d\rangle=0. \tag{41}$$

Substituting results (35) and (40) into Eq. (41) gives

$$\sum_{i=0}^{n}\int_0^T y_i(t)\int_0^{T-t}\Phi(T-t,\tau)[d(\tau)\delta(\tau)+B(\tau)u_0(\tau)]\,d\tau\,dt$$
$$+\sum_{i=0}^{n}\int_0^T v_i(\tau)u_0(T-\tau)\,d\tau+\int_0^T e(\tau)\,d(T-\tau)\,d\tau=0. \tag{42}$$

The first term on the left-hand side of Eq. (42) may be rearranged to yield

$$\int_0^T d\tau\Big[\int_0^T dt\sum_i y_i(t)\Phi(T-t, T-\tau)\delta(T-\tau)+e(\tau)\Big]\,d(T-\tau)$$
$$+\int_0^T d\tau\Big[\int_0^T dt\sum_i y_i(t)\Phi(T-t, T-\tau)B(T-\tau)+\sum_i v_i(\tau)\Big]u(T-\tau)=0. \tag{43}$$

Equation (43) must be true for arbitrary $d(T-\tau)$ and $u(T-\tau)$. Hence, the defining equations of $\chi^\perp$ are

$$\int_0^T \sum_i y_i(t)\Phi(T-t, 0)\, dt = -e(T), \qquad e(\tau)=0, \qquad \text{a.e. on } [0, T],$$

$$\int_0^T \sum_i y_i(t)\Phi(T-t, T-\tau)B(T-\tau) = -\sum_i v_i(\tau), \qquad \text{a.e. on } [0, T].$$

References

1. FENCHEL, W., *Convex Cones, Sets, and Functions*, Princeton University, Princeton, New Jersey, Mathematics Department, Mimeographed Lecture Notes, 1951.
2. ROCKAFELLAR, R. T., *Convex Analysis*, Princeton University Press, Princeton, New Jersey, 1970.
3. PETERSON, E. L., *Geometric Programming*, SIAM Review, Vol. 18, pp. 1–51, 1976.
4. ROCKAFELLAR, R. T., *Conjugate Convex Functions in Optimal Control and the Calculus of Variations*, Journal of Mathematical Analysis and Applications, Vol. 32, pp. 174–222, 1970.
5. VINTER, R. B., *Application of Duality Theory to a Class of Composite Cost Control Problems*, Journal of Optimization Theory and Applications, Vol. 13, pp. 436–460, 1974.
6. VAN SLYKE, R. M., and WETS, R. J. B., *A Duality Theory for Abstract Mathematical Programs with Applications to Optimal Control Theory*, Journal of Mathematical Analysis and Applications, Vol. 22, pp. 679–706, 1968.
7. MOSSINO, J., *An Application of Duality to Distributed Optimal Control Problems with Constraints on the Control and the State*, Journal of Mathematical Analysis and Applications, Vol. 50, pp. 223–242, 1975.
8. JEFFERSON, T. R., and SCOTT, C. H., *Generalized Geometric Programming Applied to Problems of Optimal Control, II, Computational Aspects* (to appear).
9. ROCKAFELLAR, R. T., *Integrals Which Are Convex Functionals*, Pacific Journal of Mathematics, Vol. 24, pp. 525–539, 1968.
10. ROCKAFELLAR, R. T., *Convex Functionals and Duality*, Contributions to Nonlinear Functional Analysis, pp. 215–236, Academic Press, New York, New York, 1971.

8

Projection and Restriction Methods in Geometric Programming and Related Problems[1,2]

R. A. ABRAMS[3] AND C. T. WU[4]

Abstract. Mathematical programming problems with unattained infima or unbounded optimal solution sets are dual to problems which lack *interior points*, e.g., problems for which the Slater condition fails to hold or for which the hypothesis of Fenchel's theorem fails to hold. In such cases, it is possible to project the unbounded problem onto a subspace and to restrict the dual problem to an affine set so that the infima are not altered. After a finite sequence of such projections and restrictions, dual problems are obtained which have bounded optimal solution sets and *interior points*. Although results of this kind have occasionally been used in other contexts, it is in geometric programming (both in the original posynomial form and the generalized form) where such methods appear most useful. In this paper, we present a treatment of dual projection and restriction methods developed in terms of dual generalized geometric programming problems. Analogous results are given for Fenchel and ordinary dual problems.

1. Introduction

Duffin, Peterson, and Zener, in Ref. 1, call a geometric programming problem degenerate if there is a dual variable which has the value zero in

[1] This research was supported in part by Grant No. AFOSR-73-2516 from the Air Force Office of Scientific Research and by Grant No. NSF-ENG-76-10260 from the National Science Foundation.

[2] The authors wish to express their appreciation to the referees for several helpful comments.

[3] Associate Professor, Graduate School of Business, University of Chicago, Chicago, Illinois.

[4] Assistant Professor, School of Business Administration, University of Wisconsin—Milwaukee, Milwaukee, Wisconsin.

every feasible dual solution. They show that this happens iff the transformed primal problem (when consistent) has an unattained infimum or an unbounded optimal solution set. They then show that, by simply eliminating dual variables and corresponding primal terms, a reduced form is obtained which has a bounded optimal solution set and a positive feasible solution of the dual. The dual situation was studied in Ref. 2 where it is shown that the dual has an unbounded optimal solution set or unattained suprema when the primal is not superconsistent. In this case, a reduced superconsistent form is easily obtained by eliminating the set of constraints which hold as equalities for all feasible solutions. In linear programming, a similar characterization is given by Williams (Ref. 3) and Shefi (Ref. 4). In Ref. 5, Luenberger presents Shefi's results. They show that a linear programming problem with an optimal solution has an unbounded optimal solution set iff some dual variable is zero in all feasible dual solutions. A reduced form can be obtained by eliminating such variables and the corresponding primal constraints; but, in contrast to the geometric programming case, this does not seem to be of any particular theoretical or computational value in linear programming.

The analogous situation in convex programming was first characterized by Rockafellar (Ref. 6) where, for Fenchel dual problems, he showed that a problem is not strongly consistent iff there is a direction in which the dual objective function is nonincreasing but not constant. He then used this fact to obtain an equivalent pair of dual problems which enabled him to prove his principal duality theorem. In Ref. 7, Peterson gives similar characterizations for various dual problems in general convex programming, and some of the characterizations are also discussed by Wu in Ref. 8. In Ref. 9, a projection method for reducing general convex programs with unattained infima is given.

Although reduced forms similar to the above can be obtained for any dual pairs of convex programming problems, they appear to be particularly useful in geometric programming (Refs. 1–2) and related problems, e.g. (Refs. 10–11) when the problem structure makes such reductions relatively easy to carry out. In this paper, we develop reduction procedures for generalized geometric programming problems. An advantage of this formulation is that, in the context of generalized geometric programming, the reduced problems have a particularly simple and transparent form, whereas in many special cases and alternate general forms the details obscure the basic simplicity of what is taking place. After developing the generalized geometric programming case, we show how the well-known degeneracy results of posynomial geometric programming can be obtained as a special case. We also indicate the form of analogous results for some other general dual pairs.

2. Notation and Preliminaries

Our notation will be standard as used, for example, in Refs. 1, 12. We review some of the basic notation and concepts.

R^n is n-dimensional Euclidean space.

For $x \in R^n$ and $y \in R^n$, (x, y) is the standard inner product of x and y, i.e.,

$$(x, y) = y^T x.$$

$[W]$ denotes the subspace generated by W.

The relative interior of a set C is denoted by ri C and the closure by $\bar{C}$.

The epigraph of a function $f: C \to R$, where $C \subset R^n$, is

$$\text{epi}\, f = \{(x, \alpha) \in R^{n+1} : f(x) \leqq \alpha, x \in C\}.$$

A convex function is closed if epi f is closed or equivalently if all its level sets are closed, i.e., if all sets

$$\mathscr{L}_\alpha = \{x : f(x) \leq \alpha\}$$

for all $\alpha \in R$ are closed.

If $f: C \to R$ is convex, its domain of definition may be extended to all of R^n by defining $f(x) = +\infty$ for $x \notin C$. Also, define

$$\text{dom}\, f \equiv \{x : f(x) < \infty\}.$$

A convex set C recedes in the direction of a vector y if, for every $x \in C$, $x + \lambda y \in C$ for all $\lambda \geq 0$. To simplify terminology, such a vector y will be called a direction of recession of C. Directions of recession are sometimes referred to as directions of infinity, e.g., Ref. 13.

The recession cone of C, $0^+ C$, is the set of all directions of recession of C.

A vector y is a direction of recession of a convex function $f(x)$ if y is a direction of recession of all nonempty level sets of f, i.e., of the sets

$$\mathscr{L}_\alpha = \{x : f(x) \leq \alpha\}$$

for all $\alpha \in R$. The set of all such vectors is denoted $0^+ f$. Thus, $v \in 0^+ f$ implies that, for each $x \in \text{dom}\, f$, $f(x + \lambda v)$ is a nonincreasing function of λ for $\lambda \geq 0$.

If both y and $-y$ are directions of recession of f, then y is said to be a direction of constancy of f.

For any $L \subset R^n$,

$$L^\perp = \{x \in R^n : (z, x) = 0 \text{ for } z \in L\}$$

is the orthogonal complement of L, and

$$L^+ = \{x \in R^n : (z, x) \geq 0 \text{ for } z \in L\}$$

is the polar or positive dual cone of L.

The vector sum of A, $B \subset R^n$ is defined as

$$A + B = \{z \in R^n : z = x + y, x \in A, y \in B\}.$$

A basic result concerning convex functions is the following (see, e.g., Refs. 12, 13). Let f be a closed convex function and D a closed convex set. If f and D do not have a common direction of recession, then the infimum of f over D is attained on a bounded set.

An optimization problem will be said to have a direction of recession if its objective function and the feasible set have a common direction of recession. A problem is said to have a direction of constancy if it has a direction of recession v such that $-v$ is also a direction of recession. The set of all directions of recession (constancy) of a problem is the recession cone (constancy space) of the problem.

3. Reduction of Generalized Geometric Programming Problems

Let f be a closed convex function with effective domain C, and let X be a closed convex cone. Let g be the conjugate transform of f, D the effective domain of g, and let Y be the positive dual cone of X. Consider the dual generalized geometric programming problems presented in Ref. 14, viz.,

$$(\mathrm{P_g}) \quad \underset{x \in C \cap X}{\text{minimize}}\, f(x), \qquad (\mathrm{D_g}) \quad \underset{y \in D \cap Y}{\text{minimize}}\, g(y).$$

The notation $\inf(\mathrm{P_g})$ and $\inf(\mathrm{D_g})$ will be used to designate the infimum of f over $C \cap X$ and that of g over $D \cap Y$, respectively. If

$$C \cap X \neq \varnothing,$$

$(\mathrm{P_g})$ is said to be consistent. If

$$\operatorname{ri} C \cap \operatorname{ri} X \neq \varnothing,$$

$(\mathrm{P_g})$ is said to be strongly consistent [the consistency and strong consistency of $(\mathrm{D_g})$ are similarly defined], and it is well known (e.g., Refs. 12 and 14) that this condition and $\inf(\mathrm{P_g}) > -\infty$ are sufficient for

$$\inf(\mathrm{P_g}) + \inf(\mathrm{D_g}) = 0.$$

We now investigate the case in which one of the problems $(\mathrm{P_g})$ or $(\mathrm{D_g})$ is not strongly consistent. The following theorem was given by Rockafellar (Ref. 6) for Fenchel dual problems which can be trivially specialized to the above problems (P_g) and (D_g). We state the theorem in terms of the above (P_g) and (D_g). A somewhat different characterization is given by Peterson in Ref. 7. Because some details of the proof will be needed later, we give a particularly simple proof which we base on the method of Ref. 7.

Theorem 3.1. (D_g) is not strongly consistent iff (P_g) has a nonzero direction of recession which is not a direction of constancy.

Proof. By definition, (D_g) is not strongly consistent if

$$\text{ri } D \cap \text{ri } Y = \varnothing.$$

This is the well-known (Ref. 12) necessary and sufficient condition for the *proper separation* of the sets D and Y, i.e., for the existence of a hyperplane

$$H_\alpha^v = \{y : (v, y) = \alpha\},$$

which separates D and Y but does not contain both D and Y. Because Y is a cone, we may take $\alpha = 0$. Thus, (D_g) is not strongly consistent iff there is a nonzero v such that:

$$(v, y) \geq 0 \qquad \text{for all } y \in Y, \tag{1}$$

$$(v, y) \leq 0 \qquad \text{for all } y \in D, \tag{2}$$

and either there is a $y^0 \in Y$ such that $(v, y^0) > 0$ or there is a $y^0 \in D$ such that $(v, y^0) < 0$. Now, (1) holds iff $v \in X$, and (2) holds iff (see Theorem 14.2 of Ref. 12) $v \in 0^+ f$. Therefore, (1) and (2) hold iff v is a direction of recession of (P_g). The *either/or* condition is equivalent to $-v \notin X$ or $-v \notin 0^+ f$. Therefore, (D_g) is not strongly consistent iff there is a v which is a direction of recession of (P_g) such that $-v$ is not a direction of recession. □

We see from the above proof that v is a direction of constancy of (P_g) iff the hyperplane H_0^v contains both D and Y.

To motivate the development of the reduced form, we consider a simple example in R^2. Let

$$X = \{x : x_1 \geq 0\},$$

and let

$$f(x) = 1/x_1, \qquad C = \{x \in R^2 : x_1 > 0\}.$$

Note that $\left[\begin{smallmatrix}0\\1\end{smallmatrix}\right]$ is a direction of constancy and that $\left[\begin{smallmatrix}1\\0\end{smallmatrix}\right]$ is a direction of recession which is not a direction of constancy. The dual problem (D_g) has objective function

$$g(y) = -2\sqrt{-y_1},$$

with

$$D = \{y : y_1 \leq 0,\ y_2 = 0\}$$

and cone

$$Y = \{y : y_2 = 0,\ y_1 \geq 0\}.$$

As indicated in the proof of Theorem 3.1, the direction of constancy of (P_g), viz., $\left[\begin{smallmatrix}0\\1\end{smallmatrix}\right]$, defines a hyperplane (the y_1-axis) that contains both Y and D. The direction of recession $\left[\begin{smallmatrix}1\\0\end{smallmatrix}\right]$, which is not a direction of constancy, defines a hyperplane (the y_2-axis) which separates, but does not contain, D and Y.

The direction of constancy and the direction of recession will be treated in the same way. We consider the direction of recession. The objective function is nonincreasing in the direction $\left[\begin{smallmatrix}1\\0\end{smallmatrix}\right]$ and *moving in that direction* can only help. Thus, we replace $1/x_1$ by its limit as $x_1 \to +\infty$. This removes the dependence on x_1, and we will replace the original feasible set with its projection on the orthogonal complement of $\left[\begin{smallmatrix}1\\0\end{smallmatrix}\right]$; i.e., we replace (P_g) by the problem of minimizing $f(x) \equiv 0$ on the x_2-axis. We will see that the dual operation is the restriction of the dual problem (D_g) to the orthogonal complement of $\left[\begin{smallmatrix}1\\0\end{smallmatrix}\right]$, i.e., to the y_2-axis. Then, the dual problem will be to minimize $-2\sqrt{-y_1}$ with the feasible set consisting of the point $\{0\}$. Proceeding in this way for the general problem, we obtain after a finite number of such steps and in the absence of a duality gap, dual problems which are *equivalent* to the original pair but which have nonempty bounded optimal solution sets and are strongly consistent.

We now return to the general case. When one of the problems has a direction of recession as in the example, we replace the objective function by the function obtained by calculating the limit in the direction of recession, i.e., we replace $f(x)$ by $\lim_{\lambda \to \infty} f(x + \lambda v)$, where v is any vector in the relative interior of the recession cone of the problem. An equivalent characterization will be convenient for finding the dual operation.

Proposition 3.1. Let h be a convex function, and let S be a convex cone contained in the recession cone of h. Let $v \in \operatorname{ri} S$, and let $P_{S^\perp}$ be the orthogonal projector on $S^\perp$. Then, for any $x \in S^\perp$,

$$\lim_{\lambda \to +\infty} h(x + \lambda v) = \inf\{h(z) : P_{S^\perp} z = x\} \equiv P_{S^\perp} h(x). \tag{3}$$

Proof. For any $x \in S^\perp$ and z such that $P_{S^\perp} z = x$, there is a $w \in [S]$ such that $z = x + w$. Therefore, we have

$$P_{S^\perp} h(x) = \inf_w \{h(x + w) : w \in [S]\}. \tag{4}$$

Any $w \in [S]$ may be written as a linear combination of vectors in S, e.g.,

$$w = \sum \alpha_i s_i, \qquad s_i \in S, \qquad \text{for all } i.$$

Let $\beta_i = -\alpha_i$ if $\alpha_i < 0$ and $\beta_i = 0$ otherwise. Then,

$$w + \sum \beta_i S_i \equiv \bar{w} \in S;$$

and, since h is nonincreasing in directions in S,

$$h(x+\bar{w}) \leq h(x+w).$$

Therefore, in the right-hand side of (4), we may replace $[S]$ with S, i.e.,

$$P_{S^\perp} h(x) = \inf_w \{h(x+w) : w \in S\}.$$

Let $v \in \text{ri } S$. Then, for any $w \in S$ and for sufficiently large λ,

$$v - (1/\lambda) w \in S.$$

Thus,

$$h(x+\lambda v) = h(x + w + \lambda(v - w/\lambda)) \leq h(x+w),$$

$$P_{S^\perp} h(x) \geq \lim_{\lambda \to +\infty} h(x+\lambda v).$$

It is obvious that, for any $x \in S^\perp$ and $v \in \text{ri } S$,

$$\lim_{\lambda \to +\infty} h(x+\lambda v) \geq \inf\{h(x+w) : w \in S\}.$$ □

Let S be the recession cone of (P_g); i.e., S is the set of all recession directions of (P_g). The reduced form of (P_g) is then defined to be the result of taking the limit of the objective function in a direction $v \in \text{ri } S$ and projecting the constraint set on $S^\perp$, i.e., the reduced form of (P_g) is

(P'_g) minimize $P_{S^\perp} f(x)$,

subject to $x \in P_{S^\perp} C \cap P_{S^\perp} X$.

Note that, because $S \subset 0^+ C \cap 0^+ X$,

$$P_{S^\perp} C \cap P_{S^\perp} X = P_{S^\perp}(C \cap X)$$

follows from Lemma 3.1 of Ref. 9.

Theorem 3.2. $\inf(P'_g) = \inf(P_g)$.

Proof. We first assume that $\inf(P_g)$ is finite. For any $\epsilon > 0$, there is a $z \in C \cap X$ such that

$$f(z) < \inf(P_g) + \epsilon.$$

Define $x = P_{S^\perp} z$. Then,

$$x \in P_{S^\perp} C \cap P_{S^\perp} S$$

and

$$P_{S^\perp} f(x) = \inf\{f(z) : P_{S^\perp} z = x\} \leq f(z).$$

Therefore,

$$\inf(\mathrm{P}'_g) \leq \inf(\mathrm{P}_g).$$

If $\inf(\mathrm{P}_g) = -\infty$, similar reasoning shows that $\inf(\mathrm{P}'_g) = -\infty$.

Now assume that $\inf(\mathrm{P}'_g)$ is finite. For any $\epsilon > 0$, there is an $x \in P_{S^\perp} C \cap P_{S^\perp} X$ such that

$$P_{S^\perp} f(x) < \inf(\mathrm{P}'_g) + \epsilon/2.$$

The definition of $P_{S^\perp} f(x)$ implies that there is a $z \in C \cap X$ such that $P_{S^\perp} z = x$ and

$$f(z) < P_{S^\perp} f(x) + \epsilon/2.$$

Therefore,

$$f(z) \leq \inf(\mathrm{P}'_g) + \epsilon,$$

which implies, when $\inf(\mathrm{P}'_g)$ is finite, that

$$\inf(\mathrm{P}'_g) \geq \inf(\mathrm{P}_g).$$

The same reasoning shows that

$$\inf(\mathrm{P}'_g) = -\infty$$

implies that

$$\inf(\mathrm{P}_g) = -\infty.$$

Finally, it is clear that (P_g) is inconsistent iff (P'_g) is inconsistent. □

The dual of (P'_g) has a particularly simple form which we now determine. We first calculate $(P_{S^\perp} X)^+$ and the conjugate of $P_{S^\perp} f$.

Lemma 3.1. Let $P_{S^\perp}$ be an orthogonal projector on the subspace $S^\perp$. Then,

$$(P_{S^\perp} X)^+ = (X^+ \cap S^\perp) + [S].$$

Proof.

$$\begin{aligned} y \in (P_{S^\perp}X)^+ &\Leftrightarrow (y, z) \geq 0, \qquad \text{for all } z \in P_{S^\perp}X, \\ &\Leftrightarrow (y, P_{S^\perp}x) \geq 0, \qquad \text{for all } x \in X, \\ &\Leftrightarrow (P_{S^\perp}y, x) \geq 0, \qquad \text{for all } x \in X, \\ &\Leftrightarrow P_{S^\perp}y \in X^+ \\ &\Leftrightarrow y \in (X^+ \cap S^\perp) + [S], \end{aligned}$$

where the third equivalence follows because $P_{S^\perp}$ is self-adjoint and the last equivalence follows by the orthogonal decomposition of y onto $[S]$ and $S^\perp$.

Lemma 3.2. The conjugate of $P_{S^\perp}f(x)$ is $f^*[P_{S^\perp}(y)]$, and its effective domain is $D \cap S^\perp + [S]$.

Proof. Using the fact that $P_{S^\perp}$ is self-adjoint, the expression for $(P_{S^\perp}f)^*$ follows directly from Theorem 16.3 of Ref. 12. The formula for the effective domain is obvious. □

It is also possible to obtain Lemma 3.1 from Lemma 3.2 by using the indicator function on the set X.

The dual of problem (P'_g) is, therefore,

(DP'_g) $$\text{minimize } g(P_{S^\perp}(y)), \quad y \in \{D \cap S^\perp + [S]\} \cap \{Y \cap S^\perp + [S]\}.$$

The function $g(P_{S^\perp}(y))$ is clearly constant along $[S]$ and, for $y \in S^\perp$,

$$g(P_{S^\perp}(y)) = g(y).$$

Therefore, we may eliminate $[S]$ and replace (DP'_g) with

(D'_g) $$\text{minimize } g(y), \quad y \in (D \cap S^\perp) \cap (Y \cap S^\perp).$$

Note the simple form of (D'_g). It is just (D_g) restricted to the subspace $S^\perp$. We call (D'_g) the reduced form of (D_g). (D'_g) may also be obtained directly as the dual of the problem obtained from (P'_g) by extending the domain of $P_{S^\perp}f$ to the whole space and defining

$$P_{S^\perp}f(x) = P_{S^\perp}f(P_{S^\perp}x)$$

for $x \notin S^\perp$.

Lemma 3.3. $S^\perp \supset D \cap Y$.

Proof. Let $v \in S$. If v is a direction of recession of (P_g) which is not a direction of constancy, the proof of Theorem 3.1 shows that $v^\perp$ separates D and Y, and hence that $v^\perp \supset D \cap Y$. According to the sentence following the proof of Theorem 3.1, if v is a direction of constancy of (P_g), $v^\perp \supset D \cup Y$. Therefore, in either case, $v^\perp \supset D \cap Y$. Thus, $y \in D \cap Y$ implies that $(y, v) = 0$ for any $v \in S$. Hence, $S^\perp \supset D \cap Y$. □

An immediate consequence of Lemma 3.3 is the following theorem.

Theorem 3.3. $\inf(D_g) = \inf(D'_g)$.

Finally we show that the reduced dual is the dual of the reduced primal in $S^\perp$.

Theorem 3.4. In the vector space $S^\perp$, (D'_g) is the dual of (P'_g).

Proof. That $Y \cap S^\perp$ is the dual cone of $P_{S^\perp}X$ in $S^\perp$ follows easily from the proof of Lemma 3.1, just by adding the condition $y \in S^\perp$ to each equivalence. Restricting the domain of the conjugate transform of $P_{S^\perp}f$ to $S^\perp$ does not change its value on $S^\perp$. Thus, the objective function is identical to that of (DP'_g), which is equal to that of (D'_g) on $S^\perp$. □

Although we have shown that (D'_g) is the dual of (P'_g), (P'_g) is not necessarily the dual of (D'_g), because the function $P_{S^\perp}f$ may fail to be closed and the cone $P_{S^\perp}X$ may also fail to be closed. In this case, as in Ref. 9, it may be advantageous to modify the problem by replacing both $P_{S^\perp}f$ and $P_{S^\perp}X$ by their closures, which we denote $\overline{P_{S^\perp}f}$ and $\overline{P_{S^\perp}X}$. Symmetric dual problems are then obtained. The dual of (D'_g) is, therefore,

$$(\bar{P}'_g) \qquad \text{minimize } \overline{P_{S^\perp}f}(x),$$
$$x \in \operatorname{dom} \overline{P_{S^\perp}f} \cap \overline{P_{S^\perp}X}.$$

From the definition of the closure of a convex function and the fact that the feasible set of $(\bar{P}'_g)$ contains the feasible set of (P'_g), it follows that

$$\inf(P'_g) \geq \inf(\bar{P}'_g).$$

Theorem 3.5. If $\inf(P'_g)$ is unequal to $\inf(\bar{P}'_g)$, then $\inf(P_g)$ is unequal to $\inf(D_g)$, i.e., the original dual pair has a duality gap.

Proof. In view of the preceding sentence, if the infima are unequal, we must have

$$\inf(\bar{P}'_g) < \inf(P'_g).$$

Because $(\bar{P}'_g)$ is the dual of (D'_g), we must have

$$\inf(\bar{P}'_g) + \inf(D'_g) \geq 0;$$

therefore,

$$\inf(P'_g)+\inf(D'_g)>0.$$

But

$$\inf(P'_g)=\inf(P_g),$$

by Theorem 3.2, and

$$\inf(D'_g)=\inf D_g,$$

by Theorem 3.3. Therefore,

$$\inf(P_g)+\inf(D_g)>0,$$

as was to be proven. □

It may happen that (D'_g) is not strongly consistent, in which case $(\bar{P}'_g)$ must have a nonzero recession cone S_1. Then, (D'_g) may be restricted to $S_1^\perp$ and $(\bar{P}'_g)$ may be projected into $S_1^\perp$. Continuing in this way (D_g) may, in a finite number of steps, be reduced to a strongly consistent problem, the dual of which will have a nonempty bounded optimal solution set. The reduced *closed* primal will have infimum equal to the infimum of the original problem iff the original dual pair has no duality gap.

Examples of problems requiring more than one step for reduction are given in Refs. 2, 9, and 15. If the only nonlinearities present are faithfully convex functions, a method given in Ref. 15 can be used to obtain a strongly consistent problem in a number of steps which is less than or equal to the number of explicit nonlinear constraints.

4. Posynomial Geometric Programming

In this section, we indicate how the previous results specialize to what is known (Refs. 1–2) as degeneracy and reduction to canonical form in posynomial geometric programming. Consider the transformed primal problem (p. 82 of Ref. 1):

$$(P_p) \qquad \text{minimize } \log \sum_{i\in J(0)} c_i \exp(x_i),$$

$$\text{subject to } \sum_{i\in J(k)} c_i \exp(x_i)\le 1, \qquad k=1,\ldots,p,$$

$$x\in R(A),$$

where $c_i>0$, for all i, A is the exponent matrix of the original problem, and

$R(A)$ denotes the range space of A. Its dual is

$$(\mathrm{D}_p) \qquad \text{minimize} \sum_{i \in J(0)} y_i \log(y_i/c_i) + \sum_{k=1}^{p} \sum_{i \in J(k)} y_i \log(y_i/c_i\lambda_k),$$

$$\text{subject to} \sum_{i \in J(k)} y_i = \lambda_k, \qquad k = 1, \ldots, p,$$

$$\sum_{i \in J(0)} y_i = 1,$$

$$A^T y = 0, \qquad y \geq 0.$$

To utilize the generalized geometric programming formulation, define

$$f(x) = \log \sum_{i \in J(0)} c_i \exp(x_i),$$

$$C = \left\{ x \in R^n : \sum_{i \in J(k)} c_i \exp(x_i) \leq 1,\ k = 1, \ldots, p \right\},$$

$$X = R(A).$$

Then, the remaining function and sets may be determined to be

$$g(y) = \sum_{i \in J(0)} y_i \log(y_i/c_i) + \sum_{k=1}^{p} \sum_{i \in J(k)} y_i \log\left(y_i/c_i \textstyle\sum_{J(k)} y_j\right),$$

$$D = \left\{ y \in R^n : y_i \geq 0,\ i = 1, \ldots, n,\ \sum_{J(0)} y_i = 1 \right\},$$

$$Y = N(A^T).$$

To apply Theorem 3.1, we first note that neither (P_p) nor (D_p) can have a direction of constancy. In (P_p), a direction of recession cannot have positive components; and, in (D_p), a direction of recession cannot have negative components. Therefore, if v is a direction of recession, $-v$ cannot be one; hence, no direction of constancy can exist. As a result, Theorem 3.1, when applied to (P_p) and (D_p), states that one problem has a nonzero direction of recession iff the other is not strongly consistent. (P_p) is said to be superconsistent if there is a feasible point which satisfies all nonlinear constraints strictly. Strong consistency in (P_p) is easily seen to be equivalent to superconsistency, and strong consistency in the dual is equivalent to the existence of a strictly positive dual solution. Thus, we obtain the well-known conditions for degeneracy.

When (P_p) or (D_p) has a direction of recession, the reduced form (P_g^r) may be obtained by projecting the objective function and constraint sets. Details are given in Refs. 2 and 9. We therefore consider only the reduced form (D'_g). First, suppose that (P_p) has a recession cone S. The recession cone S of (P_p) is contained in $R(A)=X$. Therefore,

$$S^{\perp} \supset Y \qquad \text{and} \qquad S^{\perp} \cap Y = Y.$$

As shown in Ref. 9, every $v \in \operatorname{ri} S$ is of the form

$$v = -\sum_{i \in U} \alpha_i e_i,$$

where e_i is the ith unit vector, $\alpha_i \geq 0$, and

$$U = \{i : y \in D \cap Y \Rightarrow y_i = 0\}.$$

Because $y \in D \cap S^{\perp}$ implies $y \geq 0$, and $(v, y) = 0$ for all $v \in S$, we conclude that $y_i = 0$ for $i \in U$. Substituting zero for such y_i yields the reduced form of Ref. 1.

Finally, consider the case in which (D_p) has a recession cone S and (P_p) is not superconsistent. As above, $S \subset Y$ and hence $S^{\perp} \cap X = X$. The form of the vectors in S is given in Ref. 2. Once this is known, the calculation of $C \cap S^{\perp}$ is a straightforward but tedious matter. The result is that

$$C \cap S^{\perp} = \Big\{ x \in R^n \colon \sum_{i \in J(k)} c_i \exp(x_i) \leq 1,\ k \in \{1, 3, \ldots, p\}/W,\ x_i = \hat{x}_i,$$
$$i \in J(k),\ k \in W \Big\},$$

where W is the set of constraints that hold as equalities for all feasible solutions and the $\hat{x}_i$ are the unique feasible values of the x_i in those constraints. See Ref. 2 for more detailed explanation of W and $\hat{x}_i$. Hence, the reduced form is obtained by explicitly substituting the unique values of x_i, $i \in J(k)$, $k \in W$.

5. Reduction of Other General Dual Pairs

In this section, we give results analogous to those of Section 3 for two of the more common dual pairs, namely, the dual problems of Fenchel and the ordinary nonlinear programming primal problem and its Lagrangian dual. It is pointed out in Ref. 14 that the generalized geometric programming formulation of duality is equivalent to Rockafellar's general dual problems and to the dual problems of Fenchel. In Ref. 16, it is shown that Fenchel's dual problems and the ordinary Lagrangian dual are equivalent. Thus, the reduced form of one can be derived from any of the others. However, in the case of the ordinary nonlinear programming problem, a direct approach is simpler and more intuitive.

Dual Problems of Fenchel. Let f and g be closed convex and concave functions with effective domains F and G, respectively. Let G^* and F^* be the effective domains of the conjugate transformation of g and f. The dual problems of Fenchel are:

$$(\mathrm{P}_F) \qquad \underset{x \in F \cap G}{\text{minimize}} \{f(x) - g(x)\},$$

$$(\mathrm{D}_F) \qquad \underset{y \in G^* \cap F^*}{\text{maximize}} \{g^*(y) - f^*(y)\}.$$

Theorem 3.1, which was first given for Fenchel dual problems, holds exactly as stated, with (P_g) and (D_g) replaced by (P_F) and (D_F). The reduced forms are similar to those of Section 3. Let S be the recession cone of (P_F); the reduced forms of (P_F) and (D_F) are

$$(\mathrm{P}_F^R) \qquad \underset{x \in P_{S^\perp}(F \cap G)}{\text{minimize}} P_{S^\perp}(f-g)(x),$$

$$(\mathrm{D}_F^R) \qquad \underset{y \in F^* \cap G^* \cap (S^\perp + y^0)}{\text{maximize}} g^*(y) - f^*(y),$$

where y^0 is any point in the hyperplane separating F^* and G^*. Note that $S^\perp + y^0$ is an affine subset of the separating hyperplane which contains y^0, and this affine set contains $F^* \cap G^*$. The analogues of all results of Section 3 apply. One difficulty in the above reduced form is that a direction of recession of $f-g$ need not be a direction of recession of either f or g. Thus, Proposition 3.1 is not directly applicable. We take care of this difficulty as follows.

Proposition 5.1. Let v be a direction of recession of $f-g$, and let y^0 be any point in the hyperplane with normal v which separates F^* and G^*. Then, v is a direction of recession of both $f(x)-(y_0, x)$ and $g(x)-(y_0, x)$.

Proof. Assume that H_α^v is a separating hyperplane of F^* and G^* such that

$$(v, y) \le \alpha \qquad \text{for all } y \in F^*,$$

$$(v, y) \ge \alpha \qquad \text{for all } y \in G^*,$$

$$(v, y^0) = \alpha.$$

Then, for $\lambda \geq 0$,

$$f(x+\lambda v)-(y_0, x+\lambda v) = \sup_{y \in F^*} \{(x+\lambda v, y) - f^*(y)\} - (y_0, x+\lambda v)$$

$$\leq \sup_{y \in F^*} \{(x, y) - f^*(y)\} + \lambda \sup_{y \in F^*} (v, y) - (y_0, x+\lambda v)$$

$$\leq f(x) + \lambda\alpha - (y_0, x) - \lambda\alpha$$

$$= f(x) - (y_0, x).$$

The other half is proven similarly. □

It is now possible to replace $f-g$ by $[f(x)-(x, y_0)]-[g(x)-(x, y_0)]$ and to use Proposition 3.1 to calculate the projected functions separately; i.e., it follows that

$$P_{S^\perp}(f-g)(x) = P_{S^\perp}[f(x)-(x, y^0)] - P_{S^\perp}[g(x)-(x, y^0)],$$

and each of the projected functions may be calculated by taking limits.

Reduction of the Ordinary Lagrangian Dual Problem. Let $f^0(x)$, $f^1(x), \ldots, f^m(x)$ be convex functions, and let C be a convex set such that

$$\operatorname{dom} f^i \supset C, \qquad i = 1, \ldots, m.$$

We consider the *ordinary* nonlinear programming problem

$$(\mathrm{P}_0) \qquad \text{minimize } f^0(x),$$
$$\text{subject to } f^i(x) \leq 0, \qquad i = 1, 2, \ldots, m,$$
$$x \in C,$$

and the dual

$$(\mathrm{D}_0) \qquad \text{maximize} \inf_{\substack{y \geq 0 \\ x \in C}} \left\{ f^0(x) + \sum_{i=1}^{m} y_i f^i(x) \right\}.$$

Although, as pointed out above, (P_0) and (D_0) may be considered as a special case of the generalized geometric or Fenchel dual problems, we feel that a direct approach is far more enlightening in this case. To characterize *degeneracy*, first define the (concave) dual function in the usual way, i.e.,

$$h(y) = \inf_{x \in C} \left\{ f^0(x) + \sum_{i=1}^{m} y_i f^i(x) \right\},$$

$$T = \{y \geq 0 : h(y) > -\infty\}.$$

Then, we consider (D_0) to be the problem of maximizing $h(y)$ over T. Note that, since y must be nonnegative in T, (D_0) cannot have a direction of constancy.

Theorem 5.1. There is no $x \in C$ such that $f^i(x) < 0$, $i = 1, 2, \ldots, m$, iff $h(y)$ has a direction of recession.

Proof. If there is no $x \in C$ that satisfies $f^i(x) < 0$, $i = 1, \ldots, m$, the theorem of Fan, Glickberg, and Hoffman (FGH; see, e.g., Theorem 21.1 of Ref. 12) implies the existence of a $0 \leq v \neq 0$ such that

$$\sum_{i=1}^{m} v_i f^i(x) \geq 0 \qquad \text{for all } x \in C.$$

Note that the assumption $\operatorname{dom} f^i \supset C$ at the beginning of the section is needed to apply the FGH theorem. Then, for $\lambda \geq 0$,

$$h(y + \lambda v) = \inf_{x \in C} \left\{ f^0(x) + \sum_{i=1}^{m} y_i f^i(x) + \lambda \sum_{i=1}^{m} v_i f^i(x) \right\}$$

$$\geq h(y) + \lambda \inf_{x \in C} \left\{ \sum_{i=1}^{m} v_i f^i(x) \right\}.$$

The last term is nonnegative. Therefore,

$$h(y + \lambda v) \geq h(y),$$

and v is a direction of recession of the concave function $h(y)$.

Conversely, suppose that, for some $x^0 \in C$, $f^i(x^0) < 0$, $i = 1, \ldots, m$, and suppose that $v \neq 0$ is a direction of recession of $h(y)$. Then,

$$\sum_{i=1}^{m} v_i f^i(x^0) < 0.$$

It follows that, for $\lambda > 0$,

$$h(y + \lambda v) = \inf_{x \in C} \left\{ f(x) + \sum_{i=1}^{m} y_i f^i(x) + \lambda \sum_{i=1}^{m} v_i f^i(x) \right\}$$

$$\leq \inf_{x \in C} \left\{ f(x) + \sum_{i=1}^{m} y_i f^i(x) + \lambda \sum_{i=1}^{m} v_i f^i(x^0) \right\} < h(y),$$

which contradicts the fact that v is a direction of recession. □

We see from the proof of the theorem that a direction of recession of the dual is the vector shown to exist by the FGH theorem. Partition the constraints of (P_0) into the following two sets:

$$I_1 = \{k : \exists x \in C \ni f^i(x) \leq 0,\ i = 1, \ldots, m, \text{ and } f^k(x) < 0\},$$

$$I_2 = \{k : f^i(x) \leq 0,\ x \in C,\ i = 1, \ldots, m \Rightarrow f^k(x) = 0\}.$$

For each $k \in I_1$, let $x^k \in C$ satisfy $f^i(x^k) \leq 0$, $i = 1, \ldots, m$, and $f^k(x^k) < 0$. Denote the number of elements in I_1 by η. Then, the feasible vector

$$\bar{x} = (1/\eta) \sum_{k=1}^{\eta} x^k$$

satisfies the constraints of I_1 strictly. Assuming that I_2 is not empty, the FGH theorem implies that

$$\sum_{i=1}^{m} v_i f^i(\bar{x}) \geq 0.$$

But $f^i(\bar{x}) = 0$ for $i \in I_2$ and $f^i(\bar{x}) < 0$ for $i \in I_1$. Therefore, $v_i = 0$ for $i \in I_1$.

The restriction analogous to that of Section 3 is to require that the values of $\{f^1(x), f^2(x), \ldots, f^m(x)\}$ lie in the orthogonal complement of the recession cone of D_0, i.e., to require that $f^k(x) = 0$ for each k such that $v_k > 0$. For many classes of functions, e.g., strictly faithfully convex, $f^k(x) = 0$ for $x \in C$ may be replaced by the requirement that some components of x take on a unique value as in Section 4 or that x satisfy a set of linear equations (e.g., Ref. 17). The projected dual function $P_{S^\perp} h(y)$ may be calculated by taking the limit of $h(y)$ in the direction of v. If $f^k(x) \neq 0$, the limit of $h(y)$ in the direction will be minus infinity. Thus, the result of the projection in D_0 is also to require that $f^k(x) = 0$ for k such that $v_k > 0$.

The existence of an interior point in the dual problem D_0 is characterized by the following theorem which we state without proof.

Theorem 5.2. There does not exist a positive vector $y > 0$ in the dual feasible set T iff P_0 has a direction of recession in which one of the constraints is unbounded below.

If the dual has no positive feasible vector, a reduced form of the primal is obtained by taking limits of all functions in a direction in the relative interior of the recession cone. Constraints which approach minus infinity may be dropped because, in the limit, they will be satisfied for any $x \in C$. The restriction of D_0 is obtained by simply deleting the components of the dual vectors y that must be zero and the corresponding constraints.

References

1. DUFFIN, R. J., PETERSON, E. L., and ZENER, C., *Geometric Programming—Theory and Applications,* John Wiley and Sons, New York, New York, 1967.
2. ABRAMS, R., *Consistency, Superconsistency and Dual Degeneracy in Geometric Programming,* Operations Research, Vol. 24, pp. 325–335, 1976.

3. WILLIAMS, A. C., *Complementary Theorems for Linear Programming*, SIAM Review, Vol. 12, pp. 135–137, 1970.
4. SHEFI, A., *Reduction of Linear Inequality Constraints and Determination of All Feasible Extreme Points*, Stanford University, PhD Dissertation, 1969.
5. LUENBERGER, D. G., *Introduction to Linear and Nonlinear Programming*, Addison-Wesley Publishing Company, Reading, Massachusetts, 1973.
6. ROCKAFELLAR, R. T., *Convex Functions and Dual Extremum Problems*, Harvard University, PhD Dissertation, 1963.
7. PETERSON, E. L., *Fenchel's Hypothesis and the Existence of Recession Directions in Convex Programming*, Northwestern University, Center for Mathematical Studies in Economics and Management Science, Discussion Paper No. 152, 1976.
8. WU, C. T., *Reduction and Restriction Methods for Simplifying and Solving Nonlinear Programming Problems*, Northwestern Universty, PhD Dissertation, 1975.
9. ABRAMS, R., *Projections of Convex Programs with Unattained Infima*, SIAM Journal on Control, Vol. 13, pp. 706–718, 1975.
10. PETERSON, E. L., and ECKER, J. G., *Geometric Programming: Duality in Quadratic Programming and Lp-Approximation, III, Degenerate Programs*, Journal of Mathematical Analysis and Applications, Vol. 24, pp. 365–383, 1970.
11. ABRAMS, R., *Degenerate Quadratic Programming and Lp-Approximation Problems*, Journal of Mathematical Analysis and Applications, Vol. 55, No. 2, 1976.
12. ROCKAFELLAR, R. T., *Convex Analysis*, Princeton University Press, Princeton, New Jersey, 1969.
13. STOER, J., and WITZGALL, C., *Convexity and Optimization in Finite Dimensions, I*, Springer-Verlag, New York, New York, 1970.
14. PETERSON, E. L., *Symmetric Duality for Generalized Unconstrained Geometric Programming*, SIAM Journal of Applied Mathematics, Vol. 19, pp. 487–526, 1970.
15. ABRAMS, R. A., and KERZNER, L., *A Simplified Test for Optimality*, Journal of Optimization Theory and Applications (to appear).
16. MAGNANTI, T. L., *Fenchel and Lagrange Duality Are Equivalent*, Mathematical Programming, Vol. 7, pp. 253–258, 1974.
17. BEN-TAL, A., BEN-ISRAEL, A., and ZLOBEC, S., *Characterization of Optimality in Convex Programming Without a Constraint Qualification*, Journal of Optimization Theory and Applications, Vol. 20, pp. 417–437, 1976.

9

Transcendental Geometric Programs[1]

G. Lidor[2] and D. J. Wilde[3]

Abstract. This paper treats a class of posynomial-like functions whose variables may appear also as exponents or in logarithms. It is shown that the resulting programs, called transcendental geometric programs, retain many useful properties of ordinary geometric programs, although the new class of problems need not have unique minima and cannot, in general, be transformed into convex programs. A duality theory, analogous to geometric programming duality, is formulated under somewhat more restrictive conditions. The dual constraints are not all linear, but the notion of *degrees of difficulty* is maintained in its geometric programming sense. One formulation of the dual program is shown to be a generalization of the chemical equilibrium problem where correction factors are added to account for nonideality. Some of the computational difficulties in solving transcendental programs are discussed briefly.

1. Introduction and Formulation

Prototype geometric programs deal with a class of positive algebraic functions called posynomials (Ref. 1, p. 2). This paper presents an extension to geometric programming, dealing with functions whose variables may appear also in exponents or logarithms. The resulting class of programs is called *transcendental geometric programs* or simply *transcendental programs*.

The study of transcendental programs was motivated primarily by the potential applications of primal transcendental programs. However, our

[1] This research was partially supported by the National Institute of Health Grant No. GM-14789; Office of Naval Research under Contract No. N00014-75-C-0276; National Science Foundation Grant No. MPS-71-03341 A03; and the US Atomic Energy Commission Contract No. AT(04-3)-326 PA # 18.

[2] Assistant Professor, Department of Computer Sciences, The City College, New York, New York.

[3] Professor, Department of Mechanical Engineering, Stanford University, Stanford, California.

results extend also to dual problems, of interest in computing chemical equilibria. We demonstrate that dual transcendental programs can represent nonideal chemical systems.

Our main results show that the duality theory of geometric programming applies also to transcendental programs in a somewhat weaker and more restrictive form. In particular, the notion of degrees of difficulty and its implications for problems with zero degrees of difficulty are shown to hold also for transcendental programs. In the remainder of this section, we formulate the primal transcendental program (PTP). Section 2 examines the necessary conditions for minima of PTP and establishes a lower bound to it by means of the geometric inequality. The dual transcendental program (DTP) and its relation to the primal program are presented in Section 3, followed by some properties of transcendental programs in Section 4. Section 5 discusses computational aspects and presents general approaches to solution techniques. In Section 6, we show that dual transcendental programs are closely related to chemical equilibrium in nonideal systems.

We consider positive functions in positive variables $x \in R_+^m$ and $t \in R_+^p$, where the subscript + indicates the positive orthant. The distinction between the *algebraic variables* x and the *transcendental variables* t is mostly for convenience. Algebraic variables are those appearing only in the usual posynomial forms, as shown below. The *transcendental functions* considered here have the form

$$g_k(x, t) \triangleq \sum_{j \in \langle k \rangle} u_j(x, t),$$

where $\langle k \rangle$ denotes a subset of consecutive integers of $\{1, 2, \ldots, n\}$ and the *terms* $u_j(x, t)$ are defined by

$$u_j(x, t) \triangleq C_j \left[\prod_{i=1}^{m} x_i^{a_{ij}} \right] \prod_{l=1}^{p} [t_l^{b_{lj}} \exp(d_{lj} t_l)].$$

Here, C_j are positive constants, and a_{ij}, b_{lj}, d_{lj} are fixed real numbers for $i = 1, 2, \ldots, m$, $j = 1, 2, \ldots, n$, $l = 1, 2, \ldots, p$.

We note that, if either $d_{lj} = 0$ for all l and j or $b_{lj} = 0$ for all l and j, then the problem reduces to a standard geometric program. In the latter case, this reduction is achieved by defining

$$x_{m+l} = \exp(t_l).$$

The *primal transcendental geometric program* is as follows.

Program PTP

$$\text{minimize } g_0(x, t),$$

$$\text{subject to } g_k(x, t) \leq 1, \qquad k = 1, 2, \ldots, K,$$

$$x > 0, \qquad t > 0.$$

This problem can be characterized by (i) a positive n-vector of coefficients C, (ii) real exponent matrices $A \in R^{m\times n}$, $B \in R^{p\times n}$, $D \in R^{p\times n}$, and (iii) a partition of the integer set $\{1, 2, \ldots, n\}$ defining the index sets $\langle k \rangle$, $k = 0, 1, \ldots, K$.

Before examining the necessary conditions for optimality we give several examples of problems leading to transcendental programs.

Example 1.1. *Logarithmic Functions.* Let

$$u_j(x) \triangleq C_j \prod_{i=1}^{m} [x_i^{a_{ij}}(\log x_i)^{b_{ij}}],$$

where $x_i > 1$, $i = 1, 2, \ldots, m$. Define

$$t_i \triangleq \log x_i, \qquad i = 1, 2, \ldots, m.$$

Then

$$u_j(t) = C_j \prod_{i=1}^{m} [t_i^{b_{ij}} \exp(a_{ij}t_i)].$$

Example 1.2. *Posynomial Functions in Exponents.* Let

$$v_j(x) \triangleq C_j \prod_{i=1}^{m} x_i^{a_{ij}},$$

a posynomial term, and consider constraints of the form

$$v_s(x) \cdot \exp\left\{\sum_{j=1}^{n} v_j(x)\right\} \le 1.$$

Let

$$t_1 \triangleq \sum_{j=1}^{n} v_j(x).$$

The constraint may be rewritten as

$$v_s(x) \cdot \exp(t_1) \le 1,$$

with the additional constraint

$$\sum_{j=1}^{n} t_1^{-1} v_j(x) \le 1.$$

Example 1.3. Smooth curves on a log–log plot can give rise to functions like:

$$v(x) \triangleq x^{-p(x)},$$

where $p(x)$ is a posynomial term. Let

$$t_1 \triangleq \log x, \qquad t_2 \triangleq p(x).$$

To minimize $v(x)$, solve the problem:

$$\text{minimize } (t_1^{-1} t_2^{-1}),$$
$$\text{subject to } x^{-1}\exp(t_1) \leq 1, \qquad p(x)^{-1} t_2 \leq 1.$$

We shall assume at the outset that program PTP is consistent and that a positive minimizing point exists so that the infimum operation can always be replaced by the minimum.

2. Necessary Conditions for Minima

Necessary conditions for a minimum in the primal program (PTP) are derived through a straightforward application of Kuhn–Tucker conditions (Ref. 2) and the logarithmic derivatives.

Proposition 2.1. Let problem PTP be superconsistent. If $(x^*, t^*) > 0$ is a (local) minimum for problem PTP, there exist nonnegative multipliers $(\delta_1, \delta_2, \ldots, \delta_n)$ and $(\lambda_0, \lambda_1, \ldots, \lambda_K)$ such that:[4]

$$A\delta = 0, \tag{1}$$

$$(B + TD)\delta = 0, \tag{2}$$

$$\sum_{j \in \langle k \rangle} \delta_j = \begin{cases} 1 = \lambda_0, & k = 0, \\ \lambda_k, & k = 1, 2, \ldots, K, \end{cases} \tag{3}$$

$$\delta \geq 0, \qquad \lambda \geq 0.$$

Here A, B, D are the matrices in PTP, δ and λ are the multipliers vectors, and T is a $p \times p$ diagonal matrix with

$$T_{ll} = t_l^*.$$

Proof. Conditions (1) and (3) are the standard dual constraints derived in the usual way as described by Duffin, Peterson, and Zener (Ref. 1, p. 117). Note that PTP is assumed to be superconsistent, i.e., having an *interior feasible point.* To prove (2), let the Lagrangian function be

$$L(x, t, \mu) \triangleq \mu_0 g_0(x, t) + \sum_{k=1}^{K} \mu_k [g_k(x, t) - 1],$$

[4] Equations (1) represent algebraic orthogonality conditions. Equations (2) represent transcendental orthogonality conditions. Equations (3) are normality conditions. Finally, $\delta \geq 0$ and $\lambda \geq 0$ constitute nonnegativity conditions.

where $\mu_k \geq 0$ are the Lagrange multipliers, with $\mu_0 \triangleq 1$. Since $L(x^*, t^*, \mu^*) > 0$ at the optimal point, we can replace the Kuhn–Tucker condition

$$\partial L(x^*, t^*, \mu^*)/\partial t_l = 0, \qquad l = 1, 2, \ldots, p,$$

by

$$\partial \log[L(x^*, t^*, \mu^*)]/\partial t_l = 0.$$

Written explicitly, we have

$$[t_l^*/L(x^*, t^*, \mu^*)]\left[\sum_{k=0}^{K} \mu_k^* \cdot \sum_{j\in\langle k\rangle} \{(1/t_l^*)b_{lj}u_j(x^*, t^*) + d_{lj}u_j(x^*, t^*)\}\right] = 0. \tag{4}$$

Define

$$\delta_j = \mu_k \cdot u_j(x^*, t^*)/g_0(x^*, t^*), \qquad k = 0, 1, \ldots, K, \qquad j \in \langle k\rangle.$$

Substituting in (4) and recalling that, by complementary slackness,

$$L(x^*, t^*, \mu^*) = g_0(x^*, t^*),$$

we obtain

$$\sum_{j=1}^{n} (b_{lj}\delta_j + t_l d_{lj}\delta_j) = 0, \qquad l = 1, 2, \ldots, p. \tag{5}$$

These are conditions (2). □

We note that the transcendental orthogonality conditions are linear in the *primal variables* t; furthermore, each of the p constraints (5) has only one transcendental variable; thus, if δ were known, t could be easily computed via (5).

The following example demonstrates that these necessary conditions lead to a definition of *degrees of difficulty* equivalent to the definition for posynomial programs, with the same striking result for problems of zero degrees of difficulty, namely, a complete solution without any optimization.

Example 2.1

$$\text{minimize } g_0(x, t) \triangleq x_1^{-1}x_2^2 t + x_1 t^{-2} \exp(-t),$$

$$\text{subject to } g_1(x, t) \triangleq x_2^{-1} \exp(2t) \leq 1,$$

$$x_1, x_2, t > 0.$$

The corresponding necessary conditions are:

$$-\delta_1 \quad +\delta_2 \qquad\qquad = 0, \tag{6-1}$$

$$2\delta_1 \qquad\qquad -\delta_3 = 0, \tag{6-2}$$

$$\delta_1 \quad -(2+t)\delta_2 \quad +2t\delta_3 = 0, \tag{7}$$

$$\delta_1 \quad +\delta_2 \qquad\qquad = 1, \tag{8}$$

$$\delta_1, \delta_2, \delta_3 \geq 0.$$

Equations (6) and (8) have the unique solution

$$\delta_1^* = 0.5, \qquad \delta_2^* = 0.5, \qquad \delta_3^* = 1.$$

With these values, Eq. (7) yields

$$t^* = \tfrac{1}{3}.$$

The constraint $g_1(x, t)$ is tight, since $\delta_3 > 0$, whence

$$x_2^* = \exp(2/3) = 1.94773.$$

$\delta_1^* = \delta_2^*$ implies

$$x_1^{-1} x_2^2 t^* = x_1 t^{*-2} \exp(-t^*),$$

so that

$$x_1^* = [x_2^{*2} t^* / t^{*-2} \exp(t^*)]^{1/2} = 0.44282,$$

and finally

$$g_0(x^*, t^*) = 5.71134. \qquad \square$$

We can now establish a lower bound on $g(x, t)$ by applying the arithmetic–geometric mean inequality (Ref. 3).

Lemma 2.1. For any feasible (x, t) and any $(\delta(t), \lambda(t))$ satisfying the conditions of Proposition 2.1,

$$g_0(x, t) \geq v(\delta, \lambda) \triangleq \left[\prod_j (C_j/\delta_j)^{\delta_j}\right]\left[\prod_k \lambda_k^{\lambda_k}\right]\left[\prod_l (-\eta_l/e\zeta_l)^{\eta_l}\right]. \tag{9}$$

Here,

$$\eta_l = \sum_{j=1}^{n} b_{lj}\delta_j, \qquad \zeta_l = \sum_{j=1}^{n} d_{lj}\delta_j, \tag{10}$$

and e is the base of natural logarithms (2.718 . . .).

Proof. The proof is analogous to the proof of the main lemma of geometric programming (Ref. 1, p. 114). For feasible (x, t) and λ, we have

$$\prod_{k=1}^{K} [g_k(x, t)]^{\lambda_k} \leq 1.$$

Therefore,

$$g_0(x, t) \geq \prod_{k=0}^{K} [g_k(x, t)]^{\lambda_k},$$

but

$$[g_k(x, t)]^{\lambda_k} \geq \left[\prod_{j \in \langle k \rangle} (u_j(x, t)/\delta_j)^{\delta_j} \right] \lambda_k^{\lambda_k},$$

where

$$x^x = x^{-x} \triangleq 0 \qquad \text{when} \qquad x = 0.$$

Thus,

$$g_0(x, t) \geq \prod_{k=0}^{K} \left\{ \left[\prod_{j \in \langle k \rangle} (u_j(x, t)/\delta_j)^{\delta_j} \right] \lambda_k^{\lambda_k} \right\}.$$

The right-hand side with conditions (1) and (2) yields

$$\left[\prod_{j=1}^{n} (C_j/\delta_j)^{\delta_j} \right] \left[\prod_{k=0}^{K} \lambda_k^{\lambda_k} \right] \left[\prod_{j} \left(\prod_{l=1}^{p} t_l^{b_{lj}} \exp(d_{lj} t_l) \right)^{\delta_j} \right]$$

$$= \left[\prod_{j} (C_j/\delta_j)^{\delta_j} \right] \left[\prod_{k} \lambda_k^{\lambda_k} \right] \left[\prod_{l} t_l^{\eta_l} \exp(\zeta_l t_l) \right].$$

But, by (5) and (10),

$$t_l = -\eta_l/\zeta_l, \qquad l = 1, 2, \ldots, p,$$

whence

$$t_l^{\eta_l} \exp(\zeta_l t_l) = (-\eta_l/\zeta_l)^{\eta_l} \exp(-\eta_l) = (-\eta_l/e\zeta_l)^{\eta_l},$$

completing the proof. □

3. Dual Problem

One is tempted, in view of Lemma 2.1, to eliminate t from the necessary conditions (2) and construct a *dual objective function* to be maximized; namely, we have the following problem.

Problem DPL

$$\text{maximize } v(\delta,\lambda) \triangleq \prod_j (C_j/\delta_j)^{\delta_j} \prod_k \lambda_k^{\lambda_k} \prod_l (-\eta_l/e\zeta_l)^{\eta_l},$$

$$\text{subject to } A\delta = 0,$$

$$\sum_{j\in\langle k\rangle} \delta_j = \begin{cases} 1, & k=0, \\ \lambda_k, & k=1,2,\ldots,K, \end{cases}$$

$$\delta \geq 0, \qquad \lambda \geq 0,$$

$$\text{where} \qquad \eta_l = \sum_j b_{lj}\delta_j, \qquad \zeta_l = \sum_j d_{lj}\delta_j.$$

Unfortunately, this *dual problem* does not serve as a *global lower bound* on $g_0(x, t)$, since feasible values of δ and λ do not necessarily satisfy conditions (2) for an arbitrary feasible t. The above dual program is in this sense not a *pure dual program*, since it implicitly incorporates primal variables t. The elimination of t from (2) does not take into account the added requirement that t be *primal feasible* or equivalently $-\eta_l/\zeta_l$ must be primal feasible. The following proposition establishes the equivalence of transcendental and ordinary posynomial programs when t is fixed.

Proposition 3.1. For any fixed feasible value of t, the primal program PTP and the dual program DPL reduce to a pair of posynomial primal and dual programs when conditions (2) are satisfied.

Proof. Consider a fixed $\bar{t}$ and define the modified coefficients

$$\bar{C}_j = C_j \prod_l \bar{t}_l^{b_{lj}} \exp(d_{lj}\bar{t}_l).$$

The primal problem reduces to a primal geometric program with coefficient vector $\bar{C}$. The dual geometric objective function is then

$$\begin{aligned} v(\delta,\lambda) &= \left[\prod_j (\bar{C}_j/\delta_j)^{\delta_j}\right]\left[\prod_k \lambda_k^{\lambda_k}\right] \\ &= \left[\prod_j (C_j/\delta_j)^{\delta_j}\right]\left[\prod_k \lambda_k^{\lambda_k}\right]\left[\prod_j \left(\prod_l \bar{t}_l^{b_{lj}} \exp(d_{lj}\bar{t}_l)\right)^{\delta_j}\right] \\ &= \left[\prod_j (C_j/\delta_j)^{\delta_j}\right]\left[\prod_k \lambda_k^{\lambda_k}\right]\left[\prod_l \bar{t}_l^{\eta_l} \exp(\zeta_l t_l)\right] \\ &= \left[\prod_j (C_j/\delta_j)^{\delta_j}\right]\left[\prod_k \lambda_k^{\lambda_k}\right]\left[\prod_l (-\eta_l/e\zeta_l)^{\eta_l}\right], \end{aligned}$$

where the last expression follows from the same argument used in Lemma 2.1.

With Proposition 3.1, we may now prove the equivalent of the *main lemma* of geometric programming.

Lemma 3.1. Let (x, t) be feasible for PTP, and let $\delta \triangleq \delta(t)$, $\lambda \triangleq \lambda(t)$ satisfy the necessary conditions (1)–(3) and nonnegativity. Then,

$$g_0(x, t) \geq v(\delta, \lambda). \tag{11}$$

Moreover, under these conditions,

$$g_0(x, t) = v(\delta, \lambda)$$

iff

$$\delta_j(t) = \begin{cases} u_j(x, t)/g_0(x, t), & j \in \langle 0 \rangle, \\ \lambda_k(t) u_j(x, t), & j \in \langle k \rangle, \quad k = 1, 2, \ldots, K. \end{cases} \tag{12}$$

Proof. The proof is completely analogous to the proof of the main lemma of geometric programming, where $g_k(x, t)$ now replaces the posynomial function $g_k(x)$ (Ref. 1, p. 114). □

The validity of (11) and (12) is based on the fact that the same t appears in both the primal and the dual problems. In order to let t vary, we observe that, for fixed $\bar{t}$,

$$g_0(x, \bar{t}) \geq v(\delta(\bar{t}), \lambda(\bar{t})).$$

Thus,

$$\min_x g_0(x, \bar{t}) \geq \max_{\delta,\lambda} v(\delta(\bar{t}), \lambda(\bar{t})),$$

whence

$$\min_t \min_x g_0(x, t) \geq \min_t \max_{\delta,\lambda} v(\delta(t), \lambda(t)),$$

where the minimum on the left is subject to the constraints of PTP and the optimization on the right is subject to the necessary conditions (1)–(3) and primal feasibility of t, that is, there exists an x such that (x, t) is primal feasible. If the latter condition is removed, then clearly the right-hand side represents a lower bound on $g_0(x, t)$. However, this lower bound may be *strictly* lower than the minimal value of $g_0(x, t)$. The last inequality gives rise to a dual program involving a *saddle point* of $v(\delta, \lambda, t)$.

Program DTP

$$\min_{t} \max_{\delta,\lambda} v(\delta, \lambda, t) \triangleq \left[\prod_{j=1}^{n} (C_j/\delta_j)^{\delta_j}\right]\left[\prod_{k=1}^{K} \lambda_k^{\lambda_k}\right]\left[\prod_{j}\left\{\prod_{l} t_l^{b_{lj}} \exp(d_{lj}t_l\delta_j)\right\}\right],$$

$$\text{subject to } t \text{ is primal feasible,} \tag{13}$$

$$A\delta = 0, \tag{14}$$

$$(B + TD)\delta = 0, \tag{15}$$

$$\sum_{j\in\langle k\rangle} \delta_j = \begin{cases} \lambda_k, & k = 1, 2, \ldots, K, \\ 1, & k = 0, \end{cases} \tag{16}$$

$$\delta, \lambda \geq 0, \tag{17}$$

where A, B, T, D are the matrices defined by program PTP and the necessary conditions (1)–(3).

With conditions (13) and (15), one may question the usefulness of this dual program. Before treating this question, we prove the duality theorem for transcendental geometric programs. We note first that, for any *fixed* t^* satisfying the constraints of both PTP and DTP, the corresponding programs, denoted by PTP(t^*) and DTP(t^*), are a pair of primal and dual geometric programs as a direct consequence of Proposition 3.1. Since the transcendental orthogonality conditions (5) are explicitly included in DTP, one may rewrite Proposition 3.1 as follows.

Proposition 3.2. For any fixed feasible t^*, programs PTP(t^*) and DTP(t^*) reduce to a pair of posynomial primal and dual programs, respectively.

Theorem 3.1. Suppose that the primal program PTP is superconsistent and attains its constrained minimum at a feasible point (x^*, t^*). Then, (i) the corresponding dual program DTP is consistent and has a feasible optimal solution $(\delta^*, \lambda^*, t^*)$; (ii) $g_0(x^*, t^*) = v(\delta^*, \lambda^*, t^*)$; (iii) the relations (12) hold at $(x^*, t^*, \delta^*, \lambda^*)$; and (iv) if $(\delta^*, \lambda^*, t^*)$ is optimal for the dual program DTP, each minimizing vector x for PTP(t^*) satisfies the system of equations

$$u_j(x^*, t^*) = \begin{cases} \delta_j^* v(\delta^*, \lambda^*, t^*), & j \in \langle 0\rangle, \\ \delta_j^*/\lambda_k^*, & j \in \langle k\rangle, \end{cases}$$

where k ranges over the integers for which $\lambda_k^* > 0$.

Proof. (i) DTP is consistent, since (x^*, t^*) must satisfy the necessary conditions for a minimum of $g_0(x, t)$ (Proposition 2.1). We only have to show that t^* is optimal also for DTP. t^* is clearly feasible for DTP by Proposition 2.1; so, suppose it is not optimal. Then, there exists $t'>0$ such that

$$\min_{x} g_0(x, t') \geq g_0(x^*, t^*) \geq v(\delta^*, \lambda^*, t^*) > \max_{\delta,\lambda} v(\delta, \lambda, t'). \tag{18}$$

For the fixed, feasible t', we may apply Proposition 3.2 and the geometric programming duality theory to show that there exist vectors

$$x'>0, \qquad (\delta', \lambda') \geq 0$$

optimizing PTP(t') and DTP(t'), respectively, such that

$$g_0(x', t') = v(\delta', \lambda', t'),$$

contrary to (18); hence, t^* is optimal for DTP.

Statements (ii), (iii), and (iv) follow directly from Proposition 3.2 and the posynomial geometric programming duality theorem (Ref. 1, p. 117). □

The reader may justly ask how the condition of primal feasibility of t could be incorporated in a *dual problem* without attaching the whole primal constraint set to the problem. This restriction on t serves to ensure *global duality*, namely, the dual provides a lower bound to the global minimum of the primal program, but the theory sacrifices much in elegance to obtain it. In some situations, t may not be severely restricted by the primal constraints, for instance, in unconstrained problems or in problems where each t_l appears only once in the constraints. In these cases, when $t>0$, primal feasibility is always guaranteed. From a practical point of view, the duality theory presented here is useful both in terms of computational algorithms, to be discussed later, and in terms of economic interpretation and sensitivity analysis, in much the same way that applies to posynomial duality.

4. Some Properties of Transcendental Programs

Extension to Exponents with an Arbitrary Base. The original formulation where all transcendental functions are of the form $\exp(d_{lj}t_l)$ can be easily extended to any base, so that programs with terms like $\alpha^{d_{lj}t_l}$ are also transcendental geometric programs. All that is required is a transformation of the matrix D, since

$$\alpha^{d_{lj}t_l} = \exp[d_{lj}t_l \log \alpha] = \exp(\tilde{d}_{lj}t_l),$$

where

$$\tilde{d}_{lj} \triangleq d_{lj} \log \alpha.$$

In general, one may have a different base for each term and variable, for example,

$$C_1\alpha_{11}^{t_1}\alpha_{21}^{t_2}\alpha_{31}^{t_3}+C_2\alpha_{21}^{-t_1}\alpha_{22}^{-t_2}\alpha_{23}^{-t_3}, \qquad \text{etc.}$$

The problem is equivalent to problem PTP, where the matrix D is replaced by $\tilde{D}$ defined by

$$\tilde{D}=(\tilde{d}_{lj})=d_{lj}\log\alpha_{lj}.$$

Degrees of Difficulty in Transcendental Programs. It was already pointed out in Example 2.1 that the dual transcendental program has a more restrictive set of constraints. The algebraic orthogonality conditions

$$A\delta=0$$

and the normality constraint still determine the dimension of the dual space. Indeed, if

$$n=m+1 \qquad \text{and} \qquad \text{rank } A=m,$$

these constraints have a unique solution (which may or may not be feasible even if $\delta>0$). We define, therefore, the degree of difficulty (dd) in the same way as for ordinary geometric programs. If

$$dd=0$$

and the dual program DTP is feasible, the solution is unique, as demonstrated in Example 2.1.

Recovering Primal Variables from the Dual Solution. As noted earlier, t need not appear explicitly in the dual problem, although it implicitly appears in the dual objective function and constraints. Given the optimal solution δ^*, we can find t^* by

$$t_l^*=\Big(-\sum_j b_{lj}\delta_j^*\Big)\Big/\Big(\sum_j d_{lj}\delta_j^*\Big)=-\eta_l^*/\zeta_l^*. \tag{19}$$

The variables x_i^* may now be computed by Theorem 3.1(iv) by first incorporating the values of t^* into the coefficients in the primal as described in Propositions 3.1 and 3.2. Thus,

$$\log\bar{C}_j+\sum_i a_{ij}\log x_i^*=\begin{cases}\log[\delta_j^*v(\delta^*,\lambda^*,t^*)], & j\in\langle 0\rangle,\\ \log[\delta_j^*/\lambda_k^*], & j\in\langle k\rangle, \quad k>0,\end{cases}$$

and the computation of x^* reduces to a solution of a linear system of equations.

Second-Order Conditions. Since the necessary conditions apply to any stationary point and not only to global (or local) minima, computational methods based on these conditions, namely dual methods, may terminate at so-called equilibrium points (Refs. 4, 5) or *quasiminima*. To study such solutions, we show here that second-order conditions may provide additional constraints to ensure true minima. We are mostly interested in the Hessian matrix of $v \triangleq v(\delta, \lambda, t)$, restricted to those values which satisfy the necessary conditions.

For simplicity, we shall look at the logarithm of v, rather than v itself. Let

$$V = \log v.$$

We have

$$V = \sum_j \delta_j(\log C_j - \log \delta_j) + \sum_k \lambda_k \log \lambda_k + \sum_j \sum_l (B_{lj} \log t_l + d_{lj} t_l)\delta_j.$$

second partial derivatives of V are

$$\partial^2 V/\partial\delta_r\, \partial\delta_s = \begin{cases} 1/\lambda_k - 1/\delta_r, & \text{if } r = s,\ r \in \langle k \rangle, \\ 1/\lambda_k, & \text{if } r \in \langle k \rangle \text{ and } s \in \langle k \rangle,\ r \neq s, \\ 0, & \text{otherwise,} \end{cases} \tag{20}$$

$$\partial^2 V/\partial\delta_j\, \partial t_l = b_{lj}/t_l + d_{lj} = -b_l\eta_l/\zeta_l + d_{lj}, \tag{21}$$

$$\partial^2 V/\partial t_r\, \partial t_s = \begin{cases} -\eta_r/t_r^2 = -\zeta_r^2/\eta_r, & r = s, \\ 0, & \text{otherwise,} \end{cases} \tag{22}$$

where

$$\eta_l = \sum_j b_{lj}\delta_j, \qquad \zeta_l = \sum_j d_{lj}\delta_j.$$

Once λ is eliminated, the Hessian may be described in partitioned form:

$$H(\delta, t) \triangleq \begin{pmatrix} H_{\delta\delta} & H_{\delta t} \\ H_{t\delta} & H_{tt} \end{pmatrix},$$

where

$$H_{\delta\delta} \in R^{n \times n}, \qquad H_{\delta t} \in R^{n \times p}, \qquad H_{t\delta} \in R^{p \times n}, \qquad H_{tt} \in R^{p \times p}.$$

$H_{\delta\delta}$ is the usual Hessian of the logarithm of dual geometric programs and is thus negative definite for all $\delta > 0$ (Ref. 1). For (δ, t) to be a saddle point, H_{tt} must be positive definite (t is strictly positive); so,

$$-\eta_l/t_l^2 > 0 \qquad \text{for all } l.$$

Thus,

$$\eta_l < 0 \qquad \text{for all } l; \tag{23}$$

and, since

$$t_l = -\eta_l/\zeta_l > 0,$$

we must also have

$$\zeta_l > 0 \qquad \text{for all } l. \tag{24}$$

To guarantee convergence of dual algorithms to a saddle point, one may add the last two restrictions to the dual constraints in the form

$$\eta_l \leq -\epsilon, \qquad \zeta_l \geq \epsilon, \tag{25}$$

where ϵ is a small positive constant.

Although the existence of a saddle point in the dual problem is required to ensure that the corresponding primal point is a global minimum, it need not be unique. The primal problem may have more than one point where the global minimum value of the objective function is attained. By Theorem 3.1, for every such point (x^*, t^*) there is a corresponding saddle point $(\delta^*, \lambda^*, t^*)$ in the dual.

Condensation of Transcendental Programs. Condensation techniques proved effective in linearizing posynomial programs (Refs. 6, 7). A similar approach applies also to transcendental programs. Let

$$g_k(x, t) \triangleq \sum_{j\in\langle k\rangle} u_j(x, t),$$

and let $(x^0, t^0) > 0$ be fixed. Define the weights

$$w_j^0 \triangleq w_j(x^0, t^0) \triangleq u_j(x^0, t^0)/g_k(x^0, t^0).$$

Then,

$$g_k(x, t) \geq \prod_{j\in\langle k\rangle} [u_j(x, t)/w_j^0]^{w_j^0} \qquad \text{for all } x, t > 0.$$

The right-hand side is a single term posynomial–exponential function, called the *condensation of* $g_k(x, t)$ *at* (x^0, t^0), defined by

$$\tilde{g}_k(x, t, x^0, t^0) \triangleq \tilde{C}_k \prod_i x_i^{\alpha_{ik}} \prod_l t_l^{\beta_{lk}} \exp(t_l \gamma_{lk}),$$

where

$$\tilde{C}_k \triangleq \prod_{j\in\langle k\rangle} (C_j/w_j^0)^{w_j^0}, \qquad \alpha_{ik} \triangleq \sum_{j\in\langle k\rangle} a_{ij} w_j^0,$$

$$\beta_{lk} \triangleq \sum_{j\in\langle k\rangle} b_{lj} w_j^0, \qquad \gamma_{lk} \triangleq \sum_{j\in\langle k\rangle} d_{lj} w_j^0.$$

Taking logarithms, we obtain

$$\log \tilde{g}_k = \log \tilde{C}_k + \sum_i \alpha_{ik} \log x_i + \sum_l (\beta_{lk} \log t_l + \gamma_{lk} t_l).$$

A condensed transcendental program can thus be viewed as a general linear–logarithmic problem, though this form is somewhat different from the linear–logarithmic problems treated by Clasen (Refs. 8, 9).

It can be easily verified that the following relations hold (Ref. 10).

(i) $\tilde{g}_k(x^0, t^0, x^0, t^0) = g_k(x^0, t^0)$,

(ii) $\nabla \tilde{g}_k(x^0, t^0, x^0, t^0) = \nabla g_k(x^0, t^0)$,

(iii) $\tilde{g}_k(x, t, x^0, t^0) \leq g_k(x, t) \qquad$ for all $(x, t) > 0$.

These properties of condensed transcendental programs suggest algorithms based on linearization similar to Dembo's method (Ref. 7) or modified versions thereof (see Lidor, Ref. 10).

5. Computational Aspects

Computationally, transcendental programs are inherently more difficult than prototype geometric programs or even signomial programs (Ref. 5) for the same reasons that render the theory more complex. Due to the possibility of multiple optima and stationary points which are not even local optima, most algorithms can guarantee at best convergence to some local optimum. It would seem that the dual program DTP, with its unorthodox *primal feasibility* and transcendental orthogonality constraints, does not offer all of the advantages of the linearly constrained prototype dual program. No attempt was made in this work to devise or test computational algorithms. Convex cutting-plane algorithms based on condensation will fail due to lack of the *nesting property*. However, condensation is still useful in approximating problems with negative coefficients in the same way that applies to complementary geometric programs (Ref. 4). Proposition 3.2 suggests a method to reduce the problem to a sequence of geometric programs by fixing t at each iteration, and improving it from one iteration to the next. Improvement of t may turn out to be quite complicated and the approach is not guaranteed to converge. Perhaps, existing techniques for solving general nonconvex programs would prove superior to any approach based on prototype geometric programming algorithms; still, the existence of a dual saddle point could be employed in computations.

6. Nonideal Chemical Systems and the Dual Problem

It has been shown by Avriel (Ref. 12) and Passy and Wilde (Ref. 13) that the problem of finding the equilibrium composition of reacting chemical species in an *ideal solution* is equivalent to the dual program of prototype geometric programming. We shall show here that some ways of treating nonideal systems lead to dual transcendental programs.

The free energy F of a chemical system with δ_j moles of species j at a fixed temperature T and pressure P is given by

$$F(T, P, \delta) = \sum_j \delta_j \mu_j(T, P, \delta),$$

where $\mu_j(T, P, \delta)$ is the *chemical potential* or the *partial molar free energy* of species j. Under fixed pressure and temperature, we may write (Ref. 14)

$$\mu_j = c_j + \log a_j,$$

where c_j is some standard value and a_j is the *activity* of species j. In principle, all of the nonideality in the system may be expressed in the values of a_j. In the ideal case,

$$a_j = \hat{\delta}_j \triangleq \delta_j / \lambda_k, \qquad j \in \langle k \rangle.$$

The variable $\hat{\delta}_j$ represents the *mole fraction* of species j. In the nonideal case, without loss of generality, the nonideality may be put into the *activity coefficient* $\gamma_j(\delta)$ by defining

$$a_j = \gamma_j(\delta) \cdot \hat{\delta}_j,$$

so that

$$F(\delta) = \sum_j \delta_j [c_j + \log \gamma_j(\delta) + \log \hat{\delta}_j]. \tag{26}$$

In practical applications, $\gamma_j(\delta)$ is expressed by complex empirical relations. Theoretical considerations (see Ref. 10) require that $F(\delta)$ be a homogeneous function of degree one, indicating that $\gamma_j(\delta)$ is homogeneous of degree zero or

$$\gamma_j(\delta) = \gamma_j(\hat{\delta}),$$

a function of *concentration*. Several relations of this type appear in the chemical literature. Among the better known are the virial equations and the Wilson equation (Ref. 15). We shall see that relations of this form may be obtained by appropriate selection of the matrices B and D in the transcendental program. Since the primal variables t have no meaning in this context, we consider problem DPL, where t has been eliminated using the transcendental orthogonality conditions. We have

$$\begin{aligned} F(\delta) &= -\log v(\delta, \lambda) \\ &= \sum_j \delta_j (c_j + \log \delta_j) - \sum_k \lambda_k \log \lambda_k - \sum_l \eta_l \log(-\eta_l / e \cdot \zeta_l), \end{aligned}$$

where

$$\eta_l = \sum_j b_{lj}\delta_j, \qquad \zeta_l = \sum_j d_{lj}\delta_j, \qquad c_j = -\log C_j,$$

$$\sum_l \eta_l \log(-\eta_l/e\zeta_l) = \sum_l \sum_j b_{lj} \log(-\eta_l/e\zeta_l)\cdot \delta_j = \sum_j \left[\sum_l b_{lj} \log(-\eta_l/e\zeta_l)\right]\delta_j.$$

Let

$$\tau_l(\delta) = -\eta_l/e\zeta_l.$$

The last expression becomes

$$\sum_j \left[\sum_l b_{lj} \log\{\tau_l(\delta)\}\right]\delta_j,$$

whence

$$F(\delta) = \sum_j \delta_j\Big(c_j + \log \hat{\delta}_j - \sum_l b_{lj} \log \tau_l\Big).$$

Comparing this with Eq. (26), we see that

$$\gamma_j(\delta) = \prod_l \tau_l(\delta)^{-b_{lj}}.$$

We can now examine structures of B and D that will yield meaningful interpretation for $\gamma_j(\delta)$. Note that the number of rows p in B and D can be arbitrarily chosen (the number of columns equals the number of species n).

Example 6.1. $p = K$ = the number of chemical phases in the system. Let

$$b_{lj} = \begin{cases} -1, & j \in \langle l \rangle, \\ 0, & \text{otherwise.} \end{cases}$$

In this case, each row l in B represents a phase and has -1's for all species in that phase. Similarly, let

$$d_{lj} = \begin{cases} f_l/e, & j \in \langle l \rangle, \quad f_l > 0, \\ 0, & \text{otherwise,} \end{cases}$$

where f_l are given constants.

Then,

$$\gamma_j(\delta) = \prod_l \left[-\Big(\sum_j b_{lj}\delta_j\Big)\Big/\Big(e\sum_j d_{lj}\delta_j\Big)\right]^{-b_{lj}}$$

$$= \Big(\sum_{j\in\langle l\rangle} \delta_j\Big)\Big/\Big(f_l \sum_{j\in\langle l\rangle} \delta_j\Big) = 1/f_l, \qquad j \in \langle l \rangle.$$

Chemically, this $\gamma_j(\delta)$ introduces a fixed correction factor for all species in a given phase. Although we have not found such factors in the literature, this situation may arise in a solution with a dominant polar solvent.

Example 6.2. $p - n$ – number of species. Let

$$b_{lj} = \begin{cases} b_{jj}, & l = j, \\ 0, & \text{otherwise.} \end{cases}$$

B is an $n \times n$ diagonal matrix. Let

$$d_{lj} = \begin{cases} d_l, & l \in \langle k \rangle \text{ and } j \in \langle k \rangle, \quad k = 1, 2, \ldots, K, \\ 0, & \text{otherwise.} \end{cases}$$

D is an $n \times n$ matrix with block angular form. Each block corresponds to one phase and each row in the block corresponds to one species in that phase. For example,

$$D = \left[\begin{array}{ccc|ccc|ccc} d_1 & d_1 \cdots & d_1 & & & & & & \\ d_2 & d_2 & d_2 & & 0 & & & 0 & \\ d_{n_1} & d_{n_1} & d_{n_1} & & & & & & \\ \hline & & & d_{n_1+1} & \cdots & d_{n_1+1} & & & \\ & 0 & & & & & & 0 & \\ & & & d_{n_2} & & d_{n_2} & & & \\ \hline & & & & & & d_{n_2+1} & \cdots & d_{n_2+1} \\ & 0 & & & 0 & & & & \\ & & & & & & d_{n_3} & & d_{n_3} \end{array}\right].$$

For species j in phase k, we obtain

$$\gamma_j(\delta) = [(1/ed_l)(-b_{jj}\delta_j/\lambda_k)]^{-b_{jj}}, \qquad j, l \in k.$$

Define

$$\beta_j = -b_{jj}/ed_l, \qquad \rho_j = -b_{jj}.$$

Then

$$\gamma_j(\delta) = \beta_j(\hat{\delta}_j)^{\rho_j}, \qquad \log \gamma_j(\delta) = \log \beta_j + \rho_j \log \hat{\delta}_j.$$

This is a simple linear correction to the *ideal* $\log \hat{\delta}_j$. By selecting more complex structures for B and D, more elaborate correction terms may be obtained.

Remark 6.1. In the chemical case, the dual is not a saddle-point problem; moreover, primal feasibility of the (implicit) t is not required. In this case, the primal problem is a saddle-point problem.

Remark 6.2. The *dual nature* of chemical equilibrium problems suggests that they can be solved more efficiently by dual methods. The corresponding primal program is still useful in sensitivity analysis and in gaining insight into the nature of equilibrium solutions (Refs. 10, 16).

7. Conclusions

We have shown that the existing theory of posynomial geometric programming provides most of the tools for handling the more complex and more general class of transcendental geometric programs. Practical problems may demonstrate a smoother and more regular behavior of transcendental functions than that which may be anticipated by theory. Our results lead the way to the development of techniques for solving such problems.

References

1. DUFFIN, R. J., PETERSON, E. L., and ZENER, C., *Geometric Programming*, John Wiley and Sons, New York, New York, 1967.
2. KUHN, H. W., and TUCKER, A. W., *Nonlinear Programming*, Proceedings of the Second Berkeley Symposium on Mathematics and Probability, Edited by T. Neyman, University of California Press, Berkeley, California, 1950.
3. HARDY, G. H., LITTLEWOOD, J. E., and POLYA, G., *Inequalities*, Cambridge University Press, Cambridge, England, 1964.
4. AVRIEL, M., and WILLIAMS, A. C., *Complementary Geometric Programming*, SIAM Journal of Applied Mechanics, Vol. 19, pp. 125–141, 1970.
5. PASSY, U., and WILDE, D. J., *Generalized Polynomial Optimization*, SIAM Journal of Applied Mechanics, Vol. 15, pp. 1344–1356, 1967.
6. DUFFIN, R. J., *Linearizing Geometric Programs*, SIAM Review, Vol. 12, pp. 211–227, 1970.
7. DEMBO, R. S., *Solution of Complementary Geometric Programs*, Technion—Israel Institute of Technology, M.S. Thesis, 1972.
8. CLASEN, R. J., *The Linear Logarithmic Programming Problem*, The RAND Corporation, Research Memorandum No. RM-3707-PR, 1963.
9. CLASEN, R. J., *The Numerical Solution of the Chemical Equilibrium Problem*, The RAND Corporation, Research Memorandum No. RM-4345-PR, 1965.
10. LIDOR, G., *Chemical Equilibrium Problems Treated by Geometric and Transcendental Programming*, Stanford University, Technical Report No. 75–8, Department of Operations Research, 1975.
11. ZANGWILL, W. I., *Nonlinear Programming: A Unified Approach*, Prentice-Hall, Englewood Cliffs, New Jersey, 1969.
12. AVRIEL, M., *Topics in Optimization: Block Search, Applied and Stochastic Geometric Programming*, Stanford University, PhD Thesis, 1966.

13. PASSY, U., and WILDE, D. J., *Mass Action and Polynomial Optimization*, Journal of Engineering Mathematics, Vol. 3, pp. 325–335, 1969.
14. DENBIGH, K., *The Principles of Chemical Thermodynamics*, Cambridge University Press, Cambridge, England, 1961.
15. PRAUSNITZ, J. M., ECKERT, C. A., DRYE, R. V., and O'CONNEL, J. P., *Computer Calculations for Multicomponent Vapor–Liquid Equilibria*, Prentice-Hall, Englewood Cliffs, New Jersey, 1967.
16. BIGELOW, J. H., and SHAPIRO, N. Z., *Sensitivity Analysis in Chemical Thermodynamics*, The RAND Corporation, Report No. P-4628, 1971.

10

Solution of Generalized Geometric Programs[1,2]

M. AVRIEL,[3] R. DEMBO,[4] AND U. PASSY[5]

Abstract. A cutting plane algorithm for the solution of generalized geometric programs with bounded variables is described and then illustrated by the detailed solution of a small numerical example. Convergence of this algorithm to a Kuhn–Tucker point of the program is assured if an initial feasible solution is available to initiate the algorithm. An algorithm for determining a feasible solution to a set of generalized posynomial inequalities which may be used to find a global minimum to the program as well as test for consistency of the constraint set, is also presented. Finally an application in optimal engineering design with seven variables and fourteen nonlinear inequality constraints is formulated and solved.

1. Introduction

This paper describes an algorithm for the solution of generalized geometric programs with bounded variables. Convergence of the algorithm to a Kuhn–Tucker solution (usually a local minimum) of the problem is assured if an initial feasible solution is available to initiate the algorithm.

In Section 2 the geometric programming theory necessary for the understanding of this paper is presented. Section 3 describes a cutting plane algorithm for the solution of regular geometric programs and gives a method

---

[1] Reproduced with permission, from International Journal for Numerical Methods in Engineering, Vol. 9.

[2] The authors are indebted to Dr. A. C. Williams whose ideas inspired the derivation of the algorithm presented in this work. This research was supported in part by the Gerard Swope Research Fund.

[3] Faculty of Industrial Engineering and Management, Technion—Israel Institute of Technology, Haifa, Israel.

[4] Faculty of Chemical Engineering, Technion—Israel Institute of Technology, Haifa, Israel.

[5] Faculty of Industrial Engineering and Management, Technion—Israel Institute of Technology, Haifa, Israel.

of combining this algorithm with an existing generalized geometric programming algorithm in an efficient manner. A method for finding a solution to a set of signomial inequalities is discussed in Section 4 and is demonstrated in Section 5, together with the algorithms of Section 3, with a numerical example. Finally, in Section 6, an engineering design example is developed and solved using the above methods.

2. Background and Definitions

2.1. Generalized Geometric Programming. We define a generalized geometric program (GGP) as the following nonlinear mathematical programming problem:

Minimize

$$P_0(\bar{x}) - Q_0(\bar{x}) \tag{1}$$

subject to

$$P_k(\bar{x}) - Q_k(\bar{x}) \leq 1, \qquad k = 1, \ldots, K, \tag{2}$$

$$0 < x_j^{\mathrm{LB}} \leq x_j \leq x_j^{\mathrm{UB}}, \qquad j = 1, \ldots, N, \tag{3}$$

where $P_k(\bar{x})$ and $Q_k(\bar{x})$, $k = 0, 1, \ldots, K$ are posynomials of the general form:

$$P_k(\bar{x}) = \sum_{i=1}^{I_k} u_{ik} = \sum_{i=1}^{I_k} c_{ik} x_1^{a_{1ik}} x_2^{a_{2ik}} \cdots x_N^{a_{Nik}}, \tag{4}$$

$$Q_k(\bar{x}) = \sum_{l=1}^{L_k} v_{lk} = \sum_{l=1}^{L_k} d_{lk} x_1^{b_{1lk}} x_2^{b_{2lk}} \cdots x_N^{b_{Nlk}}. \tag{5}$$

Each

$$u_{ik} = c_{ik} \prod_{i=1}^{N} x_j^{a_{jik}} \tag{6}$$

and

$$v_{lk} = d_{lk} \prod_{j=1}^{N} x_j^{b_{jlk}} \tag{7}$$

are called *posynomial* terms, or *monomials*.

The exponents a_{jik} and b_{jlk} are arbitrary real constants whereas the coefficients c_{ik} and d_{lk} are given *positive* numbers. Note that the variables x_j are assumed to be bounded. It is also assumed that the minimum value of the objective function is positive.

A regular geometric program (GP) is defined as above and such that no "Q" posynomials are present. A GP may be transformed into a convex program using a simple logarithmic transformation (Ref. 2). This property of a GP is a very important one since it implies that every local optimum is a global optimum for the problem. A GGP, however, does not possess this property.

A GGP may always be transformed into an equivalent problem, with a linear objective function, by the addition of one variable and one constraint to the original problem. We will therefore consider the following problem as having the most general, or standard GGP, form:

Minimize

$$x_0 \tag{8}$$

subject to

$$P_k(x) - Q_k(x) \leq 1, \qquad k = 0, 1, \ldots, K, \tag{9}$$

$$0 < x_j^{\mathrm{LB}} \leq x_j \leq x_j^{\mathrm{UB}}, \qquad j = 0, 1, \ldots, N, \tag{10}$$

where

$$P_0(\bar{x}) - Q_0(\bar{x}) \leq x_0, \tag{11}$$

or

$$\frac{P_0(\bar{x})}{x_0} - \frac{Q_0(\bar{x})}{x_0} = P_0(x) - Q_0(x) \leq 1, \tag{12}$$

$$x = (x_0, x_1, \ldots, x_N). \tag{13}$$

2.2. The Arithmetic–Geometric Inequality. The classical inequality stating that the weighted arithmetic mean of positive numbers $\omega_1, \omega_2, \ldots, \omega_N$ is greater than or equal to the geometric mean may be written as follows:

$$\sum_{i=1}^{N} \omega_i \geq \prod_{i=1}^{N} \left(\frac{\omega_i}{\epsilon_i}\right)^{\epsilon i}, \tag{14}$$

where

$$\sum_{i=1}^{N} \epsilon_i = 1, \tag{15}$$

$$\epsilon_1 \geq 0, \qquad i = 1, \ldots, N. \tag{16}$$

Equality holds in (14) if and only if

$$\frac{\omega_1}{\epsilon_1}=\frac{\omega_2}{\epsilon_2}\cdots=\frac{\omega_N}{\epsilon_N}. \tag{17}$$

It is upon this inequality that large portions of geometric programming theory (Ref. 2) are based. In the following section a further application of this inequality is developed.

2.3. Condensed Programs. Given a set of nonnegative weights ϵ_i, such that $\sum_i \epsilon_i = 1$, and any posynomial

$$g(x)=\sum_i \omega_i(x)=\sum_i \theta_i \prod_{j=0}^{N} x_j^{\phi_{ij}}, \tag{18}$$

we define a condensed posynomial, formed at a point $\tilde{x}>0$, as

$$g(x, \tilde{x})=\prod_i \left(\frac{\omega_i(x)}{\epsilon_i(\tilde{x})}\right)^{\epsilon_i(\tilde{x})} \tag{19}$$

$$=\theta(\tilde{x}) \prod_{j=0}^{N} x_j^{\phi_j(\tilde{x})}, \tag{20}$$

where

$$\theta(\tilde{x})=\prod_i \left(\frac{\theta_i}{\epsilon_i(\tilde{x})}\right)^{\epsilon_i(\tilde{x})} \tag{21}$$

and

$$\phi_j(\tilde{x})=\sum_i \epsilon_i(\tilde{x})\phi_{ij}. \tag{22}$$

An important result of the above definition is that $g(x, \tilde{x})$ is a monomial. For a given $\tilde{x}>0$ we will choose the set of weights

$$\epsilon_i(\tilde{x})=\frac{\omega_i(\tilde{x})}{g(\tilde{x})}. \tag{23}$$

It is easily seen that these weights satisfy conditions (15) and (16) and, therefore, as a direct consequence of the arithmetic–geometric inequality we have

$$g(x, \tilde{x})\leq g(x) \tag{24}$$

for any positive x and $\tilde{x}$.

Substituting (23) in (19) gives

$$g(x, \tilde{x})=\prod_i \left(\frac{\omega_i(x)}{\omega_i(\tilde{x})} g(\tilde{x})\right)^{\epsilon_i(\tilde{x})} \tag{25}$$

Now, for $x = \tilde{x}$ we have

$$g(x, \tilde{x}) = \prod_i g(\tilde{x})^{\epsilon_i(\tilde{x})} = g(\tilde{x})^{\sum_i \epsilon_i(\tilde{x})} = g(\tilde{x}). \tag{26}$$

Thus, the condensed posynomial equals the original posynomial at the point of condensation $\tilde{x}$.

Condensation may be used both as an analytical (Refs. 1 and 2) and a computational aid and a number of algorithms for the solution of GGP's and GP's have been developed (Refs. 3–5) using condensation. One of these, an algorithm developed by Avriel and Williams (Ref. 3) is presented in the next section.

2.4. An Algorithm for Solving a GGP. The algorithm described here is due to Avriel and Williams (Ref. 3). Similar algorithms have also been proposed by Passy (Ref. 4) and Pascual and Ben-Israel (Ref. 5). For reasons that will be made obvious in Section 4, we will refer to the algorithm in this section as "Phase 2."

Consider the kth constraint of a GGP

$$P_k(x) - Q_k(x) \leq 1. \tag{27}$$

This may be rewritten as

$$\frac{P_k(x)}{1 + Q_k(x)} \leq 1. \tag{28}$$

Let $Q_k(x, x^{(p)})$ denote the *monomial* obtained by condensing the posynomial $1 + Q_k(x)$ at the point $x^{(p)}$. The following program, obtained by substituting $Q_k(x, x^{(p)})$ for $1 + Q_k(x)$ in GGP, will be referred to as $\mathrm{GP}^{(p)}$:

Minimize

$$x_0 \tag{29}$$

subject to

$$\frac{P_k(x)}{Q_k(x, x^{(p)})} \leq 1, \qquad k = 0, 1, \ldots, K, \tag{30}$$

$$0 < x_j^{\mathrm{LB}} \leq x_j \leq x_j^{\mathrm{UB}}, \qquad j = 0, 1, \ldots, N. \tag{31}$$

This program has the following properties:

(a) $\mathrm{GP}^{(p)}$ is a regular GP since the functions $P_k(x)/Q_k(x, x^{(p)})$ are posynomials. (A posynomial divided by a monomial is itself a posynomial.)

(b) Any point x^F satisfying the constraints of $\mathrm{GP}^{(p)}$ will satisfy the constraints of GGP. This can be observed by using the condensation inequality (24) derived in Section 2.3, i.e.,

$$\frac{P_k(x^F)}{1+Q_k(x^F)} \leq \frac{P_k(x^F)}{Q_k(x^F, x^{(p)})} \leq 1. \tag{32}$$

(c) Inequality (32) implies that the feasible set of $GP^{(p)}$ is entirely contained in GGP and therefore the optimal solution to $GP^{(p)}$ will be a *feasible* (but not necessarily optimal) point for GGP.

Now consider the sequence of $GP^{(p)}$ problems where $GP^{(0)}$ is constructed using a point feasible for GGP, and $GP^{(p)}$, $p = 1, 2, \ldots$, is constructed using the optimal solution to $GP^{(p-1)}$. Avriel and Williams (Ref. 3) showed that the sequence of optimal solutions to $GP^{(p)}$, $p = 0, 1, 2, \ldots$, converges to a point satisfying the Kuhn–Tucker necessary conditions for optimality of GGP, provided that certain mild regularity conditions are satisfied.

2.5. Linearizing a GP. In this section we demonstrate how GP may be approximated by a linear program (LP) using condensation (Ref. 1). The constraints and objective function of the GP are condensed to monomials and after effecting a logarithmic transformation the problem becomes an LP, suitable for solution by some variant of the simplex method.

Consider the GP:

Minimize

$$x_0 \tag{33}$$

subject to

$$g_k(x) \leq 1, \qquad k = 0, 1, \ldots, K, \tag{34}$$

$$0 < x_j^{\mathrm{LB}} \leq x_j \leq x_j^{\mathrm{UB}}, \qquad j = 0, 1, \ldots, N, \tag{35}$$

where

$$g_k(x) = \sum_{i=1}^{I_k} \theta_{ik} \prod_{j=0}^{N} x_j^{\phi_{ijk}}. \tag{36}$$

In Section 2.3 we showed that the posynomials $g_k(x)$ can be approximated by *monomials* of the form

$$g_k(x, \tilde{x}) = \theta_k(\tilde{x}) \prod_{j=0}^{N} x_j^{\phi_{jk}(\tilde{x})}, \tag{37}$$

where

$$g_k(x, \tilde{x}) \leq g_k(x), \tag{38}$$

with equality if and only if $x = \tilde{x}$.

We now consider the condensed program $\overline{GP}$:

Minimize

$$x_0 \tag{39}$$

subject to

$$g_k(x, \tilde{x}) \leq 1, \qquad k = 0, 1, \ldots, K, \tag{40}$$

$$0 < x_j^{\text{LB}} \leq x_j \leq x_j^{\text{UB}}, \qquad j = 0, 1, \ldots, N. \tag{41}$$

It is obvious from inequality (38) that a point x^F satisfying the GP constraints (34) will also satisfy the $\overline{GP}$ constraints (40), i.e.,

$$g_k(x^F, \tilde{x}) \leq g_k(x^F) \leq 1, \qquad k = 0, 1, \ldots, K. \tag{42}$$

In general, the converse will only be true when $x^F = \tilde{x}$. This implies that the feasible set of program GP is contained in that of $\overline{GP}$ and, therefore, the solution to $\overline{GP}$ will not, in general, be feasible for GP. Program $\overline{GP}$ will now be shown to be equivalent to a linear program.

The natural logarithmic function $\ln y$ is monotonic increasing and is defined for $y > 0$. Therefore, the following program will be equivalent to $\overline{GP}$.

Minimize

$$\ln x_0 \tag{43}$$

subject to

$$\ln g_k(x, \tilde{x}) = \ln \theta_k(\tilde{x}) + \sum_{j=0}^{N} \phi_{jk}(\tilde{x}) \ln x_j \leq 0, \qquad k = 0, 1, \ldots, K, \tag{44}$$

$$\ln x_j^{\text{LB}} \leq \ln x_j \leq \ln x_j^{\text{UB}}, \qquad j = 0, 1, \ldots, N. \tag{45}$$

This program is a linear program in the variables $\ln x_j$, $j = 0, 1, \ldots, N$, however, it is not in a convenient form for direct application of the simplex method (Ref. 6) since the variables in $\ln x_j$ may take on negative values. We therefore define new variables z_j

$$z_j = \ln x_j - \ln x_j^{\text{LB}}, \qquad j = 0, 1, \ldots, N, \tag{46}$$

and

$$z_j^{\text{UB}} = \ln x_j^{\text{UB}} - \ln x_j^{\text{LB}} = \ln(x_j^{\text{UB}}/x_j^{\text{LB}}), \qquad j = 0, 1, \ldots, N. \tag{47}$$

Substituting into (45) yields the program:

Minimize

$$z_0 + \ln x_0^{\text{LB}} \tag{48}$$

subject to

$$\bar{g}_k(z, \tilde{x}) \leq 0, \qquad k = 0, 1, \ldots, K, \tag{49}$$

$$0 \leq z_j \leq z_j^{\text{UB}}, \qquad j = 0, 1, \ldots, N, \tag{50}$$

where

$$\bar{g}_k(z, \tilde{x}) = \ln \theta_k(\tilde{x}) + \sum_{j=0}^{N} \phi_{jk}(\tilde{x}) \ln x_j^{\text{LB}} + \sum_{j=0}^{N} \phi_{jk}(\tilde{x}) z_j. \tag{51}$$

Noting that

$$\ln \theta_k(\tilde{x}) + \sum_{j=0}^{N} \phi_{jk}(\tilde{x}) \ln x_j^{\text{LB}} = \ln g_k(x^{\text{LB}}, \tilde{x}), \tag{52}$$

we may rewrite the above program as:

Minimize

$$z_0 \tag{53}$$

subject to

$$\sum_{j=0}^{N} \phi_{jk}(\tilde{x}) z_j \leq -\ln g_k(x^{\text{LB}}, \tilde{x}), \qquad k = 0, 1, \ldots, K, \tag{54}$$

$$0 \leq z_j \leq z_j^{\text{UB}}, \qquad j = 0, 1, \ldots, N, \tag{55}$$

where $\ln x_0^{\text{LB}}$, a constant, is omitted from the objective function. We will refer to this program as LP($\tilde{x}$) since it is a regular LP, with upper bounded variables, constructed using the point $\tilde{x}$. Program LP($\tilde{x}$) can be efficiently solved using a modified version of the dual simplex method (Ref. 6), which accounts for upper-bounded variables implicitly. Details are given by Dembo (Ref. 7).

3. Details of the GP and GGP algorithms

3.1. An Algorithm for the Solution of a GP.

Consider the GP:

Minimize

$$x_0 \tag{56}$$

subject to

$$g_k(x) \leq 1, \qquad k = 0, 1, \ldots, K, \tag{57}$$

$$0 < x_j^{\text{LB}} \leq x_j \leq x_j^{\text{UB}}, \qquad j = 0, 1, \ldots, N. \tag{58}$$

Our algorithm proceeds as follows:

Step 1. Using an arbitrary starting point, x^0, linearize the GP as described in Section 2.5 and form LP(x^0). Set $M = 1$.

Step 2. Solve LP(x^{M-1}). Call solution z^M and compute x^M by Eq. (46).

Step 3. Evaluate the GP constraints (57) at x^M. If

$$g_k(x^M) \leq 1 + \epsilon, \qquad k = 0, 1, \ldots, K, \tag{59}$$

where ϵ is some small predetermined positive number, then x^M is optimal, otherwise define

$$g_l(c) = \max_{0 \leq k \leq K} \{g_k(x^M) : g_k(x^M) > 1\}. \tag{60}$$

Step 4. Condense $g_l(x)$ at x^M to obtain $g_l(x, x^M)$ which, in turn, is transformed into the linear constraint

$$\bar{g}_l(z, x^M) \leq 0. \tag{61}$$

Add this constraint to those of LP(x^{M-1}) and name the new LP problem LP(x^M). Increment M and return to Step 2.

In Step 1 the GP is approximated by a linear program LP(x^0), for which highly efficient algorithms have already been developed. If the point x^M, obtained by solving LP(x^{M-1}), lies outside the region described by $g_k(x) \leq 1$, $k = 0, 1, \ldots, K$, then Step 4 generates a modified LP problem that excludes $z^M = \ln x^M$ from its feasible region. Thus a series of LP's with progressively smaller feasible regions are solved until a point x^M obtained from one of these problems satisfies (59), at which stage the algorithm terminates.

This type of algorithm is known as a "cutting-plane" algorithm (see Ref. 8) and the constraints generated in Step 4 are known as "cuts," since they cut off part of the feasible region of the approximating linear program at each iteration.

In order to see that this cut does not cut off any section of the feasible region of GP, we observe from (42) that for any point, x^F, feasible for GP we have

$$g_l(x^F, x^M) \leq g_l(x^F) \leq 1, \tag{62}$$

i.e., x^F will also be feasible for the cut.

At each iteration M of the above algorithm we are required to solve the linear programming problem LP(x^{M-1}). However, the problem LP(x^M), solved at iteration $M+1$, differs from LP(x^{M-1}) only in that it has an additional constraint. Use of the dual simplex method enables the transformation from LP(x^{M-1}) to LP(x^M) to be carried out in such a manner that the optimal solution to LP(x^{M-1}) is used as a starting point for the solution of LP(x^M). Thus, only a modest amount of computation is required in moving from one iteration to the next. The problem however, may become large if many iterations are required for its solution, since all cuts are stored in the simplex tableau until the solution to the GP is attained.

3.2. A Possible Acceleration Technique. In Section 2 we showed how to solve a GGP by solving a series of GP subproblems. We also showed in Section 3.1 how to solve each GP subproblem using a cutting-plane algorithm. Thus, by simply combining the above two algorithms, we have a complete algorithm for the solution of a GGP. However, there is possibly a more efficient way of combining the above two algorithms, whereby convergence to a GGP solution is accelerated. This acceleration technique is based on the following observations pertaining to the above two algorithms:

(a) The sequence of optimal $\mathrm{GP}^{(p)}$ problem solutions is feasible for the GGP and thus each such solution $x_0^*(GP^{(p)})$ is greater than or equal to the optimal solution to the GGP, i.e.,

$$x_0^*(\mathrm{GP}^{(p)}) \geq x_0^*(\mathrm{GGP}). \tag{63}$$

(b) The sequence of optimal $\mathrm{LP}(x^M)$ solutions (cutting-plane iterations) converging to a particular $\mathrm{GP}^{(p)}$ solution is not feasible for the $\mathrm{GP}^{(p)}$ and thus

$$x_0^*(\mathrm{LP}(x^M)) \leq x_0^*(\mathrm{GP}^{(p)}). \tag{64}$$

At some stage, during the course of solution to $\mathrm{GP}^{(p)}$, the current optimal solution $x^*(\mathrm{LP}(x^M))$ may be feasible for GGP. This point may have a lower objective function value than the solution to $\mathrm{GP}^{(p)}$ itself and usually it will serve as a "better" point than the $\mathrm{GP}^{(p)}$ optima, for the formation of $\mathrm{GP}^{(p+1)}$.

This method may be summarized as follows: The GGP problem is solved by a series of $\mathrm{GP}^{(p)}$ approximations. Each $\mathrm{GP}^{(p)}$ may not be solved completely and the first GGP feasible point, calculated during the solution to a $\mathrm{GP}^{(p)}$ is used as the point about which $\mathrm{GP}^{(p+1)}$ is formed. The example in Section 5 demonstrates this method.

It is worth noting that this technique does not necessarily guarantee to accelerate the convergence rate of our GGP algorithm. In most practical examples studied convergence was improved considerably. However, there were cases where poorer convergence was obtained using this technique.

4. Solving a Set of Generalized Posynomial Inequalities

The algorithm, referred to as "Phase 2" in Section 2.4 of this paper, requires as its starting point a feasible solution to the following generalized posynomial constraints:

$$\frac{P_k(x)}{1+Q_k(x)} \leqslant 1, \qquad k=0, 1, \ldots, K. \tag{65}$$

In many cases, solving these inequalities, i.e., determining a value for x that satisfies (65), may be as difficult as the solution of the GGP problem itself. It is important, therefore, to have an efficient algorithm which yields a solution to (65). Our method uses the GGP algorithm itself for this purpose and because of the similarities to the corresponding method in linear programming for finding an initial feasible solution, we refer to the algorithm described below as "Phase 1."

Consider the following generalized geometric program, called GGP(ω), formed from the GGP problem of Section 2:

Minimize

$$\prod_{k=0}^{K} \omega_k \tag{66}$$

subject to

$$\frac{P_k(x)}{1+Q_k(x)} \leqslant \omega_k, \qquad k=0, 1, \ldots, K, \tag{67}$$

$$\omega_k \geqslant 1, \qquad k=0, 1, \ldots, K, \tag{68}$$

$$0 < x_j^{\mathrm{LB}} \leqslant x_j \leqslant x_j^{\mathrm{UB}}, \qquad j=0, 1, \ldots, N. \tag{69}$$

The reason for introducing GGP(ω) is made apparent by the theorem below, the proof of which is elementary and will be omitted.

Theorem 4.1. The point $x = x^*$ satisfies the inequalities (65) if and only if the optimal solution to GGP(ω) is (x^*, ω^*), where $\prod_{k=0}^{K} \omega_k^* = 1$.

If every local optimum of problem GGP(ω) is global, then the solution of GGP(ω), using our algorithm described in Section 2, may be guaranteed to yield a feasible point to the inequalities (65).

In the case where GGP(ω) has local optima that are not global we know that the desired solution to (65) is one of them, however, convergence to this particular solution is not guaranteed. We summarize the steps of the Phase 1 algorithm below:

Step 1. Let x^0 be any point satisfying

$$x_j^{\mathrm{LB}} \leqslant x_j^0 \leqslant x_j^{\mathrm{UB}}, \qquad j=0, 1, \ldots, N. \tag{70}$$

Define

$$\omega_k^0 = \max\left\{\frac{P_k(x^0)}{1+Q_k(x^0)}, 1\right\}, \qquad k = 0, 1, \ldots, K. \tag{71}$$

The point (x^0, ω^0), where $\omega^0 = (\omega_0^0, \omega_1^0, \ldots, \omega_K^0)$, is thus a feasible solution for GGP(ω).

Step 2. Consider the values of ω_k^0 for $k = 0, 1, \ldots, K$.

(a) If $\omega_k^0 = 1$ for all k, then Phase 1 terminates, and x^0 solves (65).

(b) If for at least one value of k, $\omega_k^0 > 1$, then we solve GGP(ω) using our Phase 2 algorithm with the initial point (x^0, ω^0).

Step 3. Examine the optimal solution (x^*, ω^*) to GGP(ω).

(a) If $\omega_k^* = 1$ for all k, then x^* will constitute a solution to (65).

(b) If for some k, $\omega_k^* > 1$, then the algorithm has failed to converge.

There is one further application of the Phase 1 algorithm. Assume that during the course of solution to a GGP the Phase 2 algorithm converged to a local but not global minimum of the problem. We could attempt to improve on this solution by constraining the objective function to a value less than that attained previously and by solving the resulting problem, using the Phase 1 method. If Phase 1 converges to a feasible solution of the restricted problem this solution may be used as a starting value for the phase 2 algorithm, which will then converge to a "better" local minimum. This aspect will be demonstrated in the following section.

5. An Illustrative Example

In order to demonstrate the Phase 1 and Phase 2 algorithms, we solve the following problem, sketched in Fig. 1.

Minimize

$$x_1 \tag{72}$$

subject to

$$P_1(x) - Q_1(x) = x_1/4 + x_2/2 - x_1^2/16 - x_2^2/16 \leq 1, \tag{73}$$

$$P_2(x) - Q_2(x) = x_1^2/(6x_2) + x_2/6 + 7/(3x_2) - x_1/x_2 \leq 1. \tag{74}$$

Note that since the objective function is linear, the problem is in the standard form of GGP. Assume that

$$1.0 \leq x_1 \leq 5.5, \tag{75}$$

$$1.0 \leq x_2 \leq 5.5. \tag{76}$$

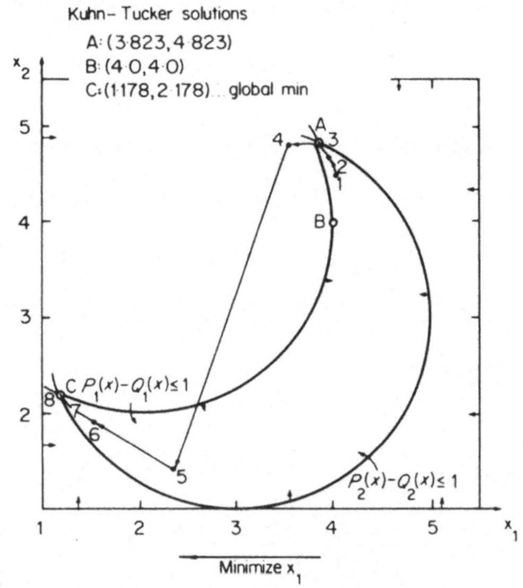

Fig. 1. Solution to a numerical example.

Our demonstration takes the following form:

(a) Starting at the feasible solution, $x_1 = 4$; $x_2 = 4.5$, we use the Phase 2 algorithm to converge to a local minimum.

(b) We then use the Phase 1 algorithm to find a feasible point, with a lower objective function value than the above local minimum.

(c) The feasible point calculated in (b) is used as a starting point for the Phase 2 algorithm, which then converges to an improved local minimum (global minimum) of the problem.

5.1. Phase 2 Algorithm, Starting Point $x_1 = 4$; $x_2 = 4.5$. We first rewrite the constraints (73) and (74) in the form of (28), that is

$$\frac{P_1(x)}{1+Q_1(x)} = \frac{x_1/4 + x_2/2}{1 + x_1^2/16 + x_2^2/16} \leqslant 1 \tag{77}$$

$$\frac{P_2(x)}{1+Q_2(x)} = \frac{x_1^2/(6x_2) + x_2/6 + 7/(3x_2)}{1 + x_1/x_2} \leqslant 1. \tag{78}$$

Now consider the condensation of posynomials $1 + Q_1(x)$ and $1 + Q_2(x)$ at the point $\tilde{x} = (\tilde{x}_1, \tilde{x}_2)$,

$$(1+Q_1)(x, \tilde{x}) = \left(\frac{1}{\epsilon_{11}}\right)^{\epsilon_{11}} \left(\frac{x_1^2/16}{\epsilon_{21}}\right)^{\epsilon_{21}} \left(\frac{x_2^2/16}{\epsilon_{31}}\right)^{\epsilon_{31}}, \tag{79}$$

$$(1+Q_2)(x, \tilde{x}) = \left(\frac{1}{\epsilon_{12}}\right)^{\epsilon_{12}} \left(\frac{x_1/x_2}{\epsilon_{22}}\right)^{\epsilon_{22}}, \tag{80}$$

where the weights ϵ_{ik} are computed at $\tilde{x}$ according to (23). For example,

$$\epsilon_{11}=\frac{1}{1+Q_1(\tilde{x})}, \qquad \epsilon_{21}=\frac{x_1^2/16}{1+Q_1(\tilde{x})}. \tag{81}$$

Substituting $x^{(0)}=\tilde{x}=(4, 4.5)$, we have

$$(1+Q_1)(x, x^{(0)})=0.435x_1^{0.612}x_2^{0.775} \tag{82}$$

$$(1+Q_2)(x, x^{(0)})=1.997x_1^{0.471}x_2^{-0.471}. \tag{83}$$

Thus $GP^{(0)}$ will be the following problem:

Minimize

$$x_1$$

subject to

$$g_1(x)=\frac{P_1(x)}{(1+Q_1)(x, x^{(0)})}=0.574x_1^{0.388}x_2^{-0.755}+1.148x_1^{-0.612}x_2^{0.225}\leqslant 1, \tag{84}$$

$$g_2(x)=\frac{P_2(x)}{(1+Q_2)(x, x^{(0)})}=0.083x_1^{1.529}x_2^{-0.529}+0.083x_1^{-0.471}x_2^{1.471}$$

$$+1.169x_1^{-0.471}x_2^{-0.529}\leqslant 1, \tag{85}$$

$$1.0\leqslant x_1\leqslant 5.5, \tag{86}$$

$$1.0\leqslant x_2\leqslant 5.5. \tag{87}$$

We now proceed with the solution to $GP^{(0)}$ using the cutting-plane algorithm described in Section 3.1.

Step 1. The problem is linearized at $x^{(0)}$. Consider the condensed constraints,

$$g_1(x, x^{(0)})=1.720x_1^{-0.305}x_2^{-0.083}\leqslant 1, \tag{88}$$

$$g_2(x, x^{(0)})=0.516x_1^{0.166}x_2^{0.277}\leqslant 1. \tag{89}$$

Using the transformations of (46) we have

$$z_1=\ln x_1-\ln x_1^{\mathrm{LB}}=\ln x_1-\ln(1)=\ln x_1, \tag{90}$$

$$z_2=\ln x_2-\ln x_2^{\mathrm{LB}}=\ln x_2-\ln(1)=\ln x_2, \tag{91}$$

and we obtain problem $LP^{(0)}$:

Minimize

$$z_1$$

subject to

$$\bar{g}_1(z, x^{(0)}) = -0.305z_1 - 0.082z_2 \leq -0.542, \tag{92}$$

$$\bar{g}_1(z, x^{(0)}) = 0.166z_1 + 0.277z_2 \leq 0.661, \tag{93}$$

$$0 \leq z_1 \leq \ln 5.5, \tag{94}$$

$$0 \leq z_2 \leq \ln 5.5. \tag{95}$$

Step 2. The solution to the linear program $\text{LP}^{(0)}$ is $z_1^{(1)} = 1.350$, $z_2^{(1)} = 1.579$, and hence from (90) and (91), $x_1^{(1)} = 3.858$, $x_2^{(1)} = 4.852$.

Values of GPP constraints (73), (74) at this point are

$$P_1(x^{(1)}) - Q_1(x^{(1)}) = 0.989 < 1, \tag{96}$$

$$P_2(x^{(1)}) - Q_2(x^{(1)}) = 1.005 > 1. \tag{97}$$

Step 3. Constraint (74) is violated so we therefore linearize $g_2(x)$ at $x^{(1)} = (3.858, 4.852)$.

$$g_2(x, x^{(1)}) = 0.498x_1^{0.097}x_2^{0.369} \leq 1. \tag{98}$$

Hence,

$$\bar{g}_2(z, x^{(1)}) = 0.097z_1 + 0.369z_2 \leq 0.697. \tag{99}$$

Step 4. Inequality (99) is added to problem $\text{LP}^{(0)}$ to obtain $\text{LP}^{(1)}$. The solution to $\text{LP}^{(1)}$ is:

$$z_1^{(2)} = 1.354, \qquad z_2^{(2)} = 1.566. \tag{100}$$

Hence,

$$x_1^{(2)} = 3.872, \qquad x_2^{(2)} = 4.786, \tag{101}$$

and

$$P_1(x^{(2)}) - Q_1(x^{(2)}) = 0.992 < 1, \tag{102}$$

$$P_2(x^{(2)}) - Q_2(x^{(2)}) = 0.998 < 1. \tag{103}$$

This point thus satisfies the GGP constraints and may be used to form the problem $\text{GP}^{(1)}$. It is interesting to note that the above solution is not optimal for $\text{GP}^{(0)}$. The optimal solution to $\text{GP}^{(0)}$ is $x_1 = 3.876$; $x_2 = 4.766$ and requires approximately two more cutting-plane iterations to be reached (depending on the accuracy required).

Convergence to a local minimum (3.823, 4.823) of the GGP problem (72)–(76) is shown in Table 1.

Table 1. Convergence to local minimum using Phase 2.

Location in Fig. 1	Phase 2 iteration	Number of cuts	Next approximating point	Comments
1	0	—	(4.0, 4.5)	
2	1	1	(3.872, 4.786)	
3	2	0	(3.824, 4.824)	
3	3	0	(3.823, 4.823)	Local optimum

5.2. Use of Phase 1. We choose to solve the problem below in order to obtain a feasible point with a lower objective function value than the above local minimum:

Minimize

$$\omega_1\omega_2 \tag{104}$$

subject to

$$P_1(x) - Q_1(x) \leq \omega_1, \tag{105}$$

$$P_2(x) - Q_2(x) \leq \omega_2, \tag{106}$$

$$1 \leq x_1 \leq 3.5, \qquad 1 \leq x_2 \leq 5.5,$$

$$1 \leq \omega_1 \leq 2.5, \qquad 1 \leq \omega_2 \leq 2.5,$$

starting at the point $(x_1, x_2) = (3.5, 4.823)$, which violates the constraint $P_1(x) - Q_1(x) \leq 1$. Note that the starting point $x_1 = 3.5$, $x_2 = 4.823$, $\omega_1 = 2.5$, $\omega_2 = 1$ satisfies (105) and (106).

Convergence of this problem to the feasible solution $(x_1, x_2) = 2.330$, 1.384) is shown in Table 2.

5.3. Convergence to the Global Minimum. Convergence to the global minimum (1.178, 2.178), using the Phase 2 algorithm, and the starting solution obtained in Section 5.2 is shown in Table 3.

Table 2. Phase 1 converging to a feasible solution.

Location in Fig. 1	Phase 1 iteration	Number of cuts	Next approximating point	Comments
4	0	—	(3.5, 4.823)	Not feasible
5	1	1	(2.330, 1.384)	Feasible

Table 3. Convergence to a global minimum.

Location in Fig. 1	Phase 2 iteration	Number of cuts	Next approximating point	Comments
5	0	—	(2.330, 1.384)	Feasible
6	1	1	(1.574, 1.961)	
7	2	2	(1.261, 2.125)	
8	3	0	(1.178, 2.176)	
8	4	0	(1.178, 2.178)	Global optimum

In this example there is at least one more point, marked B in Fig. 1, which satisfies the Kuhn–Tucker necessary conditions for optimality. This point is *not* a local minimum, but under very special and degenerate conditions (which almost never occur in practice) the GGP algorithm could converge to it (Ref. 3).

6. An Application in Optimal Engineering Design

In this section we demonstrate the applicability of our Phase 2 algorithms, discussed in the preceding sections, to the solution of an optimal engineering design problem. Our example is taken from the petroleum industry and has been previously described and solved by Bracken and McCormick (Ref. 16), using the penalty function method, and by Sauer, Colville, and Burwick (Ref. 15), using a method developed by Colville (Ref. 13).

6.1. Description of the Process. A simplified process flow diagram of an alkylation process is given in Fig. 2. It consists of a reactor into which olefin feed and isobutane make-up are introduced. Fresh acid is added to

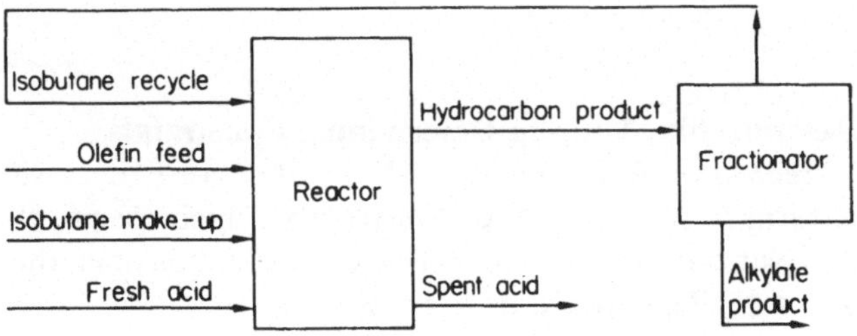

Fig. 2. Simplified alkylation process flow diagram.

catalyze the reaction and spent acid is withdrawn. The hydrocarbon product from the reactor is fed into a fractionator and the isobutane withdrawn from the top of the fractionator is recycled back to the column. Alkylate product is withdrawn from the bottom of the fractionator. The following simplifying assumptions are made:

(a) The olefin fed is pure butylene.
(b) Isobutane make-up and isobutane recycle are pure isobutane.
(c) Fresh acid strength is 98 per cent by weight.

6.2. Process Variables. Payne (Ref. 14) discusses the process variables and their relationships with each other. Some of these relationships are material balances, while some are correlations between variables within certain ranges, described by linear or nonlinear regressions. The variables considered in the model of the process are the following:

x_1 = olefin feed (barrels per day)
x_2 = isobutane recycle (barrels per day)
x_3 = acid addition rate (thousands of pounds per day)
x_4 = alkylate yield (barrels per day)
x_5 = isobutane make-up (barrels per day)
x_6 = acid strength (weight percent)
x_7 = motor octane number
x_8 = external isobutane-to-olefin ratio
x_9 = acid dilution factor
x_{10} = F–4 performance number

Values to be taken on by the variables are all bounded from above and below. The "independent" variables x_1, x_2, and x_3, i.e., those which can be directly controlled, and the "dependent" variables, x_4 and x_5, have limitations imposed on them by the capacity of the plant and/or the economic situation under analysis. For example, only 2000 barrels per day of olefin feed, x_1, may be available for processing. The dependent variables x_6, x_7, x_8, x_9, and x_{10} have bounds that are directly related to the process. Values for these bounds are given in Table 4.

6.3. Relationships Used in Determining Constraints

6.3.1. Regression Relationships. We will express regression relationships in the form of two inequality constraints, which specify the range for which these relationships are valid. For example, consider the regression equation $Y = f(z)$. This would be expressed as:

$$d_l Y \leq f(z) \leq d_u Y, \tag{107}$$

Table 4. Lower and upper bounds on variables, and starting values.

Variable	Lower bound	Upper bound	Starting value
x_1, olefin feed (barrels per day)	1	2,000	1,745
x_2, isobutane recycle (barrels per day)	1	19,200	12,000
x_3, acid addition rate (thousands of pounds per day)	1	120	110
x_4, alkylate yield (barrels per day)	0	5,000	3,048
x_5, isobutane make-up (barrels per day)	1	2,000	1,974
x_6, acid strength (weight percent)	85	93	89.2
x_7, motor octane number	90	95	92.8
x_8, external isobutane-to-olefin ratio	3	12	8
x_9, acid dilution factor	1.2	4	3.6
x_{10}, F–4 performance number	145	162	145

or

$$f(z) \leq d_u Y, \tag{108}$$

$$-f(z) \leq -d_l Y. \tag{109}$$

The deviation parameters d_l and d_u establish the percentage difference of the estimated value from the true value. Values for these deviation parameters assumed below, correspond with those used by Payne (Ref. 14).

The alkylate yield x_4 is a function of the olefin feed x_1 and the external isobutane-to-olefin ratio x_8. Nonlinear regression analysis, holding reactor temperature between 80° and 90°F and reactor acid by weight percent strength between 85 and 93, yields:

$$\tfrac{99}{100} x_4 \leq x_1(1.12 + 0.13167 x_8 - 0.00667 x_8^2) \leq \tfrac{100}{99} x_4. \tag{110}$$

The motor octane number x_7 is a function of the external isobutane-to-olefin ratio x_8 and the acid strength by weight per cent x_6. Nonlinear regression analysis, using the same reactor conditions as for x_4, yields:

$$\tfrac{99}{100} x_7 \leq 86.35 + 1.098 x_8 - 0.038 x_8^2 + 0.325(x_6 - 89) \leq \tfrac{100}{99} x_7. \tag{111}$$

The acid dilution factor x_9 may be expressed as a linear function of the F–4 performance number x_{10}:

$$\tfrac{9}{10} x_9 \leq 35.82 - 0.222 x_{10} \leq \tfrac{10}{9} x_9, \tag{112}$$

Similarly, the F–4 performance number x_{10} may be expressed as a linear function of the motor octane number x_7:

$$\tfrac{99}{100} x_{10} \leq -133 + 3x_7 \leq \tfrac{100}{99} x_{10}. \tag{113}$$

6.3.2. Mass Balances. The isobutane makeup x_5 may be determined by a volumetric reactor balance. Assume the volumetric shrinkage is 0.22 volume per volume of alkylate yield. The balance is then

$$x_4 = x_1 + x_5 - 0.22x_4, \tag{114}$$

or

$$x_5 = 1.22x_4 - x_1. \tag{115}$$

The acid dilution factor x_9 may be derived from an equation expressing acid addition rate x_3 as a function of alkylate yield x_4, acid dilution factor x_9, and acid strength by weight percent x_6. We have

$$1000x_3 = \frac{x_4 x_9 x_6}{98 - x_6}. \tag{116}$$

Rearranging, we obtain

$$x_9 = \frac{98000x_3}{x_4 x_6} - \frac{1000x_3}{x_4}. \tag{117}$$

The external isobutane-to-olefin ratio x_8 is given by

$$x_8 = \frac{x_2 + x_5}{x_1}. \tag{118}$$

Using (115) and rearranging we obtain

$$x_2 = x_1 x_8 - 1.22x_4 + x_1. \tag{119}$$

The process constraints are thus given by (110), (111), (112), (113), (115), (117), and (119), together with the bounds imposed on each specific variable. These constraints must hold simultaneously for the process to be in balance.

Constraints (115), (117), and (119) are equalities and therefore, before we can convert them into GGP constraints, they must be replaced by an equivalent set of inequalities. Methods for accomplishing this are discussed in the literature (Ref. 11). We propose to use the above equations to *eliminate* the variables x_5, x_9, and x_2 from the inequality constraints (110)–(113). However, since x_5, x_9, and x_2 are all bounded from above and below, so are the functions equal to these variables. Thus for each variable we eliminate, we must add two inequality constraints to the problem constraint set. For example, in order to eliminate x_2 by using (119) we must add the following two constraints:

$$x_1 x_8 - 1.22x_4 + x_1 \leq x_2^{UB}, \tag{120}$$

$$x_1 x_8 - 1.22x_4 + x_1 \geq x_2^{LB}. \tag{121}$$

This procedure transforms the constraint set into a set composed solely of inequalities.

6.4. Process Constraints. The set of constraints described in Section 6.3 is now rewritten in GGP form:

x_4 regression (110)

$$0.00667x_1x_8^2+0.99x_4+1-1.12x_1-0.13167x_1x_8\leq 1, \qquad (122)$$

$$1.12x_1+0.13167x_1x_8+1-0.00667x_1x_8^2-1.0101x_4\leq 1. \qquad (123)$$

x_7 regression (111)

$$0.038x_8^2+0.99x_7-0.325x_6-1.098x_8-56.425\leq 1, \qquad (124)$$

$$58.425+1.098x_8+0.325x_6-0.038x_8^2-1.0101x_7\leq 1. \qquad (125)$$

x_9 regression (112)

$$0.222x_{10}+88200x_3x_4^{-1}x_6^{-1}-900x_3x_4^{-1}-34.82\leq 1, \qquad (126)$$

$$36.82+1111.11x_3x_4^{-1}-108780x_3x_4^{-1}x_6^{-1}-0.222x_{10}\leq 1. \qquad (127)$$

x_{10} regression (113)

$$134+0.99x_{10}-3x_7\leq 1, \qquad (128)$$

$$3x_7-132-1.0101x_{10}\leq 1. \qquad (129)$$

Eliminated variable x_5 (115)

$$1.22x_4-x_1-1999\leq 1, \qquad (130)$$

$$x_1+2-1.22x_4\leq 1. \qquad (131)$$

Eliminated variable x_9 (117)

$$24500x_3x_4^{-1}x_6^{-1}-250x_3x_4^{-1}\leq 1, \qquad (132)$$

$$1000x_2x_4^{-1}+2.2-98000x_3x_4^{-1}x_6^{-1}\leq 1. \qquad (133)$$

Eliminated variable x_2 (119)

$$x_1x_8+x_1-1.22x_4-19199\leq 1, \qquad (134)$$

$$1.22x_4+2-x_1-x_1x_8\leq 1. \qquad (135)$$

6.5. Profit Function. The profit function is defined in terms of alkylate product or output value minus feed and recycle costs. Operating costs not reflected in the function are assumed to be constant. The following values for the value and cost parameters are to be used in the profit function:

$$\text{Alkylate product value} = \$0.063 \text{ per octane-barrel}$$
$$\text{Olefin feed cost} = \$5.04 \text{ per barrel}$$
$$\text{Isobutane recycle costs} = \$0.035 \text{ per barrel}$$
$$\text{Acid addition cost} = \$10.00 \text{ per thousand pounds}$$
$$\text{Isobutane makeup cost} = \$3.36 \text{ per barrel}$$

The total daily profit to be maximized is

$$\text{Profit} = 0.063x_4x_7 - 5.04x_1 - 0.035x_2 - 10x_3 - 3.36x_5. \qquad (136)$$

In order to make our problem compatible with the constraints of Section 6.4 and with the GGP formulation, we adjust our profit function in the following way:

(a) First, we *minimize* negative profit instead of *maximizing* profit.

(b) Since we expect the process to yield a positive value for the profit gained, we must add a large positive constant (say 3000) to our objective function, to ensure that it remains positive.

(c) We eliminate the variables x_2 and x_5 from (136), using Eqs. (115) and (119). After adjusting the profit in the above manner, we obtain the following objective function:

Minimize

$$1.715x_1 + 0.035x_1x_8 + 4.0565x_4 + 10x_3 + 3000 - 0.063x_4x_7. \qquad (137)$$

6.6. Problem Solution. The optimization of the above alkylation process is a GGP, when written in the form: Minimize (137) subject to the constraints (122) to (135), with bounds on each variable as given in Table 4.

Initial values for process variables, taken from a nearly balanced process developed by engineers, are also given in Table 4. These values constitute a feasible solution to the above GGP and thus Phase 2 may be used directly in determining the optimal solution.

Successive iterations of the Phase 2 algorithm are shown in Table 5 and the optimal solution is presented together with initial solution and variable bounds in Table 6.

With the present bounds on the process variables the profit is seen to be \$1774 per day. This presents an increase of \$895 over the starting value.

It is interesting to note that the isobutane makeup x_5 is at its upper limit in the above optimal solution. This leads to the conclusion that an increase in availability of isobutane will lead to an increase in profit for the process. In order to demonstrate this, the problem was solved assuming an availability of 2200 barrels of isobutane per day. The modified problem yielded a profit

Table 5. GGP solution of the alkylation problem.

Iteration	Number of cutting plane iterations	Profit of the process (dollars per day)
0	—	879
1	5	1421
2	5	1751
3	6	1775
4	2	1774

Table 6. Optimal solution to the alkylation problem.

Variable	Lower bound	Optimum value	Upper bound
x_1	1	1,695.87	2,000
x_2	1	15,771.0	19,200
x_3	1	53.6631	120
x_4	1	3,029.36	5,000
x_5	1	2,000.00	2,000
x_6	85	90.1229	93
x_7	90	95.0000	95
x_8	3	10.4790	12
x_9	1.2	1.54828	4
x_{10}	145	153.535	162

of \$1953 per day, which is an increase of \$1074 over the starting value. As was expected, in this solution the isobutane makeup was at its limiting value of 2200 barrels per day.

The solution was obtained using a FORTAN IV computer program, written specifically for the solution of GGP problems and based on this article (Ref. 9). Optimal variable values agree with those found by other methods (Ref. 16). The computation time required for the above solution was approximately 0.70 sec of IBM 370/165 computer time.

7. Conclusions

We have solved a large number of engineering optimization problems using Ref. 9 and the results seem to indicate satisfactory convergence of the algorithm. In one case [Colville Test Problem 3 (Ref. 12)] where computation time was compared with other nonlinear programming methods ours ranked among the fastest. Probably the most outstanding feature of our algorithm is its ability to handle problems with "loose" constraints at the optimal solution. Other methods, for the solution of this class of problems (see, e.g., Ref. 10), rely on the assumption that all problem constraints will be active at the optimal solution. Upper and lower bounds on the variables that occur in most practical applications are also handled in an efficient manncr.

References

1. DUFFIN, R. J., *Linearizing Geometric Programs*, SIAM Review Vol. 12, pp. 211–237, 1970.
2. DUFFIN, R. J., PETERSON, E. L., and ZENER, C., *Geometric Programming*, John Wiley and Sons, New York, New York, 1967.
3. AVRIEL, M., and WILLIAMS, A. C., *Complementary Geometric Programming*, SIAM Journal on Applied Mathematics, Vol. 19, pp. 125–141, 1970.
4. PASSY, U., *Generalized Weighted Mean Programming*, SIAM Journal on Applied Mathematics, Vol. 20, pp. 763–778, 1971.
5. PASCUAL, L. T., and BEN-ISRAEL, A., *Constrained Maximization of Posynomials by Geometric Programming*, Journal of Optimization Theory and Applications, Vol. 5, 73–86, 1970.
6. DANTZIG, G. B., *Linear Programming and Extensions*, Princeton University Press, Princeton, 1963.
7. DEMBO, R. S., *Solution of Complementary Geometric Programming Problems*, M.Sc. Thesis, Technion, Haifa, 1972.
8. KELLEY, J. E., *The Cutting Plane Method for solving Convex Programs*, SIAM Journal on Applied Mathematics, Vol. 8, pp. 703–712, 1960.
9. DEMBO, R. S., *GGP—A Program for Solving Generalized Geometric Programs, Users Manual*, Department of Chemical Engineering, Technion, Haifa, 1972.
10. BLAU, G., and WILDE, D. J., *A Lagrangean Algorithm for Equality Constrained Generalized Polynomial Optimizations*, AIChE Journal, Vol. 17, pp. 235–245, 1971.
11. BLAU, G., and WILDE, D. J., *Optimal System Design by Generalized Polynomial Programming*, Canadian Journal of Chemical Engineering, Vol. 47, pp. 317–326, 1969.
12. COLVILLE, A. R., *A Comparative Study of Nonlinear Programming Codes*, IBM NYSC Report 320–2949, 1968.
13. COLVILLE, A. R., *Process Optimization Program for Non-Linear Programming*, IBN NYSC, 1964.
14. PAYNE, R. E., *Alkylation—What You Should Know About This Process*, Petroleum Refiner, Vol. 37, pp. 9, 316–319, 1958.
15. SAUER, R. N., COLVILLE, A. R., and BURWICK. C. W., *Computer Points the Way to More Profits*, Hydrocarbon Processing and Petroleum Refiners, Vol. 43, pp. 2, 84–92, 1964.
16. BRACKEN, J., and MCCORMICK, G. P., *Selected Applications in Nonlinear Programming*, John Wiley and Sons, New York, New York, 1968.

11

Current State of the Art of Algorithms and Computer Software for Geometric Programming[1,2]

R. S. Dembo[3]

Abstract. This paper attempts to consolidate over 15 years of attempts at designing algorithms for geometric programming (GP) and its extensions. The pitfalls encountered when solving GP problems and some proposed remedies are discussed in detail. A comprehensive summary of published software for the solution of GP problems is included. Also included is a numerical comparison of some of the more promising recently developed computer codes for geometric programming on a specially chosen set of GP test problems. The relative performance of these codes is measured in terms of their robustness as well as speed of computation. The performance of some general nonlinear programming (NLP) codes on the same set of test problems is also given and compared with the results for the GP codes. The paper concludes with some suggestions for future research.

1. Introduction

Ever since its inception, geometric programming (GP) has been somewhat of an outcast in the mainstream of mathematical programming

---

[1] An earlier version of this paper was presented at the ORSA/TIMS Conference, Chicago, 1975.

[2] This work was supported in part by the National Research Council of Canada, Grant No. A-3552, Canada Council Grant No. S74-0418, and a research grant from the School of Organization and Management, Yale University. The author wishes to thank D. Himmelblau, T. Jefferson, M. Rijckaert, X. M. Martens, A. Templeman, J. J. Dinkel, G. Kochenberger, M. Ratner, L. Lasdon, and A. Jain for their cooperation in making the comparative study possible.

[3] Assistant Professor of Operations Research, School of Organization and Management, Yale University, New Haven, Connecticut.

literature. Indeed, to this very day many prominent members of the mathematical programming community regard GP as a highly specialized curiosity and a "dead" area for research. The reason for this is twofold. Firstly, the historical development of computational procedures and duality theory in GP has, for the most part, taken place outside of the accepted state-of-the-art procedures in nonlinear programming (NLP). Secondly, there has been a failure on the part of many mathematical programmers to realize that a wide variety of important practical problems (for example, optimal engineering design problems) may be effectively modelled using geometric programming.

It is precisely because of its applicability to optimal engineering design that GP has been enthusiastically accepted by the engineering community. In fact, engineers have had almost an exclusive hand in the development of GP software. To some extent, which has been detrimental to GP in terms of improving its image in mathematical programming circles, mainly because GP software development has, as a result, lagged far behind general NLP software development. This is particularly true with regard to software for solving the linearly constrained dual GP. To be more precise, apart from work currently in progress (Ref. 1), to this author's knowledge there is no published algorithm for the linearly constrained dual that implements some specialized version of one of the latest numerically stable techniques for linearly constrained nonlinear programming, as discussed in Gill and Murray (Ref. 2). Matrix factorization is virtually unheard of in GP circles.

There has been, however, one fortunate byproduct of the above phenomenon. Since GP software developers were to a large extent not prejudiced by mathematical programming folklore, the implementation and testing of many algorithms that would otherwise have been shunned by "respectable" mathematical programmers has been carried out by GP researchers. A good example of this is Kelley's cutting-plane algorithm for convex programs. The method is purported to be at best geometrically convergent (see Wolfe, Ref. 3), numerically unstable, and definitely not an algorithm to be recommended for the solution of convex programming problems. Though theoretically there is definitely justification for the above hypothesis, there is no computational evidence to show that Kelley's algorithm does in fact perform worse (or better) than existing algorithms for convex programming.

It is shown in Section 5 and confirmed by Rijckaert and Martens (Ref. 4) that one of the most efficient[4] and robust[5] software packages currently

[4] In terms of standardized CPU time.

[5] By robust, we mean that the code will succeed in solving the majority of problems for which it was designed, to within prescribed tolerance limits.

available for solving GP's is a simple application of Kelley's cutting-plane algorithm. Furthermore, the algorithm has proved to be so successful that there are at least four known software packages based on it (Refs. 5–8) and in each case the respective authors have claimed excellent results.

The purpose of this paper is not only to summarize the current state-of-the-art of GP software but to identify the main sources of difficulty in designing such software and to point to directions for future research. To this end, we will discuss the following topics: solving GP's using general-purpose NLP software (Section 2); factors influencing the choice between primal-based and dual-based algorithms (Section 3); published extensions to signomial programming (Section 4); computational comparison of some of the above software (Section 5) and analysis of the results (Section 6); and conclusions and suggestions for future research (Section 7).

2. Solving GP's Using General-Purpose NLP Software

Contrary to popular belief, one cannot in general simply solve geometric programming problems using general NLP software without taking certain necessary precautions. To show this, we consider the primal and dual programs separately.

2.1. Solving Primal Geometric Programs. A primal geometric program (PGP) may be defined as the following nonlinear programming problem:

$$\text{(PGP)} \qquad \underset{x}{\text{minimize}} \qquad g_0(x) = \sum_{j\in J_0} c_j \prod_{i=1}^{m} x_i^{a_{ij}}, \tag{1}$$

$$\text{subject to} \qquad g_k(x) = \sum_{j\in J_k} c_j \prod_{i=1}^{m} x_i^{a_{ij}} \leq 1, \qquad k = 1, 2, \ldots, p, \tag{2}$$

$$x_i > 0, \qquad i = 1, 2, \ldots, m, \tag{3}$$

where (i) the sets J_k, $k = (0, 1, 2, \ldots, p)$, number terms in the objective function J_0 and the constraints J_k, $k = 1, 2, \ldots, p$, and (ii) the parameters c_j and a_{ij} are real constants with the restriction that $c_j > 0$ for

$$j \in \bigcup_{k=0}^{p} J_k = \{1, 2, \ldots, n\}.$$

The PGP has a number of special features that are important from the point of view of algorithmic design. They are the following.

(a) Derivatives of the objective and constraint functions of any order are available explicitly. Furthermore, they are relatively cheap to compute once function evaluations have been made. For example, in a posynomial with q terms, only q multiplications and q divisions are required to compute the first derivative of the function with respect to some variable, once the values of these terms are known.

(b) The problem is convex in the variables $\log x$, and thus may be solved by any convex programming algorithms that account for feature (c) below.

(c) Observe that the primal variables x_i, $i = 1, 2, \ldots m$, are constrained to be *strictly positive*. Thus, the feasible region of PGP may not be compact and strictly speaking we should write "seek the infimum of" in place of "minimize" in the above programming problem. When a primal variable goes to zero or to some negative value, some of the terms in one or more posynomial functions might become undefined (for example, terms that contain the variable raised to a negative power).

The above characteristics of the primal indicate that any general NLP software may be used to solve GP problems provided that care is taken to avoid negative and, in some cases, zero values of the primal variables. One simple way of accomplishing this is to bound these variables from below using some small positive value. However, this approach may run into difficulty in cases where the GP is degenerate (see Duffin, Peterson, and Zener, Ref. 9) and some terms do go to zero in the optimal solution.

Since analytical derivatives are available and are relatively cheap to compute, it seems reasonable to expect gradient-based methods to be suitable for specialization to solving PGP problems. It is also not difficult to write a suitable front-end to any general gradient-based code that will read the term coefficients c_j, $j = 1, 2, \ldots, n$, the exponent matrix a_{ij}, $i = 1, 2, \ldots, m$ and $j = 1, 2, \ldots, n$, and numbers defining the sets J_k, $k = 0, 1, \ldots, p$, and from this data compute function and gradient values. Furthermore, since under a simple transformation the primal is convex, the above discussion applies to convex programming software as well.

Geometric programming is also amenable to solution via separable programming techniques. The separable primal geometric program (SPGP) is given below.

$$\text{(SPGP)} \qquad \underset{y;z}{\text{minimize}} \qquad \sum_{j \in J_0} c_j \exp(y_i), \tag{4}$$

$$\text{subject to} \qquad \sum_{j \in J_k} c_j \exp(y_j) \leq 1, \qquad k = 1, 2, \ldots, p \tag{5}$$

$$y_j - \sum_{i=1}^{m} a_{ij} z_i = 0, \qquad j = 1, 2, \ldots, n. \tag{6}$$

Equivalence of this separable formulation of the primal to the primal program PGP is easily recognized if we let

$$z_i = \log x_i, \qquad i = 1, 2, \ldots, m,$$

and define the constants c_j and a_{ij} and the sets J_k, $k = 0, 1, 2, \ldots, p$, as in PGP.

The SPGP formulation of the primal is a problem in $n + m$ variables as opposed to the m-variable formulation PGP. However, the addition of these variables results in a problem with a very special structure that may be exploited by special-purpose algorithms. Also, the above formulation would allow the solution of GP problems by widely available mathematical programming software systems such as MPSX (Ref. 10).

To this author's knowledge, there have only been two attempts at devising special-purpose algorithms for the solution of SPGP, namely Codes 1 and 3 (Appendix). Computational experience with these codes, both of which are based on linearization methods, is not encouraging. However, this should not be taken as conclusive evidence that solving SPGP is a poor way to approach the solution of GP problems. An algorithm more in keeping with the state-of-the-art in NLP would be to use a Newton-type method with an active constraint set strategy for maintaining feasibility of the nonlinear inequality constraints (5) and a projection method for handling the linear equality constraints. What makes this approach so attractive is the fact that the Hessian of a Lagrangian involving the constraints (5) would be a positive definite diagonal matrix, a fact that could surely be exploited computationally.

Another approach that is also in keeping with current practice in NLP would be to incorporate the nonlinear constraints (5) into an augmented Lagrangian and minimize this Lagrangian with respect to the linear equality constraints (6).

2.2. Solving Dual Geometric Programs. The dual geometric program (DGP) as defined by Duffin, Peterson, and Zener (Ref. 9) is the following linearly constrained nonlinear programming problem:

$$\text{(DGP)} \quad \text{maximize} \quad v(\delta) = \prod_{j=1}^{n} (c_j/\delta_j)^{\delta_j} \prod_{k=1}^{p} \lambda_k^{\lambda_k}, \tag{7}$$

$$\text{subject to} \quad \sum_{j \in J_0} \delta_j = 1, \tag{8}$$

$$\sum_{k=1}^{p} \sum_{j \in J_k} a_{ij}\delta_j = 0, \qquad i = 1, 2, \ldots, m, \tag{9}$$

$$\delta_j \geq 0, \qquad j = 1, 2, \ldots n, \tag{10}$$

where $$\lambda_k = \sum_{j\in J_k} \delta_j, \qquad k = 1, 2, \ldots, p. \tag{11}$$

A subtle but important point is that the λ_k's are not treated as independent variables in the problem and the relationships in (11) are not treated as constraints, but rather as definitions. Whereas from a theoretical viewpoint this distinction might appear to be a case of semantics, it is extremely bad practice computationally. *In fact, it is the view of this author that the explicit elimination of the λ variables and the explicit formation of the reduced dual problem have been the singular most important factors in the failure to design efficient and numerically stable software for the dual problem.* These statements will be justified in the discussion below.

The reduced dual geometric program (RDGP) is obtained by eliminating $m+1$ *basic variables* from the program DGP and expressing them in terms of $d = n-(m+1)$ *nonbasic variables*.[6] This results in the following dual program in the variables r_i, $i = 1, 2, \ldots, d$:

(RDGP) $\underset{r}{\text{maximize}}$

$$\hat{v}(r) = K_o\left[\prod_{i=1}^{d} K_i^{r_i}\right]\left[\prod_{j=1}^{n} \delta_j(r)^{-\delta_j(r)}\right]\left[\prod_{k=1}^{p} \lambda_k(r)^{\lambda_k(r)}\right] \tag{12}$$

subject to

$$\delta_j(r) = b_j^{(0)} + \sum_{i=1}^{d} r_i b_j^{(i)} \geq 0, \qquad j = 1, 2, \ldots, n, \tag{13}$$

where

$$\lambda_k(r) \triangleq \left[\sum_{j\in J_k} b_j^0\right] + \sum_{i=1}^{d} r_i\left[\sum_{j\in J_k} b_j^{(i)}\right], \qquad k = 1, 2, \ldots, p, \tag{14}$$

$$K_i \triangleq \prod_{j=1}^{n} c_j^{b_j^{(i)}}, \qquad i = 0, 1, \ldots, d. \tag{15}$$

The reduced dual was strongly emphasized in Duffin, Peterson, and Zener (Ref. 9) as being a computationally useful formulation of the dual GP. Unfortunately, the theoretical exposition in Ref. 9 was taken far too literally by researchers who were attempting to design algorithms for the dual. Without exception, every dual-based code known to this author (see Appendix) *explicitly computes and stores* the basis vectors $b^{(i)}$, $i = 0, 1, \ldots, m$, using Gaussian elimination or some related technique. This is completely *contrary to accepted practice in nonlinear programming* (Ref. 2, Chapter 2). A far more stable approach numerically would be to carry out

[6] The quantity d is sometimes referred to as the degree of difficulty of a GP.

the reduction procedure implicitly by storing a matrix Z whose columns span the null space of the equality constraint coefficient matrix of DGP (Refs. 2-2). The matrix Z can be computed and stored explicitly or in product form by performing an orthogonal triangulation of the coefficient matrix. Incidentally, this same factorization can be used efficiently to find a stable least-square solution to the primal–dual optimality relationships (Ref. 11) in order to recover the optimal primal variables. For details on this approach the reader should consult (Refs. 1–2).

We feel that algorithms for the dual should be based on the following separable dual geometric program (SDGP), that is equivalent to DGP:

$$\text{(SDGP)} \quad \underset{\delta,\lambda}{\text{maximize}} \quad V(\delta, \lambda) = \sum_{j=1}^{n} \delta_j \log(c_j/\delta_j) + \sum_{k=1}^{p} \lambda_k \log \lambda_k, \tag{16}$$

$$\text{subject to} \quad A_o\delta = 1, \tag{17}$$

$$A\delta = 0, \tag{18}$$

$$B\delta - \lambda = 0, \tag{19}$$

$$\delta \geqq 0. \tag{20}$$

The constraints (17)–(19) are a matrix representation of (9)–(11). Here,

$$\dim(A_0) = 1 \times n, \qquad \dim(A) = m \times n,$$

$$\dim(B) = p \times n,$$

where n = number of primal terms, m = number of primal variables, and p = number of primal constraints.

Notice that the above dual program has $n+p$ variables as opposed to n variables in the formulation given in (8)–(11). Also, the above formulation is a convex program whereas the original is not.

At first, it might seem ridiculous to increase both the number of variables and the number of constraints. However, this results in a problem with a very special structure. Firstly, the objective function is separable, and hence has a *diagonal Hessian* which can be utilized efficiently in a Newton-type algorithm for the dual problem (Ref. 1). The additional constraints should cause no consternation either. It is easy to construct algorithms which take implicit account of them. In fact, in Ref. 1 it is shown how the above problem may be solved by an algorithm in which the major computational effort involves either recurring an $m \times m$ *or* an $(n-m-1)\times(n-m-1)$ matrix at each iteration, depending on which is smaller.

The programs DGP, RDGP, and SDGP have a number of important characteristics, some of which preclude the direct application of NLP software for finding a numerical solution.

A major source of difficulty is the fact that the dual objective function is not differentiable with respect to the dual variables at points where they take on the value zero. To see this, for example, note that

$$\partial V/\partial \delta_j = \log(c_j/\delta_j) - 1, \tag{21}$$

which is undefined at $\delta_j = 0$. This fact (combined with the fact that, at an optimal solution, if for any $j \in J_k$, $\delta_j^* = 0$, then $\delta_j^* = 0$ for all $j \in J_k$) will cause general NLP software to fail if applied directly to these dual programs.

There are simple-minded remedies to the *nondifferentiability problem*, some of which are given below.

(a) Bound the variables such that $\delta_j \geq \epsilon > 0$, $j = 1, 2, \ldots, n$. This overcomes the differentiability problem, but causes other numerical problems such as an ill-conditioned Hessian at points near the solution. Also, most algorithms will tend to *zig-zag* between the constraints $\delta_j \geq \epsilon$, $j \in J_k$, until they have convinced themselves that $\delta_j^* = \epsilon$ for all $j \in J_k$.

More important is that simply zeroing all $\delta^* = \epsilon$ to $\delta^* = 0$, as is done in a large number of dual codes (see Appendix), will result in *infeasibilities* in the dual equality constraints (8) and (9). This could cause *poor estimates of the primal variables* to be computed using the primal–dual optimality conditions (Ref. 43). This author has long felt that this is precisely why it is often stated (see, for example, Ref. 12) that highly accurate dual solutions are required to obtain an even moderately accurate primal optimal solution. Thus, if the $\delta \geq \epsilon$ bounds are used, one must ensure that feasibility of the dual constraints is restored when the appropriate dual variables are zeroed.

The choice of an ϵ is also difficult. For example, in Problem 1 (Ref. 5) there are a substantial number of dual variables whose optimal value is less than 10^{-8}.

(b) A better approach than the one above is to approximate the jth term in the dual objective function by a quadratic[7] at points for which $\delta_j \leq \epsilon$. This is done as follows:

$$\delta_j \log(c_j/\delta_j) \simeq \alpha\delta_j^2 + \beta\delta_j, \qquad 0 \leq \delta \leq \epsilon, \tag{22}$$

where

$$\alpha = -1/\epsilon, \tag{23}$$

$$\beta = \log(c_j/\epsilon) + 1. \tag{24}$$

The advantage of the above approximation is that the objective function $V(\delta, \lambda)$ so defined will be continuous and differentiable, since α and β are chosen so that the derivatives and function values of $\delta_j \log(c_j/\delta_j)$ and

[7] The quadratic approximation presented here arose out of a series of discussions the author had with L. Lasdon and M. Saunders in an attempt to apply their general-purpose codes to the dual GP.

$\alpha\delta_j^2 + \beta\delta_j$ are equal at $\delta_j = \epsilon$. Also, the two functions are equal at $\delta_j = 0$; however, their gradients differ at this point. The quadratic approximation has a gradient of β at $\delta_j = 0$, whereas as δ_j tends to zero the gradient of $\delta_j \log(c_j/\delta_j)$ tends to infinity.

Here, the choice of ϵ is not as difficult to make as in the bounding method discussed above. A balance must be struck between making ϵ too small, in which case the Hessian matrix will become ill-conditioned (since the contribution of this term will be the diagonal element $-2/\epsilon$), and making ϵ too large, in which case the gradient at $\delta_j = 0$ [namely, $\beta = \log(c_j/\epsilon) + 1$], will be too small to approximate the true behavior of the objective function at $\delta_j = 0$. A limited amount of experimentation with the method has shown that a value of $\epsilon = 10^{-5}$ seems to suffice.

It should be noted that the value of ϵ could be chosen dynamically by the algorithm under consideration. For example, in a Newton-type algorithm, ϵ could be set to its minimum value such that the condition number of the Hessian of the objective function $V(\delta, \lambda)$ would not be much larger than if this particular variable were not present. Since the Hessian is diagonal with elements $-\delta^{-1}$ and λ^{-1} (Ref. 13), the above criterion would result in an ϵ value that is not "very much" smaller than the smallest δ_j.

For algorithms based on an active constraint set strategy, the above quadratic approximation is only needed for variables that are exactly zero, as a means of providing the algorithm with approximate curvature information so that it can decide whether or not the active constraint $\delta_j = 0$ should be dropped from the basis at a particular iteration. Since

$$\beta = \log(c_j/\epsilon) + 1$$

will be positive for any $\epsilon < c_j e$, if for example we choose ϵ such that

$$\epsilon = \min\{0.9c_j e, 10^{-5}\}, \tag{25}$$

we are always assured that the gradient of the approximating quadratic will have the correct sign. Furthermore, this method will not be subject to the *zeroing problem* alluded to in (a) above.

We will not deal with the important topic of converting an optimal dual solution into an optimal primal solution, since this is covered in detail in Dembo (Ref. 11). It will suffice to say that the results of Ref. 11 indicate that any algorithm for the dual should compute the optimal Lagrange multipliers, ω_i^*, $i = 0, 1, 2, \ldots, m$, corresponding to the normality and orthogonality constraints (8) and (9), since they are related to the optimal primal variables x_i^* and optimal dual objective function V^* by

$$\omega_i^* = \log x_i^*, \qquad i = 1, 2, \ldots, m \tag{26}$$

$$\omega_0^* = 1 - V^*. \tag{27}$$

Thus, an optimal solution of the primal can be computed to the same degree of accuracy as the optimal dual multipliers. As is mentioned in Dembo (Ref. 11), (27) provides us with a useful check on the accuracy of the multipliers since V^* and ω_0^* may be computed independently. Also, a dual algorithm is not complete unless it provides for the case where the multipliers ω_i are not unique and a subsidiary problem (Ref. 11) might have to be solved in order to recover an optimal solution of the primal problem. Only one of the dual-based codes in the Appendix, namely CSGP, provides for such an eventuality.

3. Factors Influencing the Choice between Primal-Based and Dual-Based Algorithms

The question is often raised as to whether geometric programs should be solved using algorithms based on the dual program or by direct solution of the primal program. To ask whether the primal problem or the dual problem should be solved is an oversimplification. It would probably be more correct to ask *when* should the primal problem be solved as opposed to the dual, and vice versa. There are obvious cases where the dual program is a very much simpler problem than the corresponding primal (for example, a geometric program with zero degrees of difficulty). Similarly, it is easy to construct geometric programs where the primal problem may be very much easier to solve than the dual (for example, consider the minimization of a posynomial function of one variable with a large number of terms).

It is well known that linear programming (LP) is a special case of GP (see, for example, Duffin, Peterson, and Zener, Ref. 9). Therefore, as Templeman (Ref. 14) quite rightly points out, a special case of the above dilemma occurs in LP when one has to decide whether to solve a problem using primal-based or dual-based methods. For LP, the problem is much simpler and one can easily identify cases where a primal approach would be advantageous, and vice versa. Also, the same algorithm, namely the simplex method, may be applied to both the primal program and the dual.

In geometric programming, the decision as to whether to solve the dual program or the primal is a far less obvious one. For the general case, there seems to be no way out other than to draw on empirical evidence generated by computational comparisons such as the one described in Section 6 and also in Rijckaert and Martens (Ref. 4). There is however one special case of GP, other than LP, for which the same algorithm may be applied to both the primal and dual problems; hence, an a priori estimate of which of the two problems is easier to solve can be made with a fair degree of certainty.

Consider a pair of primal–dual GP problems in the case where there are no posynomial inequality constraints in the primal problem (this is often referred to as an unconstrained GP). Here, the primal program (UPGP) and dual program (UDGP) may both be written as convex, *separable* linearly constrained nonlinear programming problems:

$$\text{(UPGP)} \quad \underset{y,\omega}{\text{minimize}} \quad \log \sum_{j=1}^{n} c_j \exp(y_j), \tag{28}$$

$$\text{subject to} \quad y - A^T\omega = 0; \tag{29}$$

$$\text{(UDGP)} \quad \text{maximize} \quad \sum_{j=1}^{n} \delta_j \log(c_j/\delta_j), \tag{30}$$

$$\text{subject to} \quad \sum_{j=1}^{n} \delta_j = 1, \tag{31}$$

$$A\delta = 0, \tag{32}$$

$$\delta \geq 0. \tag{33}$$

Since the cost of function and derivative evaluations is roughly the same for UPGP and UDGP, the difference in computational effort required to solve them will be a function of the relative sizes (number of variables and constraints) of these dual programs, if the *same* algorithm is applied to both. In both cases, the equality constraints may be handled implicitly using projection matrices; however, the primal problem (UPGP) does have a slight edge over the dual, in that it does not possess inequality constraints. Also, the primal objective function is differentiable at all points in the primal feasible region, whereas the dual is not.

For constrained GP problems, the tables are turned. The primal problem (SPGP) is subject to the nonlinear inequality constraints (5), whereas the dual (SDGP) remains a linear constrained problem. The author's feeling is that, with few exceptions, the nonlinear inequality constraints of the primal problem make the dual (SDGP) a more attractive problem to solve, *even when the degree of difficulty* $(n-m-1)$ *is very large.* The reason for this is twofold. Firstly, nonlinear constraints are at least an order of magnitude more difficult to deal with than are linear constraints. Secondly, and this is what most researchers in the area of GP seem to be unaware of, the dual problem (SDGP) may be solved by a Newton-type algorithm where at each iteration the main amount of work involved lies in solving a square system of equations, whose dimension is either equal to the degree of difficulty of the problem $(n-m-1)$ or to the number of primal variables (m), depending on which of these two quantities is smaller. Details of such an algorithm are given in Dembo (Ref. 1).

4. Extensions to Signomial Programming

A signomial programming (SP) problem is a program of the form given in (1)–(3), (namely, PGP), in which the term coefficients may take on any real value. This type of programming problem is sometimes referred to as an algebraic program.

In general, SP problems are nonconvex, and the elegant duality theory associated with posynomial programs does not carry over to signomial programs. Attempts have been made at defining pseudo-dual problems (see Ref. 15 and Ref. 16, Chapter 5); however, the use of a pseudo-dual program as a vehicle for computing an optimal solution to a primal program PSP (that is, PGP where some $c_j < 0$) is not to be recommended unless certain safeguards are incorporated into the algorithm. This is because a local maximum of the pseudo-dual program might correspond to a *local maximum* of the primal (recall that PSP is a nonconvex *minimization* problem). Thus, if a dual approach is used to solve PSP, the algorithm must contain a built-in checking procedure to ascertain whether or not the computed stationary point of the primal problem is in fact a local minimum. If it is not, then the algorithm should invoke an alternative procedure until convergence to a local minimum is achieved. To the author's knowledge, *none* of the existing codes that solve SP problems via a pseudo-dual approach (see dual Codes 2 and 11 in the Appendix) have built-in safeguards.

Apart from the *pseudo-dual approach*, there are essentially three different ways in which algorithms for SP problems have been designed. These are discussed below.

4.1. Complementary Algorithm of Avriel and Williams. (*Ref. 17*). This algorithm solves an SP problem by solving a sequence of GP approximations. Each GP approximation is computed using posynomial condensation (Ref. 17). It was noted by Dembo (Ref. 17, also reported in Ref. 18) that the complementary algorithm may be accelerated if an exterior method[8] is used to solve the approximating GP primal.

The codes GGP, QUADGP, SIGNOPT, GEOEPS, and GEOLP (see Appendix) all use this approach to solving SP problems. Unfortunately, a feasible point is required to initiate the algorithm, and this in general means that a Phase 1 routine has to be incorporated into the code (Ref. 20).

4.2. Harmonic Method of Duffin and Peterson. (*Ref. 20*). Here too, the SP problem is solved by solving a sequence of approximating GP

[8] By *exterior method*, we mean that the sequence of points converging to an optimal solution remains infeasible until the solution is reached.

problems. Each *harmonic approximation* of Duffin and Peterson can be shown to be weaker than the complementary approximation (see, for example, Ref. 16) and in general results in an approximate GP problem whose dual has a larger degree of difficulty than in the complementary approximation. The harmonic approach, however, does have one important property that, to the author's knowledge, has never been exploited computationally. That is, the exponent matrix remains constant for every GP problem in the approximating sequence, which is not the case in the complementary algorithm. Without making specific use of this property, any algorithm based on the harmonic method will be *dominated* by one based on the condensation method, other things being equal. Jefferson's code GPROG (see Appendix) uses the harmonic approach. Bradley's code QUADGP (see Appendix) has an option to use either the harmonic algorithm or the complementary algorithm. In both these codes, the invariance of the exponent matrix is *not* used. Bradley (Ref. 21) demonstrates the obvious superiority of the condensation procedure on a number of test problems.

The same remarks in Section 4.2 regarding feasible starting points apply here also.

4.3. Direct Solution of the Signomial Programming Problem. To date there has only been one code developed to solve the SP problem directly. Rijckaert and Martens (Ref. 22) solve the nonlinear equations corresponding to the Kuhn–Tucker first-order necessary conditions for optimality of the primal SP problem (PSP). However, they do not indicate whether they have built in safeguards to ensure that they compute a local minimum of PSP and not a stationary point or a local maximum.

4.4. Convergence of the Complementary and Harmonic Algorithms. Unfortunately, it has been this author's experience that the complementary (and hence the harmonic) algorithm tends to converge linearly for most problems.[9] This makes it a poor method to use in a GP code, especially for problems with relatively few negative terms (see Test Problem 4A, Ref. 23). It is for this reason that the author feels that the best way to solve signomial problems in general is by a direct attack on the primal program written in separable form (that is, SPGP where the coefficients c_j are not all positive). This will surely be more efficient than solving a sequence of similar-sized GP problems. Some justification for this statement is given in the next section.

[9] A computational comparison of these algorithms is given in Ref. 24.

5. Numerical Comparison of Some GP Codes

This section summarizes the results of a Colville-type study (Ref. 25) that was undertaken by this author in the period from June 1974 to July 1976. The ambitious aims of the study were: (i) to identify which available GP codes were obviously superior to other in terms of computational efficiency; (ii) to test the robustness of various approaches; (iii) to isolate a good set of problems that would test various critical aspects of GP algorithms; and (iv) to answer the embarrassing question: are specialized GP codes more efficient in the solution of GP problems than good general-purpose NLP codes?

Only one of the above aims was achieved to any degree of satisfaction, namely, the study did produce a good set of test problems (Ref. 23). Our justification for this conclusion comes from the feedback from people who have actually attempted to solve these problems.[10] Their general conclusion is that the problem set contains a good mix of well-scaled, badly-scaled, easy, and difficult problems and also captures the inadequacies of various algorithmic approaches to GP. We will discuss the particular nature of each of the problems later, when the computational results are analyzed.

There is one major drawback, however, to conducting a comparative study based on a hand-picked sample of test problems. That is, very little in the way of inferences can be made as to the relative performance of the codes in question on a *different* set of problems (Ref. 27). It is precisely this sort of inference that one wishes to make; namely, since code X did better than code Y on the test problems, this will be true for a larger class of problems. Unfortunately, the methodology for designing comparative studies in mathematical programming is primitive, to say the least, and has only recently been considered as a serious topic for research.

The study reported here was conducted in the following way.[11] An attempt was made to obtain the participation of all authors of GP software that were known to the author at the time the study was conducted. Each participant was informed that the problems would be run at the particular author's home institution, on the computer for which the code was originally developed. In addition, Colville's standard timer (Ref. 25) was supplied in an attempt to standardize the CPU timing results and stopping criteria and

[10] The names and addresses of people other than those mentioned in this study who have solved the problems in (Ref. 23) is available on request from the author.

[11] As is mentioned in the text, the author is fully aware of the drawbacks of such a study (see Refs. 27–28) and cautions the reader to be wary of any conclusions drawn on the basis of the results presented here. In particular, the use of a standardized timing routine may in extreme cases make timing results meaningless.

tolerances were specified as reported in (Ref. 23). Where possible, an attempt was made to standardize the use of compilers. For example, participants using IBM machines were requested to use the FORTG compiler when compiling the timing program. In some cases, participants did not adhere strictly to the rules, and this has added some additional noise to the results.

To some extent, the experimental design did allow the participants to tune their codes to the set of problems in Ref. 23, and so the timing results in Table 4 represent the best results of each code on this set of problems, without a major alteration to the code design itself. Actually, in two cases (Refs. 8, 29) the participants redesigned the Phase I section of their codes as a result of repeated failures on some of the test problems.

An attempt was also made to include CPU timing results for some recently developed general-purpose NLP codes in order to compare with GP codes tested. Table 1 summarizes the characteristics of the general purpose NLP codes that participated in the study.

The GP codes that participated were SIGNOPT, GEOEPS-GEO-GRAD, GEOLP, GPKTC, GPROG, and GGP. A summary of their main features is given in the Appendix. Details of the computer configuration and other aspects of the timing runs for the GP participants are given in Table 2.

Important characteristics of the test problems are summarized in Table 3.

The standardized times (actual CPU time divided by Colville standard time) for all participating codes on all test problems are given in Table 4. Blank entries in the table indicate that the code did not converge to a solution. The participants were all asked to solve the problems to within the convergence and constraint tolerance criteria specified in Table 3 (see Dembo, Ref. 23). Whereas most participants met the crucial primal feasibility criterion, stopping criteria and other internal tolerances were not equivalent from one code to the next. This does add an additional degree of uncertainty in interpreting the results. Only one of the GP codes, namely GPROG, did not solve problems to the required feasibility tolerances. The results in Table 4 for GPROG refer to a primal feasibility tolerance of 0.01, whereas the required feasibility tolerances specified in Ref. 23 range from 10^{-4} to 10^{-6}.

In an attempt to measure the sensitivity of various codes to achieve different degrees of primal feasibility, two levels of feasibility tolerances were specified in Dembo (Ref. 23). Since only a few of the participants responded to a request for runs at both tolerance levels, these results are not reported here. They are reported, however, for the code GGP in Ref. 23.

Table 1. Characteristics of general-purpose NLP codes tested.

Code	Algorithm	Language	Compiler	Participants*	Institution	Computer configuration	Colville standard time
GRG	Generalized reduced gradient (Ref. 30)	FORTRAN	WATFIV (NOCHECK)	Lasdon Ratner Jain	Stanford University (currently at Case Western Reserve University)	IBM 370/168	16.83 sec using WATFIV (NOCHECK)
COMET	Penalty function method	FORTRAN	RUN	Himmelblau	University of Texas at Austin	CDC 6600	20 secs
GREG	Generalized reduced gradient (Ref. 31)	FORTRAN	RUN	Himmelblau (Abadie/Guigou)	University of Texas at Austin	CDC 6600	20 secs
GPM/ GPMNLC	Extended gradient projection method (Ref. 32)	FORTRAN	RUN	Himmelblau (Kreuser/Rosen)	University of Texas at Austin	CDC 6600	20 secs
GAPF-QL	Penalty function method modified to avoid ill-conditioning of Hessian	FORTRAN	RUN	Himmelblau (Newell)	University of Texas at Austin	CDC 6600	20 secs

* Names in parentheses refer to original authors of the code.

Table 2. Characteristics of GP codes participating in study.

Code	Language	Compiler	Participants	Institution	Computer configuration	Colville standard time
SIGNOPT	FORTRAN	Not specified	Templeman	University of Liverpool	ICL 1906A CD6 7600 combination	2.22 secs
GEOPS/GEOGRAD	FORTRAN	FORTH	Dinkel	Pennsylvania State University	IBM 370/168	8.75 secs (FORTG)
GEOLP	FORTRAN	(OPT = 2)	Kochenberger			
GPKTC	FORTRAN	FORTH (OPT = 2)	Rijckaert Martens	Katholieke Universiteit Leuven	IBM 370/158	27.73 secs (FORTG)
GPROG	FORTRAN and COMPASS	Not specified	Jefferson	University of New South Wales Australia	CYBER 72	30 secs (approx.)
GGP	FORTRAN	FORTH (OPT = 2)	Dembo	University of Waterloo (currently at Yale University)	IBM 370/158	25.30 secs (FORTG)

Table 3. Test problem characteristics (Ref. 23).

Problem	Type	Variables	Constraints*	Terms*	Required tolerances EPSCON†	EPSCGP‡
1A	GP	12	3	31	10^{-6}	10^{-4}
1B	GP	12	3	31	10^{-6}	10^{-4}
2	SP	5	6	32	10^{-5}	10^{-4}
3	SP	7	14	58	10^{-5}	10^{-4}
4A	SP	8	4	16	10^{-5}	10^{-4}
4B	SP	8	4	16	10^{-5}	10^{-3}
4C	GP	9	5	15	10^{-5}	10^{-4}
5	SP	8	6	19	10^{-5}	10^{-4}
6	SP	13	13	53	10^{-6}	10^{-4}
7	SP	16	19	85	10^{-5}	10^{-3}
8A	GP	7	4	18	10^{-6}	10^{-4}
8B	GP	7	4	18	10^{-6}	10^{-4}
8C	GP	7	4	18	10^{-6}	10^{-4}

* Does not include simple bounds on variables.
† If $g_k(x) \leq 1 + \text{EPSCON}$, $k = 1, 2, \ldots, p$, then the constraints are considered to be satisfied.
‡ Convergence tolerance, see Footnote 12.

6. Analysis of Comparative Study Results

We will analyze the results presented in Table 4 in terms of the original aims of the study. First, results for GP codes are analyzed. Later, these are compared with the results for NLP codes.

6.1. Results for GP Codes

(a) *Efficiency.* By glancing at the bracketed numbers in Table 4, we see immediately that the GP codes GPKTC and GGP stand out among the rest in terms of the speed with which the problems were solved. The code GPKTC was within 10% of the fastest standardized time for 7 of the 13 test problems, whereas GGP was within 10% of the fastest time in 8 of 13 cases. A closer scrutiny reveals that GGP seemed to do better on the larger problems and on problems with many simple bounding constraints. This is entirely consistent with the findings in Ref. 16.

An interesting result shown in Table 4 is that GGP consistently dominates GEOLP in terms of standardized times. This result is interesting because these two codes are based on the *identical mathematical algorithm.*

Table 4. Standardized times for test problems.

Problem	Type	Starting point (primal)	SIGNOPT (GP)	GRG (NLP)	COMET (NLP)	GREG (NLP)	GPM/ GPMNLO (NLP)	GAPF-QL (NLP)	GEOPS GEOGRAD (GP)	GEOLP (GP)	GPKTC (GP)	GPROG (GP)	GGP (GP)
1A	GP	NF						4.651	0.5648	1.1166	[0.0584]	0.1276	0.274
1B	GP	NF		[0.056]	1.5262	0.1197		0.5155	0.5648	1.1166	[0.0554]		0.271
2	SP	F	2.6259	0.010		0.2444	0.1349*	0.442*	0.0405	0.0325	0.0454	0.1937	[0.002]
3	SP	F			0.7497	0.1165		0.5820		0.2992	[0.0868]		[0.082]
4A	SP	NF	1.1524	0.038	0.0544	0.0503	0.1193*	0.0322		1.7721	[0.0186]	0.7730	0.280
4B	SP	NF	1.1524	0.034	NR	NR	NR	0.0329		NR	[0.0183]		0.132
4C	GP	NF	0.1650	0.052	0.0795	0.0818	0.0958	0.0404	0.0537	0.0843	0.0244		[0.021]
5	SP	NF	1.4330	0.049	0.1301*	0.0709	0.5404	1.0247	0.5057	0.2817	[0.0323]	1.6220	0.125
6	SP	NF				0.5747				0.5390	[1.3545]		[0.327]
7	SP	NF						2.4454		0.4581	0.6864		[0.240]
8A	GP	NF	3.8943	0.282	>3	0.4479	0.1684		[0.0877]	0.1695	0.2155		[0.095]
8B	GP	NF	6.0576	0.194	1.8736	0.2864	0.2942		0.1406	0.1559	0.1861		[0.095]
8C	GP	NF	31.782	0.443	0.9045	0.3043	0.2577		0.1731	0.1274	0.1405		[0.079]

Problem type: GP = Posynomial program, SP = Signomial program.
Starting Point: NF = Not feasible, F = Feasible.
Bracketed numbers indicate standardized times that are within 10% of the best time. Blank entries indicate that the code was unable to solve the particular problem.
Asterisk indicates that the error in the computed value of an optimal solution was between 5% and 10%.
NR = Not reported.

This underscores the fact that what we are testing in this study is the performance of codes and not algorithms.

The results, however, do shed some light on the underlying algorithms. Consider for example Problems 4A, 4B, and 4C. Problems 4A and 4B are both signomial problems in 8 variables, 4 signomial constraints, and 7 degrees of difficulty. The sole difference between these problems is that Problem 4A has a tighter convergence tolerance ($\text{EPSCGP} = 10^{-4}$) than Problem 4B ($\text{EPSCGP} = 10^{-3}$).[12] In Ref. 26 and Table 3, it is shown that GGP requires more than twice as much CPU time for Problem 4A as for Problem 4B. This is indicative of the sensitivity of the Avriel–Williams (Ref. 17) algorithm to this commonly used termination tolerance criterion. The code GPKTC solves signomial problems directly, and Table 4 shows that the algorithm used is not at all sensitive to such a criterion; that is, GPKTC requires roughly the same amount of CPU time to achieve optimality in Problems 4A and 4B.

An even more dramatic indication of how inefficient the Avriel–Williams procedure (Ref. 17) can be is found by examining the relative performance of GGP and GPKTC on Problems 4A and 4C. Problem 4A as mentioned above is a signomial problem. However, what was not mentioned was that *only two of* 16 *terms in the problem have negative coefficients.* Thus, Problem 4A is *almost posynomial.* Problem 4C is a GP approximation of 4A, which is obtained by *condensing-out* these two negative terms (Ref. 19). Therefore, Problem 4C is constructed to be a very similar-sized and similar-structured problem to Problem 4A but without any negative terms. For this problem (and using the same tolerance criteria as for Problem 4A) GGP is 14 times faster than for Problem 4A and is slightly faster than GPKTC. Note also that GPKTC needs approximately the same amount of CPU time for all three problems and, not unexpectedly, all the GP codes using the Avriel–Williams algorithm (namely, SIGNOPT and GEOLP) exhibit the same type of behavior as GGP on these three problems.

It is difficult to compare GEOEPS/GEOGRAD, GPROG, and SIGNOPT, since they have few data points in common. Very roughly speaking, GEOEPS/GEOGRAD seems to be faster than GPROG and SIGNOPT; when it converges, it is fairly competitive with GPKTC and GGP. The standardized times for SIGNOPT show that it can exhibit extremely slow convergence (see Problem 2, for example). The reader is

[12] The tolerance EPSCGP is defined in Ref. 23 by

$$|[g_0(x^i) - g_0(x^{i-1})]/g_0(x^{i-1})| \leq \text{EPSCGP},$$

where $g_0(x^i)$ is the objective function value at the ith *ϵ-feasible point* x^i and x^i, $i = 1, 2, \ldots$, is the sequence of points that converge to a minimum of the signomial problem. That is, if the above inequality is satisfied, then x^i is assumed to be in the vicinity of a stationary point.

asked to interpret this with caution, since it is this author's feeling that the inaccuracies introduced by using standardized timers penalize participants with very fast computers. The above tests using SIGNOPT were run on a CDC 7600 with a standardized time of 2.2 seconds. This is by far the fastest computer in the study. Fortunately, this effect is not present when comparing GGP and GPKTC, since IBM 370/158 computers were used in both cases.

(b) *Robustness.* The number of blank entries in Table 4 shows that, despite the *tuning effect* mentioned in Section 5, many codes were simply unable to solve certain problems. The most difficult problems (in terms of the number of codes that failed to solve them) were Problems 1A, 6, and 7. Problems 6 and 7 were the largest in the study.

Among the GP codes, GEOLP, GPKTC, and GGP stand out as being very robust on this set of test problems, since they never failed to converge to a solution. This is confirmed once again by the independently conducted study in Ref. 4. The codes GEOEPS/GEOGRAD and SIGNOPT each failed to converge for 5 of the 13 problems, and GPROG was the least robust, with only 4 successes out of a total of 13 problems.

6.2. Results for General NLP Codes

(a) *Efficiency.* Of the NLP codes tested, GRG ranks as the most efficient in terms of standardized CPU time and appears to do consistently better than GREG. This result is misleading and would probably be *reversed* if the timing runs for GRG were to be carried out using the FORTH (OPT = 2) compiler. The reason for this statement is graphically illustrated in Ref. 55. In Ref. 55, Problems 4C and 5 were solved by GRG using the FORTH (OPT = 2) compiler. The Colville timer, also run using FORTH (OPT = 2), yielded a standard time of 3.91 seconds for the IBM 370/168 [as compared with 16.83 using WATFIV (NOCHECK)]. The standardized times thus computed for Problems 4C and 5 were 0.109 and 0.069, as opposed to 0.052 and 0.049 using WATFIV (NOCHECK)!

The NLP codes all seemed to do badly on the most difficult problems in the set namely, Problems 1A, 6, and 7. The only NLP method that was able to solve the badly-scaled chemical equilibrium problem (Ref. 23) was the penalty method GAPF-QL. This is probably because care was taken in the coding of GAPF-QL to account for ill-conditioning (see Appendix). The only NLP code to solve Problem 6 was Abadie and Guigou's reduced gradient code GREG (Ref. 28). Since the majority of the effort in Problem 6 lies in finding a feasible solution (Phase 1), this *might* indicate that, among the general NLP codes tested, GREG has the best Phase 1 component.

On first attempt, the NLP code GAPF-QL converged only on Problems 3, 4, and 7 (Ref. 33). The values given in Table 4 therefore in some sense show the best results that could be attained by GAPF-QL on these problems and are a result of a number of trial runs.

The code GPM/GPMNLC seemed to produce the most inaccurate results; for the 3 of 7 problems solved, there was more than a 5% error in the *optimal solution value* computed. The fact that errors of this size appear in the results for COMET, GPM/GPMNLC, and GAPF-QL leads the author to suspect that the constraint tolerances specified in Dembo (Ref. 23) were not strictly adhered to when these solutions were computed. However, without direct information to the contrary, the author will assume that the values in the table are CPU times computed with the specified tolerances.

(b) *Robustness.* Among the general NLP codes, the two generalized reduced gradient codes that were tested, namely GRG and GREG, proved to be the most robust. The code GRG failed on 3 of 13 problems, whereas GREG only failed on 2 of the problems. The codes COMET and GAPF-QL each failed on 4 problems, and GPM/GPMNLC was the worst, with 5 failures and a record of inaccurate solutions for the problems that it did solve.

6.3. Performance of GP Codes versus General NLP Codes

(a) *Efficiency.* In order to demonstrate the relative efficiencies of the two sets of codes, standard times for the *best* two GP codes (GPKTC and GGP) are compared with the best time computed by any one of the 5 general NLP codes. This comparison is given in Table 5.

Table 5. Standardized times of general NLP codes versus GP codes.*

Problem	Type	Variables	Nonlinear constraints	Best NLP / GPKTC	Best NLP / GGP	Best NLP / Best GP
1A	GP	12	3	79.6	17.0	79.6
1B	GP	12	3	1.0	[0.2]	1.0
2	SP	5	6	[0.2]	5.0	5.0
3	SP	7	14	1.3	1.4	1.4
4A	SP	8	4	1.7	[0.1]	1.7
4B	SP	8	4	1.8	[0.2]	1.8
4C	GP	9	5	1.7	1.9	1.9
5	SP	8	6	1.5	[0.4]	1.5
6	SP	13	13	1.6	1.8	1.8
7	SP	16	19	1.4	4.0	4.0
8A	GP	7	4	[0.8]	1.8	1.8
8B	GP	7	4	1.0	2.0	2.0
8C	GP	7	4	1.8	3.3	3.3

* Bracketed numbers refer to problems where the best standardized NLP time was better than the standardized GP time in question.

An interesting observation is that the best GP time is *always better* than the best general NLP time. Also, each of the GP codes GPKTC and GGP outperform the best NLP result for the vast majority of the test problems. As expected, the standardized times for GGP on Problems 4A and 4B are considerably worse than those for the best NLP time, since these problems were specifically designed to demonstrate the worst features of the Avriel–Williams algorithm used in GGP.

(b) *Robustness.* The best GP codes were very much more robust on these problems than their NLP counterparts. There were, however, GP codes that did not perform as well as most of the NLP codes.

7. Conclusions and Suggestions for Future Research

Despite the drawbacks associated with the comparative study in Section 6, it is possible to draw some inferences regarding the behavior of the various codes. These inferences hold with some (imprecise) degree of certainty for the particular problem set tested. It is encouraging to note that *all* the conclusions drawn in this section are supported by the results of an independent comparative study (Ref. 4), done in a more controlled setting and using a different set of problems. In the Rijckaert–Martens study (Ref. 4), the two best GP codes (in terms of computational efficiency and robustness) were found to be GGP and GPKTC.[13] The same conclusion is evident in the results of Section 6.

Two of the general NLP codes in this study, namely GRG and GREG, are known to be among the best available general-purpose NLP codes. In the Colville study (Ref. 25) for example, Abadie and Guigou's GREG proved to be one of the most efficient and robust codes tested. Whereas these NLP codes appeared to be fairly efficient in solving our GP test problems, neither of them converged on the badly-scaled Problem 1A or the largest problem in the study, Problem 7. Also, their combined best time was equal to the best GP time for Problem 1B and was between 1.4 and 5 times slower for the remaining problems that they managed to solve.

The above result is an indication to this author that there is a need for specialized GP codes, if not for any other reason but robustness alone. This author comes to this conclusion despite the fact that he feels that even the best GP codes available (GPKTC and GGP) are in many ways primitive and lag far behind what could be achieved by specilizing current NLP technology to GP.

[13] It should be noted that the version of GGP referred to in Ref. 4 is an earlier and less efficient version than the one used here.

The GP problems SPGP and SDGP have a structure that makes them amenable to large-scale geometric programming (Ref. 1). As in LP, one can precisely define what is meant by a large sparse GP. Sparsity in the general NLP case is not easily defined.

In thc summary, the conclusions are as follows.[14]

(i) Specialized GP codes do appear to offer improvement in computation times and to be more robust than general NLP codes, on GP problems.

(ii) Algorithms for the linearly constrained dual GP do not make use of the latest available technology that has been developed for linearly-constrained NLP.

(iii) Primal-based GP codes seem to dominate the field. The author feels that this is a reflection of (ii) above and *not* because of some inherent difficulty in the dual GP.

(iv) The Avriel–Williams (Ref. 17) and Duffin–Peterson (Ref. 34) algorithms for solving signomial programs are often extremely inefficient (refer to Problems 4A, 4B, and 4C) and experience shows that they often exhibit a linear rate of convergence.

(v) A methodology for comparison of mathematical programming software is sorely needed, in order that hypotheses about algorithm and code behavior may be tested in a scientific manner. Hopefully, Ref. 27 is a step in this direction.

7.1. Suggestions for Future Research

(i) There is a pressing need for an efficient and robust dual-based GP code (if only as an intellectual challenge). The author feels that the approach that should be taken is to specialize some of the numerically stable techniques for linearly constrained NLP as described in Ref. 2. A framework for doing this already exists (see Ref. 1). In particular, any strategy that is adopted should be amenable to extensions to large-scale applications. That is, the algorithm should be able to exploit sparsity and/or problem structure.

(ii) Some theoretical developments are that definitely needed, before dual-based algorithms can be competitive, are efficient algorithms for handling *simple bounding constraints on primal variables.* It is not unreasonable to postulate that the resulting special structure in the dual problem, when simple primal bounding constraints are present, could be exploited

[14] Strictly speaking, these conclusions are only valid for the test problems solved in this study. Since, however, Rijckaert and Martens (Ref. 4) reach similar conclusions on a *different* set of problems, there is reason to believe that these concludions will be true for a wide variety of GP problems.

computationally. This is an important consideration, because simple bounding constraints are invariably present in models of real systems.

(iii) Many nonlinear programming applications (Ref. 35) are signomial problems that may be written in the form:

$$\underset{y,z}{\text{minimize}} \quad \sum_{j\in J_0} c_j \exp(y_j), \tag{34}$$

$$\text{subject to} \quad \sum_{j\in J_k} c_j \exp(y_j)=1, \qquad k=1,2,\ldots,q, \tag{35}$$

$$\sum_{j\in J_k} c_j \exp(y_j)\leq 1, \qquad k=q+1,\ldots,p, \tag{36}$$

$$y_j-\sum_{i=1}^{m} a_{ij}z_j=0, \qquad j=1,2,\ldots,n, \tag{37}$$

$$l_j\leq z_j\leq u_j, \qquad j=1,2,\ldots,n, \tag{38}$$

where the sets J_k are defined as before and the coefficients c_j, the "exponents" a_{ij}, and the variable bounds l_j and u_j are arbitrary real numbers (with $l_j<u_j$, of course).

The common approach to solving such problems (Ref. 35) has been to somehow convert Eqs. (35) to inequalities and then to apply the Avriel–Williams procedure, which converts the solution of the above signomial problem to the solution of a sequence of GP's. This strategy is fraught with difficulties and requires an experienced mathematical programmer for its implementation. It is this author's feeling that research into the development of software for signomial programs should concentrate on a direct solution of the above program. It has special features that would make either a generalized reduced gradient approach or an augmented Lagrangian approach attractive possibilities when designing a code. Naturally, if the code is to be competitive, the linear constraints (37) and (38) should be handled in some implicit fashion.

8. Appendix: Summary of GP Software Reported in the Literature

This appendix summarizes available information on GP codes that have appeared in the literature. We only include publications in which there is some evidence that the proposed algorithm has been coded and tested on a number of problems. There have been many attempts at coding GP algorithms, and it is hoped that none of these have been omitted here. Codes that are not mentioned in this section have not been omitted purposefully and were simply not known to this author at the time of writing.

Information on the codes is presented in the chronological order in which they appeared in the literature. An asterisk next to the code's name indicates that it appears in the computational study in Section 5.

Primal-Based Codes

Code 1

Name: DAP.

Author: G. V. Reklaitis (Ref. 36).

References: G. V. Reklaitis and D. J. Wilde (Refs. 37, 38, 16).

Algorithm: Solves SP directly using the differentiable algorithm of Wilde and Beightler (Ref. 39). Signomial programs solved via sequential GP approximation scheme of Duffin (Ref. 40).

Comments: No Phase 1 method reported. Reklaitis (Ref. 41) has compared DAP to the primal code GGP and in general has found GGP to be far more efficient.

Code 2

Name: *GGP**

Author: R. Dembo.

References: Dembo (Ref. 18, Ref. 5); Avriel, Dembo, and Passy (Ref. 19).

Algorithm: Solves PGP directly using a cutting-plane algorithm based on condensation. Signomial problems are solved using an accelerated AW algorithm.

Comments: Does not require a feasible starting point. Phase 1 algorithm operates on a modified problem to compute an initial feasible point if necessary. Experience with GGP has shown it to be both reliable and efficient. The code has been widely distributed and has been used to solve a large number of GP apllications. Feedback indicates that the method is robust and efficient for small- to medium-sized problems. A very similar algorithm was coded and tested as early as 1968 at Mobil Research Laboratories (Williams, Ref. 6).

Code 3

Authors: W. Gochet and Y. Smeers (Ref. 42).

Algorithm: Solves SPGP directly using a cutting-plane algorithm. Cuts are shown to be "deeper" than Kelley cuts.

Comments: No extension to signomial programs is reported. Computational experience is reported on two of Beck and Ecker's (Ref. 12) problems.

Code 4

Authors: G. S. Dawkins, B. C. McInnis, and S. K. Moonat (Ref. 43).

Algorithm: Tangential approximation method of Hartley and Hocking (Ref. 44). Operates on PGP in the variables $z = \log x$.

Comments: No extension to signomial programming reported. No details of the implementation are given and only the solution of a single problem is presented.

Code 5

Authors: J. G. Ecker and M. J. Zoracki (Ref. 45).

Algorithm: PGP is converted to a GP with at most two monomial terms in each constraint, according to the procedure outlined by Duffin and Peterson (Ref. 20). This posybinomial problem is solved using a hybrid of the tangential approximation and cutting-plane methods.

Comments: No extensions to signomial programming are given, and only a limited amount of computational experimentation is reported.

Code 6

Name: GPKTC*.

Authors: M. J. Rijckaert and X. M. Martens (Ref. 22).

Algorithm: The Kuhn–Tucker conditions for optimality of PGP are solved iteratively using a condensation procedure. This method is essentially equivalent to a Newton–Raphson algorithm for direct solution of the Kuhn–Tucker conditions expressed in terms of the variables $z = \log x$.

Comments: The code GPKTC has been extensively tested in Rijckaert and Martens (Ref. 4) and appears in the comparative study in Section 5. Experimentation with the code has shown it to be very robust and efficient especially for small GP problems. The code does not have an efficient mechanism for handling simple bounding constraints and constraints that are slack at optimality. A Phase 1 procedure is included in GPKTC and, judging from the results in Section 5, it appears to work well.

Code 7

Name: GEOLP*.

Authors: J. J. Dinkel, W. H. Elliott, and G. A. Kochenberger (Ref. 46).

References: Dembo (Ref. 18); Avriel, Dembo, and Passy (Ref. 19).

Algorithm: Essentially the same as GGP.

Comments: The authors claim to have successfully solved fairly large problems using GEOLP, and they feel (Ref. 46) that GEOLP works better than any other GP software that they have developed. It is interesting to note (see Section 6) that, despite the fact that GGP and GEOLP are based on the same algorithm, GGP seems to do consistently better than GEOLP on the test problems in Ref. 23. One possible explanation could be that GGP contains a number of algorithm refinements not contained in Refs. 18 and 19.

Code 8

Authors: M. Rammamurthy and G. H. Gallagher (Ref. 7).

References: Dembo (Ref. 18); Avriel, Dembo, and Passy (Ref. 19).

Algorithm: Essentially the same as GGP.

Comments: The authors claim to have solved a large number of civil engineering design problems using their code. Experience similar to that quoted in Code 2 above.

Dual-Based Codes

Code 1

Author: C. J. Frank.

References: Frank (Refs. 47, 48).

Algorithm: Solves the dual program DGP by applying the direct search method of Hooke and Jeeves (Ref. 49). Probably, the first published GP code. Experimentation has indicated that, in many cases, convergence of the method is extremely slow.

Comments: Code does not solve signomial problems.

Code 2

Name: GOMTRY

Authors: G. E. Blau and D. J. Wilde (Refs. 50, 51).

Algorithm: Solves the Kuhn–Tucker conditions for the dual program SDGP. Solves signomial programs in the above manner by attacking the necessary conditions for optimality of the pseudo-dual problem (Ref. 49).

Comments: GOMTRY's convergence is relatively good for small problems but experimentation (Ref. 4) shows that the code often fails to converge especially for medium-sized problems.

Code 3

Author: G. W. Westley (Ref. 52).

Algorithm: Based on the Murtagh and Sargent (Ref. 53) projection method for linearly constrained nonlinear programs. Nondifferentiability of the dual objective function is handled by placing arbitrary bounds on the dual variables.

Comments: Solves DGP; no extensions to signomial programs are reported. Bradley (Ref. 21) has tested the code extensively and reports that it often failed to solve even simple problems taken from the literature.

Code 4

Name: SIGNOPT*.

Authors: A. B. Templeman, A. J. Wilson, and S. K. Winterbottom (Ref. 54)

Algorithm: Signomial problems are solved using the Avriel and Williams (Ref. 17) procedure. The posynomial subproblems are solved by explicitly forming the reduced RDGP and solving it using a modified Fletcher–Reeves (Ref. 55) algorithm. Nondifferentiability of dual objective is handled by placing arbitrary lower bounds on the dual variables.

Comments: The code has been extensively tested by Templeman (Ref. 14) and Bradley (Ref. 21), both of whom claim a fair degree of success in solving small- to medium-size problems. However, these authors indicate that convergence can often be very slow. These findings are born out by the results in Section 6 and in Rijckaert and Martens (Ref. 4). No Phase 1 capability is included in SIGNOPT to initiate the Avriel and Williams procedure.

Code 5

Name: GPROG*.

Author: T. Jefferson (Refs. 56, 57).

Algorithm: Explicitly forms the reduced dual RDGP and solves it using a modified Newton algorithm. Nondifferentiabilities are avoided by adding slack variables to the primal in the manner of Duffin and Peterson (Ref. 58). Extension to signomials is carried out using the harmonic mean procedure of Duffin and Peterson (Ref. 28).

Comments: Unless the invariance of the dual coefficient matrix is exploited, the harmonic approach can be shown to be less desirable than the condensation algorithm of Avriel and Williams (Ref. 17). Experience with the code shows that it often fails to converge. This code and QUADGP, however, are the only ones mentioned in this paper with the capability of performing a detailed sensitivity analysis. GPROG does not contain a Phase 1 routine for signomial problems.

Code 6

Name: CSGP.

Authors: P. A. Beck and J. G. Ecker (Ref. 12).

Algorithm: The concave simplex method is applied to DGP with a modification that allows for blocks of variables to go to zero simultaneously. It is this modification that overcomes the nondifferentiability problem.

Comments: This code stands out as being the only dual-based code that attempts in a theoretically sound manner to overcome the nondifferentiability problem and to include an option to solve subsidiary problems, if they are needed, when converting to an optimal solution of the primal PGP. No provision is made in the code for signomial problems. The code has been tested extensively by Rijckaert and Martens (Ref. 4) and Beck and Ecker (Ref. 12). Experience shows that CSGP is sometimes slow relative to other codes but that it is fairly reliable.

Code 7

Authors: G. A. Kochenberger, R. E. D. Woolsey, and B. A. McCarl (Ref. 59).

Algorithm: Solves SDGP using separable programming.

Comments: No extensions to signomial programming are mentioned. Only computational experience reported is on one small problem and for this problem the method does poorly.

Code 8

Name: NEWTGP.

Author: J. Bradley (Ref. 60).

Algorithm: Explicitly reduces dual program DGP to the program RDGP. Solves the nonlinear equations resulting from the Kuhn–Tucker conditions for optimality of RDGP using a Newton–Raphson procedure. Nondifferentiability is avoided by setting a *pseudo boundary* which prevents the algorithm from hitting a $\delta \geq 0$ constraint. This is equivalent to artificially adding a $\delta \geq \epsilon$ constraint.

Comments: The code has been tested extensively by Rijckaert and Martens (Ref. 4), and the indications are that it often fails to converge; in cases where convergence is attained, the code does not compete well against the best primal methods.

Code 9

Name: GEOGRAD, GEOEPS*.

Authors: J. J. Dinkel, G. A. Kochenberger, and B. A. McCarl (Ref. 61).

Algorithm: The algorithm is essentially the same as the one used by Bradley in NEWTGP. Extension to signomials is carried out using the GEOEPS routine which executes the Avriel and Williams (Ref. 17) algorithm.

Comments: No Phase 1 method for signomials. The implementation seems to perform reasonably well when it converges. However, it is generally not competitive with the best primal methods.

Code 10

Name: QUADGP.

Author: J. Bradley (Ref. 21).

Algorithm: Solves posynomial programs by explicitly forming the reduced dual program RDGP which is then solved by successive quadratic approximations. Nondifferentiability of the dual objective function is handled by artificially bounding dual variables from below. This bound, in contrast to those mentioned previously, is dynamic and decreases rapidly from iteration to iteration. Dual variables are zeroed when an estimate of the primal constraint multiplier becomes very small. The code has options for both the Avriel and Williams (Ref. 17) and the Duffin and Peterson (Ref. 28) extensions to signomial programming, but Bradley (Ref. 21) indicates that the Avriel–Williams procedure is to be preferred.

Comments: A feasible point is required to initiate QUADGP for signomial problems. Bradley (Ref. 21) has tested QUADGP extensively. However, the only relative measure of effectiveness with the codes in Section 5 can be obtained from Problems 8A, 8B, and 8C, for which computation times are given in his thesis. An important feature of QUADGP is that it has the capability of performing sensitivity analysis.

Code 11

Names: LAM, SP, LM, NRF, NRT, NRVB, DCA.

Authors: M. J. Rijckaert and X. M. Martens (Ref. 4).

Algorithm: LAM is a linear approximation method for solving the dual and SP is based on a separable programming algorithm. LM, NRF, NRT, NRVB and DCA are all essentially based on Newton-type algorithms for solving the Kuhn–Tucker conditions of SDGP.

Comments: These codes are described and tested in (Ref. 4) and do not seem to be competitive with the best available software. In particular, they do not appear to be robust and often fail to converge on medium-sized problems.

Code 12

Author: J. R. McNamara (Ref. 62).

Algorithm: Constructs an augmented primal problem with zero degrees of difficulty. The augmented problem depends on a number of parameters and for certain realizations of these parameters the solution to the augmented dual is determined uniquely.

Comments: Solves posynomial problems only. Computational results are given for 2 trivial examples (Ref. 62), and mention is made of larger examples. The author indicates that the proposed method does not necessarily converge (Ref. 62, p. 23).

References

1. DEMBO, R. S., *Second-Order Algorithms for the Geometric Programming Dual, Part 1: Analysis*, Mathematical Programming (to appear).
2. GILL, P. E., and MURRAY, W., *Numerical Methods in Constrained Optimization*, Academic Press, New York, New York, 1974.
3. WOLFE, P., *Convergence Theory in Nonlinear Programming*, Integer and Nonlinear Programming, Edited by J. Abadie, North-Holland Publishing Company, Amsterdam, Holland, 1970.
4. RIJCKAERT, M. J., and MARTENS, X. M., *A Comparison of Generalized Geometric Programming Algorithms*, Katholieke Universiteit te Leuven, Report No. CE-RM-7503, 1975.
5. DEMBO, R. S., *GGP—A Computer Program for Solving Generalized Geometric Programting Problems*, Technion, Israel Institute of Technology, Department of Chemical Engineering, Users Manual, Report No. 72/59, 1972.
6. WILLIAMS, A. C., Private Communication, 1972.
7. RAMMAMURTHY, S., and GALLAGHER, R. H., *Generalized Geometric Programming in Light Gage Steel Design*, Paper Presented at the ORSA/TIMS Meeting, Miami, Florida, 1976.
8. DINKEL, J. J., and KOCHENBERGER, G. A., Private Communication, 1975.
9. DUFFIN, R. J., PETERSON, E. L., and ZENER, C., *Geometric Programming—Theory and Application*, John Wiley and Sons, New York, New York, 1967.
10. IBM Corporation, *Mathematical Programming System—Extended (MPSX) and Generalized Upper Bounding (GUB) Program Description*, Program No. 5734-XM4, 1972.
11. DEMBO, R. S., *Dual to Primal Conversion in Geometric Programming*, Journal of Optimization Theory and Applications, Vol. 26, No. 1, 1978.
12. BECK, P. A., and ECKER, J. G., *A Modified Concave Simplex Algorithm for Geometric Programming*, Journal of Optimization Theory and Applications, Vol. 15, pp. 189–202, 1975.
13. DEMBO, R. S., *Sensitivity Analysis in Geometric Programming*, Yale University, School of Organization and Management, Working Paper No. SOM-35, 1978.
14. TEMPLEMAN, A. B., Private Communication, 1975.
15. PASSY, U., and WILDE, D. J., *Generalized Polynomial Optimization*, SIAM Journal on Applied Mathematics, Vol. 15, pp. 1344–1356, 1967.
16. BEIGHTLER, C. S., and PHILLIPS, D. T., *Applied Geometric Programming*, John Wiley and Sons, New York, New York, 1976.
17. AVRIEL, M., and WILLIAMS, A. C., *Complementary Geometric Programming*, SIAM Journal on Applied Mathematics, Vol. 19, pp. 125–141, 1970.
18. DEMBO, R. S., *The Solution of Complementary Geometric Programming Problems*, Technion, Israel Institute of Technology, MS Thesis, 1972.
19. AVRIEL, M., DEMBO, R. S., and PASSY, U., *Solution of Generalized Geometric Programming Problems*, International Journal of Numerical Methods in Engineering, Vol. 9, pp. 141–169, 1975.

20. DUFFIN, R. J., and PETERSON, E. L., *Geometric Programming with Signomials*, Journal of Optimization Theory and Applications, Vol. 11, pp. 3–35, 1973.
21. BRADLEY, J., *The Development of Polynomial Programming Algorithms with Applications*, Dublin University, Department of Computer Science, PhD Thesis, 1975.
22. RIJCKAERTS, M. J., and MARTENS, X. M., *A Condensation Method for Generalized Geometric Programming*, Katholieke Universiteit te Leuven, Report No. CE-RM-7503, 1975.
23. DEMBO, R. S., *A Set of Geometric Programming Test Problems and Their Solutions*, Mathematical Programming, Vol. 10, pp. 192–213, 1976.
24. DINKEL, J. J., KOCHENBERGER, G. A., and MCCARL, B., *A Computational Study of Methods for Solving Polynomial Geometric Programs*, Journal of Optimization Theory and Applications, Vol. 19, pp. 233–259, 1976.
25. COLVILLE, A. R., *A Comparative Study of Nonlinear Programming Codes*, IBM, New York Scientific Center, Report No. 320–2949, 1968.
26. RATNER, M., LASDON, L. S., and JAIN, A., *Solving Geometric Programs Using GRG—Results and Comparisons*, Standford University, Systems Optimization Laboratory, Technical Report No. SOL-76-1, 1976.
27. DEMBO, R. S., and MULVEY, J. M., *On the Analysis and Comparison of Mathematical Programming Algorithms and Software*, Proceedings of the Bicentennial Conference on Mathematical Programming, Gaithersburg, Maryland, 1976.
28. HIMMELBLAU, D. M., *Applied Nonlinear Programming*, McGraw-Hill Book Company, New York, New York, 1972.
29. RIJCKAERT, M. J., Private Communication, 1975.
30. LASDON, L. S., WARREN, A. D., RATNER, M. W., and JAIN, A., *GRG System Documentation*, Cleveland State University, Technical Memorandum No. CIS-75-01, 1975.
31. ABADIE, J., and GUIGOU, J., *Numerical Experiments with the GRG Method*, Integer and Nonlinear Programming, Edited by J. Abadie, North Holland Publishing Company, Amsterdam, Holland, 1970.
32. KREUSER, J. L., and ROSEN, J. B., *GPM/GPMNLC Extended Gradient Projection Method Nonlinear Programming Subroutines*, University of Wisconsin, Academic Computer Center, 1971.
33. HIMMELBLAU, D. M., Private Communication, 1975.
34. DUFFIN, R. J., and PETERSON, E. L., *Reserved Geometric Programs Treated by Harmonic Means*, Carnegie–Mellon University, Research Report No. 71–79, 1971.
35. DEMBO, R. S., *Some Real-World Applications of Geometric Programming*, Applied Geometric Programming, Edited by C. S. Beightler and D. T. Phillips, Prentice-Hall, Englewood Cliffs, New Jersey, 1976.
36. REKLAITIS, G. V., *Singularity in Differentiable Optimization Theory: Differential Algorithm for Posynomial Programs*, Stanford University, PhD Thesis, 1969.
37. REKLAITIS, G. V., and WILDE, D. J., *A Differentiable Algorithm for Posynomial Programs*, DECHEMA Monographien, Vol. 67, pp. 503–542, 1971.

38. REKLAITIS, G. V., and WILDE, D. J., *Geometric Programming via a Primal Auxiliary Problem*, AIIE Transactions, Vol. 6, 1974.
39. WILDE, D. J., and BEIGHTLER, D. D., *Foundations of Optimization*, Prentice-Hall, Englewood Cliffs, New Jersey, 1967.
40. DUFFIN, R. J., *Linearizing Geometric Programs*, SIAM Review, Vol. 12, pp. 211–227, 1970.
41. REKLAITIS, G. V., Private Communication, 1976.
42. GOCHET, W., and SMEERS, Y., *On the Use of Linear Programs to Solve Prototype Geometric Programs*, Katholieke Universiteit te Leuven, Center for Operations Research and Econometrics, Discussion Paper No. 7229, 1972.
43. DAWKINS, G. S., MCINNIS, B. C., and MOONAT, S. K., *Solution to Geometric Programming Problems by Transformation to Convex Programming Problems*, International Journal of Solid Structures, Vol. 10, pp. 135–136, 1974.
44. HARTLEY, H. O., and HOCKING, R. R., *Convex Programming by Tangential Approximation*, Management Science, Vol. 9, pp. 600–612, 1963.
45. ECKER, J. G., and ZORACKI, M. J., *An Easy Primal Method for Geometric Programming*, Management Science, Vol. 23, pp. 71–77, 1976.
46. DINKEL, J. J., ELLIOTT, W. H., and KOCHENBERGER, G. A., *A Linear Programming Approach to Geometric Programs*, Naval Research Logistics Quarterly (to appear).
47. FRANK, C. J., *An Algorithm for Geometric Programming*, Recent Advances in Optimization Techniques, Edited by D. D. Lavi and D. D. Vogel, John Wiley and Sons, New York, 1966.
48. FRANK, C. J., *Development of a Computer Program for Geometric Programming*, Westinghouse Report No. 64-1, HO-124-R2, 1964.
49. HOOKE, R., and JEEVES, T. A., *Direct Search Solution of Numerical and Statistical Problems*, Journal of the Association for Computing Machinery, Vol. 8, pp. 212–219, 1961.
50. BLAU, G. E., and WILDE, D. J., *A Lagrangean Algorithm for Equality Constrained Generalized Polynomial Optimization*, AIChE Journal, Vol. 17, pp. 235–240, 1971.
51. KUESTER, J. L., and MIZE, J. H., *Optimization Techniques with FORTRAN Programs*, McGraw-Hill Book Company, New York, New York, 1973.
52. WESTLEY, G. W., *A Geometric Programming Algorithm*, Oak Ridge National Laboratory, Technical Report No. ORNL-4650, 1971.
53. MURTAGH, B. A., and SARGENT, R. W. H., *A Constrained Minimization Method with Quadratic Convergence*, Optimization, Edited by R. Fletcher, Academic Press, London, 1969.
54. TEMPLEMAN, A. B., WILSON, A. J., and WINTERBOTTOM, S. K., *SIGNOPT—A Computer Code for Solving Signomial Geometric Programming Problems*, University of Liverpool, Department of Civil Engineering, Research Report, 1972.
55. FLETCHER, R., and REEVES, C. M., *Function Minimization by Conjugate Gradients*, Computer Journal, Vol. 7, pp. 149–154, 1964.

56. JEFFERSON, T., *Geometric Programming, with an Application to Transportation Planning*, Northwestern University, PhD Thesis, 1972.
57. JEFFERSON, T., *Manual for the Geometric Programming Code GPROG (CDC) VERSION 2*, University of New South Wales, Australia, Mechanical and Industrial Engineering Department, Report No. 1974/OR/2, 1974.
58. DUFFIN, R. J., and PETERSON, E. L., *Geometric Programs Treated with Slack Variables*, Applied Analysis, Vol. 2, pp. 255–267, 1972.
59. KOCHENBERGER:, G. A., WOOLSEY, R. E. D., and MCCARL, B. A., *On the Solution of Geometric Programs via Separable Programming*, Operations Research Quarterly, Vol. 24, pp. 285–296, 1973.
60. BRADLEY, J., *An Algorithm for the Numerical Solution of Prototype Geometric Programs*, Institute of Industrial Research and Standards, Dublin, Ireland, 1973.
61. DINKEL, J. J., KOCHENBERGER, G. A. and MCCARL, B. A., *An Approach to the Numerical Solution of Geometric Programs*, Mathematical Programming, Vol., pp. 181–190, 1974.
62. MCNAMARA, J. R., *A Solution Procedure for Geometric Programming*, Operations Research, Vol. 24, pp. 15–25, 1976.
63. DUFFIN, R. J., and PETERSON, E. L., *The Proximity of (Algebraic) Geometric Programming to Linear Programming*, Mathematical Programming, Vol. 3, pp. 250–253, 1972.
64. SHAPLEY, M., and CUTLER, L., *Rand's Chemical Composition Program—A Manual*, The Rand Corporation, Report No. 495–PR, 1970.
65. CLASEN, R. J., *The Numerical Solution of the Chemical Equilibrium Problem*, The Rand Corporation, Report No. 4345-PR, 1965.

12

A Comparison of Computational Strategies for Geometric Programs[1]

P. V. L. N. SARMA,[2] X. M. MARTENS,[3] G. V. REKLAITIS,[4] AND M. J. RIJCKAERT[5]

Abstract. Numerous algorithms for the solution of geometric programs have been reported in the literature. Nearly all are based on the use of conventional programming techniques specialized to exploit the characteristic structure of either the primal or the dual or a transformed primal problem. This paper attempts to elucidate, via computational comparisons, whether a primal, a dual, or a transformed primal solution approach is to be preferred.

1. Introduction

Geometric programming (GP) is a body of theoretical and algorithmic results concerned with constrained optimization problems involving a class of nonlinear algebraic functions (Ref. 1). As attested by the rather broad range of applications cited in Ref. 2, the class of optimization problems considered by GP is of considerable practical significance; hence, continued research into extensions and generalizations of the pioneering duality theory developed by Duffin and Peterson (Ref. 3) is well justified. Concomitant

---

[1] The authors wish to thank Captain P. A. Beck and Dr. R. S. Dembo for making available their codes. This research was supported in part under ONR Contract No. N00014-76-C-0551 with Purdue University.

[2] Graduate Student in Chemical Engineering, School of Chemical Engineering, Purdue University, West Lafayette, Indiana.

[3] Graduate Student in Chemical Engineering, Instituut voor Chemie–Ingenieurstechniek, Katholieke Universiteit Leuven, Leuven, Belgium.

[4] Associate Professor of Chemical Engineering, School of Chemical Engineering, Purdue University, West Lafayette, Indiana.

[5] Associate Professor of Chemical Engineering, Instituut voor Chemie–Ingenieurstechniek, Katholieke Universiteit Leuven, Leuven, Belgium.

with the theoretical developments, numerous algorithms for solving geometric programs have been proposed in the optimization literature. A summary of some of these algorithms can be found in Ref. 4. Nearly all are based upon the use of conventional nonlinear programming techniques specialized to exploit the characteristic structure of either the primal GP problem, or its dual, or the transformed convex program which underlies the primal. To date, little attention has been given to numerical experimentation and comparison among the various solution alternatives. This paper proposes to take an initial step in that direction.

Given the state-of-the-art of GP solution techniques, the potential user is faced with two levels of decision. First, a decision must be made as to which of the equivalent forms of the GP problem (primal, dual, or transformed primal) is most efficient to solve for a given application problem. Having selected which formulation is most advantageous, an appropriate choice of algorithm must then be made. This paper addresses itself primarily to the former selection process. That is, it attempts to elucidate by means of computational experiments under what conditions a primal, dual, or transformed primal solution approach is to be preferred. The numerical results reported derive from the use of five different nonlinear programming algorithms specialized to exploit the properties of the respective GP forms. The test problems used are those constructed by Beck and Ecker (Ref. 5) for purposes of testing their dual-based algorithm.

In the next section, the formulation and computationally relevant properties of the three equivalent forms of GP problems are briefly summarized. In Section 3, these forms are illustrated with an example. In Section 4, the computational results are summarized and conclusions drawn.

2. Equivalent GP Problem Structures

2.1. Primal Problem. The prototype geometric programming problem (P) is as follows:

$$\begin{aligned} &\text{minimize} && g_0(x), \\ &\text{subject to} && g_m(x) \leq 1, \quad m = 1, \ldots, M, \\ & && x \geq 0, \quad x \in E^N, \end{aligned}$$

where the *posynomial* functions $g_m(x)$ are defined as

$$g_m(x) = \sum_{t=S_m}^{T_m} c_t \prod_{n=1}^{N} x_n^{a_{nt}}$$

with specified positive coefficients c_t and specified real exponents a_{nt}. The term indices t are defined consecutively as

$$S_0 = 1, \qquad S_{m+1} = T_m + 1, \qquad T_M \equiv T.$$

The above problem is in general a nonconvex programming problem which, because of the nonlinearities of the constraints, can be expected to severely tax conventional nonlinear programming codes. However, despite the apparent difficulty of the primal problem, there are structural features of the generalized posynomial functions which can be exploited to facilitate direct primal solution.

First, note that the posynomial functions are continuously differentiable to all orders over the positive orthant. Moreover, the generalized formulation of the problem functions permit direct calculation of first-, second-, and (if desired) higher-order derivatives using generalized analytic formulas. For example, the second partial derivative of a posynomial function is given by,

$$\partial^2 g_m/\partial x_i\, \partial x_j = (1/x_i x_j)\left\{\sum_{t=S_m}^{T_m} (a_{it}a_{jt})c_t \prod_{n=1}^{N} x_n^{a_{nt}}\right\}.$$

These analytic derivative formulas can be easily assembled into subroutines which can be used with any nonlinear programming code. Consequently, any code requiring (for example) analytic second derivatives can be employed without the usual inconvenience of either coding specific second derivative subroutines for each individual application problem or else using difference approximations. With posynomial programs, the user thus needs only to read the exponent matrix $A = [a_{nt}]$ and vector of term coefficients $C = [c_t]$ into generalized derivative subroutines to utilize the full power of second-order optimization techniques. The SUMT code (Ref. 6), used to solve the primal GP problem in the tests reported in Section 4, has been modified in this fashion.

A further property of the primal functions is that, with the change of variable

$$x_n = \exp(z_n), \qquad n = 1, \ldots, N$$

they are transformed to convex functions. Moreover, this convexity is preserved under logarithmic transformation (Ref. 7) so that functions $\log g_m(z)$ are also convex. This underlying convexity of the posynomial functions implies that the primal problem basically is a convex programming problem. Hence, prototype geometric programming can be viewed as a branch of convex programming, and thus any number of convex programming algorithms could be applied to the solution of the *convexified primal problem.* One approach which has received considerable attention involves *condensation* or linearization of the convexified posynomial functions (Ref. 8).

Consider the linearization of $\log g_m(z)$ about a point z^0 in terms of the variables z, that is,

$$\log \tilde{g}_m(z, z^0) \equiv \log g_m(z^0) + [1/g_m(z^0)] \sum_{n=1}^{N} \left\{ \sum_t a_{nt} c_t \exp\left(\sum_n a_{nt} z_n^0 \right) \right\} (z_n - z_n^0).$$

Since $\log g_m(z)$ is convex, it follows that

$$\log g_m(z) \leq \log \tilde{g}_m(z, z^0).$$

Consequently, the linear inequality,

$$\log \tilde{g}_m(z, z^0) \leq 0$$

will always include the region

$$g_m(z) \leq 1$$

and can, as pointed out by Duffin (Ref. 8), be safely used as part of any outer approximation algorithm which employs linear programming subproblems. The GGP code (Ref. 9), used to solve the primal GP problems in the tests reported in Section 4, is based upon the use of such a construction. It is a specialized Kelley's cutting plane algorithm (Ref. 10) in which the linear subproblems are solved using the dual simplex method for upper bounded variables. It is interesting to note that, when the above linearization is used as a cutting plane, it always yields a deeper cut than that which would be obtained if the direct linearization of $g_m(z)$ were employed (Ref. 11).

2.2. Dual Problem. As shown in Refs. 1, 3, the primal GP problem has associated with it a dual problem (D):

maximize
$$\nu(\delta) = \prod_{t=1}^{T} (c_t/\delta_t)^{\delta_t} \prod_{m=1}^{M} \lambda_m^{\lambda_m},$$

subject to
$$\sum_{t=1}^{T_0} \delta_t = 1,$$

$$A\delta = 0,$$

$$\delta \geq 0,$$

where
$$\lambda_m = \sum_{t=S_m}^{T_m} \delta_t, \qquad m = 1, \ldots, M,$$

and where by definition

$$\lim_{\delta_t \to 0} (c_t \lambda_m/\delta_t)^{\delta_t} = 0.$$

It is well known that, at their respective optima δ^* and x^*,

$$\nu(\delta^*) = g_0(x^*)$$

and that the primal and dual solutions are related by the following log–linear equations which are defined for those t with $\delta_t^* > 0$,

$$\sum_{n=1}^{N} a_{nt} \log x_n^* = \log [\delta_t^*/c_t \nu(\delta^*)], \qquad 1 \leq t \leq T_0,$$

$$= \log (\delta_t^*/c_t \lambda_m^*), \qquad S_m \leq t \leq T_m,$$

$$m = 1, \ldots, M.$$

It is further known that the logarithm of $\nu(\delta)$ is a concave function which is continuously differentiable over the positive orthant. Hence, the dual problem with logarithmic objective function is a linearly constrained concave program for which numerous general algorithms have been reported in the literature (Ref. 12).

There are, however, several disadvantages associated with the direct maximization of the dual.

(i) The dimensionality of the dual problem will always be larger than that of the primal, except for cases with $T < 2N$ solved using solution techniques which employ implicit or explicit variable elimination strategies to accommodate the linear constraints.

(ii) It is possible that the rank of the system of log–linear equations which must be solved to determine the optimal primal variable values is less than N. In such instances, subsidiary maximizations of the dual problem must be undertaken prior to the recovery of the primal solution (Refs. 1, 13).

(iii) It has been observed empirically by several investigators (Refs. 2, 5, 14) that the relationships between the primal and dual variables are very sensitive to the values of the dual variables. Hence, these must be determined very accurately in solving the dual and thus excessive computation times may be encountered with most gradient-based algorithms.

(iv) The gradient of $\log \nu(\delta)$ is not defined when any dual variable $\delta_t = 0$. Moreover, if $\delta_t^* = 0$ for some t

$$S_m \leq t \leq T_m,$$

and m, $1 \leq m \leq M$, then all dual variables δ_t^* associated with constraint m must also vanish. The vanishing of the dual variables associated with a given primal constraint m corresponds to that primal constraint being inactive at the optimum. Since in most applications the set of active constraints is not known a priori, all possible constraints must be considered in problem formulation. Thus, in most applications, $M > N$; hence, loose constraints can generally be expected to occur. The problem of vanishing dual variables is thus unavoidable and must be addressed by an effective dual-based algorithm.

The first three of the above difficulties are unavoidable consequences of the structure of the dual. The fourth, the problem of loose constraints, has been approached using two devices: by using penalized slack variables (Refs. 15–17) within general techniques for linearly constrained problems or by modifying such techniques to require simultaneous changes of all dual variables associated with a loose constraint (Ref. 13). Computational results obtained with both approaches are used in the comparison reported in Section 4.

2.3. Transformed Primal Problem. Consider again the primal problem under the one-to-one transformation of variables

$$x_n = \exp(z_m) \qquad n = 1, \ldots, N.$$

As first shown in Ref. 1 and further elaborated in Refs. 11, 18, this change of variable, together with the definition

$$\omega = A^T z + \log c,$$

leads to a transformed primal auxiliary problem (TP):

$$\text{minimize} \qquad f_0(\omega) = \sum_{t=1}^{T_0} \exp(\omega_t),$$

$$\text{subject to } f_m(\omega) = \sum_{t=S_m}^{T_m} \exp(\omega_t) \leq 1, \qquad m = 1, \ldots, M,$$

$$L(\omega - \log c) = 0,$$

where the rows of the matrix L are any set of linearly independent vectors spanning the null space of the exponent matrix A and where

$$\log c = [\log c_1, \log c_2, \ldots, \log c_t].$$

It is readily shown that this transformed primal is in general a reduced equivalent and, if A has full rank, is exactly equivalent to the primal problem (Refs, 1, 18). Given a solution ω^* of TP a primal solution can be recovered by solving the linear system,

$$\omega^* = A^T z + \log c$$

for the transformed primal variables z^* and simply exponentiating the result.

The TP problem again reveals quite clearly the convex structure underlying the primal GP problem. Moreover, the TP problem presents several advantageous features for computation (Ref. 18).

(i) It is a *separable* convex programming problem.

(ii) Each variable ω_t occurs in exactly one of the nonlinear functions.

(iii) The nonlinear functions are continuously differentiable to all orders.

(iv) The number of linear equality constraints will be equal to $T-N$.

(v) Slackness of any of the inequality constraints does not cause numerical or analytic difficulties.

(vi) Any single term constraint reduces to the single inequality, $\omega_t \leq 0$.

Note, however, that these structural advantages have been obtained at the cost of increasing problem dimensionality from N in the primal to T in the TP problem. In a sense, the TP can thus be viewed as a compromise between the primal and dual problems.

As with the convexified primal, the structural features of TP can be exploited using any of several standard convex programming algorithms. The computational results reported in Section 4 are obtained using a special form of the differential algorithm, which may be viewed as a generalized reduced gradient algorithm with an active constraint strategy (Ref. 19).

3. Example Formulation

To illustrate differences in the three equivalent formulations of the GP problem, consider the following modification of the heat exchanger design problem formulated in Ref. 20:

$$\begin{aligned} \text{minimize} \quad & g_0(x) = c_1 x_1^{-4/3} x_2^{-7/6} x_4^{-1} + c_2 x_1^{-1} x_2^{-0.2} x_3^{0.8} \\ & \qquad + c_3 x_1 x_2 x_4 + c_4 x_1 x_2^{-1.8} x_3^{-4.8} + c_5 x_1^{-1} x_2^{-1} x_3^{-1}, \\ \text{subject to} \quad & g_1(x) = c_6 x_4^{-1} + c_7 x_3 x_4^{-1} \leq 1, \\ & g_2(x) = c_8 x_4 \leq 1, \\ & g_3(x) = c_9 x_1 x_2^{-1.8} x_3^{-4.8} + c_{10} x_2^{-2} x_4^{-4} \leq 1, \end{aligned}$$

and all $x_i \geq 0$.

The objective function represents the annual cost of a heat exchanger designed for a specified heating duty. The variables x_1 through x_4 are, respectively, the length, the number of tubes, and the inside and outside diameters of the tubes. The constraints represent limitations on the tube diameters and restrictions in the allowable tube-side pressure drop. The coefficients c_1 through c_{10} are cost- and technology-dependent parameters.

As formulated above, the problem involves four variables, three constraints, and ten terms. The equivalent log dual program will be:

$$\text{maximize} \qquad \log(\nu(\delta)) = \sum_{t=1}^{5} \delta_t \log(c_t/\delta_t) + \delta_6 \log[c_6(\delta_6+\delta_7)/\delta_6]$$
$$+ \delta_7 \log[c_7(\delta_6+\delta_7)/\delta_7] + \delta_8 \log c_8$$
$$+ \delta_9 \log[c_9(\delta_9+\delta_{10})/\delta_9]$$
$$+ \delta_{10} \log[c_{10}(\delta_9+\delta_{10})/\delta_{10}],$$

$$\text{subject to} \qquad \sum_{t=1}^{5} \delta_t = 1, \qquad \delta \geq 0,$$

$$\begin{pmatrix} -4/3 & -1 & 1 & 1 & -1 & 0 & 0 & 0 & 1 & 0 \\ -7/6 & -1/5 & 1 & -1.8 & -1 & 0 & 0 & 0 & -1.8 & -2 \\ 0 & 0.8 & 0 & -4.8 & -1 & 0 & 1 & 0 & -4.8 & 0 \\ -1 & 0 & 1 & 0 & 0 & -1 & -1 & 1 & 0 & -4 \end{pmatrix} \delta = 0$$

with

$$\delta_6 = 0 \quad \text{iff} \quad \delta_7 = 0,$$
$$\delta_9 = 0 \quad \text{iff} \quad \delta_{10} = 0.$$

The dual problem involves ten variables, five linear constraints, and two conditions requiring the simultaneous vanishing of a block of dual variables corresponding to the possible inactivity of primal constraints 1 and 3.

The TP problem corresponding to the given primal program is as follows:

$$\text{minimize} \qquad f_0(\omega) = \sum_{t=1}^{5} \exp(\omega_t)$$

$$\text{subject to} \qquad f_1(\omega) = \exp(\omega_6) + \exp(\omega_6) \leq 1,$$
$$f_3(\omega) = \exp(\omega_9) + \exp(\omega_{10}) \leq 1,$$
$$\omega_8 \leq 0,$$

and the linear constraints

$$\begin{pmatrix} 0 & 0 & 1 & 0 & 1 & 0 & 1 & 0 & 0 & 0 \\ 1 & 0 & 0 & 0 & -4/3 & 0 & -4/3 & 0 & 0 & 1/12 \\ 0 & 1 & 0 & 0 & -1 & 0 & -1.8 & -0.2 & 0 & 0.4 \\ 0 & 0 & 0 & 1 & 1 & 0 & 5.8 & 0.2 & 0 & -1.4 \\ 0 & 0 & 0 & 0 & 0 & 1 & 0 & 1 & 0 & 0 \\ 0 & 0 & 0 & 0 & 1 & 0 & 5.8 & 0.2 & 1 & -1.4 \end{pmatrix} (\omega - \log c) = 0.$$

The above convex program also involves ten variables, six linear constraints, and three inequalities, two of which are nonlinear. In some sense, the need for the conditions requiring the simultaneous vanishing of a block of dual variables is avoided at the cost of having to deal with a pair of nonlinear inequalities. Note that, in both the dual and the TP cases, recovery of the primal solution requires the solution of a system of four linear equations.

4. Experimental Procedure and Results

4.1. Test Algorithms. Five codes form the basis of this study.

(i) The sequential unconstrained minimization technique (SUMT) (Ref. 6) was used for a direct minimization of the primal.

(ii) A Kelley's cutting plane algorithm (GGP) was used for the minimization of the primal as a sequence of linear programming subproblems (Ref. 9).

(iii) The concave simplex algorithm (CS) was used for the maximization of the dual problem corresponding to a form of the primal program in which penalized slack variables are added (Ref. 21).

(iv) The modified concave simplex algorithm (MCS, Ref. 13), was used for the dual of the unmodified primal.

(v) The differential algorithm for posynomial programs (DAP, Ref. 19), was used for the transformed primal.

The SUMT code uses the well-known interior-point penalty term to incorporate the problem inequality constraints into the objective function. The resulting unconstrained problem is solved using a modified Newton method for decreasing levels of the penalty factor. The SUMT code was selected for solving GP primal problems because interior penalty functions are generally considered to be quite effective for inequality constrained problems and because it could utilize the analytic second derivatives which can be readily generated in the case of posynomials.

The GGP code uses a modification of Kelley's cutting plane method (Ref. 10), with a cut deletion strategy to solve strictly posynomial GP primal problems. In the code, the linear subproblems are solved using the dual simplex method for upper bounded variables. These and other program details are given in Ref. 22. The code was chosen because it appeared to be the most efficiently written linear-programming-based code for solving the convexified primal GP problem available at the time this work was carried out.

The CS code is an implementation of the convex simplex algorithm (Ref. 23), which is an extension of the linear simplex method of linear programming to accommodate a convex objective function. The code can be used for direct maximization of the dual providing that the primal formulation includes slack variables which will force all primal constraints to be active at the optimum. As discussed in Ref. 15, to ensure that the solution of the primal problem remains unchanged, each slack variable introduced into the constraints must also be included in the objective function in the form of a nonlinear penalty term. In principle, the resulting problem must then be solved for a sequence of penalty levels (Ref. 16); however, in practice, a single pass has been found to suffice providing that a suitable choice of penalty term is made (Refs. 15, 17). Numerical experimentation has shown (Ref. 17) that satisfactory results can be obtained if, corresponding to each slack variable x_s added to the constraints, a term $bx_s^{-\beta}$ is added to the objective function, where β is set at 10^{-4} and b at 1% of the estimated optimal value of the objective function. The test problems solved using the CS code have been modified in this fashion. The CS code was used for the maximization of the modified dual, so that the merits of the slack variable approach could be assessed relative to the more sophisticated MCS approach.

The MCS code is based on a modification of the convex simplex algorithm (Ref. 23) introduced by Beck and Ecker. They revised the conventional simplex direction-generation machinery to ensure that all dual variables associated with a given constraint reach zero simultaneously if all are tending towards zero (Ref. 13). The MCS code was chosen to represent all dual-based solution techniques because it is the most sophisticated of the direct dual maximization algorithms reported to date.

The DAP code is an adaptation of the differential algorithm of Beightler and Wilde (Ref. 24). This, in turn, may be viewed as a generalized reduced gradient technique which varies one independent variable at a time and uses an active constraint strategy to accommodate inequalities. Because of the generally large number of linear constraints in the TP problem as well as because of the separable nature of the nonlinear convex inequalities, these features were deemed appropriate for this type of problem. The DAP code was used in preference to SUMT to solve the convex TP program because it could more directly utilize the constraint structure.

4.2. Test Problems. The GP test problems utilized in this study are those reported by Beck and Ecker (Ref. 5). The problems range in size from three variables and one constraint to twenty-four variables and forty-two constraints. Included are several problems with interior optima as well as several with linear constraints. All objective functions are nonlinear and nonquadratic. The problem exponent matrices range from relatively sparse matrices (problems with few cross terms) to dense matrices (severely nonlinear problems). The characteristic dimensions of the test problems are given in Table 1. For the detailed formulation of these problems, the reader is best referred to the originating reference (Ref. 5). Note, from Table 1, that the number of primal variables plus one indicates the number of rows in the dual linear constraint array. The quantity $T-N$ indicates the number of linear constraints in the TP problem. The dimension of both the dual and TP problem variables is T. Note also that the problems with letter designations are identical to the preceding problem, but with the inactive multiterm constraints deleted.

Table 1. Test problem characteristics.

					Number of constraints			Number of constraints active at the optimum		
Problem	N	T	$T-N$	ρ	Total	Multiterm	Linear	Total	Multiterm	Linear
1	3	6	3	78	2	2	0	1	1	0
2	3	6	3	78	2	2	0	0	0	0
3	3	10	7	57	1	1	0	1	1	0
4	4	21	17	28	4	4	4	2	2	2
4A	4	13	9	30	2	2	2	2	2	2
5	5	9	4	33	3	1	0	2	1	0
6	5	9	4	60	2	2	0	0	0	0
7	5	10	5	72	3	3	0	1	1	0
8	6	17	11	53	4	4	0	4	4	0
9	6	13	7	23	5	1	0	1	1	0
10	7	18	11	70	4	4	0	2	2	0
10A	7	11	4	72	2	2	0	2	2	0
11	7	18	11	33	3	3	2	3	3	2
12	7	18	11	48	3	3	1	1	1	1
12A	7	13	6	44	1	1	1	1	1	1
13	7	48	41	35	7	7	1	2	2	0
14	7	48	41	42	9	9	1	4	4	0
14A	7	18	11	70	4	4	0	4	4	0
15	8	17	9	19	7	1	0	1	1	0
16	24	81	57	8	42	5	0	22	1	0
16C	24	50	26	4	22	1	0	22	1	0

N = Number of primal variables. T = Number of terms.
ρ = Density of exponent matrix, defined as % of nonzero entries.

4.3. Procedure and Results. The code testing in this study was carried out in two stages. In the first stage, the two dual maximization codes were compared against each other to determine whether the changing of blocks of variables or the addition of penalized slack variables was more effective. Since both the CS and MCS codes were based on the same algorithm, the same initiation procedures and termination criteria could readily be used. Note however that, since in the slack variable approach all constraints are always active, the solution of subsidiary maximization problems (Ref. 13) is not required. The CPU times required to solve the test problems with the CS and MCS codes on an IBM 370/158 are shown in Table 2. The solution times for both codes include recovery of the primal solution and, in the CS

Table 2. Solution times for CS and MCS codes.

Problem	Time (sec, IBM 370/158) CS	MCS
1	2.20	0.92
2	0.71	0.61
3	2.29	1.84
4	23.9	7.01
4A	11.0	11.0
5	5.24	1.58
6	†	18.1
7	4.74	6.06
8	16.7	4.04
9	28.8	2.17
10	—	6.90
10A	1.63	2.41
11	22.6	17.8
12	86.0	22.5
12A	8.60	7.58
13	175.0	60.0
14	174.7	51.5
14A	14.9	17.01
15	28.7	3.70
16	>240*	>240*
16C	>240*	103.2

* Problem not solved within the accuracy requirements.
† Noncanonical problem. Program terminated because of underflow.

case, are based on the choice $\beta = 10^{-4}$ and $b = 0.01 \times$ (optimal objective value).

Since the MCS code appeared to be generally superior to the slack-variable-based code, the second stage of the code testing, the comparison between the primal, dual, and transformed dual approaches, was performed using just the GGP, SUMT, MCS, and DAP codes.

In order to avoid the complication of comparing computation times on different machines, the computational tests with the MCS code shown in Table 2 were repeated on the Purdue University CDC 6500. All the test problems were rerun using the MCS code with all program parameters set at recommended default values. Although there were some minor discrepancies, the general trends parallel the IBM 370 results given in Table 2 as well as those reported in Ref. 5. Notable exceptions are Problem 6, which required comparatively more time in both of our trials than in Ref. 5, and Problem 12, which required substantially less time in our trials than in Ref. 5. We have at present no satisfactory explanation for these disparities.

As is generally the case in any optimization code-rating study, valid comparisons require that three main program parameters (starting point, termination criteria, and constraint tolerances) be consistently specified. In the present case, uniform specification of these parameters is complicated by the fact that each code actually operates on a different problem in different variables but yielding the same objective function value.

Consider first the question of starting points. While GGP can procede with an arbitrary starting point, three of the codes require starting points with specific characteristics:

SUMT, an interior primal feasible point,
MCS, a strictly positive dual feasible point,
DAP, an interior TP feasible point.

Note, however, that since an interior point of TP corresponds to an interior point of the primal and will yield exactly the same objective and inequality constraint values, a point generated by DAP can be transformed and used as starting point with SUMT, as well as with GGP. However, a primal feasible point cannot, in general, be transformed to a dual feasible point; even if it could, the primal and dual problems are sufficiently different that there is no reason to expect such corresponding points to require, in any sense, equivalent computational work to reach the respective problem optimum points. Consequently, no attempts were made in this study to relate the MCS starting points to those used in the other codes. Rather, DAP was rerun several times with each problem and the average execution time reported. The starting points which were generated as an integral part of the code were saved, transformed to their primal equivalent, and then used with

SUMT and GGP. Therefore, the SUMT and GGP times reported are also averages of several runs.

Program termination criteria are also problematic in this study. The MCS code has several secondary termination criteria, e.g., the magnitude of the relative cost vector and the relative differences between the primal and dual objective function values. The former is required to be less than 10^{-6}, and the latter less than 0.5%. However, the criterion checked last is the primal constraint tolerance (less than 0.001 must be attained). This must be considered, therefore, as the main termination criterion. The GGP code, as is typically the case with cutting plane methods, terminates on a primal constraint tolerance and, thus, the same 0.001 tolerance could also be used.

The SUMT program, on the other hand, terminates intermediate problems based on gradient magnitude, but termination of the run is

Table 3. Solution times for MCS, DAP, SUMT, and GGP codes.

	Time (sec, CDC 6500)			
Problem	MCS	DAP	SUMT	GGP
1	0.43	0.75	13.3	0.21
2	0.94	0.44	12.0	0.21
3	1.02	1.38	9.68	†
4	9.37	5.30	18.8	0.53
4A	13.0	4.47	12.3	0.47
5	0.94	2.10	22.8	1.13
6	1.7*	0.82	10.6	0.91
7	6.03	1.04	9.86	0.93
8	4.71	21.62	61.7	0.85
9	1.41	1.40	14.3	1.52
10	4.17	12.4	120.0	1.81
10A	1.62	14.1	—	1.48
11	14.8	22.4	37.7	1.30
12	30.1	15.5	36.0	3.78
12A	9.6	12.5	25.1	3.30
13	74.3	53.9	107.7	5.43
14	61.8	34.5	117.0	2.37
14A	33.7	11.3	—	2.25
15	3.14	4.12	46.2	3.70
16	>300†	10.5	>300†	2.11
16C	171.4	5.70	>300†	3.54

* Primal recovery unsuccessful. The time reported is the dual solution time.
† Problem was not solved within the accuracy requirements.

naturally based on the magnitude of the penalty term. Finally, the DAP code normally terminates on the basis of the constrained derivative magnitude alone, while the constraint tolerance is simply a search rather than a termination parameter. Because both the functions and the variables involved in the three GP forms are different, specifying a uniform gradient base termination criterion is meaningless. Consequently, since the only common element is the optimal objective function value, both the DAP and the SUMT runs were terminated when a primal objective value within 0.25% of the reported optimal dual objective function value was attained. Since this is well within the 0.5% criterion used in MCS, it is felt that serious distortions in the execution times of DAP and SUMT were not introduced. It turned out in most cases that substantially better accuracy than 0.25% was actually attained at the termination point. In all cases, the constraint tolerance was set at 0.001.

Finally, it should be noted that all SUMT runs, except those for Problem 10, used an initial penalty level of 1 and successive penalty reductions by a factor of $\frac{1}{4}$. For Problem 10, convergence could be achieved only with large initial penalty factor (10^3), followed by reduction by a factor of $\frac{1}{10}$. The GGP code requires upper and lower bounds on all variables. To avoid bias in the results, these were set at 10^{10} and 10^{-10} whenever no specific bounds were given in the test problem formulation.

The results obtained under these conditions are summarized in Table 3. Note that the execution times for both MCS and DAP include both starting-point generation and recovery of the primal solution. The SUMT times and the GGP times include only the primal objective minimization itself.

4.4. Discussion of Results. From Table 2, it is apparent that the CS code with slack variables is generally less effective than the MCS code. Although the former code requires less or nearly the same time in four cases, in the remaining seventeen cases it requires distinctly more. The difference in effectiveness is especially noticeable in the cases in which the number of constraints is four or larger, that is, Problems 4, 8, 9, 13, 14, 15 and 16C. Apparently, the reason for this stems from both the increase in the dimensionality of the dual brought about by the insertion of the slack variables into the primal (one dual variable will be introduced for every primal constraint) as well as the numerical difficulties introduced by the small exponent values (i.e., 10^{-4}) and the small dual variable values which result when a constraint is active. Of these causes, the second appears the most significant. Note the tenfold increase in execution times which occurs between Problems 12A and 12 and Problems 14A and 14 and the three-fold increase between Problems 4A and 4. The corresponding increases in the MCS execution times are about a factor of three for the first two pairs and actually a decrease for the third pair. Clearly, the increase in dimension alone is not responsible, since both codes ought to react similarly to dimensionality changes and the dimensionality changes are, percentagewise, about the same. Apparently, the numerical difficulties introduced by the slack variable constructions must be at fault.

From the results of the second stage of testing, given in Table 3, it is quite evident that the execution times for the SUMT code were consistently higher and the execution times for the GGP code were generally lower than those for the remaining codes. SUMT requires the largest times even when the optimum solution is unconstrained (Problems 2 and 6) or when the tight primal constraints are linear (Problems 4, 4A and Problems 12, 12A). Since this occurs even though SUMT uses second derivatives while the other codes are effectively gradient-based, it can be concluded that the advantages offered by the convexity underlying the primal as well as the simplicity of the dual are computationally important. At the other end of the spectrum, GGP exhibits a rather consistent predominance over the other contenders. It clearly outperforms MCS in sixteen out of twenty-one cases and is comparable in four others. Similarly, it is unequivocally superior to DAP in fifteen of twenty-one cases and comparable in five others. Overall, GGP is predominant in thirteen problems; in eight of these, it outperformed its competitors by a factor of *five* or better.

From the latter cases, it is definite that this predominance stems from more than better coding or a savings in not having to generate special starting points. It seems to be related primarily to the fact that GGP can handle constraint activity or looseness implicitly without the fairly elaborate strategies required by either DAP or the dual approach. For instance, the DAP and GGP times are closest in the problems in which there are few or no tight multiterm constraints at the optimum (Problems 6, 7, 9, 15, 16c, 16) and most disparate for the problems in which there are multiple active multiterm constraints (Problems 4, 8, 10. 11, 13, 14). Similarly, the MCS and GGP times are closest with problems involving few loose multiterm constraints (Problems 5, 9, 10A, 15) and most disparate when there are multiple loose multiterm constraints (Problems 4, 7, 12, 13, 14, 16).

A second factor favoring GGP seems to be that it can exploit the underlying convexity of the primal without incurring the increase in dimensionality that takes place with the transformed primal or the dual. For instance, the computation time for DAP doubles in going from Problem 16c to 16, even though only one multiterm constraint is tight in either case and the number of primal variables is unchanged.

The distinction between the effectiveness of the MCS code versus DAP code is less pronounced. MCS predominates in nine cases and DAP in twelve. However, if a few of the low-dimensionality problems are discounted, two basic trends do emerge. First, if more than one of the multiterm constraints are loose, as is the case in Problems 4, 6, 7, 10, 12, 13, 14, and 16, then the TP-based approach proves superior (Problem 10 is the exception). On the other hand, if all the multiterm constraints are tight, then the dual-based approach is either superior (e.g., Problems 3, 5, 8, 10A,11, 12A, and 15) or else gives quite comparable performance (e.g., Problem 9). The noticeable exceptions here are Problems 4A, 14A, and 16C in which DAP predominates. The results of Problem 16C can be explained by the presence of the large number (22) of single-term constraints. In the transformed problem, the variables associated with these terms appear only in the nonnegativity conditions and in the linear constraints, but not in the objective function; hence, they can readily be adjusted to accommodate the remaining variables. In the dual formulation, the dual variables associated with single-term constraints always appear linearly in the dual objective function; hence, they are directly involved in the optimization iterations.

Trends among the runs with a given code are more difficult to establish. As a general rule, the DAP computational effort increases with both the number of active multiterm constraints and the total number of such constraints, but not markedly with variable dimensionality. The dual-based code does seem to be affected by the magnitude of $T-N$ and, as noted in Ref. 5, by the sparsity of the exponent matrix. The GGP code is not markedly affected by variable dimensionality and appears to be aided by the presence of single-term constraints (e.g., Problem 16 vs Problem 16c).

5. Conclusions

The experimental results presented in this paper indicate that the MCS approach to solving the dual is preferable to using the slack-variable device and that a direct minimization of the primal which does not exploit the convexity underlying the primal is ineffective. The results further show that the dual problem does not offer any clear computational advantages over the convexified or transformed primal and is only likely to be competitive with highly-constrained problems in which all multiterm constraints are active at the optimum. The results are not complete enough to determine whether the convexified primal will be superior to the transformed primal even if the same algorithm is applied to each. Qualitatively, it seems that the generally lower dimensionality of the convexified primal is likely to predominate over the structural separability of the transformed primal. On the whole, it appears that the main computational advantage of GP stems from the underlying convexity and that the successful computational approach will need to exploit this property directly.

References

1. DUFFIN, R. J., PETERSON, E. L., and ZENER, C., *Geometric Programming*, John Wiley and Sons, New York, New York, 1968.
2. RIJCKAERT, M. J., *Engineering Applications of Geometric Programming*, Optimization Theory in Technological Design, Edited by M. Avriel, M. Rijckaert, and D. J. Wilde, Prentice-Hall, Englewood Cliffs, New Jersey, 1972.
3. DUFFIN, R. J., and PETERSON, E. L., *Duality Theory for Geometric Programming*, SIAM Journal on Applied Mathematics, Vol. 14, pp. 1307–1349, 1966.
4. PHILLIPS, D. T., and BEIGHTLER, C. S., *Geometric Programming: A Technical State-of-the-Art Survey*, AIIE Transactions, Vol. 5, pp. 97–112, 1973.
5. BECK, P. A., and ECKER, J. G., *Some Computational Experience with a Modified Convex Simplex Algorithm for Geometric Programming*, Paper presented at the 41st ORSA National Meeting, New Orleans, Louisiana, 1972.
6. MCCORMICK, G. P., MYLANDER, W. C., and FIACCO, A. V., *Computer Program Implementing the Sequential Unconstrained Minimization Technique for Nonlinear Programming*, Research Analysis Corporation, McLean, Virginia, 1967.
7. ZANGWILL, W. I., *Nonlinear Programming: A Unified Approach*, Chapter 3, Prentice-Hall, Englewood Cliffs, New Jersey, 1969.
8. DUFFIN, R. J., *Linearizing Geometric Programs*, SIAM Review, Vol. 12, pp. 211–227, 1970.
9. DEMBO, R. S., *GGP: A Computer Program for the Solution of Generalized Geometric Programming Problems*, Users Manual, McMaster University, Hamilton, Ontario, Canada, 1975.
10. KELLEY, J. E., *The Cutting Plane Method for Solving Convex Programs*, SIAM Journal on Applied Mathematics, Vol. 8, pp. 703–712, 1960.
11. GOCHET, W., and SMEERS, Y., *On the Use of Linear Programs to Solve Prototype Geometric Programs*, Center for Operations Research, Heverlee, Belgium, Discussion Paper No. 7229, 1972.
12. REKLAITIS, G. V., and PHILLIPS, D. T., *A Survey of Nonlinear Programming*, AIIE Transactions, Vol. 7, pp. 235–257, 1975.
13. BECK, P. A., and ECKER, J. G., *A Modified Concave Simplex Algorithm for Geometric Programming*, Journal of Optimization Theory and Applications, Vol. 15, pp. 184–202, 1975.
14. BLAU, G. E., and WILDE, D. J., *Generalized Polynomial Programming*, Canadian Journal of Chemical Engineering, Vol. 47, pp. 317–326, 1967.
15. KOCHENBERGER, G. A., *Geometric Programming, Extension to Deal with Degrees of Difficulty and Loose Constraints*, University of Colorado, PhD Thesis, 1969.
16. DUFFIN, R. J., and PETERSON, E. L., *Geometric Programs Treated with Slack Variables*, Carnegie–Mellon University, Report No. 70–45, 1970.

17. RIJCKAERT, M. J., and MARTENS, X. M., *Numerical Aspects of the Use of Slack Variables in Geometric Programming,* Katholieke Universiteit te Leuven, Leuven, Belgium, Report No. CE-RM-7501, 1975.
18. REKLAITIS, G. V., and WILDE, D. J., *Geometric Programming via a Primal Auxiliary Problem,* AIIE Transactions, Vol. 6, pp. 308–317, 1974.
19. REKLAITIS, G. V., and WILDE, D. J., *A Differential Algorithm for Posynomial Programs,* DECHEMA Monographien, Band 67, pp. 503–542, 1971.
20. AVRIEL, M., and WILDE, D. J., *Optimal Condenser Design by Geometric Programming,* Industrial and Engineering Chemistry, Process Design and Development, Vol. 6, pp. 255–262, 1967.
21. VAN DESSEL, J. P., Personal Communication, 1975.
22. DEMBO, R. S., *Solution of Complementary Geometric Programming Problems,* Technion, Israel, MS Thesis, 1972.
23. ZANGWILL, W. I., *The Convex Simplex Method,* Management Science, Vol. 14, pp. 221–238, 1967.
24. BEIGHTLER, C. S., and WILDE, D. J., *Foundations of Optimization,* Prentice-Hall, Englewood Cliffs, New Jersey, 1967.

13

Comparison of Generalized Geometric Programming Algorithms[1]

M. J. RIJCKAERT[2] AND X. M. MARTENS[3]

Abstract. Numerical results are presented of extensive tests involving five posynomial and twelve signomial programming codes. The set of test problems includes problems with a pure mathematical meaning as well as problems originating from different fields of engineering. The algorithms are compared on the basis of CPU time, number of failures, preparation time, and in-core storage.

1. Introduction

In this paper, 17 computer codes are compared on a set of 24 test problems. The algorithms include posynomial and signomial programming codes. Also, one general nonlinear programming algorithm was included to provide a point of reference for the geometric programming codes. To guarantee the diversity of the test examples, problems with only a mathematical meaning were examined as well as problems originating from different fields of engineering.

Two important special features of the present paper are that many original codes, developed by the authors of particular algorithms, were used in this comparison and that all codes were actually implemented on the same computer. Hence, the trouble of comparing CPU times obtained on different computers was avoided. In addition, we made sure that all computer codes were run under comparable conditions for starting points and stopping rules.

[1] The authors wish to thank Messieurs M. Avriel, P. Beck, J. Bradley, R. Dembo, T. Jefferson, R. Sargent and A. Templeman for the possibility of using their respective codes in this study.

[2] Professor, Instituut voor Chemie–Ingenieurstechniek, Katholieke Universiteit te Leuven, Leuvem, Belgium.

[3] Doctoral Student, Instituut voor Chemie–Ingenieurstechniek, Katholieke Universiteit te Leuven, Leuven, Belgium.

It should be clear that the purpose of the present study is not to select one algorithm as the best possible. But we hope to give through this paper a valuable hint to future users about the relative advantages and disadvantages of various algorithms. To make the paper self-contained, a short review of the essential elements of geometric programming is given, together with a brief outline of the different codes used in the study.

2. Geometric Programming Problem

A function

$$f_k(t) = \sum_{i \in J\{k\}} \sigma_i c_i \prod_{j=1}^{m} t_j^{a_{ij}}, \tag{1}$$

$$t_j > 0, \qquad j = 1, \ldots, m,$$

$$a_{ij} = \text{real constants}, \qquad j = 1, \ldots, m,\ i = 1, \ldots, n,$$

$$c_i = \text{positive constants}, \qquad i = 1, \ldots, n,$$

$$\sigma_i = \pm 1, \qquad i = 1, \ldots, n,$$

is called a signomial or a generalized polynomial.

If the σ_i appearing in (1) are all positive, the function

$$h_k(t) = \sum_{i \in J\{k\}} c_i \prod_{j=1}^{m} t_j^{a_{ij}}, \tag{2}$$

$$t_j > 0, \qquad j = 1, \ldots, m,$$

$$c_i > 0, \qquad i = 1, \ldots, n,$$

$$a_{ij} = \text{real constants}, \qquad j = 1, \ldots, m,\ i = 1, \ldots, n,$$

is called a posynomial.

Posynomials can be transferred by an exponential transformation into convex functions. One-term posynomials are called monomials.

2.1. Primal Program. Geometric programming is an optimization technique dealing with the following problem:

$$\min g_0(t),$$

$$\text{s.t. } \sigma'_k g_k^{\sigma'_k} \leq 1, \qquad k = 1, \ldots, p, \tag{3}$$

$$t_j > 0, \qquad j = 1, \ldots, m,$$

$$\text{with } \sigma'_k = \pm 1,$$

where $g_k(t)$, $k=0,\ldots,p$, are either signomials (1) or posynomials (2) or a mixture of both. Problem (3) is called the primal program.

A primal program involving signomial functions can always be rearranged as a program containing only posynomial functions in the following way (Ref. 1):

$$\begin{aligned} &\min r_0(t),\\ &\text{s.t. } r_{m'}(t)\leq 1, \qquad m'=1,\ldots,M',\\ &\quad\; s_{m''}(t)\geq 1, \qquad m''=1,\ldots,M'', \end{aligned} \tag{4}$$

with $r_{m'}(t)$ and $s_{m''}(t)$ posynomials. Because of this possibility, signomial programs are also termed reversed geometric programs.

2.2. Degree of Difficulty. If n is the total number of terms in the primal program (3) and m is the number of primal variables, then the quantity

$$D=n-m-1$$

is called the degree of difficulty of the GP-problem. For many algorithms, it characterizes the size of the problem.

2.3. Dual Program. For each primal program (3), there exists an associated dual program, which in its most general form can be formulated as follows: Find a stationary point of the function

$$v(\delta,\lambda)=\sigma_0\left\{\prod_{i=1}^{n}(c_i/\delta_i)^{\sigma_i\delta_i}\prod_{k=1}^{p}\lambda_k(\delta)^{\lambda_k\sigma'_k}\right\}^{\sigma_0}, \tag{5-1}$$

$$\text{s.t. } \sum_{i\in J\{0\}}\sigma_i\delta_i=\sigma_0, \tag{5-2}$$

$$\sum_{i=1}^{n}\sigma_i a_{ij}\delta_i=0, \qquad j=1,\ldots,m, \tag{5-3}$$

$$\lambda_k\sigma'_k=\sum_{i\in J\{k\}}\sigma_i\delta_i, \quad k=1,\ldots,p, \tag{5-4}$$

$$\delta_i\geq 0, \qquad i=1,\ldots,n, \tag{5-5}$$

$$\lambda_k\geq 0, \qquad k=1,\ldots,p, \tag{5-6}$$

$$\sigma_0=\min g_0(t)/|\min g_0(t)|. \tag{5-7}$$

For signomial programs, only very weak duality relations exist between so-called equilibrium solutions of (5) and (3). These equilibrium points satisfy the KT-necessary conditions, so that the global minimum of (3) will belong to the set of points which are defined as follows: a feasible solution t to primal program (3) is termed a primal equilibrium solution if there is a feasible solution δ to dual program (5) such that

$$c_i \prod_{j=1}^{m} t_j^{a_{ij}} = \delta_i v(\delta, \lambda), \qquad i \in J\{0\},$$

$$c_i \prod_{j=1}^{m} t_j^{a_{ij}} = \delta_i / \lambda_k(\delta), \qquad i \in J\{k\}, \qquad k = 1, \ldots, p;$$

δ is then a dual equilibrium solution. For corresponding equilibrium solutions,

$$g_0(t) = v(\delta, \lambda),$$

so that the above definition equations can generally be used to derive the primal equilibrium solution from the corresponding dual solution.

If the primal program contains only straight posynomials, so that all σ_i and σ'_k are equal to $+1$, much stronger duality relations can be derived (Ref. 2).

First note that, in this case, the logarithm of the dual objective function is concave. Furthermore, the dual relations then exist between the constrained primal minimum and the constrained dual maximum.

The dual program of a primal posynomial problem is

$$\max v(\delta, \lambda) = \max \left\{ \prod_{i=1}^{n} (c_i/\delta_i)^{\delta_i} \prod_{k=1}^{p} \lambda_k(\delta)^{\lambda_k} \right\}, \tag{6-1}$$

$$\text{s.t.} \sum_{i \in J\{0\}} \delta_i = 1, \tag{6-2}$$

$$\sum_{i=1}^{n} a_{ij} \delta_i = 0, \qquad j = 1, \ldots, m, \tag{6-3}$$

$$\lambda_k = \sum_{i \in J\{k\}} \delta_i, \qquad k = 1, \ldots, p, \tag{6-4}$$

$$\delta_i \geq 0, \qquad i = 1, \ldots, n, \tag{6-5}$$

$$\lambda_k \geq 0, \qquad k = 1, \ldots, p. \tag{6-6}$$

2.4. Condensation. The inequality relations between the arithmetic, geometric, and harmonic mean have played a significant role in the theoretical and computational evolution of geometric programming.

They can be used in an approximating program to replace respectively a posynomial by a monomial (Ref. 3) or a reversed posynomial constraint

$$g(t) \geq 1$$

by a straightforward posynomial constraint (Ref. 4)

$$g''(t, \epsilon) \geq 1.$$

2.5. Equilibrium Conditions. Passy and Wilde (Ref. 5) derived an additional set of dual equality constraints, called equilibrium conditions. These are linear logarithmic equations of the form:

$$\sum_{k=0}^{p} \sum_{i \in J\{k\}} \sigma_i \nu_{id} \log(\delta_i/\lambda_k) = \sum_{i=1}^{n} \sigma_i \nu_{id} \log c_i, \qquad d = 1, \ldots, D, \tag{7}$$

with $\lambda_0 = 1$. The quantities ν_{id} are the components of D linearly independent solution vectors ν_d of the homogenized equations (5-2) and (5-3).

2.6. Loose Constraints. The complete block of dual variables associated with a primal constraint must be zero for a constraint inactive at optimality. This phenomenon might eventually make the computation of the primal variables more difficult. Furthermore, some algorithms might fail because of numerical difficulties due to vanishing dual variables.

To circumvent such difficulties, Kochenberger (Ref. 6) proposed to introduce slack variables in penalty-function manner. They cannot be introduced in the same straightforward way as in linear programming. Rijckaert and Martens (Ref. 7) investigated the numerical aspects of the introduction of slack variables. They concluded that a sufficiently close approximation to the solution of such problems can be obtained in one single iteration, provided some precautions are taken with regard to the numerical constants used in the panalty terms.

In those algorithms which cannot handle inequality constraints, slack variables were used in this penalty-function manner.

3. Algorithms Used in This Study

Geometric programming algorithms can be classified on the basis of two different criteria: (a) one can make a distinction between primal and dual codes; and (b) a distinction can also be made between techniques for strict posynomial problems and techniques for signomial problems. It is this type of classification that will be used throughout this paper.

3.1. Posynomial Algorithms. In this section, methods are discussed for minimizing a posynomial subject to posynomial constraints. All these methods make explicit use of the strong duality relations, which are valid under these circumstances. Indeed, the primal nonlinear optimization problem can be traded for a more attractive dual linearly constrained problem. The following algorithms are henceforth all based on dual maximization.

Linear Approximation Method. The dual program, which can be represented schematically as

$$\begin{aligned} &\max d(\sigma)=\max \log v(\delta), \\ &\text{s.t. } A\cdot\delta=b, \\ &\delta\geqslant 0, \end{aligned} \tag{8}$$

is approximated in a point δ^0 by the following LP-problem (Ref. 8):

$$\begin{aligned} &\max[d(\delta^0)+\nabla d(\delta^0)\cdot(\delta-\delta^0)], \\ &\text{s.t. } A\cdot(\delta-\delta^0)=b-A\cdot\delta^0 \\ &\delta\geqslant 0. \end{aligned} \tag{9}$$

Call the solution of (9) δ^1. Hence,

$$r=\delta^1-\delta^0$$

is a promising direction for improving the value of $d(\delta)$. A new point δ^2 results from

$$\delta^2=\delta^0+\tau r, \tag{10}$$

where τ is obtained from a line minimization.

The method was coded by the authors of the present paper under the name LAM, using an LP-routine based on the revised simplex method with product form of the inverse (Ref. 9) and a line minimization based on cubic interpolation (Ref. 9).

Separable Programming. Along with the linearity of the constraints set, the dual formulation has the advantage of being separable after a logarithmic transformation. These two features are exploited to full extent by using separable programming to find the dual constrained maximum (Ref. 10). Indeed, the nonlinear dual objective function is separable and can be approximated by a piecewise linear function. The resulting approximating problem can then be solved by an LP-code with restricted basis entry. Successive refinements are obtained by changing the grid points. A polyhedral approximation based on 5 points in the region [0, 2] constituted the initial approximation. The adjustment of the grid points was performed in such a way that a variable can always return to a value lying outside the *optimal line segment* of the previous approximation.

The algorithm used was coded by the authors under the name SP.

Gradient-Projection Method. The third approach is a combination of the gradient-projection method due to Rosen with a variable-metric method in order to approximate the inverse of the Hessian of the objective function. The solution of the problem

$$\min \Phi(z),$$
$$\text{s.t. } \Psi_m(z)=0, \qquad m=1,\ldots,M, \tag{11}$$

satisfies the equations

$$\Psi_m(z)=0, \qquad m=1,\ldots,M,$$
$$\nabla\Phi(z)+\sum_{m=1}^{M}\lambda_m\nabla\Psi_m(z)=0. \tag{12}$$

A sequence of points z^k is created with corresponding vectors λ^k which converges to the solution of (12). Thereto, the gradient of the objective function is projected into a linear manifold approximating the active constraints at the current point z^k. Doing so, a projection matrix is constructed based on an approximation of the inverse of the Hessian of the objective function. This approximation is updated using a secant relation, and necessary precautions are taken so that this approximation remains positive definite. This technique was applied to the dual constrained GP-program (6) using a computer code VMP developed by Sargent and Murtagh (Ref. 11).

Newton–Raphson Method. The dual program (6) is transformed, before maximization, into an unconstrained optimization problem, solved iteratively by a modified Newton–Raphson procedure. Precautions are taken to guarantee that the substituted variables remain positive. This procedure operates as follows: (i) solve the linear dual constraints for $(m+p+1)$ variables; (ii) eliminate these variables from the objective function; (iii) the resulting dual objective function is maximized by setting the first derivatives to zero, so that a set of nonlinear equations is created, which is solved by a Newton–Raphson procedure; and (iv) if any substituted dual variable becomes negative, a step reduction is imposed on the Newton–Raphson procedure.

The code NEWTGP used in our study was programmed by Bradley (Ref. 12).

Concave Simplex Method. The dual program (6) can also be solved by the concave simplex method. Some modifications are imperative, however, in order to adapt the method to the presence of inactive primal constraints. This problem has to do with the fact that, if one dual variable vanishes, the whole block of dual variables corresponding to the same primal constraint should vanish. Beck and Ecker (Ref. 13) showed that the standard concave simplex method, when started from a positive feasible dual solution can never produce a feasible point with a zero block, even though a primal constraint may be inactive. The presence of inactive constraints may also prevent the calculation of all primal variables from the dual solution. Beck and Ecker presented a modification of the concave simplex method that circumvents the above difficulties. The technical details of this method can be found in Ref. 13.

The code used in our study, which we will designate as CSGP, is the original one developed by Beck.

3.2. Signomial Algorithms

Signomial Algorithms Based on the Lagrangian. The general dual geometric program (5) is schematically represented as: Find a stationary point of $\log v(\delta)$

$$\text{s.t.} \sum_{i=1}^{n+p} a_{li}\delta_i = b_l, \qquad l = 1, \ldots, L,$$

with $L = m + p + 1$. It is equivalent to the following problem: Find a stationary point of $\mathcal{L}(\delta, \mu)$ where

$$\mathcal{L}(\delta, \mu) = \log v(\delta) + \sum_{l=1}^{L} \mu_l \left(\sum_{i=1}^{n+p} a_{li}\delta_i - b_l \right), \tag{13}$$

with $\delta \geq 0$. In (13), μ_l, $l = 1, \ldots, L$, are Lagrange multipliers associated with the original dual constraints; $\sigma_0 = 1$ [if not, a constant can be added to $g_0(t)$].

In order to simplify the notation, we make the following transformations:

$$\text{if } i \leq n \begin{cases} \delta_i = \delta_i, \\ c_i = c_i, \\ \sigma_i = \sigma_i, \qquad i = 1, \ldots, n, \end{cases}$$

$$\text{if } i > n \begin{cases} \sigma_i = \lambda_k, \\ c_i = 1, \\ \sigma_i = -\sigma'_k \quad k = 1, \ldots, p, \qquad i = n+1, \ldots, n+p. \end{cases}$$

From the KT-conditions for (13) it follows that the stationary point satisfies:

$$-\sigma_i \log \delta_i + \sum_{l=1}^{L} a_{li}\mu_l = \sigma_i(1 - \log c_i), \qquad i = 1, \ldots, n+p,$$
$$\sum_{i=1}^{n+p} a_{li}\delta_i = b_l, \qquad l = 1, \ldots, L. \tag{14}$$

If the Lagrange multipliers μ_l are eliminated from (14) a new set of equations is obtained:

$$\sum_{i=1}^{n+p} \sigma_i(\log c_i - \log \delta_i - 1)\nu_{id} = 0, \qquad d = 1, \ldots, D,$$
$$\sum_{i=1}^{n+p} a_{li}\delta_i = b_l, \qquad l = 1, \ldots, L. \tag{15}$$

The D nonlinear equations are the equilibrium equations (7).

Newton–Raphson Techniques. (a) The first algorithm LM of this series solves the larger set of equations (14) using a Newton–Raphson iteration. The main reason for selecting this set of equations is that all dual variables δ_i appear uncoupled in the nonlinear equations. The Lagrange multipliers μ_l are unrestricted in sign.

(b) Algorithm NRF applies the Newton–Raphson technique to solve Eqs. (15), hereby making use of fixed bounds imposed on each variable δ_i. When such a bound is violated, the boundary value is selected as the new approximation for this particular variable.

(c) Algorithm NRVB is equivalent to the preceding one, but the step adjustments in the Newton–Raphson iteration are reduced if the linearization of the logarithmic equations (15) is too poor. The reduction is carried out until an acceptable error for the linearization is obtained.

(d) If the number of nonlinear (logarithmic) equations exceeds the number of linear equations, a logarithmic transformation

$$y_i = \log \delta_i, \qquad i = 1, \ldots, n+p,$$

will reverse the role of linear and nonlinear equations. The algorithm NRF applies the algorithm NFR on this transformed set. The preceding algorithms have been coded by the present authors.

Lagrangian Algorithm. Here, the set of equations (14) is solved using the separability of the logarithmic equations. Indeed, for a given value of the multipliers μ_l, the dual variables δ_i are completely determined. Hence, as proposed by Blau (Ref. 14), an iterative method can be derived requiring only the inversion of a $(p+m+1)$ matrix at each iteration, instead of a $(2p+m+n+1)$ matrix as in the original LM-algorithm.

The code GOMTRY used in this study is a slightly modified version of the one published in Ref. 15.

Marquardt Algorithm. Since a problem with as many variables as equations is the limit case of the well-known least square problem, an algorithm to solve the latter problem has been included. A modified Marquardt code VA07AD (Ref. 9) was applied to the dual program (15).

Dual Condensation Algorithm. This algorithm solves the set of equations (16) using the same underlying ideas as in the primal condensation algorithm (see below).

The code DCA was written by the present authors.

Signomial Algorithms Based on Posynomial Approximations. In this paragraph, algorithms are described that approximate signomial programs by a sequence of posynomial programs.

Use of the Geometric Inequality. For signomial programs, Avriel and Williams (Ref. 16) proposed to state the original problem as follows:

$$\min t_0,$$

$$\text{s.t. } r_k(t) - s'_k(t) \leq 1, \qquad k = 1, \ldots, K,$$

with $r_k(t)$ and $s'_k(t)$ being posynomials. If

$$s_k(t) = s'_k(t) + 1,$$

the signomial constraints can be transformed into

$$r_k(t) s_k^{-1}(t) \leq 1, \qquad k = 1, \ldots, K.$$

The denominator is subsequently approximated, using the arithmetic–geometric inequality (condensation). As a result, a so-called *complementary program* is created:

$$\min t_0,$$

$$\text{s.t. } g_k(t) \leq 1, \qquad k = 1, \ldots, K',$$

$$r_k(t) s_k^{-1}(t, \epsilon) \leq 1, \qquad k = 1, \ldots, K'',$$

$$\epsilon_i > 0, \qquad i \in J\{k\},$$

$$\sum_{i \in J\{k\}} \epsilon_i = 1, \qquad k = 1, \ldots, K''.$$

Starting from a primal feasible point, a series of condensed programs is constructed, and their solution is guaranteed to converge through a series of feasible points to an equilibrium point of the original program.

(a) Avriel *et al.* (Ref. 17) proposed an algorithm based on this approach without using the dual program. Thereto, a second condensation is used to transform all posynomials into monomials, which become linear expressions after a logarithmic transformation.

Following the idea of Kelley's cutting-plane technique, a sequence of linear programs is solved, until a feasible point is obtained for the approximating posynomial program. Then, a new approximation of the signomial program is constructed until convergence to a KT-point of the original GP is obtained.

The algorithm is able to compute in a Phase-I procedure its own necessary feasible starting point. Finally, bounds on all variables need to be specified.

Our study used the original code GGP, written by Dembo.

(b) Templeman (Ref. 18) also implements the idea of complementarity; but, for each approximating posynomial program, the corresponding dual is solved, after elimination of the linear constraints, by Fletcher and Reeves' conjugate-gradient method.

Here too, the original code SIGNOPT of Templeman has been applied.

Use of the Harmonic Inequality. As mentioned earlier, each signomial program can be written as a reversed posynomial program:

$$\begin{aligned} &\min g_0(t), \\ \text{s.t. } &g_k(t) \leq 1, \qquad k = 1, \ldots, K', \\ &g'_k(t) \geq 1, \qquad k = 1, \ldots, K'', \end{aligned}$$

where all functions are posynomials. The harmonic inequality can be applied to transform the reversed constraints

$$g'_k(t) \geq 1$$

into constraints of the type

$$g''_k(t, \epsilon) \leq 1.$$

The following posynomial program then results:

$$\begin{aligned} &\min g_0(t), \\ \text{s.t. } &g_k(t) \leq 1, \qquad k = 1, \ldots, K', \\ &g''_k(t, \epsilon) \leq 1, \qquad k = 1, \ldots, K'', \\ &\epsilon > 0, \qquad \sum_{i \in J\{k\}} \epsilon_i = 1, \quad k = 1, \ldots, K''. \end{aligned}$$

For a given set of ϵ, a solution t' of the corresponding condensed program is computed, from which in turn new values for ϵ are derived (Ref. 4). Based on the harmonic–arithmetic approach, Jefferson (Ref. 19) has coded an algorithm that maximizes the dual of the approximating program by means of a modified Newton method. Here too, the dual is reduced by eliminating as many variables as dual constraints. The code calculates a feasible starting point and then performs one-dimensional searches. Based on the dual maximum, new weights ϵ are computed, and a new iteration is started. To deal with loose constraints, an alternative way of introducing slack terms is used during the computations. Jefferson's original code GPROG has been used in our study.

Primal Condensation Algorithm. This algorithm makes use of the primal geometric program. It computes a solution of the Kuhn–Tucker conditions by condensing them into monomials by the geometric–arithmetic inequality. After a logarithmic transformation, a set of linear equations results (Ref. 20). These equations are solved so that new values for the variables are obtained, from which a set of new weights can be computed if necessary.

This method was coded under the name GPKTC by the present authors.

3.3. Constrained Fletcher and Powell Method. This algorithm is a general nonlinear programming technique to minimize a nonlinear objective function, subject to nonlinear constraints. Although the study of such algorithms obviously trespasses the scope of the present paper, one such general algorithm has been included to provide a point of reference. The choice has been made quite arbitrarily and must be seen in the light of the previous remark.

Haarhoff and Buys (Ref. 21) transform the original problem into an unconstrained one, for which they use the well-known DFP-method. Because of the form of the geometric programming constraints, analytical derivatives are used.

The computer code CONMIN was developed by Haarhoff, Buys, and von Molendoff and published in Ref. 15.

4. Numerical Results

4.1. Technical Details of the Comparison. The algorithms used in the present study were all programmed in FORTRAN IV, level H, and were implemented on the same computer. This is the IBM 370/158 of the

Computer Center at the Katholieke Universiteit Leuven, on which it took 28.11 secs to execute the standard timing program proposed by Colville (Ref. 35).

All codes were executed in double precision, except SIGNOPT, for which it was stated that the necessary precautions were taken for obtaining accurate results in single precision.

The study was planned to collect the following quantitative and qualitative data for each individual algorithm: (i) number of successes and failures on a series of test problems; (ii) CPU time (or average time) to solve the selected problems without taking into account the time for input and output operations; for codes working completely on the dual, the time to recover the primal solution was also not included; (iii) in-core storage for execution; (iv) simplicity of preparation and implementation; (v) size of the problems that can be solved; (vi) behavior on real-life problems.

These data will form the basis on which the different algorithms will be evaluated. It is clear that the above list of characteristics can be extended, but the cited criteria are obviously the most important ones in comparing algorithms.

The main goal of the present study was to examine the advantages and disadvantages of the different approaches and to guide the future user in his choice of an algorithm. Eventually, where possible, a classification of algorithms will be given; but it is clear that, on the basis of sometimes quite opposing criteria, no algorithm can be found that performs in such a way that it might be proclaimed superior to all others.

Essential elements in the evaluation of the different algorithms are the choice of a starting point, the stopping rules, and the selection of the test problems. These three important factors will now be discussed individually.

Starting Points. Three different approaches have been followed, depending on the kind of algorithm used.

(a) The algorithm computes internally its own starting point that satisfies the dual linear constraints. The posynomial codes LAM, SP, VMP, CSGP, and NEWTGP belongs to this class.

(b) Dual signomial algorithms usually need a dual starting point. Based on the observation that generally the components of the dual optimal solution are smaller than 1, random numbers between 0.01 and 1 were selected as starting points. The codes belonging to this group are LM, NRF, NRT, VA07AD, DCA, NRVB.

(c) A third class of algorithms need to be initiated at a primal starting point. In this case, the components of the starting point were randomly distributed in a sphere with its center at the optimal solution and whose radius in each direction equals 90% of the corresponding component of the optimal solution. If a component computed this way violates some bounds on the variables, this bound then replaces the computed starting value. The idea behind this scheme is that, in practice, where one does not have a prior knowledge of the optimum, one will always be able to choose a starting point in this range, based on a rough, educated guess. The algorithms GOMTRY, SIGNOPT, GGP, GPROG, CONMIN, and GPKTC belong to this group.

Stopping Rule. In order to obtain comparable results, an effort must be made to apply equivalent stopping rules for conceptually different algorithms.

Primal-based algorithms were stopped if the following two conditions were satisfied simultaneously:

$$|g_k(t^i)-1|\leq 10^{-5}, \qquad \text{for each active number,}$$

$$|[g_0(t^i)-g_0(t^{i-1})]/g_0(t^{i-1})|\leq 10^{-4},$$

where the superscript indicates the iteration number. This rule was applied to GGP, GPKTC, GPROG, GOMTRY, SIGNOPT, CONMIN, and CSGP.

Although the same conditions could be applied to dual codes, we have nevertheless preferred to use the more easily manageable stopping rules:

$$|(\delta_j^i-\delta_j^{i-1})/\delta_j^{i-1}|\leq 10^{-2}, \qquad j=1,\ldots,n,$$

$$|(\lambda_j^i-\lambda_j^{i-1})/\lambda_j^{i-1}|\leq 10^{-2}, \qquad j=1,\ldots,p.$$

The numerical values appearing on the right-hand side of the last two inequalities were determined after extensive numerical experimentation in order to create stopping rules which are in general equivalent to the ones used for the primal codes.

Test Problems. The algorithms were all tested on a set of 24 problems, which can be divided into four categories: (a) posynomial problems with only equality constraints, Problems 1–4; (b) posynomial problems with equality and inequality constraints, Problems 5–8; (c) signomial problems with only equality constraints, Problems 9–19; (d) signomial problems with equality and inequality constraints, Problems 20–24.

To comply with the formulation of geometric programming, each inequality should be either replaced by two inequalities of opposite sense or, if possible, by a single inequality. Rules for determining the sense of this inequality were first outlined in Ref. 22 and further discussed in Ref. 23.

4.2. Numerical Results. The numerical results are condensed in Tables 1–3. Tables 1 and 2 give for each problem the computer time needed by a particular algorithm, divided by the minimum solution time for that

Table 1. Relative CPU times for posynomial codes.

Code	LAM	SP	VMP	NEWTGP	CSGP	Min CPU time
Preparation time	0(LP)	(−)	0	(+)	(−)	
In-core storage	(+)	(−)	(−)	0	(−)	
P1	1.48	7.28	6.89	3.47	6.30	92
P2	20.61	66.84	66.05	5.26	50.82	76
P3	2.55	73.18	27.12	3.63	26.13	132
P4	—	—	—	—	4.87	1122
P5	53.11	199.96	293.25	6.45	13.76	93
P6	4.70	59.85	—	—	—	489
P7	—	217.29	—	120.47	3.54	815
P8	—	—	—	—	10.29	6240

problem. Hence, the fastest algorithm will have a relative time 1; for this particular algorithm, the CPU time (in 10^{-3} secs) is shown in a separate column.

The results for strict posynomial codes, which solved each problem only once, are presented in Table 1. The symbol — means that no convergence was obtained for this problem with the given algorithm.

For signomial algorithms, which solved each problem 5 times from different starting points, the average on all successful runs was used to construct Table 2. Since only successful runs were taken into account, one should consider this table in conjunction with Table 3, where the number of successful runs on each problem is shown for the different algorithms.

At this point, we would like to stress that the data were recorded for random starting points and that, for virtually all algorithms, the number of failures could be substantially reduced by having the algorithm start from a so-called educated guess, rather than from a completely random point. Finally, a few remarks are needed to complete the information given in the different tables. For Problems 10, 12, 15, 16, 18, multiple equilibrium solutions were recorded. For Problems 15 and 18, some of these solutions had vanishing dual components, which caused the failure of many dual algorithms on these problems. Incorrect solutions were obtained with SIGNOPT on Problem 8 and with CSGP on Problem 6.

Several algorithms failed to converge because the accuracy required of the results was too high. This happened to GPROG and SIGNOPT on Problem 13 and for the latter code also on Problem 17. They would have converged under less severe conditions. Lowering the accuracy to the range

Table 2. Relative CPU times for signomial codes.

Code	LM	NRF	NRVB	NRT	VA07AD	DCA	GOMTRY	GGP	SIGNOPT	GPROG	GPKTC	CONMIN	
Preparation time	(+)	(+)	(+)	(+)	0	0	0(LP)	0(LP)	0	0	0	0	Min CPU time
In-core storage	(−)	0	0	0	0	0	0	(+)	(+)	(−)	(+)	(+)	
P1	3.69	1.13	1.33	1.40	2.43	1	1.55	8.48	7.40	17.50	1.06	17.56	92
P2	—	3.21	3.52	2.94	—	2.25	2.35	5.78	7.07	34.60	1	34.00	76
P3	26.80	3.50	3.20	3.25	61.93	2.52	2.27	8.33	7.07	21.51	1	34.84	132
P4	91.59	7.73	—	14.89	—	60.76	—	5.82	—	—	1	12.73	1122
P5	—	10.81	—	22.70	—	7.48	22.27	1.70	4.31	25.48	1	29.74	93
P6	17.65	2.32	2.86	3.61	1.90	4.12	—	1	1.08	19.40	9.14	—	489
P7	—	9.05	14.11	15.07	—	8.35	—	1.18	1.00	6.95	3.53	—	815
P8	—	—	—	—	—	—	—	1.06	—	—	1.00	—	6240
P9	6.65	1.87	3.36	2.08	3.34	1.82	1.40	13.06	11.00	248.	1	15.21	47
P10	30.72	2.43	3.92	2.15	7.47	3.90	3.29	24.07	10.45	54.96	1	44.35	51
P11	35.41	3.10	6.30	4.93	24.94	6.87	4.37	13.53	14.14	51.87	1	17.33	98
P12	53.87	4.03	2.78	8.73	17.01	2.56	5.29	30.23	—	83.62	1	22.01	379
P13	—	6.34	5.48	3.44	5.44	2.84	5.28	7.52	—	117.56	1	31.62	745
P14	33.38	4.93	3.70	3.58	8.23	2.53	6.59	10.45	6.78	55.45	1	24.83	694
P15	24.62	3.95	3.33	8.24	9.43	2.86	6.51	13.98	5.06	50.46	1	26.34	518
P16	31.43	8.35	13.09	5.53	15.58	8.32	7.62	22.32	11.95	74.87	1	40.44	561
P17	48.24	10.97	12.03	14.45	21.71	6.94	10.01	8.01	13.24	61.13	1	36.36	645
P18	71.37	12.01	7.85	8.80	25.96	18.51	—	18.71	—	29.22	1	49.62	954
P19	—	6.49	10.65	7.03	20.92	3.29	1	5.93	—	—	1.06	—	1887
P20	—	—	—	—	—	—	—	5.65	—	—	1	—	2193
P21	—	—	—	—	—	—	—	1.24	—	—	1	—	2651
P22	—	—	—	—	—	—	—	1.11	—	—	1	—	4925
P23	—	—	—	811.	—	612.	—	1	—	—	15.45	—	201
P24	—	—	—	—	—	—	—	1	—	—	2.96	—	4927

Table 3. Number of successful runs for a total of 5 starting points.

Code	LM	NRF	NRVB	NRT	VA07AD	DCA	GOMTRY	GGP	SIGNOPT	GPROG	GPKTC	CONMIN
P1	5	5	5	5	5	5	5	5	5	5	5	4
P2	0	5	4	5	0	5	5	5	5	5	5	3
P3	5	5	5	5	1	5	5	5	5	5	5	5
P4	0	5	0	5	0	5	0	5	0	0	5	3
P5	0	5	0	3	0	5	4	5	5	5	5	5
P6	5	5	5	5	4	5	0	5	5	4	5	0
P7	0	5	2	3	0'	5	0	5	5	4	5	0
P8	0	0	0	0	0	0	0	5	0	0	5	0
P9	5	5	5	5	5	5	5	5	5	5	5	5
P10	5	5	5	4	3	5	4	5	5	5	5	1
P11	3	5	3	5	2	2	2	5	5	5	5	5
P12	2	5	3	4	3	5	4	3	0	3	5	4
P13	0	4	3	4	4	5	1	5	0	1	5	5
P14	5	4	5	3	2	5	2	5	5	1	5	5
P15	5	5	5	5	3	5	4	5	5	5	5	5
P16	3	2	3	2	3	5	2	5	5	2	5	5
P17	5	5	3	3	2	5	4	5	1	3	5	5
P18	4	5	3	3	1	5	0	4	0	2	5	3
P19	0	4	4	3	3	5	1	3	0	0	5	0
P20	0	3	0	0	0	0	0	5	0	0	1	0
P21	0	0	0	0	0	0	0	5	0	0	5	0
P22	0	0	0	0	0	0	0	5	0	0	5	0
P23	0	0	0	1	0	4	0	5	0	0	5	0
P24	0	0	0	0	0	0	0	5	0	0	5	0

10^{-3} to 10^{-4} was usually enough to let these two algorithms converge without any difficulty. Also, the convergence properties of CONMIN were sensitive to changes in accuracy.

We also would like to call attention to problem 12, a rather small-size problem, whose structure seems to cause difficulty for many algorithms.

The poor behavior of primal algorithms on Problem 19 is due to the fact that some of the primal variables have a large value at optimality, so that the random device used for the starting points generated in this case values which could be extremely far away from the optimal ones.

Finally, in Tables 1 and 2, we give our own experience on the sometimes quite subjective data concerning the required in-core storage and preparation time, assuming that one has to code the program by himself. This information is important for an engineer or a manager who has to solve an optimization problem and has no computer code available. The following notation was thereto adopted: (+) advantageous compared to most algorithms, 0 normal, (−) less advantageous than most algorithms. The notation 0(LP) means normal, provided an LP-code is available.

5. Conclusions

5.1. Posynomial Codes. One would expect these codes to be very efficient for their specific area of application, since they can make use to full extent of the quite impressive duality theory of geometric programming. However, in reality the situation turns out to be completely different. As shown by the results for the posynomial codes, signomial codes did remarkably better on each of these problems. This proves that to date, from a purely computational point of view, the role of the dual program has been overestimated. Its significance is, however, great for other applications, such as sensitivity analysis.

A second conclusion is that most of these algorithms tend to fail for larger problems. This is due to the fact that they are dual-based and therefore are affected by the following two important disadvantages: (i) the size of the dual program is usually much larger than the size of the corresponding primal program; and (ii) small primal terms give rise to small dual variables, which can cause considerable numerical difficulties; this is especially true for those problems where slack variables are added to active inequality constraints. Both effects are clearly illustrated in Problem 4.

The computational results also indicate that the approach of NEWTGP is the fastest one on small problems, but that CSGP is the most robust code of this group and henceforth the most reliable for larger problems. However, the latter code slows down under the presence of many equality constraints, where the modified direction-generation machinery of the concave simplex method is less useful.

VMP experiences difficulties in obtaining the required accuracy, while *SP* is obviously an unattractive way for solving posynomial programs.

5.2. Signomial Codes

Dual Codes. As already mentioned before, dual codes suffer severely from two main disadvantages. First, the size of the dual program tends to be much larger than the size of the corresponding primal. Convergence from a random starting point becomes very problematic when the size of the dual (i.e., the total number of dependent and independent dual variables) exceeds 50. For this class of algorithms, the degree of difficulty has kept its literal meaning. Second, small primal terms influence the efficiency of dual codes. A third warning concerns the usually high sensitivity of primal variables to inaccurate dual results.

The most important advantage of these codes, especially the Newton–Raphson-type code, is the simplicity of the required machinery, which makes the preparation time really minimal. As expected, the NRT code behaves better on problems were the number of equilibrium equations exceeds the number of dual linear constraints (see Problems 3 and 6). However, this phenomenon usually occurs together with a large size for the dual.

Marquardt's algorithm converges rather fast in the early iterations but slows down near the optimum. Therefore, it might be appropriate for locating a starting point for another dual code.

Although GOMTRY also performed quite speedily on some test problems, it is DCA which showed the best overall results. However, this remark should be seen in the light of the final conclusion that algorithms belonging to this class are clearly inferior to primal codes and certainly for larger-sized problems.

Primal Codes

General NLP Code. The required accuracy made the general nonlinear programming code used in this study fail for many test problems and increased the computation time for the ones on which it succeeded. The results clearly show that specific geometric programming codes did behave much better than this general nonlinear programming code.

Use of Harmonic Inequality. The harmonic–arithmetic inequality can be used to approximate a signomial program by a series of posynomial programs. Therefore, it should be considered as an alternative to the use of the geometric inequality. The results illustrate that, computationally, the latter approach is much more efficient. This conclusion is certainly not surprising, since the harmonic inequality yields a more conservative approximation than the geometric inequality. On the other hand, when using the harmonic inequality, the approximating exponent matrix remains unchanged during the subsequent iterations. However, this seems of minor computational importance compared to the disadvantage of a more conservative approximation.

Use of Geometric Inequality. Three algorithms make use of the geometric inequality to solve geometric programs: two of them through the idea of complementarity, and the other mainly through the technique of condensation. Globally, one can state that they produced the best results.

Furthermore, our figures indicate that the complementary problem can be more efficiently solved by the approach followed in GGP than by the one advocated in SIGNOPT. Indeed, the latter algorithm is surely quite fast, but it needs to find a dual feasible starting point for the complementary problem, an operation which could not always be performed in a straightforward way. Also, for bigger problems, failure occurred merely because of numerical difficulties (underflow, seemingly unbounded solutions, etc.). The results from GPKTC and GGP are clearly the best obtained in this study, and this is true not only with respect to the CPU times involved, but also as far as the number of failures is concerned. One should remember that the tests were performed from random starting points. The efficiency of these approaches can be attributed to a large extent to the fact that they are primal-based. Indeed, the primal problem is usually much smaller than the dual one. Also, defining the degree of difficulty for such codes is usually less meaningful.

A few differences in the behavior of these two codes are also illustrated by the computer results (Tables 4–5).

(a) GGP is more elegant in the handling of inactive constraints. In GPRTC, such constraints are converted to equality constraints by adding slack variables, an operation which increases the size of the problem.

(b) Bounds on primal variables influence the times in the same way. GGP needs such bounds, and therefore will naturally take advantage of their presence. The combined effect of (a) and (b) can be seen in Problems 23 and 24. However, when no such bounds are present due to physical considerations, they need to be introduced artificially, and the efficiency of the code will depend on the ability to produce strong bounds (see Problems 12 and 19).

(c) Due to the cutting-plane technique, the CPU time for GGP is not strictly proportional to the problem size.

(d) GPKTC's main disadvantage is its treatment of loose constraints. But, on the other hand, it behaves especially well on problems where few inequality constraints are present.

(e) A major advantage of GPKTC is the simplicity of the machinery used. Only systems of linear equations need to be solved.

Table 4. Characteristics of test problems.

Item	A	B	C	D	E	F	G	H	I
P1	4	2	0	0	2	No	6	1	24
P2	3	1	0	0	1	No	9	5	10
P3	4	1	0	0	1	No	12	7	25
P4	11	3	0	0	3	No	31	19	26
P5	4	0	3	0	3	No	8	3	27
P6	8	0	7	0	7	No	12	3	28
P7	8	0	6	1	7	No	12	3	28
P8	7	0	2	5	7	No	48	40	29
P9	2	1	0	0	1	No	5	2	24
P10	3	1	0	0	1	No	6	2	24
P11	4	2	0	0	2	No	7	2	30
P12	8	4	0	0	4	No	15	6	23
P13	8	6	0	0	6	No	19	10	31
P14	10	6	0	0	6	Yes	16	5	30
P15	10	7	0	0	7	No	15	4	30
P16	10	7	0	0	7	No	18	7	30
P17	11	9	0	0	9	No	19	7	24
P18	13	9	0	0	9	No	22	8	30
P19	8	5	0	0	5	No	28	19	14
P20	13	9	0	0	9	Yes	30	16	32
P21	10	7	0	0	7	Yes	23	12	33
P22	9	5	2	3	10	Yes	57	47	34
P23	5	0	2	4	6	Yes	21	15	35
P24	10	7	1	2	10	Yes	36	25	26

A = Number of primal variables.
B = Number of primal equality constraints.
C = Number of primal active inequality constraints.
D = Number of primal inactive inequality constraints.
E = Total number of primal constraints.
F = Bounds on the primal variables.
G = Number of primal terms.
H = Degree of difficulty.
I = Reference.

Table 5. Characteristics of codes.

Name	Brief description	Coded by	Signomial or posynomial	Primal or dual	External addition of slack variables	Reference
LAM	Linear approximation of the dual objective function	Rijckaert Martens	P	D	Yes	
SP	Separable programming	Rijckaert Martens	P	D	No	
VMP	Projected gradient method	Sargent Murtagh	P	D	Yes	
NEWTGP	NR iteration on the reduced dual	Bradley	P	D	Yes	11
CSGP	Modified concave simplex	Beck	P	D	No	12
LM	NR iteration on the dual KT conditions	Rijckaert Martens	S	D	Yes	
NRF	NR iterations on the dual Jacobian equations	Rijckaert Martens	S	D	Yes	
NRVB	Nr iteration on the dual Jacobian equations with reduced step length	Rijckaert Martens	S	D	Yes	
NRT	NR iterations on the dual Jacobian equations after logarithmic transform	Rijckaert Martens	S	D	Yes	
VA07AD	Least-square minimization (Marquardt)	Fletcher (Harwell)	S	D	Yes	9
DCA	Dual condensation algorithm	Rijckaert Martens	S	D	Yes	20
GOMTRY	Lagrangian algorithm of Blau	Garcia *et al.*	S	D–P	Yes	15
GGP	Primal condensation + cutting-plane technique	Dembo	S	P	No	26
SIGNOPT	Primal condensation + Fletcher–Reeves technique	Templeman	S	P-D	No	18
GPROG	Approximation by harmonic mean inequality +NR iteration	Jefferson	S	P-D	No	19
GPKTC	Primal condensation of KT-conditions + Picard-iteration	Rijckaert Martens	S	P	Yes	20
CONMIN	Constrained DFP-algorithm	Haarhoff *et al.*	S	P	Yes	21

6. Final Remarks

Although improvements to the functioning of the different codes remain possible, it is quite unlikely however that such changes will have much effect on the relative efficiency of the different algorithms. Hence, we can conclude that: (a) the computational advantages of the existing dual approaches have surely been overestimated; and (b) the geometric–arithmetic inequality can play a key role in solving geometric programming problems through the ideas of complementarity and condensation.

7. Appendix

7.1. Test Problems

Problem 1 (Ref. 24)

$$\min t_1^{-1},$$

$$g_1 = t_1 t_2^{-1} + 0.5 t_3^{-1} \leq 1,$$

$$g_2 = 0.01 t_3 t_4^{-1} + 0.01 t_2 \div 0.0005 t_2 t_4 \leq 1.$$

The solution is

$$t_1 = 82.847, \quad t_2 = 87.924, \quad t_3 = 8.293, \quad t_4 = 1.364, \quad g_0 = 0.01208.$$

The constraints g_1 and g_2 are derived from equality constraints.

Problem 2 (Ref. 10)

$$\min(5t_1 + 50{,}000 t_1^{-1} + 20 t_2 + 72{,}000 t_2^{-1} + 10 t_3 + 144{,}000 t_3^{-1}),$$

$$g_1 = 4 t_1^{-1} + 32 t_2^{-1} + 120 t_3^{-1} \leq 1.$$

The solution is

$$t_1 = 107.4, \quad t_2 = 84.9, \quad t_3 = 204.5, \quad g_0 = 6300.$$

The constraint g_1 is derived from an equality constraint.

Problem 3 (Ref. 29)

$$\min(592 t_1^{0.65} + 582 t_1^{0.39} + 1200 t_1^{0.52} + 370_1^{0.22} t_2^{-0.22} + 250 t_1^{0.40} t_3^{-0.40}$$

$$+ 210 t_1^{0.62} t_3^{-0.62} + 250 t_1^{0.40} t_4^{-0.40} + 200 t_1^{0.85} t_4^{-0.85}),$$

$$g_1 = 500 t_1^{-1} + 50 t_2 t_1^{-1} + 50 t_3 t_1^{-1} + 50 t_4 t_1^{-1} \leq 1.$$

The solution is

$$t_1 = 749.89, \qquad t_2 = 0.11114, \qquad t_3 = 1.46193, \qquad t_4 = 3.42481,$$
$$g_0 = 126{,}344.$$

The constraint g_1 is derived from an equality constraint.

Problem 4 (Ref. 26)

$$\min(t_1^{-0.00133172} t_2^{-0.002270927} t_3^{-0.00248546} t_4^{-4.67} t_5^{-4.671973}$$
$$\times t_6^{-0.00814} t_7^{-0.008092} t_8^{-0.005} t_9^{-0.00909} t_{10}^{-0.00088} t_{11}^{-0.0019}) \times 10^5,$$

$$g_1 = 0.05176 t_1 + 0.021864 t_2 + 0.097733 t_3 + 0.00669408 t_4 t_5 \leq 1,$$
$$g_2 = 10^{-6} t_1 + 10^{-5} t_2 + 10^{-6} t_3 + 10^{-10} t_4 + 10^{-8} t_5 + 10^{-3} t_6$$
$$+ 10^{-3} t_7 + 0.10898645 t_4 t_5 \leq 1,$$
$$g_3 = 0.00016108 t_2 t_5 + 10^{-23} t_2 t_4 t_5 + 1.9310^{-6} t_2 t_4^{-1} t_5 + 10^{-4} t_{10}$$
$$+ 10^{-6} t_1 + 10^{-5} t_2 + 10^{-6} t_3 + 10^{-10} t_4 + 10^{-8} t_5 + 10^{-3} t_6$$
$$+ 10^{-3} t_8 + 0.10898 t_4 t_5 + 1.610810^{-4} t_2 t_5 + 10^{-23} t_2 t_4 t_5$$
$$+ 1.9310^{-6} t_2 t_4^{-1} t_5 + 10^{-5} t_9 + 0.0001184 t_1 t_9 + 10^{-4} t_{11} \leq 1.$$

The solution is

$$t_1 = 2.5180708, \qquad t_2 = 2.5100911, \qquad t_3 = 7.6625640,$$
$$t_4 = 1.1741216, \qquad t_5 = 7.7709114, \qquad t_6 = 1.3019503,$$
$$t_7 = 4.2801718, \qquad t_8 = 2.7894894, \qquad t_9 = 1.7389956,$$
$$t_{10} = 1.9732197, \qquad t_{11} = 6.6389848, \qquad g_0 = 3.1681988.$$

The constraints g_1, g_2, g_3 are derived from equality constraints.

Problem 5 (Ref. 27)

$$\min(168 t_1 t_2 + 3651.2 t_1 t_2 t_3^{-1} + 3651.2 t_1 + 40{,}000 t_4^{-1}),$$
$$g_1 = 1.0425 t_1 t_2^{-1} \leq 1,$$
$$g_2 = 0.00035 t_1 t_3 \leq 1,$$
$$g_3 = 1.25 t_1^{-1} t_4 + 41.63 t_1^{-1} \leq 1.$$

The solution is

$$t_1 = 43.02, \qquad t_2 = 44.85, \qquad t_3 = 66.39, \qquad t_4 = 1.11, \qquad g_0 = 623015.$$

The constraints g_1, g_2, g_3 are active inequality constraints.

Problem 6 (Ref. 28)

$$\min(2t_1^{0.9}t_2^{-1.5}t_3^{-3} + 5t_4^{-0.3}t_5^{2.6} + 4.7t_6^{-1.8}t_7^{-0.5}t_8),$$

$$g_1 = 7.2t_1^{-3.8}t_2^{2.2}t_3^{4.3} + 0.5t_4^{-0.7}t_5^{-1.6} + 0.2t_6^{4.3}t_7^{-1.9}t_8^{8.5} \leq 1,$$

$$g_2 = 10t_1^{2.3}t_2^{1.7}t_3^{4.5} \leq 1,$$

$$g_3 = 0.6t_4^{-2.1}t_5^{0.4} \leq 1,$$

$$g_4 = 6.2t_6^{4.5}t_7^{-2.7}t_8^{-0.6} \leq 1,$$

$$g_5 = 3.1 1_1^{1.6}t_2^{0.4}t_3^{-3.8} \leq 1,$$

$$g_6 = 3.7t_4^{5.4}t_5^{1.3} \leq 1,$$

$$g_7 = 0.3t_6^{-1.1}t_7^{7.3}t_8^{-5.6} \leq 1.$$

The solution is

$$t_1 = 0.9701, \qquad t_2 = 0.1985, \qquad t_3 = 1.1216,$$

$$t_4 = 0.7841, \qquad t_5 = 1.0040, \qquad t_6 = 0.6948,$$

$$t_7 = 1.1157, \qquad t_8 = 0.9993, \qquad g_0 = 29.5985.$$

The constraints g_1, g_2, g_3, g_4, g_5, g_6, g_7 are active inequality constraints.

Problem 7 (Ref. 28)

$$\min(2t_1^{0.9}t_2^{-1.5}t_3^{-3} + 5t_4^{-0.3}t_5^{2.6} + 4.7t_6^{-1.8}t_7^{-0.5}t_8),$$

$$g_1 = 7.2t_1^{-3.8}t_2^{2.2}t_3^{4.3} + 0.5t_4^{-0.7}t_5^{-1.6} + 0.2t_6^{4.3}t_7^{-1.9}t_8^{8.5} \leq 1,$$

$$g_2 = 10t_1^{2.3}t_2^{1.7}t_3^{4.5} \leq 1,$$

$$g_3 = 0.2t_4^{-2.1}t_5^{0.4} \leq 1,$$

$$g_4 = 6.2t_6^{4.5}t_7^{-2.7}t_8^{-0.6} \leq 1,$$

$$g_5 = 3.1t_1^{1.6}t_2^{0.4}t_3^{-3.8} \leq 1,$$

$$g_6 = 3.7t_4^{5.4}t_5^{1.3} \leq 1,$$

$$g_7 = 0.3t_6^{-1.1}t_7^{7.3}t_8^{-5.6} \leq 1.$$

The solution is

$$t_1 = 0.96856, \qquad t_2 = 0.19355, \qquad t_3 = 1.1332,$$

$$t_4 = 0.78624, \qquad t_5 = 1.0001, \qquad t_6 = 0.69479,$$

$$t_7 = 1.1163, \qquad t_8 = 0.99689, \qquad g_0 = 29.595.$$

The constraints g_1, g_2, g_4, g_5, g_6, g_7 are active inequality constraints. The constraint g_3 is an inactive inequality constraint.

Problem 8 (Ref. 29)

$$\begin{aligned}
&\min(10t_1t_2^{-1}t_3^{-1}t_4+20t_1^{-1}t_4^{-1}t_5t_6+30t_2t_3t_4\\
&\quad+100t_1^{-1}t_2^{-1}t_3^{-1}t_4^{-1}t_5^{-1}t_6^{-1}t_7^{-1}+5t_1^2t_2^2t_3t_5t_6^{1.5}t_7^2\\
&\quad+50(t_3t_4t_5)^{-0.5}+25t_3^2t_4^2t_5^{-1}t_6^{-1}t_7^{-1}+10t_3^{0.5}t_4^{0.5}t_5t_6t_7),\\
&g_1=0.1t_1^2t_2^2t_3+0.05t_4t_5^{0.5}+0.15t_6^{0.5}t_7^{0.5}\leqslant 1,\\
&g_2=0.1t_1t_4t_7+0.05t_1t_2^{-1}t_3^{-1}t_5t_6t_7^{0.5}+0.05t_2^2t_3^2t_4^{-1}\\
&\quad+0.15t_1^{-0.5}t_2^{-0.3}t_3t_5^{0.5}+0.1t_5t_6+0.1t_4^2+0.2t_1t_2t_3\leqslant 1,\\
&g_3=t_1+t_2+t_3+t_4+t_5+t_6+t_7\leqslant 10,\\
&g_4=t_1^2+t_1t_2+t_1t_3+t_1t_4+t_1t_5+t_1t_6+t_1t_7\leqslant 50,\\
&g_5=t_1t_2^{-1}+t_2t_3^{-1}+t_3t_4^{-1}+t_4t_5^{-1}+t_5t_6^{-1}+t_6t_7^{-1}\leqslant 100,\\
&g_6=t_1t_3^{-2}+t_2t_4^{-2}+t_3t_5^{-2}+t_4t_6^{-2}+t_5t_7^{-2}\leqslant 10,\\
&g_7=t_1^{-0.5}t_3+t_2^{-0.5}t_4+t_3^{-0.5}t_5+t_4^{-0.5}t_6+t_5^{-0.5}t_7\leqslant 50.
\end{aligned}$$

The solution is

$$t_1=1.34186,\qquad t_2=0.99325,\qquad t_3=0.87050,$$
$$t_4=0.92359,\qquad t_5=3.14643,\qquad t_6=0.40408,$$
$$t_7=1.54767,\qquad g_0=178.478.$$

The constraints g_2 and g_6 are active inequality constraints.
The constraints g_1, g_3, g_4, g_5, g_7 are inactive inequality constraints.

Problem 9 (Ref. 24)

$$\min(3.7t_1^{0.85}+1.985t_1+700.3t_2^{-0.75}),$$
$$g_1\equiv 0.7673t_2^{0.05}-0.05t_1\leqslant 1.$$

The solution is

$$t_1=0.819,\qquad t_2=446,\qquad g_0=11.91.$$

The constraint g_1 is derived from an equality constraint.

Problem 10 (Ref. 24)

$$\min(0.5t_1t_2^{-1} - t_1 - 5t_2^{-1}),$$

$$g_1 = 0.01t_2t_3^{-1} + 0.01t_1 + 0.0005t_1t_3 \leq 1.$$

The solution is

$$t_1 = 88.310, \qquad t_2 = 7.454, \qquad t_3 = 1.311, \qquad g_0 = -83.21.$$

The constraint g_1 is derived from an equality constraint.

Problem 11 (Ref. 10)

$$\min(-t_1 + 0.4t_1^{0.67}t_3^{-0.67}),$$

$$g_1 = 0.05882t_3t_4 + 0.1t_1 \leq 1,$$

$$g_2 = 4t_2t_4^{-1} + 2t_2^{-0.71}t_4^{-1} + 0.05882t_2^{-1.3}t_3 \leq 1.$$

The solution is

$$t_1 = 8.1301, \quad t_2 = 0.6154, \quad t_3 = 0.5640, \quad t_4 = 5.6362, \quad g_0 = -5.7398.$$

The constraints g_1 and g_2 are derived from equality constraints.

Problem 12 (Ref. 23)

$$\min(-t_1 - t_5 + 0.4t_1^{0.67}t_3^{-0.67} + 0.4t_5^{0.67}t_7^{-0.67}),$$

$$g_1 = 0.05882t_3t_4 + 0.1t_1 \leq 1,$$

$$g_2 = 0.05882t_7t_8 + 0.1t_1 + 0.1t_5 \leq 1,$$

$$g_3 = 4t_2t_4^{-1} + 2t_2^{-0.71}t_4^{-1} + 0.05882t_2^{-1.3}t_3 \leq 1,$$

$$g_4 = 4t_6t_8^{-1} + 2t_6^{-0.71}t_8^{-1} + 0.05882t_6^{-1.3}t_7 \leq 1.$$

The solution is

$$t_1 = 6.4650, \qquad t_2 = 0.6674, \qquad t_3 = 1.0130,$$

$$t_4 = 5.9327, \qquad t_5 = 2.2326, \qquad t_6 = 0.5958,$$

$$t_7 = 0.4006, \qquad t_8 = 5.5273, \qquad g_0 = -6.0482.$$

The constraints g_1, g_2, g_3, g_4 are derived from equality constraints.

Problem 13 (Ref. 31)

$$\min(t_1 + t_2 + t_3),$$

$$g_1 = 833.332352 t_1^{-1} t_4 t_6^{-1} + 100 t_6^{-1} - 83{,}333.333 t_1^{-1} t_6^{-1} \leq 1,$$

$$g_2 = 1250 t_2^{-1} t_5 t_7^{-1} + t_4 t_7^{-1} - 1250 t_2^{-1} t_4 t_7^{-1} \leq 1,$$

$$g_3 = 1{,}250{,}000 t_3^{-1} t_8^{-1} + t_5 t_8^{-1} - 2{,}500 t_3^{-1} t_5 t_8^{-1} \leq 1,$$

$$g_4 = 0.0025\ t_4 + 0.0025 t_6 \leq 1,$$

$$g_5 = 0.0025 t_5 + 0.0025 t_7 - 0.0025 t_4 \leq 1,$$

$$g_6 = 0.01 t_8 - 0.01 t_5 \leq 1.$$

The solution is

$$t_1 = 579.307, \quad t_2 = 1359.97, \quad t_3 = 5109{,}97,$$

$$t_4 = 182.018, \quad t_5 = 295.601, \quad t_6 = 217.982,$$

$$t_7 = 286.416, \quad t_8 = 395.601, \quad g_0 = 7049.247.$$

The constraints g_1, g_2, g_3, g_4, g_5, g_6 are derived from equality constraints.

Problem 14 (Ref. 30)

$$\min(t_6 + 0.4 t_4^{0.67} + 0.4 t_9^{0.67}),$$

$$g_1 = t_1^{-1} t_2^{-1.5} t_3 t_4^{-1} t_5^{-1} + 5 t_1^{-1} t_2^{-1} t_3 t_5^{1.2} \leq 1,$$

$$g_2 = 0.05 t_3 + 0.05 t_2 \leq 1,$$

$$g_3 = 10_3^{-1} - t_1 t_3^{-1} \leq 1,$$

$$g_4 = t_6^{-1} t_7^{-1.5} t_8 t_9^{-1} t_{10}^{-1} + 5 t_6^{-1} t_7^{-1} t_8 t_{10}^{1.2} \leq 1,$$

$$g_5 = t_2^{-1} t_7 + t_2^{-1} t_8 \leq 1,$$

$$g_6 = t_1 t_8^{-1} - t_6 t_8^{-1} \leq 1,$$

$$g_7 = t_{10} \leq 0.1.$$

The solution is

$$t_1 = 2.0953, \quad t_2 = 12.0953, \quad t_3 = 7.9047,$$

$$t_4 = 0.4594, \quad t_5 = 0.3579. \quad t_6 = 0.4548,$$

$$t_7 = 10.4548, \quad t_8 = 1.6405, \quad t_9 = 1.1975,$$

$$t_{10} = 0.1000, \quad g_0 = 1.1436.$$

The constraints $g_1, g_2, g_3, g_4, g_5, g_6$ are derived from equality constraints. The constraint g_7 is an active inequality constraint.

Problem 15 (Ref. 30)

$$\min(0.05t_1 + 0.05t_2 + 0.05t_3 + t_9),$$
$$g_1 = 0.5t_9t_{10}^{-1} + 0.25t_{10}^{-1} \leqslant 1,$$
$$g_2 = t_7^{-1}t_{10} - 0.5t_1t_4t_7^{-1} \leqslant 1,$$
$$g_3 = t_7t_8^{-1} - 0.5t_2t_5t_8^{-1} \leqslant 1,$$
$$g_4 = t_8t_9^{-1} - 0.5t_3t_6t_9^{-1} \leqslant 1,$$
$$g_5 = 0.79681t_4t_7^{-1} \leqslant 1,$$
$$g_6 = 0.79681t_5t_8^{-1} \leqslant 1,$$
$$g_7 = 0.79681t_6t_9^{-1} \leqslant 1.$$

The solution is

$$t_1 = 0.7240, \qquad t_2 = 0.7240, \qquad t_3 = 0.7240,$$
$$t_4 = 0.2576, \qquad t_5 = 0.1771, \qquad t_6 = 0.1218,$$
$$t_7 = 0.2053, \qquad t_8 = 0.1411, \qquad t_9 = 0.0970,$$
$$t_{10} = 0.2985, \qquad g_0 = 0.2015.$$

The constraints $g_1, g_2, g_3, g_4, g_5, g_6, g_7$ are derived from equality constraints.

Problem 16 (Ref. 30)

$$\min(0.0t_1 + 0.05t_2 + 0.05t_3 + t_9),$$
$$g_1 = 0.5t_9t_{10}^{-1} + 0.25t_{10}^{-1} \leqslant 1,$$
$$g_2 = t_7^{-1}t_{10} - 0.5t_1t_4t_7^{-1} \leqslant 1,$$
$$g_3 = t_7t_8^{-1} - 0.5t_2t_5t_8^{-1} \leqslant 1,$$
$$g_4 = t_8t_9^{-1} - 0.5t_3t_6t_9^{-1} \leqslant 1,$$
$$g_5 = 0.700329t_4t_7^{-1} + 0.307795t_7 \leqslant 1,$$
$$g_6 = 0.700329t_5t_8^{-1} + 0.307795t_8 \leqslant 1,$$
$$g_7 = 0.700329t_6t_9^{-1} + 0.307795t_9 \leqslant 1.$$

The solution is

$$t_1 = 0.7295, \quad t_2 = 0.7133, \quad t_3 = 0.7030,$$
$$t_4 = 0.2653, \quad t_5 = 0.1821, \quad t_6 = 0.1241,$$
$$t_7 = 0.1979, \quad t_8 = 0.1329, \quad t_9 = 0.0893,$$
$$t_{10} = 0.2947, \quad g_0 = 0.1966.$$

The constraints g_1, g_2, g_3, g_4, g_5, g_6, g_7 are derived from equality constraints.

Problem 17 (Ref. 24)

$$\min t_3^{-1},$$
$$g_1 = 0.1t_{10} + t_7 t_{10} \leq 1,$$
$$g_2 = 10t_1t_4 + 10t_1t_4t_7^2 \leq 1,$$
$$g_3 = t_4^{-1} - 100t_7t_{10} \leq 1,$$
$$g_4 = t_{10}t_{11}^{-1} - 10t_8 \leq 1,$$
$$g_5 = t_1^{-1}t_2t_5 + t_1^{-1}t_2t_5t_8^2 \leq 1,$$
$$g_6 = t_5^{-1} - 10t_1^{-1}t_8t_{11} \leq 1,$$
$$g_7 = 10t_{11} - 10t_9 \leq 1,$$
$$g_8 = t_2^{-1}t_3t_6 + t_2^{-1}t_3t_6t_9^2 \leq 1,$$
$$g_9 = t_6^{-1} - t_2^{-1}t_9 \leq 1.$$

The solution is

$$t_1 = 7.004, \quad t_2 = 7.646, \quad t_3 = 7.112,$$
$$t_4 = 0.0125, \quad t_5 = 0.8120, \quad t_6 = 0.9558,$$
$$t_7 = 0.382, \quad t_8 = 0.358, \quad t_9 = 0.353,$$
$$t_{10} = 2.077, \quad t_{11} = 0.453, \quad g_0 = 0.1406.$$

The constraints g_1, g_2, g_3, g_4, g_5, g_6, g_7, g_8, g_9 are derived from equality constraints.

Problem 18 (Ref. 30)

$$\min t_9^{-1},$$
$$g_1 = t_1 + t_1t_{10} + t_1t_{10}t_{12} \leq 1,$$
$$g_2 = t_1^{-1}t_4t_{10}^{-1} + 0.01t_1^{-1}t_4t_{12}^{-1} + 0.01t_1^{-1}t_4 \leq 1,$$
$$g_3 = 100t_4^{-1}t_7t_{10}^{-1} \leq 1,$$

$$g_4 = t_1^{-1}t_2 + t_1^{-1}t_2t_{11} + t_1^{-1}t_2t_{11}t_{13} \leq 1,$$

$$g_5 = -t_2t_4^{-1}t_{11} + t_4^{-1}t_5 + 0.01t_4^{-1}t_5t_{11}t_{13}^{-1} + 0.01t_4^{-1}t_5t_{11} \leq 1,$$

$$g_6 = -0.01t_5t_7t_{11} + t_7^{-1}t_8 \leq 1,$$

$$g_7 = 12601t_2^{-1}t_3 \leq 1,$$

$$g_8 = -2100t_3t_5^{-1} + 26.2t_5^{-1}t_6 \leq 1,$$

$$g_9 = -21t_6t_8^{-1} + t_8^{-1}t_9 \leq 1.$$

The solution is

$$t_1 = 0.3917, \quad t_2 = 0.0937, \quad t_3 = 0.7435 \times 10^{-5},$$

$$t_4 = 0.4809, \quad t_5 = 0.6431, \quad t_6 = 0.0251,$$

$$t_7 = 0.0066, \quad t_8 = 0.0220, \quad t_9 = 0.5500,$$

$$t_{10} = 1.3823, \quad t_{11} = 2{,}3831, \quad t_{12} = 0.1234,$$

$$t_{13} = 0.3348, \quad g_0 = 1.81830.$$

The constraints g_1, g_2, g_3, g_4, g_5, g_6, g_7, g_8, g_9 are derived from equality constraints.

Problem 19 (Ref. 14)

$$\begin{aligned}\min(&2.0425t_1^{0.782} + 52.25t_2 + 192.85t_2^{0.9} + 5.25t_2^3 + 61.465t_6^{0.467}\\ &+0.01748t_4^{1.33}t_4^{-0.8} + 100.7t_4^{0.546} + 3.6610^{-10}t_3^{2.85}t_4^{-1.7}\\ &+0.00945t_5 + 1.0610^{-10}t_5^{2.8}t_4^{-1.8} + 116t_6 - 205t_6t_7 - 278t_2^3t_7),\end{aligned}$$

$$g_1 = 129.4t_2^{-3} + 105t_6^{-1} \leq 1,$$

$$g_2 = 1.0310^5t_2^3t_3^{-1}t_7t_8^{-1} + 1.210^6t_3^{-1}t_8^{-1} \leq 1,$$

$$g_3 = 4.68t_1^{-1}t_2^3 + 61.3t_1^{-1}t_2^2 + 160.5t_1^{-1}t_2 \leq 1,$$

$$g_4 = 1.79t_7 + 3.02t_2^3t_6^{-1}t_7 + 35.7t_6^{-1} \leq 1,$$

$$\begin{aligned}g_5 = &1.2210^{-3}t_3t_4^{-0.2}t_5^{-0.8}t_8 + 1.6710^{-3}t_3^{0.4}t_4^{-0.43}t_8\\ &+3.610^{-5}t_3t_4^{-1}t_8 + 2.10^{-3}t_3t_5^{-1}t_8 + 4.10^{-3}t_8 \leq 1.\end{aligned}$$

The solution is

$$t_1 = 5153.58, \quad t_2 = 6.64945, \quad t_3 = 169413,$$

$$t_4 = 743.39, \quad t_5 = 87998.7, \quad t_6 = 187.542,$$

$$t_7 = 0.124094, \quad t_8 = 29.2653, \quad g_0 = 17486.$$

The constraints g_1, g_2, g_3, g_4, g_5 are derived from equality constraints.

Problem 20 (Ref. 32)

$$\min(-0.28t_1t_6^{-1}+0.6732t_2t_6^{-1}+1.12t_3t_6^{-1}-31047.139t_6^{-1}+0.0074t_6^{-1}t_5+10).$$

$$g_1=0.73398t_4^{1.67}t_3^{-1}t_7t_{10}t_{11}\leq 1,$$

$$g_2=0.639926t_4^{-0.25}t_8t_{10}^{-1}-0.156564t_4^{0.42}t_9^{-1}t_{11}-0.1t_{10}t_{13}^{-1}\leq 1,$$

$$g_3=3809.973t_4^{-1.25}t_7^{-1}t_9^{-1}t_{10}^{-1}+0.195706t_4^{0.42}t_9^{-1}t_{11}\leq 1,$$

$$g_4=0.31254t_2^{-1}t_4^{1.25}t_9t_{10}t_{12}t_{13}^{-1}t_7\leq 1,$$

$$g_5=t_1t_4^{-1}t_7^{-1}t_8^{-1}t_9^{-1}-0.31254\,t_4^{0.25}t_{10}t_{13}^{-1}\leq 1,$$

$$g_6=0.02t_5^2t_6^{-1}t_7\leq 1,$$

$$g_7=t_{11}^{-1}t_{13}+1.25014t_4^{1.25}t_7t_9t_{10}t_{11}^{-1}-0.24466t_4^{1.67}t_7t_{10}\leq 1,$$

$$g_8=t_5^{-1}t_{12}+0.73398t_4^{1.67}t_5^{-1}t_7t_{10}t_{11}+t_5^{-1}t_{11}-t_5^{-1}t_{13}\leq 1,$$

$$g_9=t_{10}t_{12}^{-1}+0.24466t_4^{0.67}t_9^{-1}t_{10}t_{11}t_{12}^{-1}+0.15627t_4^{0.25}t_{10}^2t_{12}^{-1}t_{13}^{-1}+t_9t_{12}^{-1}+11t_{12}^{-1}t_{13}+1.5628t_4^{0.25}t_{10}t_{12}^{-1}\leq 1,$$

$$6.18\leq t_4\leq 129.53.$$

The solution is

$$t_1=13505,\quad t_2=36312,\quad t_3=3212,$$
$$t_4=110.78,\quad t_5=370411,\quad t_6=30.75,$$
$$t_7=1.1210^{-8},\quad t_8=47399,\quad t_9=147039,$$
$$t_{10}=7793,\quad t_{11}=19322,\quad t_{12}=362436,$$
$$t_{13}=14559,\quad g_0=-121.54.$$

The constraints g_1, g_2, g_3, g_4, g_5, g_6, g_7, g_8, g_9 are derived from equality constraints.

Problem 21 (Ref. 33)

$$\min(-0.063t_4t_7+5.04t_1+0.035t_2+10t_3+3.35t_5),$$

$$g_1=0.89286t_1^{-1}t_4-0.11756t_8+0.005955t_8^2\leq 1,$$

$$g_2=0.01741t_7-0.01912t_8+0.0006617t_8^2-0.0056596t_6\leq 1,$$

$$g_3=35.82t_9^{-1}-0.222t_9^{-1}t_{10}\leq 1,$$

$$g_4=0.333t_7^{-1}t_{10}+44.3333t_7^{-1}\leq 1,$$

$$g_5=1.020410^{-5}t_3^{-1}t_4t_6t_9+1.0204(10)^{-2}t_6\leq 1,$$

$$g_6 = 1.22 t_4 t_5^{-1} - t_1 t_5^{-1} \leq 1,$$

$$g_7 = t_1 t_2^{-1} t_8 - 1.22 t_2^{-1} t_4 + t_1 t_2^{-1} \leq 1,$$

$$t_1 \leq 2000, \quad t_2 \leq 19200, \quad t_3 \leq 120, \quad t_4 \leq 5000, \quad t_5 \leq 2000,$$
$$85 \leq t_6 \leq 93, \quad 90 \leq t_7 \leq 95, \quad 3 \leq t_8 \leq 12,$$
$$1.2 \leq t_9 \leq 4, \quad 145 \leq t_{10} \leq 162.$$

The solution is

$$t_1 = 1766, \quad t_2 = 18664, \quad t_3 = 95.12,$$
$$t_4 = 3087, \quad t_5 = 2000, \quad t_6 = 91.50,$$
$$t_7 = 94.83, \quad t_8 = 11.70, \quad t_9 = 2.19,$$
$$t_{10} = 151.48, \quad g_0 = -1237.55.$$

The constraints g_1, g_2, g_3, g_4, g_5, g_6, g_7 are derived from equality constraints.

Problem 22 (Ref. 34)

$$\min(2.8485 t_1 - 22.499 t_1 t_2 + 2.8952 t_1 t_3 + 0.3057 t_1 t_4 - 4.4318 t_1 t_5$$
$$+ 0.14 t_1 t_5^2 + 3.5974 t_1 t_6 + 0.05 t_1 t_7),$$

$$g_1 = 0.025616 t_1^2 t_7^{-1} + 0.293164 t_1^2 t_6 t_7^{-1} + 0.83877 t_1^2 t_6^2 t_7^{-1} \leq 1,$$

$$g_2 = 100 t_3 t_9^{-1} - 100 t_3 t_8^{0.01} t_9^{-1.01} + t_8 t_9^{-1} \leq 1,$$

$$g_3 = 0.4744 t_1^{-1} t_4 t_8^{-1} + 0.87564 t_1^{-1} t_4 t_6 t_8^{-1} + 0.012152 t_1 t_8^{-1}$$
$$+ 0.1391 t_1 t_6 t_8^{-1} + 0.3979 t_1 t_6^2 t_8^{-1} - 5.7222 t_6 \leq 1,$$

$$g_4 = 10.4351 t_1^{-1} t_4 t_5^{-1} t_9^{-1} 72.5476 t_5^{-1} + 5.6303 t_3 t_5^{-1} + 0.1279 t_4 t_5^{-1}$$
$$- 1.8459 t_6 - 133.9131 t_5^{-1} t_6 + 10.3930 t_3 t_5^{-1} t_6 + 0.2362 t_4 t_5^{-1} t_6$$
$$+ 19.2611 t_1^{-1} t_4 t_5^{-1} t_6 t_9^{-1} \leq 1,$$

$$g_5 = -4.44 t_5^{-1} + 41.04 t_2 t_5^{-1} + 5.63 t_3 t_5^{-1} + 0.1228 t_4 t_5^{-1} \leq 1,$$

$$g_6 = 3.309 \times 10^{-3} t_1 - 6.91 \times 10^{-3} t_1 t_3 - 4.858 \times 10^{-4} t_1 t_4 + 1.009 \times 10^{-2} t_1 t_5$$
$$- 1.294 \times 10^{-6} t_1^3 - 1.49 \times 10^{-5} t_1^3 t_6 - 4.237 \times 10^{-5} t_1^3 t_6^2$$
$$- 2.5322 \times 10^{-4} t_1 t_5^2 \leq 1,$$

$$g_7 = 0.4 t_6^{-1} \leq 1,$$

$$g_8 = 21.3351 t_4^{-1} - 1.8458 t_6 \leq 1,$$

$$g_9 = 0.002017 t_1 + 0.004878 t_1 t_2 + 0.005735 t_1 t_5 - 0.000744 t_1 t_3$$
$$- 0.000063 t_1 t_4 - 0.000019 t_1 t_7 \leq 1,$$

$$g_{10}=0.001817t_1+0.011287t_1t_2+0.010795t_1t_5+0.000013t_1t_7$$
$$-0.003304t_1t_3-0.000471t_1t_4-0.000471t_1t_4-0.000363t_1t_5^2\leq 1.$$

The solution is

$$t_1=11.7446,\quad t_2=0.36756,\quad t_3=0.3474,$$
$$t_4=12.277,\quad t_5=14.166,\quad t_6=0.4,$$
$$t_7=38.2906,\quad t_8=0.73187,\quad t_9=0.13111,$$
$$g_0=-375.784.$$

The constraints g_1, g_2, g_3, g_4, g_5 are derived from equality constraints.
The constraints g_6, g_7 are active inequality constraints.
The constraints g_8, g_9, g_{10} are inactive inequality constraints.

Problem 23 (Ref. 35)

$$\min(5.3578t_3^2+0.8357t_1t_5+37.2392t_1),$$
$$g_1=0.00002584t_3t_5-0.00006663t_2t_5-0.0000734t_1t_4\leq 1,$$
$$g_2=0.000853007t_2t_5+0.00009395t_1t_4-0.00033085t_3t_5\leq 1,$$
$$g_3=1330.3294t_2^{-1}t_5^{-1}-0.42t_1t_5^{-1}-0.30586t_2^{-1}t_3^2t_5^{-1}\leq 1,$$
$$g_4=0.00024186t_2t_5+0.00010159t_1t_2+0.00007379t_3^2\leq 1,$$
$$g_5=2275.1327t_3^{-1}t_5^{-1}-0.2668t_1t_5^{-1}-0.40584t_4t_5^{-1}\leq 1,$$
$$g_6=0.00029955t_3t_5+0.00007992t_1t_3+0.00012157t_3t_4\leq 1,$$
$$78\leq t_1\leq 102,\quad 33\leq t_2\leq 45,\quad 27\leq t_3\leq 45,$$
$$27\leq t_4\leq 45,\quad 27\leq t_5\leq 45.$$

The solution is

$$t_1=78,\quad t_2=33,\quad t_3=29.998,$$
$$t_4=45,\quad t_5=36.7673,\quad g_0=10127.13.$$

The constraints g_1, g_3, g_4, g_6 are inactive inequality constraints.
The constraints g_2, g_5 are active inequality constraints.

Problem 24 (Ref. 26)

$$\min(1.262626t_8+1.262626t_9+1.262626t_{10}-1.231059t_1t_8$$
$$-1.231059t_2t_9-1.231059t_3t_{10})$$
$$g_1=0.03475t_1t_4^{-1}+0.975t_1-0.00975t_1^2t_4^{-1}\leq 1,$$

$$g_2 = 0.03475 t_2 t_5^{-1} + 0.975 t_2 - 0.00975 t_2^2 t_5^{-1} \leq 1,$$

$$g_3 = 0.03475 t_3 t_6^{-1} + 0.975 t_3 - 0.00975 t_3^2 t_6^{-1} \leq 1,$$

$$g_4 = t_1 t_5^{-1} t_7^{-1} t_8 + t_4 t_5^{-1} - t_4 t_5^{-1} t_7^{-1} t_8 \leq 1,$$

$$g_5 = 0.002 t_2 t_9 + 0.002 t_5 t_8 + t_5 + t_6 - 0.002 t_1 t_8 - 0.002 t_6 t_9 \leq 1,$$

$$g_6 = t_2^{-1} t_3 t_9^{-1} t_{10} + t_2^{-1} t_6 + 500 t_9^{-1} - t_9^{-1} t_{10} - 500 t_2^{-1} t_6 t_9^{-1} \leq 1,$$

$$g_7 = 0.9 t_2^{-1} + 0.002 t_{10} - 0.002 t_2^{-1} t_3 t_{10} \leq 1,$$

$$g_8 = t_2 t_3^{-1} \leq 1,$$

$$g_9 = t_1 t_2^{-1} \leq 1,$$

$$g_{10} = 0.002 t_7 - 0.002 t_8 \leq 1,$$

$$0.1 \leq t_1 \leq 1, \qquad 0.1 \leq t_2 \leq 1, \qquad 0.9 \leq t_3 \leq 1,$$

$$t_4 \leq 0.1, \qquad 0.1 \leq t_5, \qquad 0.1 \leq t_6 \leq 0.9$$

$$500 \leq t_9, \qquad 0.1 \leq t_{10} \leq 500.$$

The solution is

$$t_1 = 0.8037, \qquad t_2 = 0.9, \qquad t_3 = 0.9,$$

$$t_4 = 0.1, \qquad t_5 = 0.1908, \qquad t_6 = 0.1908,$$

$$t_7 = 574.099, \qquad t_8 = 74.099, \qquad t_9 = 500,$$

$$t_{10} = 0.1, \qquad g_0 = 97.591034.$$

The constraints $g_1, g_2, g_3, g_4, g_5, g_6, g_7$ are derived from equality constraints.
The constraints g_8, g_9 are inactive inequality constraints.
The constraints g_{10} is an active inequality constraint.

7.2. Table of Random Numbers. These numbers are used to generate the different starting points:

0.53	0.47	0.98	0.11	0.15	0.98	0.03	0.61	0.22	0.17
0.59	0.52	0.64	0.02	0.38	0.40	0.57	0.73	0.93	0.51
0.43	0.21	0.16	0.92	0.55	0.97	0.34	0.47	0.03	0.28
0.58	0.11	0.69	0.19	0.64	0.26	0.24	0.04	0.46	0.43
0.83	0.28	0.79	0.73	0.91	0.92	0.82	0.37	0.75	0.78
0.66	0.02	0.33	0.82	0.77	0.74	0.52	0.37	0.11	0.18
0.84	0.89	0.21	0.39	0.56	0.98	0.89	0.99	0.23	0.15

7.3. Example of Starting-Point Calculation. Define

$a = 0.1 \times$ optimal value of variable,
$b = 1.9 \times$ optimal value of variable,
l.b. = lower bound for variable,
u.b. = upper bound for variable,
$u = a + (b - a) \times$ random number.

The calculation of the starting point is done as follows:

if $u <$ l.b., then initial value of variable = l.b.,
if l.b. $\leq u \leq$ u.b., then initial value of variable $= u$,
if u.b. $< u$ then initial value of variable = u.b.

For example, consider t_2 in Problem 24:

random number = 0.47, $u = 0.8514$,
initial value of $t_2 = 0.8514$.

References

1. DUFFIN, R. J., and PETERSON, E. L., *Geometric Programming with Signomials*, Journal of Optimization Theory and Applications, Vol. 11, pp. 3–35, 1973.
2. DUFFIN, R. J., PETERSON, E. L., and ZENER, C., *Geometric Programming*, John Wiley and Sons, New York, New York, 1967.
3. PASSY, U., *Generalized Weighted Mean Programming*, SIAM Journal of Applied Mathematics, Vol. 19, pp. 98–104, 1971.
4. DUFFIN, R. J., and PETERSON, E. L., *Reversed Geometric Programming Treated by Harmonic Means*, Indiana University Mathematics Journal, Vol. 22, pp. 531–550, 1972.
5. PASSY, U., and WILDE, D. J., *Mass Action and Polynomial Optimization*, Journal of Engineering Mathematics, Vol. 3, pp. 325–335, 1969.
6. KOCHENBERGER, G., *Geometric Programming: Extensions to Deal with Degrees of Difficulty and Loose Constraints*, University of Colorado, PhD Thesis, 1969.
7. RIJCKAERT, M. J., and MARTENS, X. M., *Numerical Aspects of the Use of Slack Variables in Geometric Programming*, Katholieke Universiteit Leuven, Chemical Engineering Department, Report No. CE-RM-7501, 1975.
8. ZANGWILL, W. I., *Nonlinear Programming, a Unified Approach*, Prentice Hall, Englewood Cliffs, New Jersey, 1969.
9. Harwell Subroutine Library, Theoretical Physics Division, Atomic Energy Research Establishment, Harwell, Berkshire, England, 1971.
10. KOCHENBERGER, A., WOOLSEY, R. E. D., and MCCARL, B. A., *On the Solution of Geometric Programming via Separable Programming*, Operations Research Quarterly, Vol. 24, pp. 285–296, 1973.

11. SARGENT, R. W. H., and MURTAGH, B. A., *Projection Methods for Nonlinear Programming*, Mathematical Programming, Vol. 4, pp. 245–268, 1973.
12. BRADLEY, J., *An Algorithm for the Numerical Solution of Prototype Geometric Programs*, Institute of Industrial Research and Standards, Dublin, Ireland, Report, 1973.
13. BECK, P. A., and ECKER, J. G., *A Modified Concave Simplex Algorithm for Geometric Programming*, Journal of Optimization Theory and Applications, Vol. 15, pp. 189–202, 1975.
14. BLAU, G., and WILDE, D. J., *A Lagrangian Algorithm for Equality Constrained Generalized Polynomial Optimization*, AIChE Journal, Vol. 17, pp. 235–240, 1971.
15. KUESTER, J. L., and MIZE, J. H., *Optimization Techniques with Fortran*, McGraw-Hill Book Company, New York, New York, 1973.
16. AVRIEL, M., and WILLIAMS, A. C., *Complementary Geometric Programming*, SIAM Journal on Applied Mathematics, Vol. 19, pp. 125–141, 1970.
17. AVRIEL, M., DEMBRO, R., and PASSY, U., *Solution of Generalized Geometric Programming*, International Journal of Numerical Methods in Engineering, Vol. 9, pp. 149–169, 1975.
18. TEMPLEMAN, A., *The Use of Geometric Programming Methods for Structural Optimization*, AGARD Lectures Series 17, Neuilly-sur-Seine, France, 1974.
19. JEFFERSON, T. J., *Geometric Programming with an Application to Transportation Planning*, Northwestern University, Urban Systems Engineering Center, Report, 1972.
20. RIJCKAERT, M. J., and MARTENS, X. M., *A Condensation Method for Generalized Geometric Programming*, Mathematical Programming, Vol. 11, pp. 89–93, 1976.
21. HAARHOFF, P. C., and BUYS, J. D., *A New Method for the Optimization of a Nonlinear Function Subject to Nonlinear Constraints*, Computer journal, Vol. 13, pp. 178–184, 1970.
22. BLAU, G. E., and WILDE, D. J., *Generalized Polynomial Programming*, Canadian Journal of Chemical Engineering, Vol. 47, pp. 317–326, 1969.
23. RIJCKAERT, M. J., *Engineering Applications of Geometric Programming*.Optimization and Design, Edited by M. Avriel, M. J. Rijckaert, and D. J. Wilde, Prentice Hall, Englewood Cliffs, New Jersey, pp. 195–220, 1973.
24. RIJCKAERT, M. J., *Studies in Neit-lineaire Programming*, Katholieke Universiteit Leuven, PhD Thesis, 1970.
25. ·HELLINCKX, L. J., and RIJCKAERT, M. J., *Minimization of Capital Investment for Batch Processes*, I1EC Process Design and Development, Vol. 10, pp. 422–423, 1971.
26. DEMBO, R. S., Technion, Haifa, Israel, Chemical Engineering Department, MS Thesis, 1972.
27. NEGHABAT, F., and STARK, R. M., *A Cofferdam Design Optimization*, Mathematical Programming, Vol. 3, pp. 263–276, 1972.
28. MINE, H., and OHNO, K., *Decomposition of Mathematical Programming Problems by Dynamic Programming and Its Application to Block Diagonal Geometric*

Programming, Journal of Mathematical Analysis and Applications, Vol. 32, pp. 370–385, 1970.
29. BECK, P. A., and ECKER, J. G., *Some Computational Experience with a Modified Convex Simplex Algorithm in Geometric Programming,* Rensselaer Polytechnic Institute, Report AFSC, 1972.
30. MARTENS, X. M., *Optimalisering van Stapsgewijs Uitgevoerde Processen met Behulp van Geometrische Programmering,* Katholieke Universiteit Leuven, PhD Thesis, 1971.
31. AVRIEL, M., and WILDE, D. J., *Optimal Condenser Design by Geometric Programming,* I and EC Process Design and Development, Vol. 6, pp. 256–263, 1967.
32. RIJCKAERT, M. J., and MARTENS, X. M., *Analysis and Optimization of the Williams–Otto Process by Geometric Programming,* AIChE Journal, Vol. 20, pp. 742–750, 1974.
33. BRACKEN, J. and MCCORMICK, G. P., *Selected Applications of Nonlinear Programming,* John Wiley and Sons, New York, New York, pp. 37–45, 1968.
34. RIJCKAERT, M. J., and MARTENS, X. M., Unpublished Material, 1977.
35. COLVILLE, A. R., *A Comparative Study of Nonlinear Programming Codes,* IBM New York Scientific Center, Report No. 320-2949, 1968.

14

Solving Geometric Programs Using GRG: Results and Comparisons[1]

M. RATNER,[2] L. S. LASDON,[3] AND A. JAIN[4]

Abstract. This paper describes the performance of a general-purpose GRG code for nonlinear programming in solving geometric programs. The main conclusions drawn from the experiments reported are: (i) GRG competes well with special-purpose geometric programming codes in solving geometric programs; and (ii) standard time, as defined by Colville, is an inadequate means of compensating for different computing environments while comparing optimization algorithms.

1. Introduction

This paper describes the performance of a generalized reduced gradient (GRG) algorithm in solving geometric programs. The code used, described in Refs. 1 and 2, is a general-purpose nonlinear programming (NLP) code and takes no advantage of the structure of geometric programs. First partial derivatives of the objective function and all constraint functions are required, and these are computed by simple forward difference approximations. All problem functions are expressed in power form, i.e., each term t_i has the form

$$t_i = c_i \prod_j x_j^{a_{ij}}.$$

[1] This research was partially supported by the Office of Naval Research under Contracts Nos. N00014-75-C-0267 and N00014-75-C-0865, the US Energy Research and Development Administration, Contract No. E(04-3)-326 PA-18, and the National Science Foundation, Grant No. DCR75-04544 at Stanford University; and by the Office of Naval Research under Contract No. N00014-75-C-0240, and the National Science Foundation, Grant No. SOC74-23808, at Case Western Reserve University.

[2] Programmer, Department of Academic Computing, Case Western Reserve University, Cleveland, Ohio.

[3] Professor, Department of Operations Research, Case Western Reserve University, Cleveland, Ohio.

[4] Senior Energy Modeling Analyst, Energy Center, SRI International, Menlo Park, California.

2. Problems Solved and Measures of Comparison

The geometric programs solved come from two sources: 8 problems given by Dembo in Ref. 3 and 24 problems given by Rijckaert and Martens in Ref. 4. Problem sizes are given in Table 1 below. The problems are good

Table 1. Problem size.

Problem	Number of variables	Number of constraints	Number of positive terms	Number of negative terms	Number of binding constraints at optimality
D1	12	3	31	0	3
D2	5	6	15	8	2
D3	7	14	31	13	5
D4A	8	4	14	2	4
D4B	8	4	14	2	4
D4C	8	5	16	0	5
D5	8	6	14	5	6
D6	13	13	27	12	11
D7	16	19	40	21	—
D8A	7	4	18	0	2
D8B	7	4	18	0	3
D8C	7	4	18	0	4
R1	4	2	6	0	2
R2	3	1	9	0	1
R3	4	1	12	0	1
R4	11	3	30	0	3
R5	4	3	8	0	3
R6	8	7	12	0	7
R7	8	7	12	0	6
R8	7	7	48	0	2
R9	2	1	4	1	1
R10	3	1	4	2	1
R11	4	2	6	1	2
R12	8	4	13	2	4
R13	8	6	14	5	6
R14	10	6	13	2	6
R15	10	7	12	3	7
R16	10	7	13	3	7
R17	11	9	14	5	9
R18	13	9	18	4	9
R19	8	5	26	2	5
R21	10	7	16	7	7
R22	9	10	36	21	7
R24	10	10	23	13	8

examples of small, dense, highly nonlinear NLP's. The problems with some negative terms are generalized geometric programs with signomial constraints. These may have local optima which are not global (such a point was encountered in Problem R18).

3. Measures of Comparison

In comparing GRG with the code used by Dembo in Ref. 3 (one of the better special-purpose GP codes) two measures were available—the final objective value obtained and the *standard time* required to achieve that value. Standard time is the execution time for the problem divided by the time required to execute a timing program written by Colville, Ref. 5. This program inverts a 40×40 matrix 10 times. Use of the standard time is supposed to compensate for the effects of different computing environments, e.g., machines, compilers, etc. To investigate this, we solved 4 problems on the IBM 370/168 at Stanford University using three different FORTRAN compilers: the FORTRAN H compiler (OPT = 2), the WATFIV compiler with the CHECK option, and the WATFIV compiler with the NOCHECK option. The results appear in Table 2, which gives the times required by GRG to solve four problems (with minimal printed output) divided by the time required to run the timing program. There is great variation in standard times between the three compilers, with widest variation (by factors of from 3 to 10) between WATFIV(CHECK) and the FORTRAN H compilers.

As pointed out by a referee, a large part of the variation is due to the difference in execution times of the Colville timing program. The actual execution times of Problems R2 and R9 (which are quite small) are about the same for all 3 compilers, suggesting that these runs are almost totally dominated by I/O time. One would not expect standard times to be useful in

Table 2. Standard execution times on three FORTRAN compilers.

Problem	WATFIV (CHECK)	WATFIV (NOCHECK)	IBM FORTRAN H (OPT = 2)
D4C	0.026	0.052	0.109
D5	0.025	0.049	0.069
R2	0.005	0.012	0.038
R9	0.003	0.007	0.033
Colville timing program (IBM 370/168 CPU, seconds)	41.80	16.83	3.91

such situations. Execution times of Problems D4C and D5 do vary considerably between the three compilers, however, indicating that the I/O time domination is not nearly as pronounced here, and the standard times are still in poor agreement with each other. A partial explanation for this is the fact that the standard timing program contains many indexing operations; hence, it should be more strongly affected by optimizing compilers like FORTH than an optimization code would be.

To compare with the other published results, we chose the WATFIV(NOCHECK) compiler, partly for convenience, partly because it gave the median times. In all GRG runs, there was no printing of intermediate results, but input data and final results were printed. In problems with run times less than one second, even this printing may consume a large fraction of total time.

Comparison with the Rijckaert and Martens results is difficult, since their starting points were chosen randomly, and were not published at the time the computations were made.[5] We chose our starting values so that

[5] A revised version of their paper includes the starting points.

Table 3. Starting values of variables used for the Rijckaert–Martens problems in GRG runs.

Problem	Coordinates
R1	41.0, 140, 0, 4.1, 2.1
R2	54.0, 126.0, 102.2
R3	375.0, 0.17, 0.73, 5.1
R4	1.25, 3.75, 3.8, 1.8, 3.9, 1.95, 2.14, 4.2, 0.85, 3.0, 3.3
R5	21.5, 67.0, 33.0, 1.6
R6	0.5, 0.3, 0.56, 1.1, 0.5, 1.05, 0.56, 1.5
R7	0.5, 0.3, 0.56, 1.1, 0.5, 1.05, 0.56, 1.5
R8	0.67, 1.5, 0.44, 1.38, 1.57, 0.6, 0.77
R9	0.41, 660.0
R10	44.1, 11.0, 0.65
R11	4.06, 1.23, 0.28, 2.82
R12	3.23, 1.32, 0.51, 8.95, 1.11, 0.9, 0.2, 8.3
R13	290.0, 2040.0, 2550, 0, 273.0, 148.0, 327.0, 143.0, 594.0
R14	1.05, 18.15, 3.95, 0.69, 0.18, 0.69, 5.22, 2.46, 0.60, 0.05
R15	0.36, 1.08, 0.36, 0.39, 0.09, 0.18, 0.1, 0.21, 0.05, 0.45
R16	0.36, 1.08, 0.36, 0.39, 0.09, 0.18, 0.1, 0.21, 0.05, 0.45
R17	3.5, 11.4, 3.6, 0.02, 0.4, 1.5, 0.19, 0.55, 0.18, 3.0, 0.23
R18	0.2, 0.21, 0.1, 0.96, 0.3, 0.5, 0.003, 0.04, 0.26, 2.8, 1.2, 0.24, 0.17
R19	2600.0, 9.9, 80,000.0, 1000.0, 44,000.0, 274.0, 0.06, 45.0
R21	900.0, 9000.0, 48.0, 4500.0, 1000.0, 90.0, 92.0, 6.0, 1.8, 150.0
R22	5.9, 0.5, 0.68, 6.1, 20.0, 0.2, 50.0, 0.36, 0.26
R24	0.4, 1.0, 0.9, 0.05, 0.38, 0.11, 1000.0, 37.0, 750.0, 0.2

odd-subscripted variables were one-half their optimal value, and even-subscripted variables were three halves their optimal value. The resulting points are shown in Table 3.

4. Computational Results

Table 4 shows the performance of GRG on the Dembo problems. Problem D1A was too badly scaled to attempt solution, and the code failed on Problems D6 and D7. In Problem D6, GRG terminated prematurely when no decrease in the objective function was achieved while attempting to move in the direction of steepest descent. In Problem D7, the program terminated short of feasibility at a local optimum of the Phase-I objective function.

Initial runs failed on Problem D3 also. Improved results were obtained by using an alternative pivoting strategy in computing the basis inverse. This strategy allowed pivoting on matrix elements smaller than allowed by the previous strategy if the alternative was entering a variable at a bound into the basis. This avoided degenerate bases in some cases, and allowed solution of Problem D3.

Table 5 shows the effects of varying GRG tolerances and solution strategy on Problems D8A–D8C. In the table headings, the term Pivalt refers to the alternative pivoting strategy described in the previous paragraph. TOL is the feasibility tolerance, i.e., all binding constraints must be within TOL of a bound. The terms H-reset and H-updated refer to the following. The code uses the BFS variable-metric method to minimize the reduced objective function. The strategy H-reset is to update the approximation H to the inverse Hessian matrix used by this method only when the line search terminates in an unconstrained optimum. Otherwise, it is reset to the identity matrix, and the search direction becomes the negative reduced gradient. The H-updated strategy uses the BFS update at each iteration, except those at which a basis change occurs.

As seen from the first two columns of Table 5, relaxing TOL from 10^{-6} (the Dembo value, used in the runs of Table 4) to 10^{-4} produces a reduction in total equivalent function calls from 6160 (the total of Problems D8A–D8C in Table 4) to 3456 and in total standard time from 0.919 to 0.553 units. Introducing the H-updated strategy produces further reductions of about 13% in function calls and in standard time (Columns 3–4, Table 5). Adding Pivalt but using $\mathrm{TOL} = 10^{-6}$ increases both totals significantly (Columns 5–6), but they are significantly better than those of Table 4. Finally, Columns 7–8 show the effects of incorporating both Pivalt and H-updated, and of starting with $\mathrm{TOL} = 10^{-4}$, running to termination, then

Table 4. Computational results for Dembo GP Problems, using specified constraint tolerances.

Problem	WATFIV time	Present standard time	Dembo standard time	Dembo optimum	Present optimum	n_f	n_g	n_e	Newton average*	Reason for termination
D1A			0.2747	4.8905E9	†					
D1B	0.94	0.055	0.2711	3.168213	3.169247	189	15	369	0.34	FC
D2	0.17	0.001	0.0024	10127.13	10122.44	17	6	47	1.14	KT
D3	0.96	0.057	0.0829	1227.18	1227.19	228	21	375	2.10	FC
D4A	0.65	0.038	0.2806	3.951698	3.951153	141	18	285	1.21	KT
D4B	0.58	0.034	0.1324	3.956197	3.951165	132	16	260	1.35	FC
D4C	0.89	0.052	0.0213	3.95207	3.95209	168	17	304	1.94	FC
D5	0.84	0.049	0.1255	7049.32	7049.24	174	16	302	1.88	FC
D6	F	F	0.3275	97.5910	261.16	304	21	577	2.74	ALPH-O
D7	F	F	0.2403	174.7888	‡					
D8A	4.75	0.282	0.0954	1809.762	1809.763	1398	72	1902	3.43	FC
D8B	3.28	0.194	0.0955	911.8796	911.8801	878	53	1249	2.82	FC
D8C	7.46	0.443	0.0792	543.6664	543.6681	2274	105	3009	4.04	FC

F = Failure.
FC = Fractional change in objective function less than 10^{-4} for 3 consecutive iterations.
KT = Kuhn–Tucker point found to within 10^{-4}.
ALPH-O = Premature termination: no function decrease in direction of steepest descent.
n_f = Number of function calls.
n_g = Number of gradient calls.
n_e = Equivalent function calls = $n_e + n_f + N \cdot n_g$, where N = number of variables.

* Average number of Newton iterations per attempt to solve for basic variables.
† Problem too badly scaled for GRG.
‡ Algorithm unable to find a feasible point.

Table 5. Results for Problems D8A–D8C with various GRG strategies.

Problem	$TOL = 10^{-4}$, no Pivalt, H-reset		$TOL = 10^{-4}$, no Pivalt, H-updated		$TOL = 10^{-6}$, Pivalt, H-updated		Initial $TOL = 10^{-4}$, final $TOL = 10^{-6}$, Pivalt, H-updated	
	Standard time*	Equivalent function calls	Standard time*	Equivalent function calls	Standard time*	Equivalent function calls	Standard time*	Equivalent function calls
D8A	0.235	1526	0.152	973	NA	1233	NA	1030
D8B	0.135	810	0.155	964	NA	1414	NA	1053
D8C	0.183	1120	0.170	1051	NA	1922	NA	1101
Total	0.553	3456	0.477	2988	0.768	4569	0.569	3184

* WATFIV(NOCHECK).
NA=Not available.

Table 6. Computational results* for Rijckaert and Martens GP problems.†

Problem	WATFIV time	Present standard time	RM standard time	RM optimum	Present optimum	n_f	n_g	n_e	Newton average	Reason for termination	Notes
R1	0.52	0.030	0.003	0.01210	0.01210	388	37	536	1.52	KT	‡
R2	0.14	0.008	0.002	6300	6299.7	64	9	91	1.39	FC	
R3	0.48	0.028	0.004	126,344	126,306	194	25	294	0.0	FC	
R4	0.48	0.028	0.039	3.1681	3.1442	118	9	172	1.40	FC	
R5	0.24	0.014	0.003	623,015	623,277	24	14	140	0.16	FC	
R6	1.36	0.080	0.017	29.5985	29.2282	207	20	367	1.26	FC	
R7	1.36	0.080	0.028	29.5985	29.2256	245	20	405	1.62	FC	
R8	0.93	0.055	0.221	178.478	181.370	253	21	400	3.15	FC	
R9	0.11	0.006	0.001	11.91	11.9002	29	5	39	0.70	KT	
R10	0.18	0.010	0.001	−83.21	−83.26	124	13	163	1.74	FC	
R11	0.25	0.014	0.003	−5.7398	−5.7398	109	13	161	1.08	FC	
R12	0.89	0.052	0.013	−6.0482	−6.0483	219	25	419	1.26	FC	
R13	1.26	0.075	0.026	7049.24	7082.93	361	28	577	2.08	FC	
R14	6.11	0.363	0.024	1.1436	1.1436	1819	112	2939	3.15	FC	
R15	1.10	0.065	0.018	0.2015	0.20566	272	18	452	2.37	KT	
R16	1.16	0.069	0.019	0.1966	0.1966	302	19	492	2.53	KT	
R17	4.03	0.239	0.022	0.1406	0.1406	939	46	1445	1.99	FC	‡
R18	F	F	0.034	1.81818				Could not find feasible point			
R19	1.51	0.089	0.067	17,486.0	17,485.9	311	27	527	1.94	KT	
R21	1.62	0.096	0.094	−1237.55	−1252.8	260	33	590	1.04	FC	
R22	1.68	0.100	0.175	−375.784	−379.96	241	27	484	2.03	FC	
R24	1.05	0.062	0.175	97.591	97.569	140	18	320	0.75	KT	

* The feasibility tolerance used was 10^{-4}, in contrast to the stricter tolerance of 10^{-5} used by Rijckaert and Martens. This difference would tend to bias results in favor of GRG.

† Problem R23 was the same as D2, so it was not rerun. Problem R20 had an unresolved typographical error.

‡ Tolerance controlling termination had to be tightened by a factor of 10.

starting at that point with TOL = 10^{-6} (initially infeasible, so in Phase I), and running to termination again. This produces results almost as good as those with TOL = 10^{-4} throughout.

The performance of GRG (including both Pivalt and H-updated) on the Rijckaert–Martens problems is shown in Table 6. The column dealing with reported standard time contains the best standard time reported by Rijckaert and Martens (Ref. 6) in a comparison of eleven special-purpose codes for geometric programming and one general purpose code. GRG was generally slower than the best code and missed the true optimum by one-to-two percent in Problems R8, R13, and R15. Otherwise, GRG solved all these problems satisfactorily.

Note that GRG is competitive or superior in its standard time on the larger problems Problems R19 through R24. Since all times except two are on the order of 1 second, the printing of some output by GRG (which may consume a large fraction of run time in these cases), plus the previously mentioned difficulties with using standard times, imply that these comparisons must be viewed with some caution.

An enhancement of the GRG code, described in Ref. 6, uses quadratic extrapolation to compute initial estimates of basic variables prior to solution of the nonlinear constraint equations in contrast to tangent vector extrapolation (Ref. 1) used in the runs described above. Some of the Dembo and Rijckaert–Martens problems were used in tests to compare the two extrapolation schemes. The results, displayed in Table 7 (which exhibits

Table 7. Performance of GRG using tangent vector vs. quadratic extrapolation.

Problem	Equivalent function calls (TV)	Equivalent function calls (QUAD)	Newton average (TV)	Newton average (QUAD)	Standard time (TV)	Standard time (QUAD)
D2	47	47	1.14	1.14	0.0107	0.0107
D4A	285	268	1.24	0.91	0.0452	0.0446
D5	277	257	2.43	2.00	0.0446	0.0440
D8A	1902	1488	4.26	3.16	0.3072	0.2496
R6	383	369	1.96	1.64	0.0879	0.0868
R8	400	331	3.51	2.54	0.1872	0.0529
R12	419	346	1.31	0.98	0.0600	0.0529
R14	1120	822	2.66	2.29	0.1907	0.1467
R15	452	433	2.50	2.14	0.0719	0.0707
R16	492	451	2.69	1.89	0.0755	0.0766
Total	5777	3500	2.84*	2.20*	1.0808	0.8354

* Average values for each column.

small discrepancies with the results in Tables 4–5, owing to minor differences in tolerances and strategies used), show the superiority of quadratic extrapolation for these problems. In the table headings, TV represents tangent-vector extrapolation and QUAD represents quadratic extrapolation.

5. Conclusions

Conclusions to be drawn from these experiments are as follows.

(i) Uncritical use of *standard time* (as defined by Colville in Ref. 1) to compare algorithms can be misleading. I/O time and the effect of optimizing compilers are two confounding factors. Further experiments on larger problems are needed to assess these factors and to develop improved procedures.

(ii) GRG, representing the class of general-purpose NLP algorithms, competes well with special-purpose geometric programming codes in solving geometric programs. GRG failed on two of the most difficult of Dembo's problems, so this implementation of GRG may not be as robust for GP's as the better GP codes. However, some failures may be due to the use of numerical derivatives, and analytic derivatives may be computed quite simply, as discussed later. Improved implementations of GRG are currently being developed. The authors hope to report on their performance in the near future.

(iii) Certain modifications in solution strategy can strongly affect the performance of GRG. Among these are the following: when the approximate Hessian matrix is reset; the logic used in basis inversion to decide when a variable at bound is to enter the basis; and the order of extrapolation (linear or quadratic) used to obtain initial estimates of the basic variables.

(iv) Certain parameter settings strongly affect GRG performance, in particular, the tolerance used to determine which constraints are binding and the tolerance used to terminate the algorithm.

Of course, strictly speaking, these conclusions are valid only for the subset of GP problems solved here. We believe, however, that they will also hold in future experiments.

In closing, we note some things left undone but worth doing. GRG could easily be made more convenient and efficient on geometric programs by coding a special subroutine to compute first partial derivatives. This would use the fact that, if the ith term in the program is

$$t_i = c_i \prod_{j=1}^{n} x_j^{a_{ij}},$$

then

$$\partial t_i/\partial x_k = a_{ik}t_i/x_k.$$

Hence, if the terms are stored when computing the constraint values and objective value, their partial derivatives are available with little additional effort. This would reduce the time required to compute the gradient of a function from the time required in these runs (nt_f, where t_f is the time required to evaluate the function and n is the number of variables) to little more than t_f. Special input subroutines could be coded to enable the user to specify the problem by inputting only (a) the constants c_i, (b) the exponent matrix a_{ij}, and (c) which terms appear in which problem functions. Currently, all problem functions must be coded directly. These enhancements would transform GRG into a *special-purpose* geometric programming code.

Some additional experiments appear useful. Geometric programs can be transformed into exponential form by the change of variables

$$x_j = \exp(y_j),$$

which transforms the ith term into

$$t_i = c_i \exp\left(\sum_j a_{ij}y_j\right).$$

Evaluation of t_i then requires only one transcendental computation, rather than one for each fractional a_{ij}. In addition, y_j is a free variable (if x_j has no upper bound), and the problem functions become convex if all c_i are positive. Some problems should be solved using both forms, to see which yields smallest solution times. In addition, tests of GRG and some good geometric programming codes should be run on the same computer, in order to remove the factor of standard times from obscuring the comparisons.

References

1. LASDON, L. S., WAREN, A., JAIN, A., and RATNER, M. W., *Design and Testing of a Generalized Reduced Gradient Code for Nonlinear Programming*, ACM Transactions on Mathematical Software, Vol. 4, pp. 34–50, 1978.
2. LASDON, L. S., WAREN, A., JAIN, A., and RATNER, M. W., *GRG System Documentation*, Cleveland State University, Cleveland, Ohio, Computer and Information Science Department, Technical Memorandum No. CIS-75-01, 1975.
3. DEMBO, R., *A Set of Geometric Programming Test Problems and Their Solutions*, University of Waterloo, Waterloo, Canada, Department of Management Sciences, Working Paper No. 87, 1974.

4. RIJCKAERT, M. J., and MARTENS, X. M., *A Comparison of Generalized Geometric Programming Algorithms*, Katholieke Universiteit Leuven, Leuven, Belgium, Report No. CE-RM-7503, 1975.
5. COLVILLE, A. R., *A Comparative Study of Nonlinear Programming Codes*, IBM, New York Scientific Center, Report No. 320-2949, 1968.
6. JAIN, A., *The Solution of Nonlinear Programs Using the Generalized Reduced Gradient Method*, Stanford University, Stanford, California, Department of Operations Research, Systems Optimization Laboratory, Technical Report No. SOL 76-6, 1976.

15

Dual to Primal Conversion in Geometric Programming[1,2]

R. S. DEMBO[3]

Abstract. The aim of this paper is not to derive new results, but rather to provide insight that will hopefully aid researchers involved in the design and coding of algorithms for geometric programs. The main contributions made here are: (i) a computationally useful interpretation of the Lagrange multipliers associated with the dual orthogonality constraints, (ii) a computationally useful interpretation of the Lagrange multiplier associated with the dual normality constraint, and (iii) an analysis of the much-avoided issue of subsidiary problems.

1. Introduction

One of the biggest drawbacks associated with solving geometric programming (GP) problems via the GP dual lies in the conversion of an optimal dual solution to an optimal primal solution. There are a number of reasons for this; some have to do with accuracy and numerical stability, whereas others stem from the fact that the extremality conditions relating the primal and dual at optimality (Ref. 3, p. 81) are insufficient to determine an optimal solution to the primal, in which case a subsidiary dual problem must be solved (Refs. 3, 4). These difficulties have led many GP researchers to view a direct solution of the primal as the most efficient approach to the numerical solution of GP problems. However, with the desire to solve larger and larger GP problems, the importance of numerical methods for the dual become

[1] This work was supported in part by the National Research Council of Canada, Grant No. A3552.

[2] The author would like to acknowledge the contribution of an anonymous referee, whose constructive criticism led to this improved version of the original paper.

[3] Assistant Professor of Operations Research, School of Organization and Management, Yale University.

more apparent. This is because sparsity can be handled more directly in the dual. Also, the linear constraints of the dual are easier to deal with than the nonlinear constraints of the primal.

This paper discusses the theory and methodology that has been developed for converting an optimal dual solution to an optimal primal solution. The relationship between the dual multipliers and the primal variables at optimality is presented in Section 2. It underscores the desirability of algorithms for the dual that compute estimates of the optimal dual multipliers. These are essentially estimates of the optimal primal variables and can be extremely useful in termination criteria. Subsidiary problems are discussed in Section 3.

2. Relationship Between Primal Variables and Dual Multipliers at Optimality

Using the notation of Ref. 3, the geometric dual problem B may be formulated as the following convex programming problem.

Problem B: Maximize

$$V(\delta)=\sum_{i=1}^{n}\delta_i\log(c_i/\delta_i)+\sum_{k=0}^{p}\lambda_k\log\lambda_k, \tag{1}$$

subject to

$$\lambda_0=1, \tag{2}$$

$$\sum_{i=1}^{n}a_{ij}\delta_i=0, \qquad j=1,2,\ldots,m, \tag{3}$$

$$\delta_i\geq 0, \qquad i=1,2,\ldots,n, \tag{4}$$

where

$$\lambda_k=\sum_{[k]}\delta_i, \qquad k=0,1,\ldots,p. \tag{5}$$

Assigning Lagrange multipliers ω_j, $j=0,1,2,\ldots,m$, to the normality and orthogonality conditions (2) and (3) yields the Lagrangian

$$L(\delta;\,\omega_0,\omega)=V(\delta)+\omega_0(\lambda_0-1)+\sum_{j=1}^{n}\omega_j\left(\sum_{i=1}^{m}a_{ij}\delta_i\right). \tag{6}$$

Theorem 2.1. (*Ref. 3*). Let δ^* be an optimal solution to the dual program B, and let

$$[\omega_0^*,\omega^*]^T=[\omega_0^*,\omega_1^*,\omega_2^*,\ldots,\omega_m^*]^T$$

be a Kuhn–Tucker vector, where ω_0^* is associated with the normality constraint (2) and ω_j^*, $j=1,2,\ldots,m$, is associated with the jth orthogonality constraint (3). Then,

$$\omega_0^* = -\log g_0(t^*), \tag{7}$$

$$\omega_j^* = \log t_j^*, \qquad j=1,2,\ldots,n, \tag{8}$$

where t^* is a primal optimal solution.[4]

Proof. It is clear that $\partial L/\partial\delta_i$ exists iff $\delta_i > 0$, in which case

$$\partial L/\partial\delta_i = \begin{cases} \log(c_i\lambda_0/\delta_i) + \omega_0 + \sum_{j=i}^{m} a_{ij}\omega_j & \text{if } i\in[0], \quad (9)\\ \log(c_i\lambda_k/\delta_i) + \sum_{j=1}^{m} a_{ij}\omega_j & \text{if } i\in[k], \quad k=1,2,\ldots,p. \quad (10)\end{cases}$$

The extremality conditions (Ref. 3, p. 81) that relate the primal geometric program (Problem A) to the above dual program (Problem B) can clearly be written as

$$0 = \begin{cases} \log(c_i/\delta_i) - V(\delta) + \sum_{j=1}^{m} a_{ij}\log t_j & \text{if } i\in[0], \quad (11)\\ \log(c_i\lambda_k/\delta_i) + \sum_{j=1}^{m} a_{ij}\log t_j & \text{if } i\in[k], \quad k=1,2,\ldots,p, \quad (12)\end{cases}$$

$\lambda_k > 0.$

Assuming that Problems A and B are canonical (Ref. 3, p. 173) and that Problem B has an optimal solution δ^*, it is known that Problem A also has at least one optimal solution t^*. Moreover, the extremality conditions (11) and (12), with $\delta = \delta^*$ and $\lambda = \lambda^*$, characterize all optimal solutions t^* to Problem A' which result from deleting the inactive constraints of Problem A [i.e., all constraints $g_k(t) \le 1$ for which $\lambda_k^* = 0$].

Under the above assumptions, it is well known that there is no duality gap, and hence

$$\log g_0(t^*) = V(\delta^*). \tag{13}$$

This, together with relationships (9), (10), (11), and (12), clearly implies that

---

[4] It should be noted that if λ_0 is eliminated from Problem B and constraint (2) is replaced by $\sum_{[0]} \delta_j = 1$, then expression (7) becomes:

$$\omega_0^* = 1 - \log g(t^*)$$

t^* is an optimal solution to problem A′ iff

$$-\log g_0(t^*) = \omega_0^*, \tag{14}$$

$$\log t_j^* = \omega_j^* \tag{15}$$

are optimal multipliers for the corresponding geometric dual Problem B′, in which case $g_k(t^*) = 1$ for each constraint in Problem A′. □

This theorem points to the importance of computing the dual multipliers when using a dual-based algorithm for GP. Two important uses these multipliers can be put to are the following.

(i) *Recovery of a Solution to the Primal.* From (15), the optimal primal variables may be recovered from

$$t_j^* = \exp(\omega_j^*), \qquad j = 1, \ldots, m. \tag{16}$$

At intermediate stages in the algorithm, (16) may be used to convert an estimate of the multiplier vector ω into first-order estimates of the primal variables. This then provides a means of simultaneously monitoring the progress toward both a primal and a dual optimal solution. Furthermore, it then becomes relatively easy to check primal feasibility criteria in the vicinity of the optimum and provides an ironclad test for optimality, namely, equality between primal and dual objective function values.

(ii) *Checking the Accuracy of a Solution.* From (16), the optimal primal and dual objective function values may be recovered from:

$$g_0(t^*) = \exp(-\omega_0^*), \tag{17}$$

$$V(\delta^*) = -\omega_0^*. \tag{18}$$

Since these relationships are only true at the optimum and depend only on ω_0^*, these provide an independent way of computing $V(\delta^*)$ and $g_0(t^*)$ which may be checked against a value computed by direct substitution. Any significant disagreement in these independently computed values is a sign that optimality has not yet been reached.

At points other than the optimum, (17) and (18) may be used to compute first-order estimates of $g_0(t)$ and $V(\delta)$. This could reduce the number of function evaluations required by the particular algorithm under consideration, especially in the vicinity of an optimal solution.

3. Subsidiary Problems

The conditions of the above theorem essentially state that the set of equations (9), (10) hold simultaneously only at Kuhn–Tucker (KT) points and are inconsistent everywhere else. Furthermore, at KT points, the

Lagrange multipliers (and hence the primal variables) are ***uniquely*** determined if the exponent matrix in Problem A′ is of rank m. This is a rather unrestrictive requirement when all the optimal dual variables are positive (that is, when Problem A = Problem A′), in view of page 83 of Ref. 3.

Subsidiary problems are needed only when the rank of the exponent matrix in Problem A′ is less than m. They have been almost totally ignored in the mathematical programming literature, except for the theoretical developments of Duffin, Peterson, and Zener (Ref. 3) and Rockafeller (Ref. 4), and the computational implementation of the former by Beck and Ecker (Ref. 1, p. 179). With the exception of the code developed in Ref. 1, the following simple *canonical* GP example will probably cause every other available dual-based GP code to fail.

Problem A: Minimize

$$g_0(t) = t_1 t_2 + t_1^{-1} t_2^{-1},$$

subject to

$$g_1(t) = 0.25 t_1^{0.5} + t_2 \leq 1,$$

$$t_1, t_2 \geq 0.$$

Problem B: Maximize

$$V(\delta) = \delta_1 \log(1/\delta_1) + \delta_2 \log(1/\delta_2) + \delta_3 \log(0.25/\delta_3)$$
$$+ \delta_4 \log(1/\delta_4) + \lambda_0 \log \lambda_0 + \lambda_1 \log \lambda_1,$$

subject to

$$\lambda_0 = 1,$$

$$\delta_1 - \delta_2 + 0.5\delta_3 = 0,$$

$$\delta_1 - \delta_2 + \delta_4 = 0,$$

$$\delta_1, \delta_2, \delta_3 \geq 0,$$

where

$$\lambda_0 = \delta_1 + \delta_2,$$

$$\lambda_1 = \delta_3 + \delta_4.$$

The unique optimal solution to the dual is

$$\delta^* = [1/2, 1/2, 0, 0] \qquad \text{and} \qquad V(\delta^*) = \log(2).$$

In keeping with our notation in Theorem 2.1, the primal Problem A′ is the above Problem A with the constraint $g_1(t) \leq 1$ deleted. Its corresponding dual (Problem B′) is the above dual problem with the variables δ_3 and δ_4 (and hence λ_1) deleted. It is easy to see that, in this case, Problem A′ has an exponent matrix of rank $1 < m = 2$; hence, a subsidiary problem must be solved in order to compute a primal optimal solution. Exactly what this implies, in terms of the original primal Problem A, is made clear below.

The optimal multipliers (ω_1^*, ω_2^*) may be computed from Eqs. (9) and (10). In the above example, this gives

$$\omega_0^* + \omega_1^* + \omega_2^* = \log(\tfrac{1}{2}), \tag{19}$$

$$\omega_0^* - \omega_1^* - \omega_2^* = \log(\tfrac{1}{2}), \tag{20}$$

which has the solution

$$\omega_0^* = \log(\tfrac{1}{2}), \tag{21}$$

$$\omega_1^* + \omega_2^* = 0. \tag{22}$$

From (17),

$$g_0(t^*) = \exp(-\omega_0^*) = 2,$$

which equals the value of $V(\delta^*)$ obtained by substituting

$$\delta^*[\tfrac{1}{2}, \tfrac{1}{2}, 0, 0]$$

in the dual objective function. Equation (22) shows the dependence of ω_1^* on ω_2^*, and hence of t_1^* on t_2^*. Specifically,

$$\omega_1^* = -\omega_2^*, \qquad \text{and therefore} \qquad t_1^* = 1/t_2^*. \tag{23}$$

In this example, there are an infinite number of pairs of values (ω_1^*, ω_2^*) that the optimal Lagrange multipliers ω_1^* and ω_2^* may assume and still satisfy (22). Each of these pairs of values yields a corresponding pair of primal variables (t_1^*, t_2^*) obtained from the transformation in (16). Observe that, since

$$t_1^* = 1/t_2^*,$$

the primal objective value is

$$g_0(t^*) = 2 = \log V(\delta^*)$$

for each and every pair (t_1^*, t_2^*) derived from a solution to (22). However, not all of these possible pairs of values will satisfy the primal constraint

$$g_1(t) = 0.25t_1^{0.5} + t_2 \le 1.$$

For example, a generalized inverse solution to

$$\omega_1^* + \omega_2^* = 0$$

would be

$$\omega_1^* = \omega_2^* = 0,$$

implying

$$t_1^* = t_2^* = 1.$$

With $t_1^* = t_2^* = 1$,

$$g_0(t^*) = \log V(\delta^*) = 2;$$

however,

$$g_1(t^*) > 1,$$

and hence the above point is *not feasible*. This is precisely what gives rise to subsidiary problems. The extremality conditions are insufficient to determine the optimal dual multipliers (and hence the optimal primal variables) uniquely. The problem then is to find a set of multipliers, satisfying the extremality conditions, that will generate a *feasible* and therefore optimal primal solution.

A general approach to subsidiary problems is given below. Let

$$T \triangleq \{k | \lambda_k^* > 0\}.$$

this is equivalent to saying that T is the set of active primal constraints at the optimum. Essentially, the *subsidiary problem* that must be solved is the Phase-1 problem: Find a $t^* \geq 0$ such that

$$g_0(t^*) = \log V(\delta^*), \tag{24}$$

$$g_k(t^*) = 1, \qquad k \in T, \tag{25}$$

$$g_k(t^*) \leq 1, \qquad k \notin T. \tag{26}$$

Since, by Theorem 2.1, (24) and (25) hold for *any* solution to the extremality conditions (11) and (12), an *equivalent* subsidiary (Phase 1) problem may be obtained by replacing (24) and (25) with the corresponding extremality equations. This gives the following problem.

Problem SP: Find a $t \geq 0$ such that

$$c_i \prod_{j=1}^{m} t_j^{a_{ij}} = \begin{cases} \delta_i^* \exp[V(\delta^*)], & i \in [0], \qquad\qquad (27) \\ \delta_i^*/\lambda_k^*, & i \in [k], \quad k \in T, \qquad (28) \end{cases}$$

$$g_k(t) \leq 1, \qquad k \notin T. \tag{29}$$

Problem SP is more convenient than the one given by (24), (25), and (26), since the posynomial constraints (24) and (25) have been decomposed into equivalent *single-term constraints* (27) and (28). At this point, it is instructive to show how the methods of Duffin, Peterson, and Zener (Ref. 3, p. 179) and Rockafeller (Ref. 4) differ in their approach to solving the above Phase-1 problem. Duffin, Peterson, and Zener (Ref. 3) require the solution of a sequence of subsidiary problems. The reason for this is that each of their subsidiary problems is essentially a problem of the following form.[5]

Problem SPDPZ: Minimize

$$t_0, \tag{30}$$

subject to

$$c_i \prod_{j=1}^{m} t_j^{a_{ij}} = \begin{cases} \delta_i^* \exp[V(\delta^*)], & i \in [0], \quad (31) \\ \delta_i^*/\lambda_k^*, \quad i \in [k], & k \in T, \quad (32) \end{cases}$$

$$g_k(t) \le t_0, \qquad k \notin T, \tag{33}$$

$$t_0,\ t_i \ge 0, \qquad i = 1, \ldots, m. \tag{34}$$

Problem SPDPZ guarantees that at least *one* of the constraints

$$g_k(t) \le t_0, \qquad k \in t,$$

will be active at its optimum. This increases the number of linearly independent equations in (27) and (28) by the number of terms in the constraint that has just become active. However, it is still possible that the augmented extremality conditions constitute an undetermined system, in which case the above procedure is repeated. Since at each iteration *at least one* additional constraint becomes active, only a finite number of subsidiary problems have to be solved. The sequence of problems eventually results in a system of equations that is sufficient to determine an optimal primal solution t^*, that is, a solution that satisfies the system (24), (25), and (26).

It is worth noting that the above approach was developed for canonical geometric programs (Ref. 3, p. 173) and might fail on degenerate geometric programs. Consider the following example.

Problem A: Minimize

$$g_0(t) = t_1 t_2 + t_1^{-1} t_2^{-1},$$

subject to

$$g_1(t) = 2t_1^2 \le 1,$$

$$t_1,\ t_2 \ge 0.$$

[5] The analysis in Ref. 3 is based entirely on the dual problem; it is part of an existence proof and was not originally intended for computational purposes. We have considered the primal here, since it provides insight into the true computational *Phase 1* nature of subsidiary problems.

Problem B: maximize

$$V(\delta)=\delta_1\log(1/\delta_1)+\delta_2\log(1/\delta_2)$$
$$+\delta_3\log(2/\delta_3)+\lambda_0\log\lambda_0+\lambda_1\log\lambda_1,$$

subject to

$$\lambda_0=1,$$
$$\delta_1-\delta_2+2\delta_3=0,$$
$$\delta_1-\delta_2=0,$$
$$\delta_1,\ \delta_2,\ \delta_3\geq 0,$$

where

$$\lambda_0=\delta_1+\delta_2,$$
$$\lambda_1=\delta_3.$$

This primal–dual pair is degenerate since, in every feasible solution to the dual, $\delta_3=0$. The optimal solution is

$$\delta^*=[\tfrac{1}{2},\tfrac{1}{2},0],$$

which yields the same optimal multiplier equations (21) and (22) as in the previous example. The subsidiary problem SPDPZ would be as follows.

Problem SPDPZ: Minimize

$$t_0,$$

subject to

$$t_1,\ t_2=1,$$
$$t_1^{-1}t_2^{-1}=1,$$
$$2t_1^2\leq t_0,$$
$$t_0,\ t_1,\ t_2\geq 0.$$

In this example, the optimal solution to Problem SPDPZ is unbounded (its dual is inconsistent); hence, it cannot be used to recover an optimal solution of Problem A. This difficulty is not encountered with Rockafeller's subsidiary problem below.

Rockafeller's approach (Ref. 4) is to solve a single subsidiary problem (an approach that is computationally oriented and is not intended to also provide an existence proof). That is, he constructs the following *single problem* that *uniquely* determines a solution to (24), (25), and (26).

Problem SPR: Minimize

$$(\tfrac{1}{2}) \sum_{j=1}^{m} t_j^2, \tag{35}$$

subject to

$$(24), (25), (26), \text{ and } t \geq 0.$$

The rationale behind his choice of objective function can be explained by the fact that it adds m linearly independent columns to the exponent matrix, which definitely increases its rank to m and thereby results in a set of extremality conditions that *uniquely* determines a vector t satisfying (24), (25), and (26). It is this uniqueness as well as the assumed consistency of the primal that results in Problem SPR yielding an *optimal* solution to Problem A satisfying *all* of the constraints (24), (25), and (26).

From a computational point of view, Problem SPR is probably better than solving a sequence of Problems SPDPZ. There are some other factors worth noting.

(i) There are many different objective functions that would have the same desired effect as the one in Problem SPR. Once again, it is not obvious which of these would be most efficient computationally.

(ii) In the special case where all of the constraints $g_k(t) \leq 1$, $k \notin T$, are single terms, the Phase-1 Problem SP is linear in $\log t$ and may be solved using linear programming techniques.

References

1. BECK, P. A., *A Modified Concave Simplex Algorithm for Geometric Programming*, Journal of Optimization Theory and Applications, Vol. 15, pp. 189–202, 1975.
2. DEMBO, R. S., *Sensitivity Analysis in Geometric Programming*, Yale University, School of Organization and Management, Working Paper No. SOM-35, 1978.
3. DUFFIN, R. J., PETERSON, E. L., and ZENER, C., *Geometric Programming*, John Wiley and Sons, New York, New York, 1967.
4. ROCKAFELLER, R. T., *Convex Programming Problems with Linearly Constrained Duals*, Nonlinear Programming, Edited by J. B. Rosen, *et al.*, Academic Press, New York, New York, 1970.

16

A Modified Reduced Gradient Method for Dual Posynomial Programming[1,2]

J. G. ECKER,[3] W. GOCHET,[4] AND Y. SMEERS[5]

Abstract. In this paper, we present a method for solving the dual of a posynomial geometric program based on modifications of the reduced gradient method. The modifications are necessary because of the numerical difficulties associated with the nondifferentiability of the dual objective function. Some preliminary numerical results are included that compare the proposed method with the modified concave simplex method of Beck and Ecker of Ref. 1.

1. Introduction

Beck and Ecker (Ref. 1) presented an algorithm for the solution of the dual of a posynomial programming problem. The algorithm is in essence Zangwill's convex simplex method (Ref. 2), with some important modifications required because of two basic problems: (i) the nondifferentiability of the objective function at a point where some of the variables are equal to zero; and (ii) the fact that each *block of dual variables* must have all

[1] This research was supported in part by the National Science Foundation, Grant No. MCS75-09443-A01.

[2] The authors sincerely thank P. Vandeputte for his computing assistance. We also wish to thank T. Magnanti and unknown referees for helpful criticism.

[3] Associate Professor, Department of Mathematical Sciences, Rensselaer Polytechnic Institute, Troy, New York.

[4] Associate Professor, Department of Applied Economics, Katholieke Universiteit Leuven; and Center for Operations Research and Econometrics, Catholic University of Leuven, Leuven, Belgium.

[5] Associate Professor, Department of Engineering, Universite Catholique de Louvain, and Center for Operations Research and Econometrics, Catholic University of Louvain, Louvain, Belgium.

components positive or all components equal to zero at optimality. Both these problems will be more fully discussed in Section 2.

Essentially, the modifications in Beck and Ecker (Ref. 1) are based on the following: whereas in the convex simplex method only one independent (nonbasic) variable is allowed to change at every iteration, Beck and Ecker show how a complete block of variables can be allowed to move to or away from zero at certain iterations. As was expected, computational experience has shown that the iterations involving block changes result in significantly greater increases in the objective function value than the (standard convex simplex) iterations involving a change in a single nonbasic variable.

In this paper, we show how Wolfe's reduced gradient method (Ref. 3) can be modified to overcome the difficulties (i) and (ii) above. The reduced gradient method allows for changes in all nonbasic variables at each iteration, and we will show below how to modify the method so that block changes can be permitted at certain iterations. An important advantage of this approach (over using the convex simplex method as the underlying algorithm) is that, on those iterations not involving block changes, the reduced gradient method permits all the nonbasic variables to be changed simultaneously. In developing the required modification, we use some of the basic ideas of Beck and Ecker (Ref. 1). A main difference, however, lies in the criteria that are used for a block of variables to move to or away from zero. In particular, with respect to moving a block of variables away from zero, a direction will be constructed that gives the largest increase in the objective function value, at least locally. The procedure of Beck and Ecker uses a direction of strict increase, in general different from the locally best direction. For a short but clear description of the reduced gradient method the reader is referred to Luenberger (Ref. 4).

2. Dual of a Posynomial Geometric Program

A posynomial geometric program can be stated as follows.

Program (P):
$$\inf \sum_{i=m_0}^{n_0} c_i \prod_{j=1}^{m} t_j^{a_{ij}},$$

subject to

$$\sum_{i=m_k}^{n_k} c_i \prod_{j=1}^{m} t_j^{a_{ij}} \leq 1, \qquad k = 1, 2, \ldots, p,$$

$$t \equiv (t_1, \ldots, t_q) > 0,$$

where

$$m_0 = 1, \qquad m_1 = n_0 + 1, \qquad m_2 = n_1 + 1, \ldots, m_p = n_{p-1} + 1, \qquad n_p = n.$$

The coefficients c_i are fixed constants and positive.

Associated with Program (P) is the following dual program, expressed here with the concave logarithmic form of the objective function.

Program (D): $$\max V(x) = \sum_{k=0}^{p} \sum_{[k]} x_i \log(c_i \Lambda_k / x_i),$$

subject to

$$A^T x = 0, \qquad \sum_{[0]} x_i = 1, \qquad x \geq 0,$$

where

$$[k] = \{m_k, m_k + 1, \ldots, n_k\}$$

and will be called *block k*,

$$\Lambda_k = \sum_{[k]} x_i,$$

$A = n \times m$ matrix of exponents of Program (P), and $x_i \log(c_i \Lambda_k / x_i)$ is defined to be zero when $x_i = 0$.

It will be assumed that Program (P) satisfies the Slater condition (superconsistent) and that the objective function attains its constrained minimum value. Unless (P) is totally degenerate, the second condition will always hold for an easily computed transformed program of (P). See Duffin, Peterson, and Zener (Ref. 5). Under these assumptions, Program (D) has a finite solution, and a primal solution can in most cases be obtained from the dual solution by solving a system of linear equations [Duffin, Peterson, and Zener (Ref. 5)]. Exceptionally, one more (subsidiary) geometric program may have to be solved of usually vastly lower dimension [Duffin, Peterson, and Zener (Ref. 5), and Rockafellar (Ref. 6)]. In this paper, however, we restrict ourselves to a solution procedure for Program (D).

It is clear that the objective function of (D) is nondifferentiable at a point where $x_i = 0$ for some i. Furthermore, it is a property of any finite optimal solution that, for a given k, either $x_i = 0$, all i in $[k]$, or $x_i > 0$, $i \in [k]$ [see Duffin, Peterson, and Zener (Ref. 5)]. Provided the gradient of the objective function exists, its components G_{x_i} can be computed as

$$G_{x_i}(x) = \log(c_i \Lambda_k / x_i), \qquad i \in [k], \qquad k = 0, 1, \ldots, p.$$

As is usual for reduced gradient methods, we make the nondegeneracy assumption that, for every feasible point, there exists a strictly positive basis.

3. Justification of the Modifications of the Reduced Gradient Method

The modifications required to apply the reduced gradient method to Program (D) are basically twofold.

(i) Since an optimal solution may be at a point for which a block of variables x_i, $i \in [k]$, equals zero, a straightforward application of the reduced gradient method will usually lead to ill-conditioned gradients and zigzagging in the neighborhood of $x_i = 0$, $i \in [k]$. Numerical experience by the authors has shown that the method in most cases fails to converge, because of these reasons.

Zigzagging occurs because a variable never becomes zero along any reduced gradient direction, unless that direction happens to point to the origin of the block, in which case all variables in the block could become zero simultaneously. This fact will be proved in Section 5.

Therefore, provided that a certain condition is satisfied at a particular iteration in the algorithm, the direction for a nonbasic block of variables x_i, $i \in [k]$, will not be determined by the reduced gradient, but by the ray from the current point x_i, $i \in [k]$, to the origin $x_i = 0$, $i \in [k]$ [Step 1(c) in Section 4]. The condition will guarantee that the direction to the origin of the block is a direction of increase.

(ii) Given that a block of variables x_i, $i \in [k]$, is zero at a certain iteration, the reduced gradient is not defined for those variables, and a procedure must be developed that finds a direction of increase, provided that one exists. The direction of best increase locally for such a block is given in Step 1(a) of the algorithm, and the proof of this result is given in Section 5.

4. Modified Reduced Gradient Method

In this section, we give a description of the reduced gradient method with the modifications that are necessary to overcome the difficulties mentioned above. The theoretical justification for these modifications will be given in Section 5.

Assume that, at the beginning of an iteration, the $m + 1$ largest variables are the basic variables, denoted by y, and the remaining variables, the nonbasic variables are denoted by z (ties are broken arbitrarily). By the nondegeneracy assumption, y is strictly positive. Also, assume that a nonbasic variable is zero only if all the variables in the same block are nonbasic and equal to zero. These assumptions can be satisfied easily for a starting vector (y, z), since there exists a dual feasible point which is strictly positive for any problem satisfying the conditions imposed on Program (P)

[Duffin, Peterson, and Zener (Ref. 5)]. Finding such a dual feasible point, which is strictly positive, requires (at most) the solution of one linear program. Moreover, the algorithm proceeds in such a way that these assumptions remain satisfied (see Section 5). Assume that the system of dual constraint equations is rewritten as

$$y + Tx = b,$$

where y is the $(m+1)$-vector of basic variables, z is the $(n-m-1)$-vector of nonbasic variables, $T = (t_{ij})$ is an$(m+1)\times(n-m-1)$ matrix, and b is an $(m+1)$-vector of constraints. Furthermore, let

$$K = \{0, 1, \ldots, p\},$$

$$K_0 = \{k | x_i \text{ is nonbasic and zero for every } i \in [k]\},$$

$$K_1 = \{k | x_i \text{ is nonbasic and positive for every } i \in [k]\},$$

$$M_i = \sum_j G_{y_j}(x) t_{ji}.$$

One iteration of our modified reduced gradient algorithm can then be described as follows.

Step 1. (a) Let $\Delta z_i = c_i \exp(-M_i)$ all $i \in [k]$ and $k \in \bar{K}_0$, where

$$\bar{K}_0 = \left\{ k \in K_0 | \sum_{[k]} c_i \exp(-M_i) > 1 \right\}.$$

(b) Let $\Delta z_i = 0$, all $i \in [k]$, and $k \in K_0 \backslash \bar{K}_0$.
(c) Let $\Delta z_i = -z_i$, all $i \in [k]$ and $k \in \bar{K}_1$, where

$$\bar{K}_1 = \{k \in K_1 | r_i \equiv G_{z_i}(x) - M_i < 0, \text{ all } i \in [k]\}.$$

(c′) An alternative here is to set $\Delta z_i = -z_i$, $i \in [k]$ and $k \in K_1$, whenever

$$-p_k z_{[k]}^T r_{[k]} \geq q_k r_{[k]}^T r_{[k]},$$

with q_k a fixed number, $0 \leq q_k \leq 1$, and the scalar

$$p_k = \|r_{[k]}\| / \|z_{[k]}\|,$$

where

$$r_{[k]} = (r_i), \qquad i \in [k] \qquad \text{and} \qquad z_{[k]} = (z_i), \qquad i \in [k].$$

(d) Let $\Delta z_i = r_i = G_{z_i}(x) - M_i$ for $i \in [k]$, $k \in K \backslash \{K_0 \cup K_1\}$.

Step 2. If $\Delta z_i = 0$, all $i \in [k]$ and $k \in K$, stop; the current point is optimal. Otherwise, compute

$$\Delta y = -T \Delta z.$$

Step 3. Find α_1, α_2, α_3 such that
(a) $\alpha_1 = \max\{\alpha | y + \alpha\Delta y \geq 0\}$,
(b) $\alpha_2 = \max\{\alpha | z + \alpha\Delta z \geq 0\}$,
(c) $\alpha_3 = \max\{V(x + \alpha\Delta x) : 0 \leq \alpha \leq \min(\alpha_1, \alpha_2)\}$.
(d) Let $\bar{x} = x + \alpha_3\Delta x$.

Step 4. Pivot in the system

$$y + Tx = b,$$

so that the $m+1$ largest components of $\bar{x}$ constitute the basic variables. Update the index sets K_0 and K_1, and return to Step 1.

5. Theoretical Justification of the Algorithm

Steps 1(a) and 1(b) provide a direction for all nonbasic variables currently at zero (remember that a nonbasic variable is zero only if all the variables in the same block are zero). These steps are justified through the following theorem.

Theorem 5.1. Assume that, at the current iteration, a block of nonbasic variables z_i, $i \in [k]$, equals zero. Then directions of increase for that block exist iff

$$\sum_{[k]} c_i \exp(-M_i) > 1,$$

and the direction of best increase locally is given by

$$\Delta z_i = c_i \exp(-M_i), \qquad i \in [k].$$

Proof. For any $\Delta z_i > 0$, $i \in [k]$, the directional derivative of the objective function at $z_i = 0$, $i \in [k]$, is well defined and given by

$$\sum_{[k]} \Delta z_i \log\left[c_i\left(\sum_{[k]} \Delta z_j / \Delta z_i\right)\right].$$

Hence, the first-order approximation at $z_i = 0$, $i \in [k]$, to the change in the objective function value is given by

$$C(\Delta z_{[k]}) \equiv \sum_{[k]} \Delta z_i \left\{\log\left[c_i\left(\sum_{[k]} \Delta z_j / \Delta z_i\right)\right] - M_i\right\},$$

so $\Delta z_i > 0$, $i \in [k]$, is a direction of increase iff $C(\Delta z_{[k]}) > 0$. Since $C(\Delta z_{[k]})$ is positively homogeneous in Δz_i, $i \in [k]$, the direction of best increase is obtained by solving the problem

$$\max \sum_{[k]} C(\Delta z_{[k]}),$$

subject to

$$\sum_{[k]} \Delta z_i = 1, \qquad \Delta z_i > 0, \qquad i \in [k].$$

Neglecting the positivity conditions, the solution can easily be obtained as

$$\Delta z_i^* = c_i \exp(-M_i) \Big/ \left[\sum_{[k]} c_i \exp(-M_i) \right], \qquad i \in [k],$$

which is strictly positive. The value of $C(\Delta z_{[k]})$ for $\Delta z_i = \Delta z_i^*$ equals $\log \sum_{[k]} c_i \exp(-M_i)$. This shows that directions of increase exist iff

$$\sum_{[k]} c_i \exp(-M_i) > 1. \qquad \square$$

Theorem 5.1 justifies a so-called block increase away from zero, since all variables in such a block will move away from zero. If no directions of increase exist for a block currently at zero, then Step 1(b) of the algorithm keeps those variables at zero.

We now justify Step 1(c) of the algorithm. It was already pointed out in Section 3 that no variable can go to zero unless all variables in the same block go to zero. In particular, a nonbasic variable of block k will never go to zero when the direction specified by the reduced gradient (or any other direction, for that matter) is used, unless by chance that direction happens to be the direction

$$\Delta z_i = -z_i,$$

pointing towards the origin of block k. Since this is very unlikely and remembering that, at an optimal solution, either all variables in a block or none are zero, Step 1(c) provides a direction for all variables in a block to move to zero simultaneously, whenever a certain condition is satisfied. The condition in Step 1(c) requires that the reduced gradients for all variables in a nonbasic block be negative. This means that the objective function value increases when all the variables in that block are decreased along the reduced gradient direction. The direction specified in Step 1(c) also decreases all variables in that block, but is different from the reduced gradient direction. However, it can easily be seen to be a direction of increase, since

$$\sum_{[k]} (-z_i) r_i > 0 \qquad \text{whenever } k \in \bar{K}_1.$$

This procedure is suggested in the modified convex simplex method of Beck and Ecker. (Ref. 1).

The alternative Step 1(c′) has the same objective. It states that the direction to zero for a block of variables will be selected whenever the inner product between that direction and the reduced gradient (being a measure of the increase in the objective function locally when the direction $-z_i$ is selected) is at least a fraction of the norm of the reduced gradient (being a measure of the increase in the objective function locally when the direction r_i is selected). The scalar p_k is a normalization constant to make a meaningful comparison possible. The extreme case $q_k = 1$ corresponds to always selecting the reduced gradient as the direction in Step 1(c), while $q_k = 0$ corresponds to selecting $\Delta z_i = -z_i$ provided it is a direction of increase. The remaining steps follow exactly the reduced gradient method and need no further explanation.

It remains to be shown that no variable, basic or nonbasic, will go to zero without all variables in the same block going to zero simultaneously.

Theorem 5.2. Under the assumptions stated so far, application of the algorithm in Section 4 never leads to one or more variables becoming zero without all the variables in the same block becoming zero simultaneously.

Proof. The only place in the algorithm where a variable could conceivably go to zero, without the corresponding block of variables going to zero, is in Step 3(c) where the maximum of the objective function is searched in the direction Δx For $0 \leq \alpha < \alpha_3$ and a given Δx, it holds that

$$dV(x)/d\alpha = \sum_{k \in \bar{K}_0} \sum_{[k]} \Delta z_i \log\left[c_i\left(\sum_{[k]} \Delta z_j/\Delta z_i\right)\right]$$

$$+ \sum_{i \in N \setminus K_o} G_{z_i}(x + \alpha\, \Delta x)\, \Delta z_i + \sum_j G_{y_j}(x + \alpha\, \Delta x)\, \Delta y_j$$

where N is the set of nonbasic variables. $dV(x)/d\alpha$ is a continuous function of α for $0 \leq \alpha < \alpha_3$; and, since Δx is a direction of increase, it holds that

$$[dV(x)/d\alpha]_{\alpha=0} > 0.$$

Suppose now that, as $\alpha \to \alpha_3$, there exist proper subsets S_k of $[k]$, $k \in K' \subseteq K$, such that

$$z_i \to 0, \qquad i \in S_k,$$

$$y_j \to 0, \qquad j \in S_k.$$

It is possible, of course, that S_k contains either only basic variables y_j or only nonbasic variables z_i. It is immediate from the expressions of $G_{z_i}(\cdot)$ and $G_{y_j}(\cdot)$ that

$$G_{z_i}(x+\alpha\,\Delta x)\to+\infty \qquad \text{as } \alpha\to\alpha_3 \text{ for } i\in S_k, \qquad k\in K'\subseteq K,$$

$$G_{y_i}(x+\alpha\,\Delta x)\to+\infty \qquad \text{as } \alpha\to\alpha_3 \text{ for } j\in S_k, \qquad k\in K'\subseteq K.$$

Since for variables x_i going to zero it must be the case that $\Delta x_i<0$, it follows that

$$\sum_{k\in K'}\sum_{S_k} G_{z_i}(x+\alpha\,\Delta x)\,\Delta z_i+\sum_{k\in K'}\sum_{S_k} G_{y_j}(x+\alpha\,\Delta x)\,\Delta y_j\to-\infty \qquad \text{as } \alpha\to\alpha_3.$$

All other terms in the expression for $dV(x)/d\alpha$ are bounded because $\min(\alpha_1, \alpha_2)$ is finite, since at least one variable is assumed to go to zero. Hence,

$$dV/d\alpha\to-\infty \qquad \text{as } \alpha\to\alpha_3.$$

This contradicts the fact that the maximum of V in the direction Δx occurs at $\alpha=\alpha_3$. Hence, no subsets S_k as defined above can exist. □

6. Some Computational Experience

An experimental code for the above algorithm has been written to obtain a preliminary idea of its convergence behavior. Using this code, we have solved the test problems of Beck and Ecker (Ref. 1) including those problems that required subsidiary problems. The results of this investigation are summarized in Table 1 below. Because of the experimental nature of the code, only the number of iterations required to solve the test problems by the modified concave simplex algorithm of Beck and Ecker and by the method proposed in this paper are compared. No definite conclusions should be drawn from this comparison, since the codes differ in the line search and the selection criteria of the basis. However, the stopping criterion was taken the same in both cases, and it can also be expected that the amount of work to be done at every iteration will not be significantly different provided identical line searches are used.

The problems numbers in Table 1 below are the same as in Beck and Ecker (Ref. 1). Problems 5 and 6 are problems that require the use of subsidiary problems. So far, our experimental code does not explicitly have a subroutine to handle subsidiary problems. However, both problems could be solved by our code, but one can probably construct counterexamples where it would fail.

Table 1. Number of iterations required.

Problem	Modified concave simplex method	Proposed method
1	2	2
2	28	52
3	149	91
4	11	4
5	63	8
6	18	12
7	83	27
8	13	6
9	37	165
10	169	14
11	118	123
12	459	62
13	419	317
14	26	7

7. Conclusions

Modifications to the reduced gradient method have been developed that allow for the solution of the dual of a posynomial programming problem. In order to overcome the difficulties associated with the nondifferentiability of the objective function, we have shown that on certain iterations blocks of variables can be allowed to move to or away from zero. On the iterations not involving block changes, standard reduced gradient iterations are performed. The preliminary results of our computational study show that the method proposed here is probably competitive with the modified concave simplex method [Beck and Ecker. (Ref. 1)].

Considerable work has been done by others in developing efficient reduced gradient codes; see, for example, Lasdon, Fox, and Ratner. (Ref. 7). A final goal in our work is to adapt one of these codes (which has proved to be efficient on other problems) to our proposed method for solving the dual posynomial programming problem.

References

1. BECK, P. A., and ECKER, J. G., *A Modified Concave Simplex Algorithm for Geometric Programming*, Journal of Optimization Theory and Applications, Vol. 15, pp. 189–202, 1975.

2. ZANGWILL. W. I., *The Convex Simplex Method*, Management Science, Vol. 14, pp. 221–238, 1967.
3. WOLFE, P., *Methods of Nonlinear Programming*, Nonlinear Programming, Edited by J. Abadie, John Wiley and Sons (Interscience Publishers), New York, New York, 1967.
4. LUENBERGER, D. G., *Introduction to Linear and Nonlinear Programming*, Addison-Wesley Publishing Company, Reading, Massachusetts, 1973.
5. DUFFIN, R. J., PETERSON, E. L., and ZENER, C., *Geometric Programming Theory and Applications*, John Wiley and Sons, New York, New York, 1967.
6. ROCKAFELLAR, R. T., *Some Convex Programs Whose Duals are Linearly Constrained*, Nonlinear Programming, Edited by J. B. Rosen, O. L. Mangasarian, and K. Ritter, Academic Press, New York, New York, 1970.
7. LASDON, L. S., FOX, R. L., and RATNER, M. W., *Nonlinear Optimization Using the Generalized Reduced Gradient Method*, Case Western Reserve University, Cleveland, Ohio, Department of Operations Research, Technical Memorandum No. 325, 1973.

17

Global Solutions of Mathematical Programs with Intrinsically Concave Functions

U. PASSY[1]

Abstract. An implicit enumeration technique for solving a certain type of nonconvex program is described. The method can be used for solving signomial programs with constraint functions defined by sums of quasi-concave functions and other types of programs with constraint functions called intrinsically concave functions. A signomial-type example is solved by this method. The algorithm is described together with a convergence proof. No computational results are available at present.

1. Introduction

The question of under what conditions a local minimum is also a global one is associated with the notion of generalized concave (convex) functions (Refs. 1–3). A point set defined by $\{x \mid x \in X;\ f(x) \geq a\}$, where X is some convex subset of E^n, (i) is convex if f is quasiconcave, Ref. 3, and (ii) can be transformed into a convex set (see Section 4) if f is a generalized concave function in the sense of Ref. 2.

While the class of concave functions is closed under addition, the class of generalized concave functions is not. It is well known that, if f_1 and f_2 are quasiconcave, the function $h = f_1 + f_2$ is not necessarily quasiconcave. Point sets defined by the sum of generalized convex functions, e.g., $\{x \mid x \in X;\ h(x) \geq a\}$, are not convex; common mathematical programming methods, if applied to this type of program, will usually locate only a stationary point.

In the present work, an algorithm for locating the global solution of such nonconvex programs is described, and a solved example is included. The first part of this work deals with quasiconcave functions. Only in the latter sections are generalized concave functions analyzed.

---

[1] Associate Professor, Faculty of Industrial Engineering and Management, Technion—Israel Institute of Technology, Haifa, Israel.

Consider the following nonconvex mathematical programming problem, called P.I.:

$$\min_{x\in D} f_0(x), \qquad f_0(x)=\sum_{i=1}^{n} c_i x_i. \tag{1}$$

The constraint set D is defined by the following relations:

$$D=X\bigcap_{l=1}^{m} P(l), \tag{2}$$

where X is a given convex and compact set in E^n and

$$P(l)=\{x\,|\,x\in E^n, f_l(x)\geq 1\}. \tag{3}$$

The functions f_l are given by

$$f_l(x)=\sum_{j=1}^{I(l)} g_{lj}(x), \tag{4}$$

where g_{lj} are positive continuous quasiconcave functions defined over X, i.e.,

$$g_{lj}(x)>0, \qquad \text{for all } x\in X. \tag{5}$$

2. Preliminary Results

Associated with P.I. is the following problem:

$$\min_{x\in Y} f_0(x). \tag{6}$$

The constraint set Y is defined through the following sets. Let

$$\mathcal{U}(l)=\left\{U(l)=(u_{l1},\ldots,u_{lI(l)})\,\middle|\,\sum_{j=1}^{I(l)} u_{lj}=1,\, u_{lj}>0\right\}, \qquad l=1,\ldots,m,$$

be a set of weights. Then

$$S(U(l))=\{x\,|\max_{1\leq j\leq I(l)}\{g_{lj}(x)/u_{lj}\}\geq 1\}, \qquad U(l)\in\mathcal{U}(l),$$

$$S(l)=\bigcap_{U(l)\in\mathcal{U}(l)} S(U(l)),$$

$$Y=X\bigcap_{l=1}^{m} S(l). \tag{7}$$

Lemma 2.1. For $l=1,\ldots,m$, the two point sets $X\cap P(l)$ and $X\cap S(l)$ are equal, i.e.,

$$X\cap P(l)=X\cap S(l).$$

Proof. First, it is necessary to cite a result associated with generalized means (Ref. 4). Given two sets

$$\{(w_1,\ldots,w_n)|\, w_i\in E^1,\, w_i>0\},$$

$$U=\left\{(u_1,\ldots,u_n)|\, u_i\in E^1,\, u_i>0,\, \sum_{i=1}^{n} u_i=1\right\}, \tag{8}$$

then

$$\min_{u\in U}\max_{1\le i\le n}\{w_i/u_i\}=w_l/u_l^*,$$

where

$$u_l^*=w_l\Big/\sum_{j=1}^{n} w_j.$$

The proof can be found in Ref. 4.

(a) Let $x\in X\cap P(l)$; then,

$$\sum_{j=1}^{I(l)} g_{lj}(x)\ge 1.$$

Define

$$u_{lj}^*=g_{lj}(x)/f_l(x). \tag{9}$$

Clearly,

$$U^*(l)=(u_{l1}^*,\ldots,u_{lI(l)}^*)\in\mathscr{U}(l)$$

and

$$g_{lj}(x)\ge u_{lj}^*>0 \qquad \text{for all } j.$$

Hence,

$$\max_{1\le j\le I(l)}\{g_{lj}(x)/u_{lj}^*\}\ge 1,$$

and therefore $x\in S(l)\cap X$.

(b) Let $x\in X\cap S(l)$ and $x\notin X\cap P(l)$, so $f_l(x)<1$. Choose $U^*(l)$ as in part (a), Eq. (9). Recall that the g_{lj} are all positive functions. In this case,

$$\max_{1<j<I(l)}\{g_{lj}(x)/u_{lj}^*\}=\sum_{j=1}^{I(l)} u_{lj}^*(g_{lj}(x)/u_{lj}^*)=\sum_{j=1}^{I(l)} g_{lj}(x)<1,$$

contradicting the assumption that $x\in S(l)\cap X$. □

Corollary 2.1. The point sets D and Y [Eqs. (2) and (6)] are equal, i.e., $D = Y$. This follows from Lemma 2.1 above and the definitions of Y and D.

Corollary 2.2. The following two programs are identical:

$$\min_{x \in D} f_0(x) = \min_{x \in Y} f_0(x).$$

This is an immediate consequence of the previous corollary. A method for solving P.I. can now be described.

Step (0). Minimize $f_0(x)$ [Eq. (1)] over $Y^{(0)} = X$.

Step (k). Let $x^{(k-1)}$ solve subproblem $k-1$, and let

$$f_q(x^{(k-1)}) = \min_{1 \le l \le m} f_l(x^{(k-1)}).$$

If

$$f_q(x^{(k-1)}) \ge 1,$$

terminate; otherwise choose a vector $U^{(k)}(q)$ such that

$$\max_{1 \le j \le I(q)} \{g_{qj}(x^{(k-1)})/u_{qj}^{(k)}\} < 1.$$

There are various ways to select such a vector. One satisfactory formula follows from Eq. (9) and is given by

$$u_{qj}^{(k)} = g_{qj}(x^{(k-1)})/f_q(x^{(k-1)}), \qquad j = 1, \ldots, I(q). \tag{10}$$

Minimize $f_0(x)$ over $Y^{(k)}$, where

$$Y^{(k)} = Y^{(k-1)} \cap S(U^{(k)}(q)). \tag{11}$$

Start the next step.

The sets $Y^{(k)}$ generate a sequence $\{Y^{(k)}\}$ with

$$Y \subseteq Y^{(k)} \subseteq Y^{(k-1)}. \tag{12}$$

Thus,

$$f_0^{(k)} \equiv f_0(x^{(k)}) \le f_0(x^{(k+1)}) \equiv f_0^{(k+1)},$$

and $\{f_0^{(k)}\}$ is a bounded monotonic nondecreasing sequence.

Lemma 2.2. Let $x^{(k)}$ be the sequence of points generated by the algorithm if $U^{(k)}(q)$ is calculated by Eq. (10). Then, (a) any convergent subsequence $\{x^i\}_{i \in J} \to x^*$ converges to a point satisfying

$$f_0(x^*) = \lim_{k \to \infty} f_0^{(k)};$$

(b) the point x^* is feasible $x^* \in Y = D$; and (c) the point x^* is an optimal solution of P.I.

Proof. (a) Let $\{x^{(i)}\}_{i \in J} \to x^*$. The sets $Y^{(i)}$ are compact, since $Y^{(0)} = X$. Therefore, since $f_0(x)$ is continuous,

$$\lim_{k \in J} f_0(x^{(k)}) = f_0(x^*).$$

Since $\{f_0(x^{(k)})\}$ is a bounded monotonic nondecreasing sequence, then

$$f_0(x^*) = \lim_{k \in J} f_0(x^{(k)}) = \lim_{k \to \infty} f_0^{(k)} = f_0^*.$$

(b) Let $\{x^{(k)}\}_{k \in J} \to x^*$ be any convergent subsequence. Define m sequences

$$K(l) = \{k \mid k \in J, f_l(x^{(k)}) = \min_{1 \le i \le m} f_i(x^{(k)})\}, \qquad l = 1, \ldots, m.$$

Let

$$L1 = \{l \mid K(l) \text{ is finite}\},$$

$$L2 = \{l \mid K(l) \text{ is infinite}\}.$$

If $l \in L1$, then there exists an N such that

$$f_l(x^{(k)}) \ge \min_{i \in L2} f_i(x^{(k)}), \qquad k > N.$$

If $l \in L2$, then

$$\max_{1 \le j \le I(l)} \{g_{lj}(x^{(k)})/u_{lj}^{(k)}\} \ge 1,$$

since

$$x^{(k)} \in Y^{(k)} \subseteq S(U^{(k)}(q)).$$

Moreover, since $Y^{(k)} \subseteq Y^{(k-1)}$,

$$\max_{1 \le j \le I(l)} g_{lj}(x^{(k)})/u_{lj}^{(n)} \ge 1, \qquad n \le k.$$

In the limit,

$$\lim_{k \in K(l)} \{\max_{1 \le j \le I(l)} \{g_{lj}(x^{(k)})/u_{lj}^{(n)}\}\} = \max_{1 \le j \le I(l)} \{g_{lj}(x^*)/u_{lj}^{(n)}\} \ge 1, \qquad n = 1, \ldots.$$

In particular, for $n \in K(l)$, since f_l and g_{lj} are continuous and positive

$$1 \le \lim_{n \in K(l)} \{\max_{1 \le j \le I(l)} \{g_{lj}(x^*)/u_{lj}^{(n)}\}\}$$

$$= \lim_{n \in K(l)} \max_{1 \le j \le I(l)} \{g_{lj}(x^*)/[g_{lj}(x^{(n)})/f_l(x^{(n)})]\} = f_l(x^*).$$

The last equality follows from Eq. (10). Hence,

$$f_l(x^*) \geq 1 \qquad \text{for } l \in L2,$$

and

$$f_i(x^*) \geq f_j(x^*) \qquad \text{for } i \in L1 \text{ and for } j \in L2.$$

(c) For any k, we have

$$Y^{(k)} \supseteq Y = D \qquad \text{and} \qquad f_0(x^{(k)}) \leq \min_{x \in Y} f_0(x).$$

Also,

$$f_0(x^*) = \lim_{k \to \infty} f_0^{(k)} \leq \min_{x \in Y} f_0(x).$$

Therefore, $f_0(x^*)$ is a solution for P.I. □

At each iteration of the previously described algorithm, the solution of a nonconvex program defined by f_0 and $Y^{(k)}$ is required. A method for solving these programs is now described.

For simplicity and to avoid unnecessarily cumbersome notation, assume that

$$m = 1, \qquad P(1) \equiv P,$$

i.e., the constraint set is given by

$$D = X \cap P \qquad \text{and} \qquad P = \left\{ x \,\middle|\, \sum_{i=1}^{K} g_i(x) \geq 1 \right\},$$

where

$$K = I(1).$$

Extension for $m > 1$ can easily be done. At the beginning, $k = 0$ and $f_0(x)$ is minimized over X, which is a convex and compact set. This program is therefore a convex program and can be solved numerically by various methods. After evaluating $U^{(1)}$ by Eq. (10) (the second index q is not required, since $m = 1$) the process proceeds one step ahead and $k = 1$.

If $f(x^0) \geq 1$ the algorithm terminates; otherwise, a new constraint is adjoined to the constraint set, given by the conditional expression

$$\max_{1 \leq i \leq K} \{g_i(x)/u_i^{(1)}\} \geq 1,$$

which can be explicitly written as

$$g_1(x) \geq u_1^{(1)}, \qquad \text{or } g_2(x) \geq u_2^{(1)}, \qquad \ldots, \text{or } g_K(x) \geq u_K^{(1)}. \tag{13}$$

Since the g_i's are quasiconcave, $Y^{(1)}$ is a union of K convex sets and is generally not convex. It is given by

$$Y^{(1)} = \bigcup_{i=1}^{K} \{x \mid x \in X, g_i(x) \geq u_i^{(1)}\}.$$

The optimum at this stage can be located after solving K convex subproblems

$$P(1, i): \min\{f_0(x) \mid x \in X, g_i(x) \geq u_i^{(1)}\}, \qquad i = 1, \ldots, K.$$

The first index in $P(1, i)$ indicates the iteration number 1, and the second index corresponds to the ith term in the conditional constraint given by Eq. (13). The number of conditional constraints increases as the process evolves. At the jth iteration, there are j conditional constraints:

$$g_1(x) \geq u_1^{(1)} \quad \text{or } g_2(x) \geq u_2^{(1)} \ldots \text{or } g_K(x) \geq u_K^{(1)}, \tag{14-1}$$

$$g_1(x) \geq u_1^{(2)} \quad \text{or } g_2(x) \geq u_2^{(2)} \ldots \text{or } g_K(x) \geq u_K^{(2)}, \tag{14-2}$$

$$\vdots$$

$$g_1(x) \geq u_1^{(j)} \quad \text{or } g_2(x) \geq u_2^{(j)} \ldots \text{or } g_K(x) \geq u_K^{(j)}. \tag{14-j}$$

Each conditional constraint is a union of convex sets, the intersection of which is again a union of convex sets; and $Y^{(j)}$ is therefore a union of a finite number of convex sets. The solution for the jth iteration can therefore be found after solving a finite number of convex subproblems. Although this number may be large, only a small portion of it must be explicitly solved. This is best illustrated by an example; but, before we proceed further, some additional notation is required.

The constraint set at iteration j is characterized by a $j \times K$ dimensional matrix with elements $\{u_i^{(q)}\}$, $q = 1, \ldots, j$ and $i = 1, \ldots, K$. Thus, $Y^{(j)}$ differs from $Y^{(j-1)}$ by its last row. This matrix is described by the tabeleau given in Table 1.

Table 1. Tableau generated by the algorithm.

Iteration	$g_1(x) \geq$	$g_2(x) \geq$	$\cdots$	$g_K(x) \geq$
1	u_1^1	u_2^1	$\cdots$	u_K^1
2	u_1^2	u_2^2	$\cdots$	u_K^2
$\vdots$	$\vdots$	$\vdots$		$\vdots$
j	u_1^j	u_2^j	$\cdots$	u_K^j
$\vdots$	$\vdots$	$\vdots$		$\vdots$

Since $Y^{(j)}$ is a union of a finite number (say R) convex subsets, each of them is characterized by a set of indices $Q_r(j)$, $r=1, \ldots, R$, where

$$Q_r(j) \subseteq C(j) = \{(i, l) \mid 1 \leq i \leq j, 1 \leq l \leq K\}. \tag{15}$$

These subsets, denoted by $P(Q_r(j))$, are given by

$$P(Q_r(j)) = \{x \mid x \in X, g_l(x) \geq u_l^i, (i, l) \in Q(j)\}. \tag{16}$$

Each convex subset $P(Q_r(j))$ is associated with a convex subproblem, i.e.,

$$\min_{x \in P(Q_r(j))} f_0(x).$$

Example 2.1. Given an objective function $f_0(x)$, a convex subset $X \subseteq E^n$, and

$$f_1(x) = g_1(x) + g_2(x) + g_3(x),$$

assume that the tableau after the second iteration is given by Table 2. The constraint set $Y^{(0)}$ is convex, since $Y^{(0)} = X$. After the first iteration the $Q_r(1)$, $r = 1, 2, 3$, are

$$Q_1(1) = \{(1, 1)\}, \qquad Q_2(1) = \{(1, 2)\}, \qquad Q_3(1) = \{(1, 3)\},$$

and

$$Y^{(1)} = (P(1, 1)) \cup P((1, 2)) \cup P((1, 3)).$$

The solution of the first iteration is determined from the best value of the objective function, resulting from minimizing it over $P((1, 1))$, $P((1, 2))$, and $P((1, 3))$. After the second iteration,

$$C(2) = \{(i, l) \mid 1 \leq i \leq 2, 1 \leq l \leq 3\}.$$

The $Q_r(j)$ are given by

$$\{(1, 1); (2, 1)\}, \qquad \{(1, 1); (2, 2)\}, \qquad \{(1, 1); (2, 3)\},$$
$$\{(1, 2); (2, 1)\}, \qquad \{(1, 2); (2, 2)\}, \qquad \{(1, 2); (2, 3)\},$$
$$\{(1, 3); (2, 1)\}; \qquad \{(1, 3); (2, 2)\}, \qquad \{(1, 3); (2, 3)\}.$$

Table 2. Tableau for the example.

Iteration	$g_1(x) \geq$	$g_2(x) \geq$	$g_3(x) \geq$
1	0.100	0.300	0.600
2	0.400	0.100	0.500

Thus,

$$Y^{(2)} = P((1,1);(2,1)) \cup P((1,1);(2,2)) \cup P((1,1);(2,3)) \cup P((1,2);(2,1)) \cup P((1,2);(2,2)) \cup P((1,2);(2,3)) \cup P((1,3);(2,1)) \cup P((1,3);(2,2)) \cup P((1,3);(2,3)).$$

However, observe that

$$P((1,1);(2,1)) = P((2,1)), \qquad P((1,2);(2,1)) \subseteq P((2,1)) \subseteq Y^{(2)},$$
$$P((1,2);(2,2)) = P((1,2)), \qquad P((1,2);(2,3)) \subseteq P((1,2)) \subseteq Y^{(2)},$$
$$P((1,3);(2,3)) = P((1,3)), \qquad P((1,3);(2,2)) \subseteq P((1,3)) \subseteq Y^{(2)},$$
$$P((1,3);(2,1)) \subseteq P((2,1)) \subseteq Y^{(2)}.$$

Thus,

$$\begin{aligned} Y^{(2)} &= P((2,1)) \cup P((1,1);(2,2)) \cup P((1,1);(2,3)) \cup P((1,2)) \cup P((1,3)) \\ &= [P((1,2)) \cup P((1,3))] \cup [P((2,1)) \cup P((1,1);(2,2)) \\ &\qquad \cup P((1,1);(2,3))]. \end{aligned}$$

The solution at the second iteration is found after solving five subproblems. However, the solution of the first two can be obtained from the first iteration. Thus only three new subproblems are solved in the second iteration. It is shown that $Y^{(2)}$, given by the union of nine convex subsets, was eventually expressed as the union of only five subsets, resulting in five subproblems, only three of which have to be solved explicitly. This reduction is essential to the algorithm. The definitions and properties of these reduced sets are given in Section 3.

3. Reduced Sets

Definition 3.1. Given a K-dimensional vector $U = (u_1, \ldots, u_K)$, a pair (b, i), where $b \in E^1$ and $1 \leq i \leq K$, is said to cover U if $b \geq u_i$.

Definition 3.2. Given an $s \times K$ matrix $\{u_i^{(k)}\}$ and a subset of indices $Q_r(s) \subseteq C(s)$ [see Eq. (15)], then $Q_r(s)$ is said to be a covering set of $\{u_i^{(k)}\}$ if the induced set of numbers $\{u_l^{(i)}, l) \mid (i, l) \in Q_r(s)\}$ covers the rows $U^1, \ldots, U^i$ of the matrix $\{u_i^{(k)}\}$.

If $Q_r(s)$ is a covering set, then

$$P(Q_r(s)) \subset Y^{(s)}.$$

Definition 3.3. A covering set $Q_r(s)$ is said to be a reduced covering set if it has the following properties. There is at most one pair (i, j) of each row and column, i.e.,

$$\text{(i)}\quad (i,j)\in Q_r(s) \Rightarrow \begin{cases} \text{(a)} & (i,k)\notin Q_r(s), \qquad 1\le k\le K,\ k\neq j, \\ \text{(b)} & (l,j)\notin Q_r(s), \qquad 1\le l\le s,\ l\neq i, \end{cases}$$

$$\text{(ii)}\quad (l,i)\in Q_r(s);\ (k,j)\in Q_r(s) \Rightarrow u_j^{(l)} > u_j^{(k)} \text{ and } u_i^{(k)} > u_i^{(l)}.$$

It can be verified easily that only the reduced sets induce subproblems whose solution must be calculated explicitly. However, the solutions of some of these subproblems are known from previous iterations. Considering again Example 2.1, note that $P((1, 2))$ and $P((1, 3))$ are reduced covering sets for the second iteration. The solution of these two subproblems is found during the first iteration. A subproblem $P(Q_r(j))$, whose explicit solution must be calculated during the jth iteration, is called an essential subproblem. The essential subproblems in the second iteration (see Example 2.1) are

$$\text{(i) } P((2,1)), \qquad \text{(ii) } P((1,1);(2,2)), \qquad \text{(iii) } P((1,1);(2,3)).$$

Note that, if $Q(j)$ is a reduced covering set, then the induced subproblem $P(Q(j))$ is essential if $(j, i)\in Q(j)$ for some i.

Consider now the following finite set $\Omega(k)$

$$\Omega(k) = \{(Q(i), a_i, b_i)\}, \tag{17}$$

where $Q(i)$ is a reduced covering set for the kth iteration, $a_i \in E^n$ is the solution of the subproblem $P(Q(i))$, and $b_i \in E^1$ is the value of the objective function at that point, if a solution exists. Otherwise, let $a_i = (M, \ldots, M)$ and $b_i = M$, where M is a given symbol. The number of elements in $\Omega(k)$ is $N(k)$. This set is updated after every iteration.

The detailed description of the algorithm can now be given.

Step (0). Minimize $f_0(x)$ over $Y^{(0)}(=X)$; call the minimizer $x^{(0)}$.

Step (k).

(i) Evaluate $f_1(x^{(k-1)})$.

(ii) If $f_1(x^{(k-1)}) \geq 1$, terminate; otherwise proceed to (iii).

(iii) Evaluate $u_i^{(k)}$, $i = 1, \ldots, K$, by Eq. (10). The third index q is not required, since $m = 1$.

At this stage, an additional row, consisting of the $u_i^{(k)}$, is added to the tableau.

(iv) Delete from $\Omega(k-1)$ each element whose $Q_r(k)$ is not a covering set for the new tableau.

(v) Identify and solve all the k essential subproblems.

(vi) Adjoint the new elements (Q, a, b) obtained in (v) to the remaining elements in $\Omega(k-1)$. This new set is $\Omega(k)$. Determine $N(k)$.

(vii) Set $x^{(k)} \in \{a \mid f_0(a) = \min_{1 \le i \le N(k)} b_i\}$.
(viii) Set $k = k+1$ and go to (i).

Lemma 3.1. The algorithm converges.

Proof. The algorithm is the same as previously described (see Section 2).

If $m > 1$, then all of the results hold, but there are m tableaux of $u_i^{(j)}(l)$, $l = 1, \ldots, m$. If the number of rows in each tableau is α_l, then at the kth iteration

$$k = \sum_{l=1}^{m} \alpha_l.$$

Accordingly, step k above should be written as

(i) Let $f_j(x^{(k-1)}) = \min_{1 \le l \le m} f_l(x^{(k-1)})$.
(ii) If $f_j(x^{(k-1)}) \ge 1$ terminate; otherwise proceed to (iii).
(iii) Evaluate $u_i^{\alpha j+1}(j)$.

At this stage, an additional row of u's is added to the jth tableau.

4. Intrinsically Concave Functions

Definition 4.1. A real function g defined on a set $C \subseteq E^n$ is said to be intrinsically concave on the set C if a mapping $T: C \to E^m$ and a quasiconcave function

$$G: T(C) \to E^1 \tag{18}$$

exist such that (i) $T(C)$ is a convex subset of E^m, and T^{-1} exists, and (ii) the following two sets are equal:

$$\{z \mid z \in T(C), G(z) \ge 1\} = T(\{x \mid, x \in C, g(x) \ge 1\}).$$

Since $G(z)$ is quasiconcave, and since $T(C)$ is convex, the above set is convex. It is very difficult in general to determine whether a given function is intrinsically concave. However, examples of such functions can be found in Refs. 1, 2, 5.

Example 4.1

$$g(x) = 13 - 10x_1^{-1}x_2^{-1} - 5x_1^2 - x_1^{-1},$$

$$C = (E^2)^+, \qquad \text{the positive orthant.}$$

For this case,

$$T(C)=\left\{(z_1, z_2, z_3)\,|\, z=\begin{bmatrix}-1 & -1\\ 2 & 1\\ -1 & 0\end{bmatrix}\begin{bmatrix}\log x_1\\ \log x_2\end{bmatrix};\ x\in C\right\}.$$

$T(C)$ is a subspace of E^3 and

$$G(z)=13-10\exp(z_1)-5\exp(z_2)-\exp(z_3)$$

is a concave function.

$$\{z\,|,\ z\in T(C),\ G(z)\geq 1\}=T(\{x\,|\,x\in C,\ 13-10x_1^{-1}x_2^{-1}-5x_1^2x_2-x_1^{-1}\geq 1\}).$$

One can recognize that $g(x)$ is given in this example by the numerical constant 13 minus a posynomial (Ref. 6).

Example 4.2. (*Ref.* 5). A positive function $g(x)$ defined on a convex set $C\subseteq E^n$ is said to be L-concave if, for each $x^1, x^2\in C$ and $0\leq\lambda\leq 1$,

$$g((1-\lambda)x^1+\lambda x^2)\geq(g(x^1))^{1-\lambda}(g(x^2))^{\lambda}.$$

If g is L-concave, the function $\log(g)$ is concave. In this case,

$$T(C)=C \qquad \text{and} \qquad G(z)=1+\log[g(z)].$$

Example 4.3. (*Ref.* 1). A real function g defined on an open convex set $C\subseteq E^n$ is said to be r-concave if, for any $x^1, x^2\in C$, one has

$$g(\lambda x^1+(1-\lambda)x^2)\geq\begin{cases}\log\{\lambda\exp[rg(x^1)]+(1-\lambda)\exp[rg(x^2)]\}^{1/r}, & r\neq 0,\\ \lambda g(x^1)+(1-\lambda)g(x^2), & \text{if } r=0.\end{cases}$$

Clearly, if g is r-concave (and $r>0$) the function $\exp[rg(x)]$ is concave. In this case,

$$T(C)=C \qquad \text{and} \qquad G(x)=\exp[r(g(x)-1)].$$

Point sets defined by intrinsically concave functions can be transformed into convex point sets. However, point sets defined by a sum of intrinsically concave functions cannot, in general, be transformed into convex sets (see Section 1).

Definition 4.2. A finite set of functions $\{g_i\}$, $i=1,\ldots,P$, defined on a set $C\subseteq E^n$ is said to be intrinsically concave on C if a mapping $T: C\to E^m$ and a set of quasiconcave functions $G_i: T(C)\to E^1$, $i=1,\ldots,P$, exist such that (i) $T(C)$ is a convex set in E^m, and T^{-1} exists, and (ii) the following P pairs of sets are equal

$$\{z\,|,\ z\in T(C),\ G_i(z)\geq 1\}=T(\{x\,|,\ x\in C,\ g_i(x)\geq 1\}), \qquad i=1,\ldots,P. \tag{19}$$

All these sets are convex, since the G_i are quasiconcave and $T(C)$ is convex.

The generalized P.I. problem is defined as problem P.I., given by Eqs. (1)–(5), with the following differences. The g_{lj} are continuous positive and intrinsically concave functions on X, and the composite function

$$F_0(\cdot) \equiv f_0 T^{-1}(\cdot) \tag{20}$$

is a convex function over $T(X)$.

The results given in Lemmas 2.1, 2.2, Corollaries 2.1, 2.2, and the algorithm (Section 2) hold for any positive function and not only for quasiconcave functions. However, as was shown before, there exists no method that is capable of handling the subproblems induced by the algorithms, except for the quasiconcave case. In this special case, the difficulty was resolved, since $Y^{(j)}$ is a union of convex sets.

If the g_{qi} are intrinsically concave, then each nonconvex subproblem can be transformed into an equivalent convex one. Namely, if $P(Q(j))$ induces the following subproblem [Eq. (15)]:

$$P(Q(j)): \min\{f_0(x) \mid x \in X, g_{qi}(x) \geq u_i^l(q), (q, l, i) \in Q(j)\},$$

which is not convex, then the following program:

$$\tilde{P}(Q(j)): \min\{F_0(z) \mid z \in T(X); G_{qi}(z) \geq u_i^l(q); (q, l, i) \in Q(j)\}$$

is, by construction, a convex program. Note that the third index in $u_i^l(q)$ is required if $m > 1$. In that case, there are m different tableaux for each $f_l(x)$; q specifies the tableau $1 \leq q \leq m$; l determines the row in the qth tableau; and i specifies the column in the qth tableau $1 \leq i \leq I(q)$. As was mentioned before, the total number of rows in all the tableaux equals the iteration number.

Lemma 4.1. Let x_1 and z_2 be solutions to $P(Q(j))$ and $\tilde{P}(Q(j))$, respectively. Then,

$$f_0(x_1) = F_0(z_2).$$

Proof. The points $T(x_1)$ and $T^{-1}(z_2)$ are both feasible for $\tilde{P}(Q(j))$ and $P(Q(j))$ [see Eq. (19)]. Therefore,

$$F_0(z_2) = F_0(T^{-1}(z_2)) \nless f_0(x_1),$$

$$f_0(x_1) = f_0(T^{-1}(T(x_1)) = F_0(T(x_1)) \nless F_0(z_2).$$

Thus,

$$F_0(z_2) = f_0(x_1). \qquad \square$$

It was shown that, if g_{qi} are not quasiconcave, but merely intrinsically concave, then $P(Q(j))$ is solved indirectly by solving $\tilde{P}(Q(j))$. Observe that only the subproblems $P(Q(j))$ can be transformed into the convex program $\tilde{P}(Q(j))$.

All the results described previously in Sections 2 and 3 hold. An illustration of the method for functions which are not quasiconcave, but intrinsically concave, is given in Section 5.

5. Numerical Example (Generalized Geometric Programming)

Signomial programs with bounded variables can always be transformed into an equivalent program (Ref. 7) having the following form:

$$\min f_0(t) = t_0,$$

$$\text{s.t. } f_k(t) = a_K - u_k(t) \geq 1, \qquad k = 1, \ldots, P,$$

$$u_{P+1}(t) \geq 1,$$

$$0 < \epsilon_i \leq t_i \leq M_i, \qquad i = 0, \ldots, m,$$

where a_k is a positive constant, and $u_k(t)$, $k = 1, \ldots, P+1$, are all posynomial functions given by

$$u_k(t) = \sum_{j \in J\{k\}} c_j \prod_{i=0}^{m} t_i^{\alpha_{ij}}, \qquad k = 1, \ldots, P+1,$$

where

$$J\{k\} = \{m_k, m_k + 1, \ldots, n_k\},$$

$$m_0 = n_0 = 1, \qquad m_1 = 2, \qquad m_k = n_{k-1} + 1, \qquad n_{P+1} = n.$$

Let

(i) $C = \{t \mid t \in E^{m+1}, \epsilon_i \leq t_i \leq M_i\}$,

(ii) $T(C) = \left\{ z \mid z \in E^n, z_j = \sum_{i=0}^{m} a_{ji} \log t_i, j = 1, \ldots, n, t \in C \right\}$,

where

$$a_{01} = 1,$$

$$a_{j1} = 0 \qquad \text{for } j \neq 0,$$

$$a_{ij} = \alpha_{ij} \qquad \text{for } j \in J\{k\} \text{ and } K \neq P+1,$$

$$a_{ij} = -\alpha_{ij} \qquad \text{for } j \in J\{P+1\}.$$

The set $T(C)$ is compact and convex; and, if the rank of the coefficient matrix $\{a_{ij}\}$ is m and if $n > m$, then T^{-1} exists (see Ref. 3, p. 83). Moreover,

$$\text{(i)} \qquad F_k(z) = a_k - \sum_{j \in J\{k\}} c_j \exp(z_j), \qquad k = 1, \ldots, P,$$

$$\text{(ii)} \qquad G_j(z) = 2 - c_j^{-1} \exp(z_j), \qquad j \in J\{P+1\},$$

are all concave. Therefore,

$$\{z | z \in T(C), F_k(z) \geq 1\} = T\{t|, t \in C, f_k(t) \geq 1\}, \qquad k = 1, \ldots, P,$$

$$\{z | z \in T(C), G_j(z) \geq 1\} = T\left\{ t | t \in C, c_j \prod_{i=0}^{m} t_i^{\alpha_{ji}} \geq 1 \right\}, \qquad j \in J\{P+1\}.$$

Thus, the set of functions $f_k(t)$, $k = 1, \ldots, P$, together with the set

$$\left\{ c_j \prod_{i=0}^{m} t_i^{\alpha_{ji}} \right\}_{j \in J\{P+1\}}$$

is intrinsically concave on C and

$$F_0(z) = \exp(z_1) = f_0(T^{-1}(z))$$

is a convex function.

Each of the first P constraints consists of a single intrinsically concave function, and only the last constraint $(P+1)$ is given by a sum of several intrinsically concave functions. Therefore, if the $(P+1)$th constraint is absent, the transformed problem is convex (see Ref. 6, p. 168) and is induced by a regular geometric program (not signomials). The nonconvexity of the transformed problem is caused by the last constraint.

The method described in Section 4 can handle such problems. By following the various steps of the example, it is possible to construct an outline for an algorithm suitable for signomial programs. A detailed description of the algorithm can be found in Ref. 8.

Here we solve a signomial version of the *gravel box problem* (Ref. 6):

$$\min(40t_1^{-1}t_2^{-1}t_3^{-1} + 20t_3t_2 + 40t_1t_2 + 10t_3t_1), \tag{21}$$

$$\text{s.t. } t_1 + t_2 + t_3 \geq 8, \tag{22}$$

$$10^6 \geq t_i \geq 10^{-6}, \qquad i = 1, 2, 3. \tag{23}$$

This problem is given in its standard form by

$$\min\{t_0 | t \in C;\ u(t) = 1/8(t_1 + t_2 + t_3) \geq 1;\ f_1(t) \geq 1\}, \tag{24}$$

where

$$f_1(t) = 2 - 40t_0^{-1}t_1^{-1}t_2^{-1} - 20t_0^{-1}t_3t_2 - 40t_0^{-1}t_1t_2 - 10t_0^{-1}t_1t_3, \tag{25}$$

and the set C is defined by

$$10^6 \geq t_i \geq 10^{-6}, \qquad i = 0, 1, 2, 3. \tag{26}$$

The set of functions $f_1(t)$ and $1/8t_1$, $1/8t_2$, $1/8t_3$ is a set of intrinsically concave functions. Let

$$\lambda_i = \log t_{i-1}, \qquad i = 1, \ldots, 4.$$

Then,

(i) $T(C) = \{z \mid z \in E^8, z^T = (\lambda_1, \lambda_2, \lambda_3, \lambda_4)$

$$\times \begin{bmatrix} 1 & -1 & -1 & -1 & -1 & 0 & 0 & 0 \\ 0 & -1 & 0 & 1 & 1 & -1 & 0 & 0 \\ 0 & -1 & 1 & 1 & 0 & 0 & -1 & 0 \\ 0 & -1 & 1 & 0 & 1 & 0 & 0 & -1 \end{bmatrix}, \qquad \log 10^{-6} \leq \lambda_i \leq \log 10^6 \},$$

(ii) $F_1(z) = 2 - 40 \exp(z_2) - 20 \exp(z_4) - 10 \exp(z_5)$,

(iii) $1/8t_1 \to 2 - 8 \exp(z_6)$, $\quad 1/8t_2 \to 2 - 8 \exp(z_7)$, $\quad 1/8t_3 \to 2 - 8 \exp(z_8)$,

and

$$\{z \mid z \in T(C), F_1(z) \geq 1\} = T\{t \mid t \in C, f_1(t) \geq 1\}, \tag{27}$$

$$\{z \mid z \in T(C), 2 - 8 \exp(z_{i+5}) \geq 1\} = T\{t \mid t \in C, 1/8t_i \geq 1\}. \tag{28}$$

Recall that from (i) above

$$z_{i+5} = -\log t_i, \qquad i = 1, 2, 3,$$

and the sets appearing in (27) and (28) are convex sets. The transformed objective function is given by

$$f_0(T^{-1}(z)) = F_0(z) = \exp(z_1)$$

and it is convex.

This specific example was thoroughly investigated, and it was found that it has three different stationary points (see Table 4). When the problem was solved by condensation (Ref. 7), which locates only a local solution of generalized geometric programs, the solution obtained was dependent on the starting point, as illustrated in Table 3. As can be seen, the global minimum is 121.41, the difference between the various solutions is significant, and a method such as condensation is not sufficiently good for finding a global optimum.

The global method presently described is graphically illustrated as a tree; see Fig. 1. The subproblems $P(Q(m))$, derived from the tableaux (up to

Table 3. Numerical solution of the example by the condensation method.

Trial number	Initial point $t_1^{(0)}$	$t_2^{(0)}$	$t_3^{(0)}$	Solution obtained t_1^*	t_2^*	t_3^*	Value of objective function
I	0.3	8.0	0.5	0.2167	7.3687	0.4155	186.11
II	0.5	0.3	8.0	0.5201	0.2539	7.2260	121.41
III	8.0	0.5	0.3	7.1446	0.1729	0.6835	147.83

iteration 14) are the nodes of the tree. The circled numbers indicate the node associated with the subproblem that is a solution of the kth iteration [Step (vii) of the algorithm]. The starting node of the tree represents the solution $f_0 = 100$ to the unconstrained problem $Y^{(0)} = C$. The arcs emanating from a given node k are all the $k+1$ essential subproblems [Step (v) of the algorithm]. The three arcs amanating from node 0 correspond to the three essential subproblems associated with the first iteration:

$$P\{(1, 1)\}: t \in C, \qquad f_1(t) \geq 1, \qquad 1/8t_1 \geq u_1^{(1)},$$
$$P\{(1, 2)\}: t \in C, \qquad f_1(t) \geq 1, \qquad 1/8t_2 \geq u_2^{(1)},$$
$$P\{(1, 3)\}: t \in C, \qquad f_1(t) \geq 1, \qquad 1/8t_3 \geq u_3^{(1)}.$$

The explicit constraints for these subproblems are given in the first row of Table 4. The solution to the first iteration equals min{112.78, 135.357, 103} and it is written under node No. 1. As was mentioned before, there are various methods for choosing $U^{(k)}$. The implemented computer program is slightly modified. While $U^{(k)}$ for $k > 1$ is evaluated by Eq. (10), $U^{(1)}$ is set equal to $u_j^1 = 1/K$, where K is the number of terms in the constraint equation. For the specific example,

$$u_1^1 = \tfrac{1}{3}, \qquad u_2^1 = \tfrac{1}{3}, \qquad u_3^1 = \tfrac{1}{3}.$$

Table 4. Tableau at the third iteration for the example.†

j		$i = 1$		$i = 2$		$i = 3$
1		$t_1 \geq 2.666$	or	$t_2 \geq 2.666$	or	$t_3 \geq 2.666$
2	and	$t_1 \geq 1.6314$	or	$t_2 \geq 0.7113$	or	$t_3 \geq 5.667$
3	and	$t_1 \geq 1.208$	or	$t_2 \geq 1.43$	or	$t_3 \geq 5.361$

† Note that this tableau was written as $t_i \geq 8u_i^j$, instead of $1/8t_i \geq u_i^j$ as in Table 1.

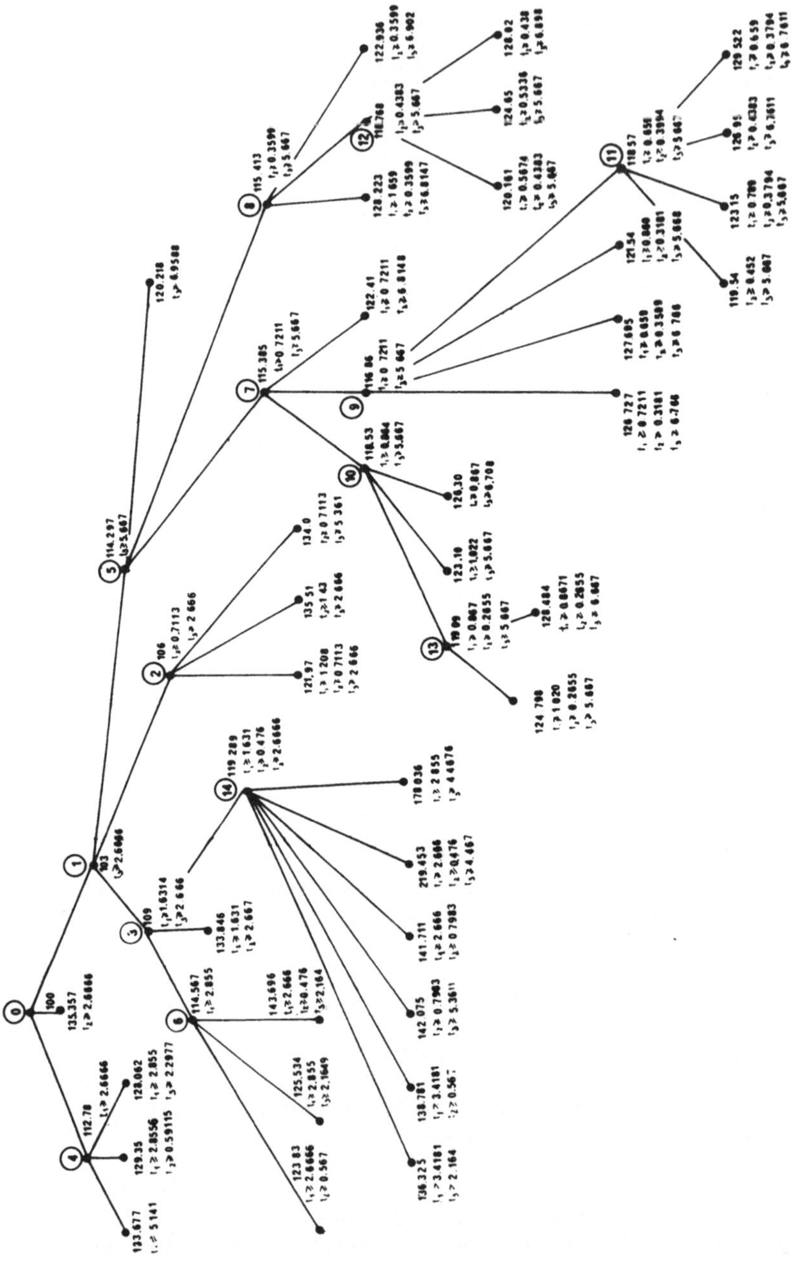

Fig. 1. Tree for the numerical example.

The optimal solution of the various subproblems $P(Q(m))$ is written directly under each node. Under this number are written all the *additional constraints* which define that particular subproblem [they are additional, since $t \in C$ and $f_1(t) \geq 1$ are included in every subproblem]. The number of subproblems $N(k)$, associated with a given iteration k, equals the number of arcs connecting nodes $0, \ldots, k-1$ to an unnumbered node or to a node numbered by an integer not less than k. So, at the second iteration, there are two arcs leaving node 0 [the arc (0, 1) is not counted] and three arcs leaving node 1, making the total number of subproblems $N(2) = 5$.

Only three arcs, however, are leaving node one, thus generating only three essential subproblems for this iteration. The total number of subproblems $N(2)$ can also be found in the tableau of Table 4 by generating all the reduced covering sets of the first two rows:

$$Q_1(2) = \{(1, 1)\}, \qquad Q_2(2) = \{(1, 2)\}, \qquad Q_3(2) = \{(2, 3)\},$$

$$Q_4(2) = \{(2, 1); (1, 3)\}, \qquad Q_5(2) = \{(3, 2); (1, 3)\}.$$

Similarly, at the 12th iteration there are $N(12) = 26$ subproblems, but there are only four essential subproblems—the ones denoted by the arcs leaving node 11.

References

1. AVRIEL, M., *Solution of Certain Nonlinear Programs Involving R-Convex Functions*, Journal of Optimization Theory and Applications, Vol. 11, pp. 159–174, 1973.
2. BEN-TAL, A., *On Generalized Means and Generalized Convexity*, Journal of Optimization Theory and Applications, Vol. 21, pp. 1–13, 1977.
3. MANGASARIAN, O. L., *Nonlinear Programming*, McGraw-Hill Book Company, New York, New York, 1969.
4. HARDY, G. H., LITTLEWOOD, J. E., and POLYA, G., *Inequalities*, Cambridge University Press, Cambridge, England, 1952.
5. MANGASARIAN, O. L., and KLINGER, A., *Logarithmic Convexity and Geometric Programming*, Journal of Mathematical Analysis and Applications, Vol. 24, pp. 388–408, 1968.
6. DUFFIN, R. J., PETERSON, E. L., and ZENER, C., *Geometric Theory and Application*, John Wiley and Sons, New York, New York, 1967.
7. PASSY, U., *Condensing Generalized Polynomials*, Journal of Optimization Theory and Applications, Vol. 9, pp. 221–237, 1972.
8. TSHERNIAK, D., *The Global Solution of Signomial Programming*, Technion, Haifa, Israel, MS Thesis, 1977.

18

Interval Arithmetic in Unidimensional Signomial Programming[1]

L. J. MANCINI[2] AND D. J. WILDE[3]

Abstract. This paper applies an interval arithmetic version of Newton's method to unidimensional problems in signomial programming. Unidimensional dual problems occur in engineering design problems formulated as a signomial program with a single degree of difficulty. Unidimensional primal problems are of interest, since many multidimensional search procedures involve unidimensional searches. The interval arithmetic method is guaranteed to generate all the local optima.

1. Introduction

In the last decade, signomial programming (or generalized geometric programming) has received much attention (Refs. 1–4) because many engineering design problems can be expressed as signomial programs. This paper applies an interval arithmetic version of Newton's root-finding method (Refs. 5 and 6) to unidimensional dual and primal problems in signomial programming. Section 2 presents the interval analysis required for the paper. The interval Newton method and its advantages over the ordinary Newton method are reviewed in Section 3. To account for multiple zeros, Section 4 presents a new combined version of the interval Newton–bisection method. Section 5 applies the combined method to unidimensional dual problems, that is, signomial dual programs with a single

[1] The authors are grateful to the National Science Foundation for support through a Graduate Fellowship and Grant No. GK-41301.

[2] Research Engineer, Physics, Mathematics, and Computer Science Department, Shell Development Company, Houston, Texas.

[3] Professor, Department of Mechanical Engineering, Stanford University, Stanford, California.

degree of difficulty. In this case, the combined method generates all the dual Kuhn–Tucker points. Many an engineering design problem can be formulated as a signomial program with a single degree of difficulty. An example design of a chemical plant is given at the end of Section 5. Finally, Section 6 applies the combined method to unidimensional primal signomial programs. One-dimensional primal problems are interesting because many multidimensional search procedures involve one-dimensional searches. In this case, the combined method is guaranteed to generate all the local optima, if any, in any given interval.

2. Interval Analysis

Interval arithmetic (Refs. 5 and 7) is a generalization of ordinary arithmetic in which the basic elements are closed intervals of the real line. Given two intervals

$$\bar{z} = [a, b] \quad \text{and} \quad \bar{y} = [c, d],$$

interval arithmetic is defined by

$$\begin{aligned} \bar{z} + \bar{y} &\equiv [a, b] + [c, d] \equiv [a + c, b + d], \\ \bar{z} - \bar{y} &\equiv [a, b] - [c, d] \equiv [a - d, b - c], \\ \bar{z} \cdot \bar{y} &\equiv [a, b] \cdot [c, d] \equiv [\min(ac, ad, bc, bd), \max(ac, ad, bc, bd)], \\ \bar{z}/\bar{y} &\equiv [a, b]/[c, d] \equiv [a, b][1/d, 1/c], \end{aligned} \tag{1}$$

where $\bar{z}/\bar{y}$ is defined only if $0 \notin [c, d]$. The degenerate interval

$$\bar{z} \equiv [z, z]$$

will not be distinguished from the real number z. The width and midpoint are two commonly used scalar functions of intervals defined by

$$w(\bar{z}) \equiv w([a, b]) \equiv b - a,$$

$$m(\bar{z}) \equiv m([a, b]) \equiv (a + b)/2.$$

On the computer, *rounded interval arithmetic* (Refs. 5 and 8) is used in place of the exact version (1) to bound roundoff error. Although rounded intervals might be slightly larger, they are guaranteed to contain the exact result. Special machine-language programs (Refs. 9 and 10) have been written for rounded interval arithmetic.

If $f: R \to R$ is continuous, then an *interval extension* of f on $\bar{z}$ is an interval-valued function $\bar{f}$ which operates on intervals and satisfies

$$\bar{f}(z) = f(z) \qquad \text{for all } z \in \bar{z},$$

$$\bar{f}(\bar{u}) \subseteq \bar{f}(\bar{y}) \qquad \text{for all } \bar{u} \subseteq \bar{y} \subseteq \bar{z}.$$

Interval extensions are not unique, and

$$\{f(y) | y \in \bar{y}\} \subseteq \bar{f}(\bar{y}) \qquad \text{for all } \bar{y} \subseteq \bar{z}. \tag{2}$$

The *best* interval extension, or united extension (Ref. 5), is that function $\bar{f}$ in which equality holds in (2). If f is a rational function, then the natural extension of f is obtained by replacing z by $\bar{z}$ and performing the ordinary arithmetic operations in interval arithmetic.

3. Unidimensional Interval Newton Method

Suppose that all the solutions to

$$f(z) = 0 \tag{3}$$

are desired, where $f: R \to R$ is continuously differentiable. Moore (Ref. 5) was the first to present an interval version of Newton's method for finding a solution to (3). Assume that a continuous interval extension $\bar{f}'$ of the first derivative of f is available. If for some interval $\bar{z}^0$,

$$0 \notin \bar{f}'(\bar{z}^0), \tag{4}$$

then the *interval Newton operator* (INO)

$$\bar{N}(z, \bar{z}) \equiv z - f(z)/\bar{f}'(\bar{z})$$

is well defined for all $z \in \bar{z} \subseteq \bar{z}^0$. Suppose that it is not known if $\bar{z}^0$ contains a zero z^* of f. If (4) holds, then there is at most one zero of f in $\bar{z}^0$. In this case, the INO possesses three elegant existence properties (Refs. 5 and 7):

(i) if $z^* \in \bar{z}$, then $z^* \in \bar{N}(z, \bar{z})$ for all $z \in \bar{z} \subseteq \bar{z}^0$; (5)

(ii) if $\bar{z} \cap \bar{N}(z, \bar{z}) = \emptyset$ for any $z \in \bar{z} \subseteq \bar{z}^0$, then $z^* \notin \bar{z}$; (6)

(iii) if $\bar{N}(z, \bar{z}) \subseteq \bar{z}$ for any $z \in \bar{z} \subseteq \bar{z}^0$, then $z^* \in \bar{N}(z, \bar{z})$. (7)

The *interval Newton method* (INM), starting with $\bar{z}^0$ satisfying (4) and a point $z^0 \in \bar{z}^0$, generates an interval sequence $\{\bar{z}^n\}$ and a sequence of points $\{z^n\}$ defined by (Ref. 6)

$$\bar{z}^{n+1} = \bar{z}^n \cap \bar{N}(z^n, \bar{z}^n), \quad z^{n+1} \in \bar{z}^{n+1}, \qquad n = 0, 1, 2, \ldots.$$

The ordinary Newton method begins with a point z^0 and generates a sequence $\{z^n\}$ defined by

$$z^{n+1} = N(z^n), \qquad n = 0, 1, 2, \ldots, \tag{8}$$

where the ordinary Newton operator is given by

$$N(z) = z - f(z)/f'(z),$$

and it is assumed in (8) that $0 \neq f'(z^n)$ for all n. In the following optimization applications of the INM, the problem is to find the positive zeros of certain unidimensional functions. Since these functions are undefined for non-positive arguments, the iteration (8) cannot be directly applied, since there is no guarantee that z^{n+1} will be positive. The ordinary Newton method could be modified to

$$z^{n+1} = \max[\not{p}, N(z^n)], \qquad n = 0, 1, 2, \ldots, \tag{9}$$

where $\not{p}$ is some *appropriate* small positive number. In the INM, no such difficulty is present. If the INM is initiated with a positive interval $\bar{z}^0$, then $\bar{z}^n \subseteq \bar{z}^0$ is positive for all n; and $f(z^n)$ is well defined, since $z^n \in \bar{z}^n$.

Under certain conditions the INM never fails. If $z^* \in \bar{z}^0$ and (4) holds, then from Nickel (Ref. 6) the INM *always* converges with at least a superlinear rate. In addition, if f' is rational and $w(\bar{z}^0)$ is sufficiently small, then Moore (Ref. 5) showed that convergence is quadratic. Furthermore, if $z^* \in \bar{z}^0$, then by (5) $z^* \in \bar{z}^n$ for all n, so that $\{\bar{z}^n\}$ is a sequence of *error bounds* (Ref. 6) converging to z^*. This sequence of error bounds generated in the INM can be used as a stopping criterion.

4. Combined Interval Newton Method

The INM will find any zero of f in an interval $\bar{z}^0$ satisfying (4). To overcome this restriction and to account for multiple zeros of f, Moore (Ref. 5) suggested combining the INM with an *interval bisection method.* Moore's idea was programmed by Dargel, Loscalzo, and Witt (Ref. 8) to find error bounds on the real zeros of unidimensional rational functions. They studied only rational functions because they have natural extensions. However, their method finds error bounds on all zeros, in a given interval, of any function f having continuous interval extensions of f and f'.

The combined method performs interval bisections and stores the intervals on a list until (4) is satisfied, when the INO can be applied. Since Moore's combined method does not use the existence property (7) of the INO, a new combined INM is presented which directly incorporates (7); see Fig. 1.

Step 0. Place $\bar{z}^0$ in the IMZ list (intervals which *might* contain zeros).

Step 1. Check if IMZ is empty. If so, print the IDZ list (intervals which *do* contain a zero). Otherwise, pick some $\bar{z}$ from IMZ.

Step 2. Evaluate $\bar{f}(\bar{z})$. If $0 \notin \bar{f}(\bar{z})$, then $\bar{z}$ cannot contain a zero of f. Discard $\bar{z}$ and return to Step 1. Otherwise, $\bar{z}$ might contain a zero of f.

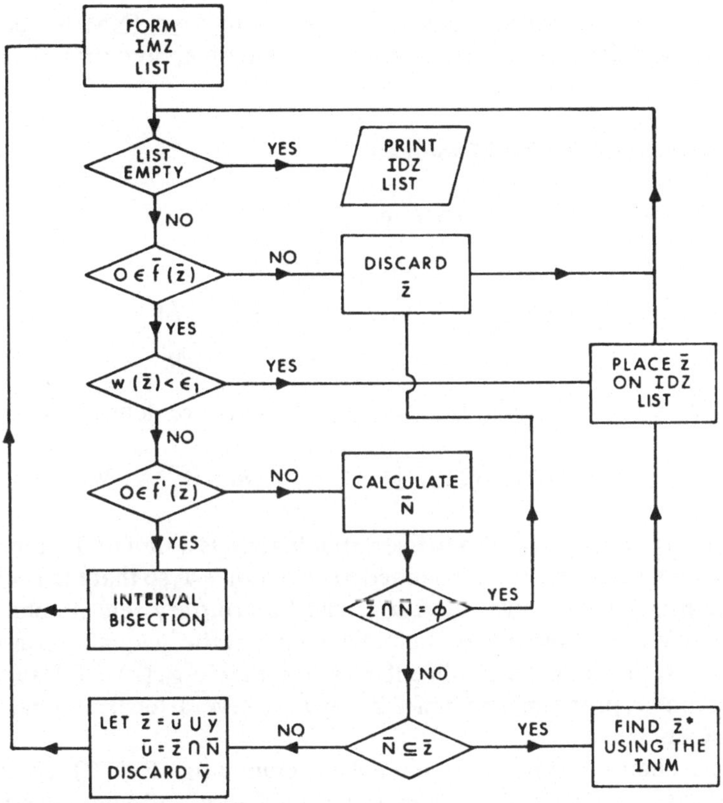

Fig. 1. Combined interval Newton method.

Step 3. If $w(\bar{z}) < \epsilon_1$, place $\bar{z}$ on IDZ and assign it a termination type 0. This signifies that a *small* interval has been found on which the interval extension of f contains zero (Ref. 8).

Step 4. Evaluate $\bar{f}'(\bar{z})$. If $0 \in \bar{f}'(\bar{z})$, then the INO cannot be applied. Bisect $\bar{z}$, place the bisected intervals on IMZ, and return to Step 1. Otherwise, $\bar{z}$ contains at most one zero z^* of f.

Step 5. Compute $\bar{N} \equiv \bar{N}(m(\bar{z}), \bar{z})$. Since $w(\bar{z}) \geq \epsilon_1$, it is easy to show (Ref. 11) that $\bar{z} \not\subseteq \bar{N}$. Therefore, there are three possible outcomes.

(a) $\bar{z} \cap \bar{N} = \emptyset$. By (6), $z^* \notin \bar{z}$. Discard $\bar{z}$ and return to Step 1.

(b) $\bar{N} \subseteq \bar{z}$. By (7), $z^* \in \bar{N}$. Use the INM to find an error bound $\bar{z}^*$ on z^* such that $w(\bar{z}^*) < \epsilon_2$. Convergence to z^* is guaranteed. Place $\bar{z}^*$ on IDZ with termination type 1 to signify verification of a zero, and return to Step 1.

(c) Here, $\bar{z}$ and $\bar{N}$ have a nonempty intersection $\bar{u}$. Let $\bar{z} = \bar{u} \cup \bar{y}$. By (5), if $z^* \in \bar{z}$ then $z^* \in \bar{u}$. Discard $\bar{y}$, place $\bar{u}$ on IMZ, and return to Step 1.

In the remaining sections, the above combined INM is applied to unidimensional problems in signomial programming, both primal and dual.

5. Unidimensional Dual Problems

The primal signomial program is

$$\begin{aligned} &\min g_0(x), \\ \text{s.t. } &\sigma_m g(x)^{\sigma_m} \leq 1, \qquad m = 1, \ldots, M, \\ &0 < x_n, \qquad n = 1, \ldots, N, \end{aligned} \tag{10}$$

where the $1+M$ signomial functions $g_m : R_+^N \to R$ are defined as

$$g_m(x) \equiv \sum_{t=1}^{T_m} \sigma_{mt} c_{mt} \prod_{n=1}^{N} x_n^{a_{mtn}}, \qquad m = 0, \ldots, M. \tag{11}$$

Each g_m is a function of N positive variables and is a sum of T_m terms. The *term signum functions* σ_{mt} are defined to be $+1$ or -1, so that the *coefficients* c_{mt} are positive. The *exponents* a_{mtn} can be arbitrary real numbers. The *constraint signum functions* σ_m define the sense of the inequality constraints. If $\sigma_m = +1$, then $g_m(x) \leq 1$; while if $\sigma_m = -1$, then $-g_m(x) \geq 1$. Usually (10) is not a convex program, and hence it may have several local minima, or even none at all.

The duality theory of signomial programming (Ref. 2) associates a nonnegative dual variable δ_{mt} with each primal term, and a nonnegative dual variable λ_m with each primal constraint. The dual linear constraints are

$$\sum_{t=1}^{T_0} \sigma_{0t} \delta_{0t} = \sigma_0, \tag{12}$$

$$\sum_{m=0}^{M} \sum_{t=1}^{T_m} \sigma_{mt} a_{mtn} \delta_{mt} = 0, \qquad n = 1, \ldots, N, \tag{13}$$

$$\lambda_m = \sigma_m \sum_{t=1}^{T_m} \sigma_{mt} \delta_{mt}, \qquad m = 1, \ldots, M, \tag{14}$$

where $\sigma_0 (= \pm 1)$ is a signum function associated with g_0. Now, (12)–(13) are $1+N$ linear conditions in T unknowns, where

$$T \equiv \sum_{m=0}^{M} T_m$$

is the total number of primal terms. Any well-formulated problem has

$$T \geq 1 + N,$$

and the *number of degrees of difficulty* is the difference $T-(1+N)$. It is assumed here that

$$T=N+2,$$

and hence the signomial program has *one degree of difficulty*.

It is also assumed that the primal program is either unconstrained or has all constraints active. In this case, all the dual variables are positive (Ref. 11), and $(\tilde{\delta}, \tilde{\lambda})$ is a *dual Kuhn–Tucker stationary point* (DKT) iff it is feasible [(12)–(14)] and satisfies the equilibrium condition (Ref. 3)

$$\prod_{m=0}^{M} \prod_{t=1}^{T_m} (\delta_{mt}/\lambda_m)^{\sigma_{mt}v_{mt}} = K. \tag{15}$$

In the above $\lambda_0 \equiv 1$, $v \in R^{N+2}$ is a solution to the homogeneous counterpart of (12)–(13); that is,

$$\begin{aligned} \sum_{t=1}^{T_0} \sigma_{0t}v_{0t} &= 0, \\ \sum_{m=0}^{M} \sum_{t=1}^{T_m} \sigma_{mt}a_{mtn}v_{mt} &= 0, \qquad n=1,\ldots,N, \end{aligned} \tag{16}$$

and the *equilibrium constant* K is defined as

$$K \equiv \prod_{m=0}^{M} \prod_{t=1}^{T_m} (c_{mt})^{\sigma_{mt}v_{mt}}.$$

Given a DKT stationary point $(\tilde{\delta}, \tilde{\lambda})$, its corresponding primal Kuhn–Tucker stationary point $\tilde{x}$ is easily obtained by solving any N linearly independent equations chosen from among the following $N+2$ equations:

$$\sum_{n=1}^{N} a_{0tn} \log x_n = \log[\tilde{\delta}_{0t}\sigma_0\, d(\tilde{\delta}, \tilde{\lambda})/c_{0t}], \qquad t=1,\ldots,T_0, \tag{17}$$

$$\sum_{n=1}^{N} a_{mtn} \log x_n = \log[\tilde{\delta}_{mt}/\tilde{\lambda}_m c_{mt}], \qquad m=1,\ldots,M, \qquad t=1,\ldots,T_m,$$

where $d(\delta, \lambda)$ is given by

$$d(\delta, \lambda) = \sigma_0\left[\prod_{m=0}^{M} \prod_{t=1}^{T_m} (c_{mt}\lambda_m/\delta_{mt})^{\sigma_{mt}\delta_{mt}}\right]^{\sigma_0}. \tag{18}$$

The dual variables δ can be partitioned and reindexed by a single subscript, so that (12)–(14) can be written as

$$\begin{aligned} W\delta_J + X\delta_I &= \begin{bmatrix} \sigma_0 \\ \underset{\sim}{0} \end{bmatrix}, \\ S\delta_J + U\delta_I - \lambda &= \underset{\sim}{0}, \end{aligned} \tag{19}$$

where $\delta_J \in R^{1+N}$, $\delta_I \in R$, and W is the nonsingular coefficient matrix of δ_J in (12)–(13). It is assumed without loss of generality that δ_J corresponds to the first $1+N$ indices. The term signum functions σ_{mt} and homogeneous solution v in (16) are indexed in a similar manner. However, the reindexed term signum functions are denoted by s_i, $i=1,\ldots,N+2$, to avoid confusion with the constraint signum functions σ_m. Note that the homogeneous solution v is given in the reindexed form by $(-W^{-1}X, 1)$.

Using (19), δ_I can be considered as the independent dual variable, since it determines the dual solution by

$$\begin{aligned} \delta_i &= h_i + v_i\delta_I, \qquad i=1,\ldots,1+N,\\ \lambda_m &= k_m + l_m\delta_I, \qquad m=1,\ldots,M, \end{aligned} \tag{20}$$

where

$$h \equiv W^{-1}\binom{\sigma_0}{0}, \qquad k \equiv Sh, \qquad l \equiv S(-W^{-1}X) + U.$$

Defining $h_{N+2} \equiv 0$ and taking the logarithm of (15) implies that $(\tilde{\delta}, \tilde{\lambda})$ is a DKT stationary point iff $\tilde{\delta}_I$ is a zero of the unidimensional *equilibrium function e* given by

$$e(\delta_I) = \sum_{i=1}^{N+2} s_i v_i \log(h_i + v_i\delta_I) - \sum_{m=1}^{M} q_m \log(k_m + l_m\delta_I) - \log K, \tag{21}$$

where the constants q_m are defined in the original indexing as

$$q_m \equiv \sum_{t=1}^{T_m} \sigma_{mt} v_{mt}, \qquad m=1,\ldots,M.$$

To apply the combined INM to e given by (21) it is necessary to have continuous interval extensions of e and e'. Using the monotonicity of the logarithm function, the united extension of each log term in (21) for $\bar{\delta}_I = [a, b]$ is

$$\alpha_1 \overline{\log}(\alpha_2 + \alpha_3\bar{\delta}_I) \equiv \begin{cases} [\alpha_1 \log(\alpha_2+\alpha_3 a), \alpha_1 \log(\alpha_2+\alpha_3 b)], & \text{if } \alpha_1\alpha_3 \geq 0,\\ [\alpha_1 \log(\alpha_2+\alpha_3 b), \alpha_1 \log(\alpha_2+\alpha_3 a)], & \text{if } \alpha_1\alpha_3 < 0, \end{cases} \tag{22}$$

for appropriate constants α_1, α_2, α_3. It is assumed in (22) that all the arguments for the logarithm function are positive. An interval extension $\bar{e}(\bar{\delta}_I)$ is evaluated by calculating the united extension of each log term and summing, using interval arithmetic. The continuity of $\bar{e}$ follows from Theorems 4.1 and 4.2 by Moore (Ref. 5).

Differentiating (21) yields

$$e'(\delta_I) = \sum_{i=1}^{N+2} s_i v_i v_i (h_i + v_i \delta_I)^{-1} - \sum_{m=1}^{M} q_m l_m (k_m + l_m \delta_I)^{-1}.$$

Since e' is rational, the natural extension can be used. Therefore, the combined INM can be used to find all the zeros of e. Using (20), all the dual stationary points can be determined; and the global solution (δ^*, λ^*) is that DKT stationary point with the smallest value of $d(\delta, \lambda)$ given by (18) (Ref. 2). Then, the global primal solution x^* satisfies

$$g_0(x^*) = d(\delta^*, \lambda^*),$$

and x^* can be calculated from (δ^*, λ^*) using (17).

Example 5.1. Consider the design of a hypothetical chemical plant (Fig. 2) which is an expanded version of a problem proposed by Wilde and Beightler (Ref. 4). The primal variables are x_1 = reactor temperature, x_2 = reactor pressure, x_3 = weight fraction of a catalyst used, and x_4 = weight fraction of product leaving the reactor. The cost functions are given in Table 1. The product concentration x_4 is a direct function of the temperature x_1 expressed as

$$x_4 = 0.1 x_1^{0.25}. \tag{23}$$

Using (23), the minimum cost design is the solution to an unconstrained signomial program with

$$g_0(x) \equiv 3.18(10^{-4}) x_1^{1.1} x_2^{0.6} + 114.3 x_2^{-1} x_3^{-1} + 2.28 x_3 + 0.015 x_2 - 3.38 x_1^{0.25}. \tag{24}$$

Since $T = 5$ and $N = 3$, (24) has one degree of difficulty.

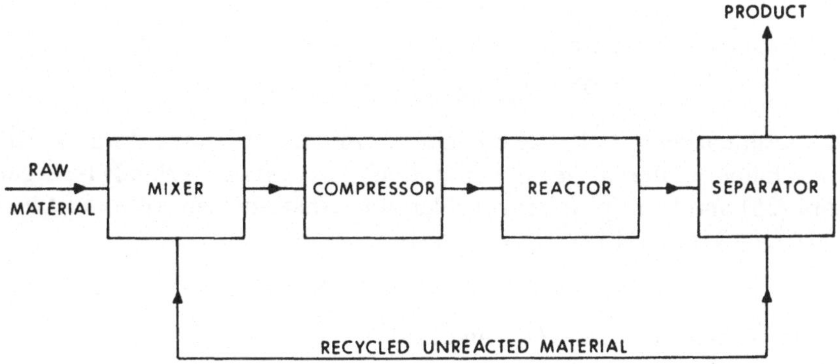

Fig. 2. Hypothetical chemical plant.

Table 1. Component cost functions for chemical plant.

Component	Cost ($/100 lb of material processed)
Reactor	$3.18(10^{-4})x_1^{1.1}x_2^{0.6}$
Separator	$114.3x_2^{-1}x_3^{-1}$
Catalyst	$2.28x_3$
Compressor	$0.015x_2$
Byproduct sales	$-33.8x_4$

Define the partitioning $\delta \rightarrow (\delta_J, \delta_I)$ as

$$\delta_J \equiv (\delta_{01}, \delta_{02}, \delta_{03}, \delta_{04}), \qquad \delta_I \equiv \delta_{05}.$$

The reduced linear system (20) is

$$\delta_i = h_i + v_i \delta_I, \qquad i = 1, \ldots, 4, \tag{25}$$

where

$$h = [0, -0.33333, -0.33333, -0.33333, 0],$$

$$v = [0.22727, 0.30303, 0.30303, 0.16667, 1].$$

The signum function σ_0 in (12) has been set to -1, since

$$\sigma_0 = \operatorname{sign}(g_0(x^*)),$$

and it is assumed that the minimum cost is negative. Note that (25) implies that $\delta_I > 2$ to maintain positivity.

The combined INM, using Zoltan's interval arithmetic package (Ref. 10), found all the zeros of e defined by (21) in

$$\bar{\delta}_I^0 = [2.01, 20]$$

with tolerances

$$\epsilon_1 = \epsilon_2 = 10^{-6}.$$

The result was a single error bound

$$\bar{\delta}_I^* = [2.3445110, 2.3445111]$$

with termination type 1. The number of bisections performed was 6, while the number of interval Newton operations was 5. It can certainly be argued from (25) and from physical considerations that any zero of e must lie in [2.01, 20]. Using (18) and (25), x^* satisfies

$$g_0(x^*) = d(\delta^*) = -7.8226,$$

and x^* is calculated using (17) and (23) as

$$x^* = [866.8, 29.94, 1.293, 0.5426].$$

6. Unidimensional Primal Problems

The unidimensional primal problem is to find all the stationary points of a signomial $g: R_+ \to R$ given by

$$g(x) = \sum_{t=1}^{T} \sigma_t c_t x^{a_t},$$

where the subscripts m and n have been dropped from (11). Equivalently, the problem is to find all zeros of g'. To apply the combined INM, continuous interval extensions of g' and g'' are required, where

$$g'(x) = \sum_{t=1}^{T} g_t(x), \qquad g''(x) = (1/x) \sum_{t=1}^{T} (a_t - 1) g_t(x),$$

$$g_t(x) \equiv \sigma_t c_t a_t x^{a_t - 1}, \qquad t = 1, \ldots, T.$$

Since the functions g_t are monotonic, their united extension on any positive interval $\bar{x} = [l, u]$ is

$$\bar{g}_t(\bar{x}) = \begin{cases} [\sigma_t c_t a_t l^{a_t-1}, \sigma_t c_t a_t u^{a_t-1}] & \text{if } \sigma_t a_t (a_t - 1) \geq 0, \\ [\sigma_t c_t a_t u^{a_t-1}, \sigma_t c_t a_t l^{a_t-1}] & \text{if } \sigma_t a_t (a_t - 1) < 0. \end{cases} \tag{26}$$

It follows from Moore's Theorems 4.1 and 4.2 (Ref. 5) and the inclusion monotonicity of interval arithmetic that

$$\bar{g}'(\bar{x}) \equiv \sum_{t=1}^{T} \bar{g}_t(\bar{x}), \qquad \bar{g}''(\bar{x}) \equiv (1/\bar{x}) \sum_{t=1}^{T} (a_t - 1) \bar{g}_t(\bar{x})$$

are continuous interval extensions of g' and g'' evaluated on $\bar{x}$.

Example 6.1. Suppose that

$$g(x) = -(8/3)x^{-1} - (1/6)x^3 + (19/6)x^2 - (34/3)x;$$

then, the united extensions $\bar{g}_t(\bar{x})$ [see Eq. (26)] are

$$\bar{g}_1(\bar{x}) = [(8/3)u^{-2}, (8/3)l^{-2}], \qquad \bar{g}_3(\bar{x}) = [(19/3)l, (19/3)u],$$

$$\bar{g}_2(\bar{x}) = [-(1/2)u^2, -(1/2)l^2], \qquad \bar{g}_4(\bar{x}) = -34/3.$$

Table 2 presents the results of the combined INM with

$$\bar{x}^0 = [0.1, 1000]$$

and

$$\epsilon_1 = \epsilon_2 = 10^{-6}.$$

Fourteen bisections and 19 interval Newton operations were performed.

Table 2. Combined interval Newton method on primal example problem.

Error bounds $\bar{x}^*$	Termination type
[0.5846508, 0.5846513]	1
[2.000000, 2.000001]	1
[10.515760, 10.515761]	1

Table 3 presents the results of applying the modified ordinary Newton's method (9) with $\not k = 0.1$. In this case, there is no guarantee that all the zeros can be found by changing the starting point x^0. Notice how a difference of only 0.001 in x^0 determined the zero to which the method converged. In each run, the procedure was terminated when

$$|g'(x^n)| < 10^{-6}.$$

Table 3. Modified ordinary Newton method on primal example problem.

N	A	B	C
0	4.1500000	4.1510000	4.1520000
1	1.0705934	1.0692110	1.0678268
2	4.1254240	4.1779889	4.2328025
3	1.1040196	1.0312252	0.94991002
4	3.2864825	7.4733179	0.10000000
5	1.7925042	14.517429	0.14804897
6	2.0020307	11.493324	0.21599735
7	2.0000000	10.608200	0.30616465
8		10.516755	0.41187205
9		10.515758	0.51001043
10			0.56872850
11			0.58385186
12			0.58464884
13			0.58465091

References

1. DUFFIN, R. J., PETERSON, E. L., and ZENER, C., *Geometric Programming*, John Wiley and Sons, New York, New York, 1966.
2. PASSY, U., and WILDE, D. J., *Generalized Polynomial Optimization*, SIAM Journal of Applied Mathematics, Vol. 15, pp. 1344–1356, 1967.
3. PASSY, U., and WILDE, D. J., Mass Action and Polynomial Optimization, Journal of Engineering Mathematics, Vol. 3, pp. 325–335, 1969.

4. WILDE, D. J., and BEIGHTLER, C. S., *Foundations of Optimization*, Prentice Hall, Englewood Cliffs, New Jersey, 1967.
5. MOORE, R. E., *Interval Analysis*, Prentice Hall, Englewood Cliffs, New Jersey, 1966.
6. NICKEL, K., *Triplex-Algol and Applications*, Topics in Interval Analysis, Edited by E. R. Hansen, Clarendon Press, Oxford, England, 1969.
7. HANSEN, E. R., Editor, *Topics in Interval Analysis*, Clarendon Press, Oxford, England, 1969.
8. DARGEL, R. H., LOSCALZO, F. R. and WITT, T. H., *Automatic Error Bounds On Real Zeros of Rational Functions*, Communications of the ACM, Vol. 9, pp. 806–809, 1966.
9. HANSEN, E. R., and SMITH, R., *A Computer Program for Solving a System of Linear Equations and Matrix Inversion with Automatic Error Bounding Using Interval Arithmetic*, Lockheed Missiles and Space Company, Report No. 4-22-66-3, 1966.
10. ZOLTAN, A. C., *Interval Arithmetic Subroutine Package for the IBM/360*, Universidad Central de Venezuela, Facultad de Ciencias, Departamento de Computacion, Report No. 69-05, 1969.
11. MANCINI, L. J., *Applications of Interval Arithmetic in Signomial Programming*, Stanford University, PhD Thesis, 1975.
12. MANCINI, L. J., and WILDE, D. J., *Signomial Dual Kuhn–Tucker Intervals*, Journal of Optimization Theory and Applications, Vol. 28, pp. 11–27, 1979.

19

Signomial Dual Kuhn–Tucker Intervals[1]

L. J. Mancini[2] and D. J. Wilde[3]

Abstract. Signomial programs are a special type of nonlinear programming problems which are especially useful in engineering design. This paper applies interval arithmetic, a generalization of ordinary arithmetic, to a dual equilibrium problem in signomial programming. Two constructive applications are considered. Application I involves uniqueness of local solutions; Application II involves existence and error bounds.

1. Introduction

If all the functions in a nonlinear programming problem are generalized polynomials, then the problem is called a signomial program. Many problems in engineering design and operations research have been formulated as signomial programs (Ref. 1). In general, signomial programs are nonconvex, and may have multiple local optima. A duality theory exists for signomial programs (Refs. 2 and 3) relating local solutions of the original, or primal, program to local solutions of a linearly constrained dual problem.

Interval arithmetic generalizes ordinary arithmetic to closed intervals of the real line (Ref. 4). Originally, algorithms using interval arithmetic were developed by numerical analysts to bound computer roundoff error (Ref. 5). Recently, it has been shown that interval analysis can be used to verify the uniqueness of solutions to nonlinear equations. Furthermore, an interval version of Newton's operator can be used to verify the existence of solutions

[1] The authors are grateful to the National Science Foundation for support through a Graduate Fellowship and Grant No. GK-41301.

[2] Supervisor, Operations Research Group, Logistics and Planning Applications Development Section, Computer Services Department, Standard Oil Company, San Francisco, California.

[3] Professor, Design Division, Department of Mechanical Engineering, Stanford University, Stanford, California.

and to compute error bounds (Ref. 5). These results have optimization applications since local solutions to nonlinear programs are related to solutions of nonlinear equations and inequalities.

This paper applies interval analysis to a dual equlibrium problem in signomial programming. It shows how the equilibrium conditions (Ref. 6) are transformed into a structure which is ideal for interval analysis. Two constructive applications are then considered. Application I deals with uniqueness. Under certain conditions, the uniqueness of a local solution to the dual signomial problem, in a certain region, is verified. This can be used in branch-and-bound algorithms for the global solution of signomial programs (Ref. 7), to discard regions from further consideration. Application II, first introduced by Robinson (Ref. 8) for general nonlinear programs, deals with existence and error bounds. Given an approximate dual solution, an interval Newton operator is used, under certain conditions, to verify the existence of an exact solution and to compute an error bound on it. Both procedures are illustrated with an example.

2. Dual Equilibrium Problem for Signomial Programs

The relationship between a signomial program and a certain dual equilibrium problem is presented here. The dual problem is a system of linear and nonlinear equations which define dual Kuhn–Tucker solutions. It will be transformed into a linear–logarithmic structure which is ideal for interval analysis.

A *primal signomial program* P can be written as

$$\min g_0(x),$$

s.t.

$$g_m(x) \leq 1, \qquad m = 1, \cdots, M,$$

where the $1+M$ *signomial functions* g_m are given by

$$g_m(x) = \sum_{t=1}^{T_m} \sigma_{mt} c_{mt} \prod_{n=1}^{N} x_n^{a_{mtn}}, \qquad m = 0, \cdots, M.$$

Each g_m is a function of N positive variables (i.e., the vector x) and is the sum of T_m terms. The term *signum functions* σ_{mt} are either $+1$ or -1, so that the corresponding term *coefficients* c_{mt} are positive. The reason for separating the coefficients from their signs will become clear when the dual problem is presented. The *exponents* a_{mtn} are arbitrary real numbers. In general, P is not a convex program, and hence it may have several local minima, or even none at all. If all signum functions σ_{mt} are $+1$, then the functions g_m are called *posynomials*, and P can be transformed into a convex program (Ref. 9). Other notations for defining P which use two subscripts with index sets and no signum functions are available (Ref. 2). This notation is used for engineering design purposes (Ref. 10).

The duality theory for signomial programs (Ref. 3) relates primal Kuhn–Tucker solutions to Kuhn–Tucker solutions of a linearly constrained dual problem. It associates a nonnegative dual variable δ_{mt} with each term in the primal program, a total of

$$T = \sum_{m=0}^{M} T_m$$

dual variables. Furthermore, a nonnegative λ_m is associated with each primal constraint function. The dual constraints are

$$\begin{aligned} &\sum_{t=1}^{T_0} \sigma_{0t}\delta_{0t} = \sigma_0, \\ &\sum_{m=0}^{M} \sum_{t=1}^{T_m} \sigma_{mt}a_{mtn}\delta_{mt} = 0, \qquad n = 1, \cdots, N, \end{aligned} \tag{1}$$

$$\sum_{t=1}^{T_m} \sigma_{mt}\delta_{mt} - \lambda_m = 0, \qquad m = 1, \cdots, M, \tag{2}$$

where a_{mtn} are the exponents and σ_{mt} are the signum functions in the primal program. The signum function σ_0 $(=\pm 1)$ on the right-hand side of the first constraint is related to the sign of g_0 and is assumed given. Note that Eqs. (2) merely define the λ variables in terms of δ. Also note that the dual constraints are independent of the primal coefficients c_{mt}, a useful property for design problems (Ref. 10).

Equations (1) are $1+N$ linear equations in the T dual variables δ. Any well-formulated problem has

$$T \geq 1+N,$$

and it can be assumed that Eqs. (1) have full row rank (see Ref. 11). The difference

$$D = T-(1+N)$$

is called the *degrees of difficulty* of the signomial program. If $D = 0$, then the unique dual solution is obtained by solving the square linear system defined by Eqs. (1) (e.g., see Ref. 12). Assume hereafter that $D > 0$.

Assume that the primal program is either unconstrained or has all strongly binding constraints. This assumption can be relaxed using results presented in Ref. 2. In this case, all dual Kuhn–Tucker solutions (δ^*, λ^*) are strictly positive (see Ref. 11). Define the vectors $v^{(r)} = (v_{mt}^{(r)})$ as the D linearly independent solutions to the homogeneous counterpart of Eqs. (1). That is, $v^{(r)}$, $r = 1, \cdots, D$, satisfy

$$\sum_{t=1}^{T_0} \sigma_{0t} v_{0t}^{(r)} = 0,$$
$$\sum_{m=0}^{M} \sum_{t=1}^{T_m} \sigma_{mt} a_{mtn} v_{mt}^{(r)} = 0, \qquad n = 1, \cdots, N, \tag{3}$$

and are easily computed using linear operations on Eqs. (1). Passy and Wilde (Ref. 6) have shown that (δ^*, λ^*) is a dual Kuhn–Tucker solution iff it is feasible [i.e., satisfies Eqs. (1)–(2)] and satisfies the *equilibrium conditions*

$$\prod_{m=0}^{M} \prod_{t=1}^{T_m} (\delta_{mt}/\lambda_m)^{\sigma_{mt} v_{mt}^{(r)}} = \prod_{m=0}^{M} \prod_{t=1}^{T_m} (c_{mt})^{\sigma_{mt} v_{mt}^{(r)}}, \qquad r = 1, \cdots, D, \tag{4}$$

where $\lambda_0 = 1$. The equilibrium conditions derive their name because of their relationship to the chemical equilibrium expression of the law of mass action. Given a dual Kuhn–Tucker solution, its corresponding primal Kuhn–Tucker solution is obtained by solving a square linear system (Ref. 3).

3. Interval Analysis

This section presents the interval analysis required for the paper. Interval arithmetic (Ref. 4) generalizes ordinary arithmetic to closed intervals of the real line. Given two intervals $\bar{u} = [a, b]$ and $\bar{w} = [c, d]$, *interval arithmetic* is defined by

$$\bar{u} + \bar{w} = [a + c, b + d],$$
$$\bar{u} - \bar{w} = [a - d, b - c],$$
$$\bar{u} \cdot \bar{w} = [\min(ac, ad, bc, bd), \max(ac, ad, bc, bd)],$$
$$\bar{u}/\bar{w} = [a, b] \cdot [1/d, 1/c],$$

where $\bar{u}/\bar{w}$ is defined only if $0 \notin [c, d]$. On the computer, *rounded interval arithmetic* is used in place of the exact version above to bound roundoff error. Rounded intervals might be slightly larger, but they are guaranteed to contain the exact result. Special machine-language programs have been written for rounded interval arithmetic (e.g., see Ref. 13).

Interval matrices are rectangular arrays with intervals as components. A square interval matrix $\bar{A}$ is nonsingular iff all ordinary matrices $A \in \bar{A}$ are nonsingular, that is,

$$0 \notin \{\det A \,|\, A \in \bar{A}\}.$$

Usually it is not possible to compute the right-hand side of the above exactly. However, an *interval determinant* $\overline{\det}\,\bar{A}$ satisfying

$$\{\det A \,|\, A \in \bar{A}\} \subseteq \overline{\det}\,\bar{A} \tag{5}$$

can be computed. If $\bar{A}$ is nonsingular, then an *interval inverse* $(\bar{A})^{-1}$ satisfies

$$\{A^{-1} \,|\, A \in \bar{A}\} \subseteq (\bar{A})^{-1}.$$

If f: $R^N \to R$ is continuous, then an *interval extension* of f, on an interval vector $\bar{u} \subset R^N$, is an interval-valued function $\bar{f}$, which satisfies

$$\bar{f}(u) = f(u) \qquad \text{for all } u \in \bar{u},$$

$$\bar{f}(\bar{w}) \subseteq \bar{f}(\bar{v}) \qquad \text{for all } \bar{w} \subseteq \bar{v} \subseteq \bar{u}.$$

Interval extensions, like interval determinants and interval inverses, are not unique; and

$$\{f(w) \,|\, w \in \bar{w}\} \subseteq \bar{f}(\bar{w}) \tag{6}$$

for all $\bar{w} \subseteq \bar{u}$. The *best* interval extension, or *united extension* (see Ref. 4), is the function $\bar{f}$ in which equality holds in Eq. (6). That is, if $\bar{f}$ is a united extension, then

$$[\min_{w \in \bar{w}} f(w), \max_{w \in \bar{w}} f(w)] = \bar{f}(\bar{w}).$$

If f is a rational function, then the *natural extension* is obtained by replacing u by $\bar{u}$ and performing the ordinary arithmetic operations in interval arithmetic. Finding good computable interval extensions, limiting the difference between the left-hand and right-hand sides of Eq. (6), is one of the key problems in interval analysis.

Suppose that f: $R^N \to R^N$ is continuously differentiable, and let u^* denote a zero of f. Assume continuous (see Ref. 4) interval extensions $\nabla \bar{f}_{ij}$ of the partial derivatives

$$\nabla f_{ij}(u) = \partial f_{i(u)}/\partial u_j, \qquad i, j = 1, \cdots, N,$$

are available on some $\bar{u} \subset R^N$. For $\bar{w}$, $\bar{v} \subseteq \bar{u}$, define

$$\nabla \bar{f}_{ij}(\bar{w}, \bar{v}) = \nabla \bar{f}_{ij}(\bar{w}_1, \cdots, \bar{w}_{j-1}, \bar{v}_j, \cdots, \bar{v}_n), \qquad i, j = 1, \cdots, N,$$

and let $\nabla \bar{f}(\bar{w}, \bar{v})$ be the interval matrix with components $\nabla \bar{f}_{ij}(\bar{w}, \bar{v})$. If $\nabla \bar{f}(\bar{u}, \bar{u})$ is nonsingular in the interval sense, then $\nabla \bar{f}(\bar{w}, \bar{v})$ is nonsingular for all $\bar{w}$, $\bar{v} \subseteq \bar{u}$; and an *interval Newton operator* $\bar{N}$ can be defined as

$$\bar{N}(w, \bar{w}) = w - \nabla\bar{f}(w, \bar{w})^{-1} \cdot f(w) \qquad \text{for all } w \in \bar{w} \subseteq \bar{u}.$$

Nickel (Ref. 5) has proven that the nonsingularity of interval Jacobians and the above interval Newton operator have some elegant properties.

Theorem 3.1. Suppose that $\nabla\bar{f}(\bar{u}, \bar{u})$ is nonsingular. Then,

(a) any zero u^* of f in $\bar{u}$ is unique;
(b) if $u^* \in \bar{u}$, then $u^* \in \bar{N}(u, \bar{u})$ for all $u \in \bar{u}$;
(c) if $\bar{u} \cap \bar{N}(u, \bar{u}) = \phi$ for any $u \in \bar{u}$, then $u^* \notin \bar{u}$;
(d) if $\bar{N}(u, \bar{u}) \subseteq \bar{u}$ for any $u \in \bar{u}$, then $u^* \in \bar{N}(u, \bar{u})$.

Note that Part (a), concerning uniqueness, only uses the nonsingularity of the interval Jacobian $\nabla\bar{f}(\bar{u}, \bar{u})$. Part (b) shows that the interval Newton operator never loses a zero. If the hypotheses in Part (d) hold, then the existence of a zero u^* of f in $\bar{u}$ is verified, u^* is contained in the interval $\bar{N}(u, \bar{u})$, and u^* is the unique zero of f in $\bar{u}$.

It cannot be overemphasized how much the usefulness of the above results depends on good interval extensions, which the sequel shows are available for the dual equilibrium problem.

4. Reduced Dual Problem

Our objective is to apply the uniqueness and existence properties of interval analysis to signomial programs. However, computable interval extensions of most nonlinear functions, like the second partial derivatives of signomial functions, are usually very inexact. That is, the difference between the left-hand and right-hand sides of Eq. (6) is large. For this reason, the dual equilibrium problem introduced in Section 2 is used. We will show that the partial derivatives of the equilibrium conditions have natural, united, interval extensions: an ideal structure for interval analysis.

For computational reasons, it is convenient to deal with a reduced dual problem. The reduction is obtained using the linearity of the dual constraints, Eqs. (1)–(2). Defined the *equilibrium constants* K_r as the left-hand sides of Eqs. (4):

$$K_r = \prod_{m=0}^{M} \prod_{t=1}^{T_m} (c_{mt})^{\sigma_{mt} v_{mt}^{(r)}}, \qquad r = 1, \cdots, D.$$

Taking the logarithms of Eqs. (4) implies that (δ^*, λ^*) is a dual Kuhn–Tucker solution if it is feasible and is a zero of the nonlinear system

$$E(\delta, \lambda) = \mathbf{0}, \tag{7}$$

where

$$E(\delta, \lambda)_r = \sum_{m=0}^{M} \sum_{t=1}^{T_m} \sigma_{mt} v_{mt}^{(r)} \log \delta_{mt} - \sum_{m=1}^{M} b_{mr} \log \lambda_m - \log K_r, \qquad r = 1, \cdots, D.$$

The constants b_{mr} in the above system are defined by

$$b_{mr} = \sum_{t=1}^{T_m} \sigma_{mt} v_{mt}^{(r)}, \qquad m = 1, \cdots, M, \qquad r = 1, \cdots, D.$$

Note that Eqs. (7) are linear in the logarithms of the dual variables.

Since Eqs. (2) merely define the λ variables in terms of δ, the λ variables can be considered dependent dual variables. Linear operations on Eqs. (1) will create $1+N$ additional dependent dual variables [i.e., there is only $T-(1+N)=D$ degrees of freedom in Eqs. (1)–(2)]. Partition δ into (z, y), where $(z_1, \cdots, z_{(1+N)})$ are the additional dependent dual variables and $(y_1, \cdots, y_D)$ are the independent dual variables. There is no loss of generality in assuming that z corresponds to the first $1+N$ components of δ. The signum functions σ_{mt} and homogeneous constants $v_{mt}^{(r)}$ are reindexed similarly. The dual constraints (1)–(2) can now be written in terms of the independent dual variables

$$\begin{aligned} z(y)_i &= h_i + \sum_{r=1}^{D} v_i^{(r)} y_r, \qquad i = 1, \cdots, 1+N, \\ \lambda(y)_m &= l_m + \sum_{r=1}^{D} d_{mr} y_r, \qquad m = 1, \cdots, M, \end{aligned} \tag{8}$$

where $v_i^{(r)}$ are the homogeneous constants and h_i, l_m, d_{mr} are derived constants.

Using Eqs. (8), the equilibrium conditions (7) can also be written in terms of the independent dual variables. That is, $(z(y^*), y^*, \lambda(y^*))$ is a dual Kuhn–Tucker solution iff y^* is a zero of $E(y)$, where

$$E(y) = E[z(y), y, \lambda(y)]. \tag{9}$$

Differentiating Eqs. (9) yields

$$\begin{aligned} \frac{\partial E(y)_r}{\partial y_s} &= \frac{\partial E[z(y), y, \lambda(y)]_r}{\partial y_s}, \qquad r = 1, \cdots, D, \qquad s = 1, \cdots, D, \\ &= \sum_{i=1}^{1+N} \frac{\sigma_i v_i^{(r)}}{z_i} \frac{\partial z_i}{\partial y_s} + \sum_{j=1}^{D} \frac{\sigma^{(1+N)+j} v_{(1+N)+j}^{(r)}}{y_j} \frac{\partial y_j}{\partial y_s} - \sum_{m=1}^{M} \frac{b_{mr}}{\lambda_m} \frac{\partial \lambda_m}{\partial y_s} \\ &= \sum_{i=1}^{1+N} \frac{\sigma_i v_i^{(r)} v_i^{(s)}}{z_i} + \frac{\sigma_{(1+N)+s} v_{(1+N)+s}^{(r)}}{y_s} - \sum_{m=1}^{M} \frac{b_{mr} d_{ms}}{\lambda_m}. \end{aligned} \tag{10}$$

The above shows that each element of the Jacobian matrix $\nabla E(y)$ is a rational function of the dual variables and thus has a natural interval extension. It is obtained by replacing y by $\bar{y}$, computing Eqs. (8) in interval arithmetic to obtain $\bar{z}(\bar{y})$ and $\bar{\lambda}(\bar{y})$, and then evaluating Eq. (10) in interval arithmetic. Since each dual variable in Eq. (10) occurs at most once, the natural extension equals the *exact* united extension (see Ref. 4). It is this property of the equilibrium conditions which makes them amenable to interval analysis.

5. Applications of Interval Analysis

Two applications of interval analysis are considered here. Both deal with the zeros y^* of the nonlinear system described by Eqs. (9). With Eqs. (8), these zeros define dual Kuhn–Tucker solutions $(z(y^*), y^*, \lambda(y^*))$. Application I involves the uniqueness of a zero (y^*) in intervals $\bar{y}$ containing it. The objective is to use the nonsingularity of the interval Jacobian to verify uniqueness in *large* intervals. This can be used in branch-and-bound algorithms for the global solutions of signomial programs, (e.g., see Ref. 7) to discard regions from further consideration. Application II, first introduced by Robinson (Ref. 8) for general nonlinear programs, involves proving the *existence* of a zero (y^*) and computing an *error bound* on it. The interval Newton operator is used here.

5.1. Application I uses Part (a) of Nickel's theorem to prove the uniqueness of a known zero (y^*) in an interval $\bar{y}$ containing it. Given $\bar{y}$, Eqs. (8) are used to compute intervals on the dependent dual variables

$$\begin{aligned} \bar{z}(\bar{y})_i &= \bar{h}_i + \sum_{r=1}^{D} \bar{v}_i^{(r)} \bar{y}_r, \qquad i = 1, \cdots, 1+N, \\ \bar{\lambda}(\bar{y})_m &= \bar{l}_m + \sum_{r=1}^{D} \bar{d}_{mr} \bar{y}_r, \qquad m = 1, \cdots, M. \end{aligned} \tag{11}$$

Note that the constants in Eqs. (8) have been replaced by intervals. This is because these constants are the result of linear operations on the dual constraint set and thus contain roundoff error. Computer programs (Ref. 5) are available which use interval arithmetic to solve a system of linear equations with automatic error bounding.

Given $(\bar{z}(\bar{y}), \bar{y}, \bar{\lambda}(\bar{y}))$, the partial derivatives (10) are evaluated in interval arithmetic, and the interval Jacobian $\overline{\nabla E}(\bar{y})$ is formed. If $\overline{\nabla E}(\bar{y})$ is nonsingular in the interval sense, then y^* is the unique zero in $\bar{y}$. The nonsingularity is determined by computing an interval determinant

$\overline{\det}\,\overline{\nabla E}(\bar{y})$ satisfying Eq. (5). Thus, y^* is the unique zero in $\bar{y}$ if

$$0 \notin \overline{\det}\,\overline{\nabla E}(\bar{y}). \tag{12}$$

5.2. Application II uses the interval Newton operator to prove the existence of a zero, y^*, and to calculate an error bound on it. All parts of Nickel's theorem are involved. Suppose that $\tilde{y}$ is an approximate zero of Eqs. (9), obtained as a result of some numerical optimization procedure. Let $\bar{y}$ be a *small* interval containing $\tilde{y}$ and hopefully the true zero (y^*). As in Application I, compute intervals on the dependent dual variables using Eqs. (11), form the interval Jacobian $\overline{\nabla E}(\bar{y})$, and determine if $\overline{\nabla E}(\bar{y})$ is nonsingular. If Eq. (12) holds, the interval Newton operator

$$\bar{N}(\tilde{y}, \bar{y}) = \tilde{y} - (\overline{\nabla E}(\bar{y}))^{-1} \cdot E(\tilde{y}) \tag{13}$$

is well defined.

Three results are possible. The desired result is to have the interval Newton operator contained in $\bar{y}$, that is, $\bar{N}(\tilde{y}, \bar{y}) \subseteq \bar{y}$. In this case, the existence of a zero (y^*) in $\bar{y}$ has been verified, y^* is in $\bar{N}(\tilde{y}, \bar{y})$, and y^* is the unique zero in $\bar{y}$. Existence is proven, and $\bar{N}(\tilde{y}, \bar{y})$ is an error bound on y^*. Another possible result is that $\bar{N}(\tilde{y}, \bar{y})$ and $\bar{y}$ have an empty intersection, that is,

$$\bar{N}(\tilde{y}, \bar{y}) \cap \bar{y} = \phi$$

In this case, $\bar{y}$ does not contain a zero. The remaining possibility is that

$$\bar{N}(\tilde{y}, \bar{y}) \cap \bar{y} = \bar{\omega} \neq \phi,$$

and $\bar{\omega}$ is not contained in $\bar{y}$. In this case, if $\bar{y}$ contains a zero (y^*), then y^* is the unique zero in $\bar{y}$ and y^* is in $\bar{\omega}$. This follows from the fact that the interval Newton operator never loses a zero.

6. Numerical Procedures

Applications I and II require two numerical procedures. First, to verify the nonsingularity of the interval Jacobian, a method for calculating an interval determinant $\overline{\det}\,\overline{\nabla E}(\bar{y})$ is required. The method used here is to diagonalize $\overline{\nabla E}(\bar{y})$, using partial pivoting in interval arithmetic, to obtain an upper triangular $\bar{U}$. Then,

$$\overline{\det}\,\overline{\nabla E}(\bar{y}) = \prod_{i=1}^{D} \bar{U}_{ii}. \tag{14}$$

It is easy to show that the above satisfies Eq. (5). In Ref. 14, Hansen and Smith give an alternate method.

Second, a method for calculating the interval Newton operator $\bar{N}(\tilde{y}, \bar{y})$ in Eq. (13) is required. Assuming that

$$0 \neq \det \nabla E(\tilde{y}),$$

the ordinary Newton operator

$$N(\tilde{y}) = \tilde{y} - \nabla E(\tilde{y})^{-1} \cdot E(\tilde{y})$$

is calculated by solving the linear system

$$\nabla E(\tilde{y})u = -E(\tilde{y})$$

for the correction vector u and then augmenting $\tilde{y}$ to obtain

$$N(\tilde{y}) = \tilde{y} + u.$$

The interval Newton operator is computed in an analogous manner. Assuming that Eq. (12) holds, $\bar{N}(\tilde{y}, \bar{y})$ is calculated by solving the interval linear system

$$\overline{\nabla E}(\bar{y})u = -E(\tilde{y}) \tag{15}$$

for the *interval correction vector* $\bar{u}$ and augmenting $\tilde{y}$ to obtain

$$\bar{N}(\tilde{y}, \bar{y}) = \tilde{y} + \bar{u}.$$

The theoretical solution set Ω to Eq. (15) is

$$\Omega = \{u \mid Au = -E(\tilde{y}), A \in \overline{\nabla E}(\bar{y})\}.$$

An interval solution $\bar{u}$ must satisfy $-E(\tilde{y}) \in \overline{\nabla E}(\bar{y})\bar{u}$, and hence contains Ω. The $\bar{u}$ with the smallest width (see Ref. 4) is desired to minimize the difference between $\overline{\nabla E}(\bar{y})\bar{u}$ and $-E(\tilde{y})$. In Ref. 14, Hansen and Smith present five methods for solving interval linear systems. Their recommended procedure (Method 4) is used here. In Ref. 15, Hansen presents a refinement procedure which may improve a given interval solution.

It cannot be overemphasized how much the usefulness of Application I and II depends on good numerical procedures for calculating interval determinants and solving interval linear systems.

7. Example

Consider the design of the hypothetical chemical plant in Fig. 1, which is an expanded version of a model proposed by Wilde and Beightler (Ref. 12).

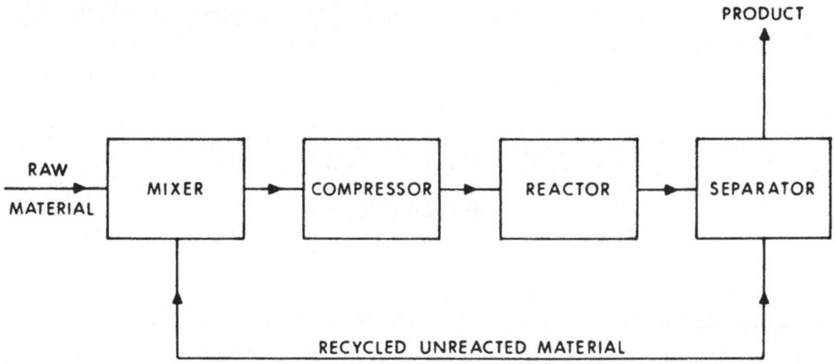

Fig. 1. Hypothetical chemical plant.

The design variables are

x_1 = reactor temperature,
x_2 = reactor pressure,
x_3 = weight fraction of a catalyst used,
x_4 = weight fraction of the product leaving the reactor.

The product x_4 is an explicit function of the temperature x_1, given as

$$x_4 = 0.1x_1^{0.25}.$$

Summing the terms in Table 1 and substituting for x_4 yields the cost function

$$g_0(x) = 0.0318x_1^{1.1}x_2^{0.6} + 11430x_2^{-1}x_3^{-1} + 228x_3 + 1.5x_2 - 25x_1^{0.25} - 495x_1^{0.2}.$$

Minimizing the above objective function is an unconstrained signomial program.

Table 1. Component costs for chemical plant.

Item	Cost ($/100 lb of material processed)
Reactor	$0.0318\,x_1^{1.1}x_2^{0.6}$
Separator	$11430\,x_2^{-1}x_3^{-1}$
Catalyst	$228\,x_3$
Compressor	$1.5\,x_2$
Recycle compressor	$250\,(1-x_4)$
Byproduct sales	$-3123.2\,x_4^{0.8}$

Since $T = 6$ and $N = 3$, the problem has two degrees of difficulty (i.e., $D = 2$). The dual constraints (1) are

$$\begin{aligned} \delta_{01} + \delta_{02} + \delta_{03} + \delta_{04} - \delta_{05} - \delta_{06} &= -1, \\ 1.1\delta_{01} - 0.25\delta_{05} - 0.2\delta_{06} &= \ \ 0, \\ 0.6\delta_{01} - \delta_{02} + \delta_{04} &= \ \ 0, \\ -\delta_{02} + \delta_{03} &= \ \ 0. \end{aligned}$$

The signum function σ_0 in Eqs. (1) is set to -1, since it is assumed the minimum cost is negative (see Ref. 3). Since the problem is unconstrained, no λ variables exist, and thus Eqs. (2) are not present.

The homogeneous vectors $v^{(r)}$ are [see Eqs. (3)]

$$\begin{bmatrix} 0.22727 & 0.30303 & 0.30303 & 0.16667 & 1 & 0 \\ 0.18182 & 0.30909 & 0.30909 & 0.20000 & 0 & 1 \end{bmatrix}^T.$$

Let the independent dual variables (y_1, y_2) be $(\delta_{05}, \delta_{06})$. Then, the reduced linear system (8) is

$$\begin{bmatrix} z_1 \\ z_2 \\ z_3 \\ z_4 \end{bmatrix} = \begin{bmatrix} 0 \\ -0.33333 \\ -0.33333 \\ -0.33333 \end{bmatrix} + \begin{bmatrix} v_1^{(1)} & v_1^{(2)} \\ v_2^{(1)} & v_2^{(2)} \\ v_3^{(1)} & v_3^{(2)} \\ v_4^{(1)} & v_4^{(2)} \end{bmatrix} \begin{bmatrix} y_1 \\ y_2 \end{bmatrix}, \tag{16}$$

where the constants $v_i^{(r)}$ are given above. Assume that the constants in Eqs. (16) are exact (i.e., no roundoff error occurred in the reduction). The partial

Table 2. Application II examples.

Case	$\tilde{y}$	$\bar{y}$	$\bar{N}(\tilde{y}, \bar{y})$, Eq. (13)
1	$\begin{bmatrix} 0.12992 \\ 1.8501 \end{bmatrix}$	$\begin{bmatrix} [0.12862, 0.13123] \\ [1.8316, 1.8687] \end{bmatrix}$	$\begin{bmatrix} [0.12727, 0.12837] \\ [1.8229, 1.8267] \end{bmatrix}$
2	$\begin{bmatrix} 0.12992 \\ 1.8501 \end{bmatrix}$	$\begin{bmatrix} [0.12472, 0.13512] \\ [1.7761, 1.9242] \end{bmatrix}$	$\begin{bmatrix} [0.12484, 0.13059] \\ [1.8160, 1.8320] \end{bmatrix}$
3	$\begin{bmatrix} 0.12772 \\ 1.8240 \end{bmatrix}$	$\begin{bmatrix} [0.12484, 0.13059] \\ [1.8160, 1.8320] \end{bmatrix}$	$\begin{bmatrix} [0.12787, 0.12794] \\ [1.8257, 1.8260] \end{bmatrix}$

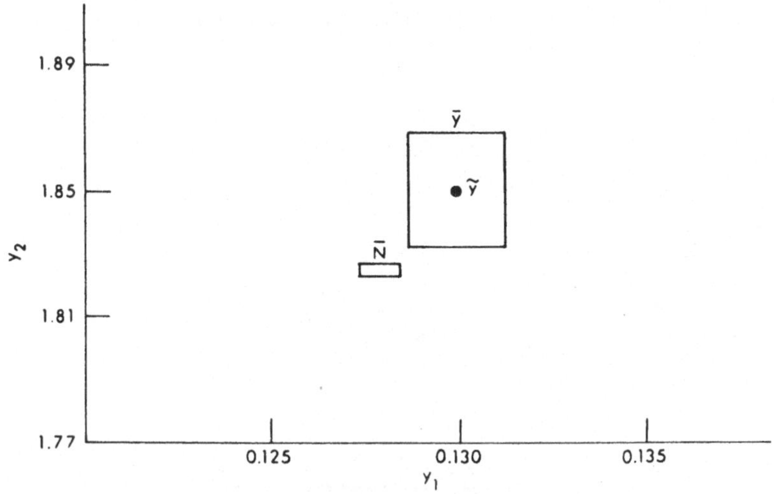

Fig. 2. Application II example, Case 1.

derivatives of the reduced equilibrium problem (10) are

$$\frac{\partial E_r}{\partial y_s}=\frac{v_1^{(r)}v_1^{(s)}}{z_1}+\frac{v_2^{(r)}v_2^{(s)}}{z_2}+\frac{v_3^{(r)}v_3^{(s)}}{z_3}+\frac{v_4^{(r)}v_4^{(s)}}{z_4}-\frac{v_{4+s}^{(r)}}{y_s}, \qquad r=1,2,\ s=1,2. \tag{17}$$

An approximate zero $\tilde{y}$ to Eqs. (9), obtained using Dembo's algorithm for signomial programs (Ref. 16), is

$$\tilde{y}=\begin{bmatrix}0.12992\\1.8501\end{bmatrix}. \tag{18}$$

First, Application II will be used to prove the existence of a zero (y^*) near $\tilde{y}$ and to give an error bound on it. Then, Application I will be used to investigate the uniqueness of y^*. All interval arithmetic operations will be performed using Zoltan's rounded interval arithmetic package (Ref. 13).

Some results implementing Application II are given in Table 2 and shown graphically in Figs. 2–4. In Case 1, an interval $\bar{y}$ of $\pm 1\%$ is placed around $\tilde{y}$ in Eq. (18). Evaluating Eqs. (16) with $\tilde{y}$ and $\bar{y}$ yields

$$z(\tilde{y})=\begin{bmatrix}0.36591\\0.27789\\0.27789\\0.058341\end{bmatrix}, \qquad \bar{z}(\bar{y})=\begin{bmatrix}[0.36225, 0.36957]\\ [0.27177, 0.28400]\\ [0.27177, 0.28400]\\ [0.054424, 0.062259]\end{bmatrix}.$$

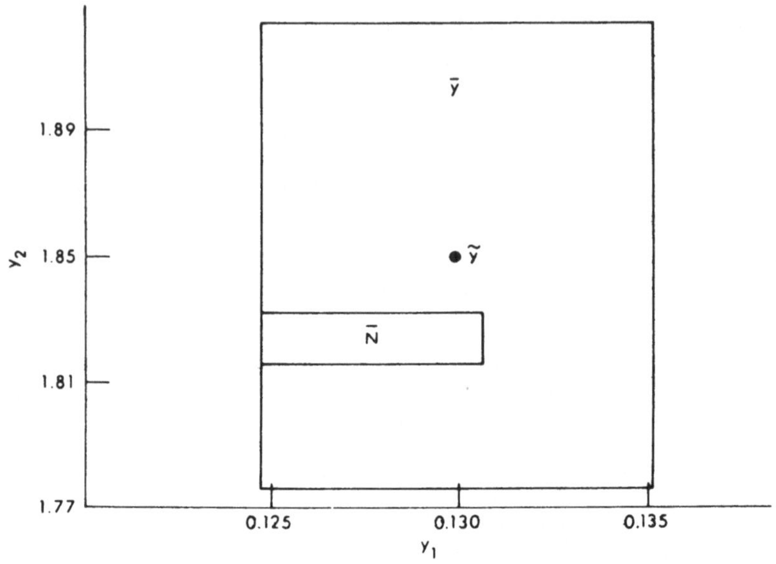

Fig. 3. Application II example, Case 2.

Evaluating Eqs. (17) using $(\bar{z}(\bar{y}), \bar{y})$ yields

$$\overline{\nabla E}(\bar{y}) = \begin{bmatrix} [-6.5421, -6.2919] & [1.3068, 1.4159] \\ [1.3068, 1.4159] & [0.85876, 0.99412] \end{bmatrix}.$$

Since $\overline{\nabla E}(\bar{y})$ is nonsingular, the interval Newton operator (13) is well defined. An interval solution $\bar{u}$ to Eqs. (15), obtained using Hansen and Smith's (Ref. 14) Method 4 is

$$\bar{u} = \begin{bmatrix} [-0.0026495, -0.0015539] \\ [-0.027147, -0.023496] \end{bmatrix}.$$

Figure 2 shows that the interval Newton operator

$$\bar{N}(\tilde{y}, \bar{y}) = \tilde{y} + \bar{u}$$

and $\bar{y}$ have an empty intersection. Thus, $\bar{y}$ does not contain a zero, and the approximate solution $\tilde{y}$ in Eq. (18) was not within one percent.

In Case 2, an interval $\bar{y}$ of $\pm 4\%$ is placed around $\tilde{y}$ in Eq. (18), and the above procedure is repeated. In this case, $\bar{N}(\tilde{y}, \bar{y}) \subseteq \bar{y}$ (see Fig. 3). Therefore, the existence of zero (y^*) has been verified, y^* is in $\bar{N}(\tilde{y}, \bar{y})$, and y^* is the unique zero in $\bar{y}$. The midpoint of $\bar{N}(\tilde{y}, \bar{y})$: (0.12772, 1.8240) can be used as a new point approximation to y^*.

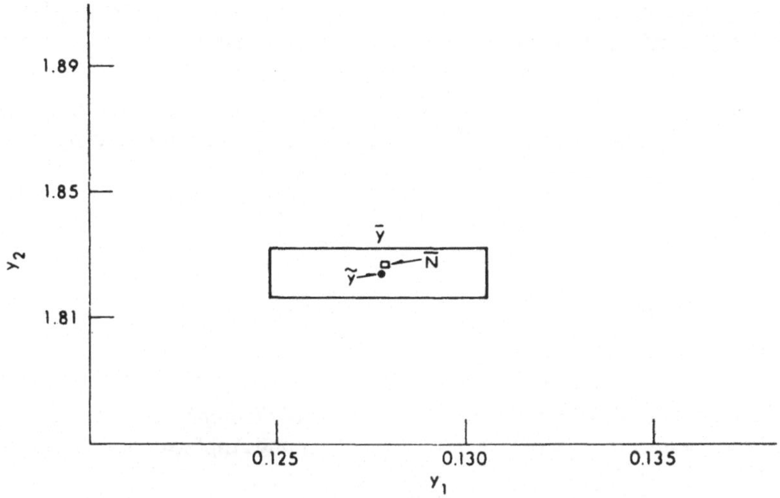

Fig. 4. Application II example, Case 3.

Since the interval Newton operator never loses a zero, an additional run is made to obtain a better error bound. Although convergence to a degenerate interval (i.e., a point) is not guaranteed (see Ref. 5), good convergence results are expected if the intervals $\bar{y}$ are small and close to y^*. In Case 3, the error bound equals the accuracy of the data in Table 1. Its midpoint

$$y^* = \begin{bmatrix} 0.12790 \\ 1.8258 \end{bmatrix} \tag{19}$$

is used as the final point approximation.

Table 3. Application I examples.

Case	Percent	$\bar{y}$	$\overline{\det}\, \overline{\nabla E}(\bar{y})$, Eq. (14)
1	±11	$\begin{bmatrix} [0.11383, 0.14198] \\ [1.6250, 2.0267] \end{bmatrix}$	[−74.273, −1.9672]
2	±12	$\begin{bmatrix} [0.11255, 0.14326] \\ [1.6067, 2.0450] \end{bmatrix}$	[−211.32, −0.96870]
3	±12.5	$\begin{bmatrix} [0.11191, 0.14390] \\ [1.5976, 2.0541] \end{bmatrix}$	[−4886.9, −0.055146]
4	±13	$\begin{bmatrix} [0.11127, 0.14454] \\ [1.5884, 2.0633] \end{bmatrix}$	[−691.39, 1670.9]

Application I is now used to investigate the uniqueness of y^* in Eq. (19). Table 3 presents some results where symmetric intervals $\bar{y}$ are placed around y^*. Interval determinants $\overline{\det}\,\overline{\nabla E}(\bar{y})$ are computed using Eq. (14). The largest interval found satisfying Eq. (12) is in Case 3. It represents a $\pm 12.5\%$ deviation around y^*. Therefore, y^* is the unique zero in

$$\bar{y} = y^* \pm 0.125 y^* = \begin{bmatrix} [0.11191, 0.14390] \\ [1.5976, 2.0541] \end{bmatrix}.$$

References

1. RIJCKAERT, M. J., and MARTENS, X. M., *A Bibliographical Note on Geometric Programming*, Journal of Optimization Theory and Applications, Vol. 26, No. 2, 1978.
2. DUFFIN, R. J., and PETERSON, E. L., *Geometric Programming with Signomials*, Journal of Optimization Theory and Applications, Vol. 11, pp. 3–35, 1973.
3. PASSY, U., and WILDE, D. J., *Generalized Polynomial Optimization*, SIAM Journal on Applied Mathematics, Vol. 15, pp. 1344–1356, 1967.
4. MOORE, R. E., *Interval Analysis*, Prentice-Hall, Englewood Cliffs, New Jersey, 1966.
5. HANSEN, E. R., and SMITH, R., *A Computer Program for Solving a System of Linear Equations and Matrix Inversion with Automatic Error Bounding using Interval Arithmetic*, Lockheed Missiles and Space Company, Report No. 4-22-66-3, 1966.
6. PASSY, U. and WILDE, D. J., *Mass Action and Polynomial Optimization*, Journal of Engineering Mathematics, Vol. 3, pp. 325–335, 1969.
7. FALK, J. E., *Global Solutions of Signomial Programs*, George Washington University, Program in Logistics, Technical Paper No. T-274, 1973.
8. ROBINSON, S. M., *Computable Error Bounds for Nonlinear Programming*, Mathematical Programming, Vol. 5, pp. 235–242, 1973.
9. DUFFIN, R. J., PETERSON, E. L., and ZENER, C., *Geometric Programming*, John Wiley and Sons, New York, 1966.
10. MANCINI, L. J., and PIZIALI, R. L., *Optimal Design of Helical Springs by Geometrical Programming*, Engineering Optimization, Vol. 2, pp. 73–81, 1976.
11. MANCINI, L. J., *Applications of Interval Arithmetic in Signomial Programming*, Stanford University, PhD Thesis, 1975.
12. WILDE, D. J., and BEIGHTLER, C. S., *Foundations of Optimization*, Prentice-Hall, Englewood Cliffs, New Jersey, 1967.
13. ZOLTAN, A. C., *Interval Arithmetic Subroutine Package for the IBM/360*, Universidad Central de Venezuela, Facultad de Ciencias, Departamento de Computacion, Report No. 69-05, 1969.
14. HANSEN, E. R., and SMITH, R., *Interval Arithmetic in Matrix Computations, Part 2*, SIAM Journal of Numerical Analysis, Vol. 4, pp. 1–9, 1967.

15. HANSEN, E. R., *On Linear Algebraic Equations with Interval Coefficients*, Topics in Interval Analysis, Edited by E. R. Hansen, Clarendon Press, Oxford, England, 1969.
16. DEMBO, R. S., *Solution of Complementary Geometric Programs*, Technion, Israel, MS Thesis, 1972.

20

Optimal Design of Pitched Laminated Wood Beams[1]

M. AVRIEL[2] AND J. D. BARRETT[3]

Abstract. The optimal design of a pitched laminated wood beam is considered. An engineering formulation is given in which the volume of the beam is minimized. The problem is then reformulated and solved as a generalized geometric (signomial) program. Sample designs are presented.

1. Introduction

The design of structures requires that members of the structure be selected with the capacity to perform under the anticipated loading conditions. The selection of member geometry is often fairly simple, once the critical loading conditions have been established. However, there are cases, particularly where the member geometry is complex, in which the selection of an optimum member configuration is difficult. In these cases, it is of interest to investigate the possibility of using optimization techniques which can predict the optimum member configuration directly. The use of direct optimization techniques could also be particularly advantageous if the maximum permitted stresses are a function of the member geometry and loading. Probably, the most common situation where this interaction between allowable stress and member geometry occurs is in column design, where the critical buckling stress is a function of length and the radius of

[1] This research was partially supported by the Office of Naval Research under Contracts Nos. N00014-75-C-0267 and N00014-75-C-0865; by the US Energy Research and Development Administration Contract No. E(04-3)-326 PA-18; and by the National Science Foundation, Grant No. DCR75-04544 at Stanford University. This work was carried out during the first author's stay at the Management Science Division of the University of British Columbia and the Systems Optimization Laboratory of Stanford University. The authors are indebted to Mr. S. Liu and Mrs. M. Ratner for their assistance in performing the computations.

[2] Professor, Faculty of Industrial Engineering and Management, Technion—Israel Institute of Technology, Haifa, Israel.

[3] Research Scientist, Western Forest Products Laboratory, Vancouver, BC, Canada.

gyration of the section. For most applications, the optimum configurations are tabulated in handbooks and the design engineer can quickly select optimum sections.

When new design criteria are developed, it is often necessary to develop new information which will ultimately be tabulated for the practicing engineer who wishes to avoid the tedious calculations required to select optimum sections. In this paper, we will investigate the possibility of using geometric programming methods to optimize the geometry of a pitched tapered beam subjected to uniformly distributed loads. The beam geometry is shown in Fig. 1 and is characterized by the span $2L$, the width b, the radius of curvature R, the heights at the center H and H_c, the height at the support H_s, the roof angle β, and the slope of the lower surface φ.

In this paper, we have chosen to optimize the design of the pitched tapered beam with respect to volume. Fox (Ref. 1) presented a computer program for minimizing the volume of beams using a technique that required a detailed understanding of the constraints and employed a rather arbitrary fixed-step search technique to move along constraint surfaces until a minimum volume was achieved. This method provides optimal designs in many cases; but, because of the arbitrary stepsizes used in the program, it can be shown that the attainment of minimum volumes is not always ensured. Since this program was developed, it has been shown by Barrett *et al.* (Ref. 2) that the tensile strength perpendicular to grain of timber is not independent of member geometry as has been previously assumed. In particular, given a set of design parameters x_i which characterize the beam geometry, the stresses and deflection are checked using formulas normally specified in building codes. Stresses and deflections so calculated must not

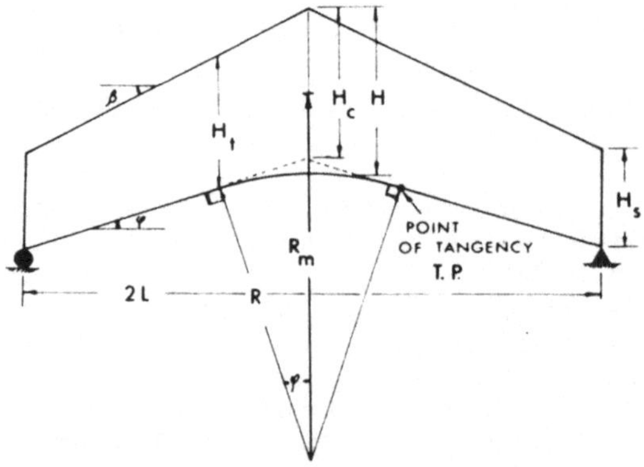

Fig. 1. Pitched tapered beam.

exceed the maximum (or allowable) values. These relationships can, therefore, be expressed in the form of inequality constraints

$$\sigma_j(x_i) \leq F_j, \tag{1}$$

where σ_j is the computed stress component and F_j is the maximum allowable value for that stress component. In addition to stress constraints, deflection constraints are usually employed to keep the deflection D of the member below a specified fraction of the span; accordingly, the constraint will have the form

$$D(x_i) \leq \alpha S, \tag{2}$$

where α is the appropriate fraction and S is the beam span.

In general, there will be other relations which control beam geometry. For example, in beam design, often the ratio of beam height to width is restricted to satisfy lateral stability requirements. In the case of the pitched tapered beam, there will be requirements for a minimum length of tangent on the beam lower surface. These constraints can be written in the form

$$f_j(x_i) \leq 1. \tag{3}$$

The allowable tensile strength perpendicular to grain for pitched tapered beams is given by

$$F_r = 324/\sqrt[5]{(bHR_m\varphi)}, \tag{4}$$

where b is the beam width, H the height at the centerline, R_m the radius to midheight, and φ the angle in radians between the centerline and the tangent point. The allowable stress according to Eq. (4) is generally lower than the 65 psi used previously, thereby requiring modification of the beam configurations which have been tabulated in Ref. 3.

The optimal beam geometry will be found by formulating the optimization problem as a generalized geometric (signomial) programming problem. In the next section, the detailed engineering formulation of the problem will be presented. The signomial programming formulation is derived in Section 3; in Section 4, the results of a few sample designs are given.

2. Engineering Considerations

Given specifications for the uniformly distributed design dead load ω_0 and snow load ω_s, the total span $2L$, and the roof angle β, the design engineer must develop a beam of adequate capacity. The capacity of a beam is assessed by evaluating stress, deflection, and geometric criteria which are normally specified by building code authorities or dictated by manufacturing requirements. For this paper, the design criteria for allowable shear and bending stress F_s and F_B and the deflection constraints are those specified by the Canadian Standard Association Standard CSA 086-1970, Ref. 4. Currently, the allowable tension perpendicular to grain stress F_r, permitted in Canada for Douglas fir, dry service condition, and normal load duration is 65 psi. It has been shown that this allowable stress is not adequately conservative for large beams; for this paper, the allowable stress shall be that recommended by Barrett *et al.* (Ref. 2).

In his design, the engineer strives to produce a minimum-cost structure consistent with the requirements for public safety and protection of property. This may be accomplished by minimizing the volume of material used in the members. Accordingly, in our optimization problem, the objective function to be minimized is the volume of the beam, subject to the design constraints developed below.

There are constraints on the bending stresses at the beam centerline and tangent point (T.P., Fig. 1), on the shear stress at the support (A, Fig. 1) and on tension perpendicular to grain stress at the centerline. The specific form of these constraints for the pitched tapered beam subjected to a uniformly distributed load is as follows.

For the bending stresses σ_B, we require

$$\sigma_B \leq F_B,$$

where F_B is the given allowable bending stress. Bending stresses must be checked at two positions, at the centerline and at the tangent point. Therefore, at the centerline, the constraint has the following form

$$\sigma_B = (6M_c/bH^2)(1 + 2.7 \tan \beta) \leq F_B, \tag{5}$$

where M_c, the bending moment at the centerline, is given by

$$M_c = \omega L^2/24. \tag{6}$$

Similarly, the bending stress at the tangent point is given by

$$\sigma_B = (6M_T/bH_T^2)(1 + 2.7 \tan \beta) \leq F_B, \tag{7}$$

where M_T, the bending moment at the tangent point, is given by

$$M_T = M_c - \omega R^2 \sin^2 \varphi/24. \tag{8}$$

Here, H, H_T = beam heights at centerline and tangent point, respectively (inches); L = half-span (inches); b = beam width (inches); β = roof angle (degrees); and ω = uniformly distributed load (pounds per foot).

The beam loading tends to increase the radius of curvature of the beam, thereby introducing stresses in the radial direction. The magnitude of the tension perpendicular to grain stress σ_r is given by

$$\sigma_r = K_r(6M_c/bH^2) \leq F_r, \tag{9}$$

where

$$K_r = A + B(H/R_m) + c(H/R_m)^2. \tag{10}$$

Here, $R_m = R + H/2 =$ radius to midheight (inches); and $A, B, C =$ functions of β, tabulated in Ref. 4.

The shear stresses τ at the support are required to satisfy the relation

$$\tau \leq F_s,$$

and this constraint is formulated as follows:

$$\tau = (3/2)(v/bH_s) \leq F_s, \tag{11}$$

where

$$v = \omega L/12. \tag{12}$$

Here, $H_s =$ beam height at the support (inches); and $v =$ shear force (pounds).

A constraint on the midspan deflection of beams is normally imposed to prevent excessive deflection which could damage ceiling materials. The maximum allowable deflection δ_{max} is usually expressed as a function of the total span. The corresponding constraint in our case is

$$\delta = (5\omega L^4/288EI)Y \leq \delta_{max}, \tag{13}$$

where (see Ref. 4)

$$Y = 0.2 + 0.8H_c/H_s, \tag{14}$$

$$I = bH_c^3/12. \tag{15}$$

Here, $E =$ modulus of elasticity (psi); and $H_c =$ height at centerline for the double-tapered component of the beam (inches).

The final constraints are constraints on geometry. The first is introduced to ensure that the tangent point (T.P., Fig. 1) is positioned at an adequate distance from the end of the beam. This constraint is required to prevent springback when the beam is released from the clamps after curing of the glue is complete. For convenience, the constraint is formulated as follows

$$L - R\sin\varphi \geq \alpha H, \tag{16}$$

where α is a given constant.

To complete the engineering formulation, it is necessary to specify the allowable values to be used in the stress and deflection constraints. These values depend on the species of wood used in the beam. If Douglas fir is used, the allowable values are as follows:

$$F_B = 2760[1 - 2000(t/R)^2], \qquad \text{psi}, \tag{17}$$

where t = lamination thickness (inches).

$$F_r = K(bHR_m\varphi)^{-0.2}, \qquad \text{psi}, \tag{18}$$

$$F_s = 190, \qquad \text{psi}, \tag{19}$$

where $K = 324$ (if φ is in radians) and

$$\delta_{\max} = 2L/\xi, \tag{20}$$

where we assume $\xi = 180$. The above formulas for F_B and F_s include a 15% increase in allowable stress for snow load conditions.

Finally, the radius of curvature R is constrained to be greater than or equal to 330 inches, so that excessive stresses will not be introduced when the beams are fabricated.

The variable portion of the half-beam volume, to be minimized, is given by (see Fig. 1)

$$V = Lb(H_s + H_c) + bR^2(\tan\varphi - \varphi). \tag{21}$$

The total half-beam volume is equal to $V - L^2 b \tan\beta$.

This concludes the engineering formulation of the optimal design problem.

3. Signomial Programming Formulation

In this section, we present a signomial programming formulation of the optimal beam design problem. The main difficulty in reformulating this problem as a signomial program is that some of the relations appearing in the preceding section are equations, whereas signomial programming constraints must be inequalities. As will be shown below, some of the equations of the engineering formulation are used to eliminate variables and others are converted into inequalities in such a way that hopefully they hold as equations at an optimum. Another minor problem arises from the appearance of trigonometric functions in the design formulas. Taylor series approximations of these functions are used below to obtain generalized polynomials as required.

Note that converting an equality constraint

$$g(x)=1$$

into two inequalities

$$g(x)\leq 1, \qquad g(x)\geq 1$$

is not practical, since the algorithm used for the numerical solution is based on the assumption that the interior of the constraint set is nonempty. For this reason, equalities can only be converted into one-sided inequalities. The sense of the inequalities is usually determined by physical or design considerations (see, for example, Ref. 5). Unfortunately, there are cases where these considerations are quite complex and cannot be observed by simple inspection of the constraints. Consequently, a trial-and-error approach is necessary. Simple examples can, however, be constructed showing that not every equality-constrained problem can be solved by converting equations into single inequalities.

Let us derive now the signomial program for the beam design in detail. The volume of the beam as given in the engineering formulation is

$$V=Lb(H_s+H_c)+bR^2(\tan\varphi-\varphi). \tag{22}$$

Using the identity

$$H_c+L\tan\varphi=H_s+L\tan\beta \tag{23}$$

and a three-term Taylor expansion of $\tan\varphi$:

$$\tan\varphi\cong\varphi+(1/3)\varphi^3+(2/15)\varphi^5, \tag{24}$$

we obtain the volume to be minimized

$$\begin{aligned} V=2LbH_c-L^2b\tan\beta+L^2b\varphi+(1/3)L^2b\varphi^3 \\ +(2/15)L^2b\varphi^5+(1/3)bR^2\varphi^3+(2/15)bR^2\varphi^5, \end{aligned} \tag{25}$$

where H_s, the beam height at the support, has been eliminated.

The variable H_c is defined by the identity

$$H_c=H+R-R/\cos\varphi. \tag{26}$$

In converting this relation into an inequality, we can ensure that, in an optimal solution, the inequality will hold as an equation by writing

$$H_c\geq H+R(1-1/\cos\varphi). \tag{27}$$

Since in (25) we try to lower the value of H_c as much as possible, the inequality in (27) will be tight in an optimal solution. Substituting

$$1/\cos\varphi - 1 \cong (1/2)\varphi^2 + (5/24)\varphi^4 + (61/720)\varphi^4 \tag{28}$$

into (27) and rearranging yields

$$H_c + (1/2)R\varphi^2 + (5/24)R\varphi^4 + (61/720)R\varphi^6 \geq H, \tag{29}$$

or

$$2R^{-1}\varphi^{-2}H - 2R^{-1}\varphi^{-2}H_c - (10/24)\varphi^2 - (122/720)\varphi^4 \leq 1. \tag{30}$$

The bending stress constraint at centerline is given by (5), (6), and (17) as

$$6M_c(1+2.7\tan\beta)/bH^2 \leq 2760[1-2000(t/R)^2], \tag{31}$$

or

$$\Gamma L^2 H^{-2} + 2000t^2 R^{-2} \leq 1, \tag{32}$$

where

$$\Gamma = (1+2.7\tan\beta)\omega/4(2760)b. \tag{33}$$

Turning now to the bending stress constraint at tangent point, we have from (6), (7), (8), and (17) that

$$6(\omega L^2/24 - \omega R^2 \sin^2\varphi/24)(1+2.7\tan\beta)/bH_T^2 \leq 2760[1-2000(t/R)^2], \tag{34}$$

or

$$\Gamma L^2 H_T^{-2} - \Gamma R^2 \sin^2\varphi H_T^{-2} + 2000t^2R^{-2} \leq 1. \tag{35}$$

The beam height at tangent point H_T is related to the other beam heights by

$$H_T = (1 - R\sin\varphi/L)H_c + (R\sin\varphi/L)H_s. \tag{36}$$

From (23) and (26), we obtain

$$H_T = H + R - R\cos\varphi - R\tan\beta\sin\varphi. \tag{37}$$

Instead of converting (37) into an inequality and guessing its sense, we substitute (37) into (35); and, by letting,

$$\sin\varphi \cong \varphi - (1/6)\varphi^3 + (1/120)\varphi^5, \tag{38}$$

$$\cos\varphi \cong 1 - (1/2)\varphi^2 + (1/24)\varphi^4, \tag{39}$$

we write the bending stress constraint at tangent point as

$$\begin{aligned}
&\Gamma L^2H^{-2}+(1/12)H^{-1}R\varphi^4+2\tan\beta H^{-1}R\varphi+(\tan\beta/60)H^{-1}R\varphi^5\\
&\quad+2000t^2R^{-2}+500t^2\tan\beta H^{-2}\varphi^5+[2(1-\Gamma-\tan^2\beta)/45]H^{-2}R^2\varphi^6\\
&\quad+\tan\beta H^{-2}R^2\varphi^3+2000t^2H^{-1}R^{-1}\varphi^2+(2000t^2\tan\beta/3)H^{-1}R^{-1}\varphi^3\\
&\quad+\tan^2\beta\, 2000t^2H^{-2}\varphi^2+[(3/4-\tan^2\beta)(2000t^2/3)]H^{-2}\varphi^4-H^{-1}R\varphi^2\\
&\quad-(\tan\beta/3)H^{-1}R\varphi^3-2000t^2\tan\beta H^{-2}\varphi^3-(\Gamma+\tan^2\beta)H^{-2}R^2\varphi^2\\
&\quad-[(3/4)-\Gamma-\tan^2\beta/3]H^{-2}R^2\varphi^4-(\tan\beta/4)H^{-2}R^2\varphi^5\\
&\quad-(2000t^2/12)H^{-1}R^{-1}\varphi^4-4000t^2\tan\beta H^{-1}R^{-1}\varphi\\
&\quad-(2000t^2\tan\beta/60)H^{-1}R^{-1}\varphi^5-[2(1-\tan^2\beta)2000t^2/45]H^{-2}\varphi^6\leq 1.
\end{aligned}\tag{40}$$

Next, we consider the constraint on tension perpendicular to grain stress. We have, by (9), (10) and (18),

$$(6M_cA/b)H^{-2}+(6M_cB/b)H^{-1}R_m^{-1}+(6M_cC/b)R_m^{-2}\leq K/(bHR_m\varphi)^{0.2},\tag{41}$$

or

$$\begin{aligned}
&(6M_cA/b^{0.8}K)H^{-1.8}R_m^{0.2}\varphi^{0.2}+(6M_cB/b^{0.8}K)H^{-0.8}R_m^{-0.8}\varphi^{0.2}\\
&\qquad\qquad+(6M_cC/b^{0.8}K)H^{0.2}R_m^{-1.8}\varphi^{0.2}\leq 1,
\end{aligned}\tag{42}$$

where

$$R_m=R+(1/2)H.\tag{43}$$

Direct substitution of (43) into (42) would yield a nonsignomial constraint and, therefore, should be avoided. Consequently, we must treat (43) as an inequality that has to hold as an equation at optimum. Since R_m has both positive and negative exponents in (42), the sense of the inequality to replace (43) cannot be determined in advance. Multiplying, however, the left-hand side of (42) by the identity $(R+H/2)/R_m$ yields the new tension-perpendicular-to-grain-stress constraint

$$\begin{aligned}
&(6M_cA/b^{0.8}K)RH^{-1.8}R_m^{-0.8}\varphi^{0.2}+(6M_cB/b^{0.8}K)RH^{-0.8}R_m^{-1.8}\varphi^{0.2}\\
&\quad+(6M_cC/b^{0.8}K)RH^{0.2}R_m^{-2.8}\varphi^{0.2}+(3M_cA/b^{0.8}K)H^{-0.8}R_m^{-0.8}\varphi^{0.2}\\
&\quad+(3M_cB/b^{0.8}K)H^{0.2}R_m^{-1.8}\varphi^{0.2}+(3M_cC/b^{0.8}K)H^{1.2}R_m^{-2.8}\varphi^{0.2}\leq 1,
\end{aligned}\tag{44}$$

where all the terms on the left-hand side are positive and the exponents of R_m are all negative. Now, we convert (43) into the inequality

$$R_m \le R + (1/2)H, \tag{45}$$

or

$$R_m R^{-1} - (1/2)HR^{-1} \le 1. \tag{46}$$

Note that the above considerations are valid if the inequality (44) is tight in the optimal solution. It may happen, however, that both (44) and (46) are strict inequalities at optimum. In this case, the sense of (46) must be reversed (such a reversal was in fact necessary in one of the cases solved).

The shear stress constraint is formulated from (11) by using (23) and (24). We obtain

$$(\omega L/8bF_s + L \tan \beta)H_c^{-1} - L\varphi H_c^{-1} - (1/3)L\varphi^3 H_c^{-1} - (2/15)L\varphi^5 H_c^{-1} \le 1. \tag{47}$$

The deflection constraint is formulated from (13), (14), (15), and (20) as

$$\eta H_c^{-3} + 4\eta H_c^{-2} H_s^{-1} \le 1, \tag{48}$$

where

$$\eta = 0.208\xi\omega L^3/Eb. \tag{49}$$

Multiplying both sides of (48) by H_s and substituting (23) yields

$$\eta H_c^{-2} - \eta L \tan \beta H_c^{-3} + \eta L \tan \varphi H_c^{-3} + 4\eta H_c^{-2} \le H_c - L \tan \beta + L \tan \varphi; \tag{50}$$

and, by (24),

$$5\eta H_c^{-3} + \eta L\varphi H_c^{-4} + (1/3)\eta L\varphi^3 H_c^{-4} + (2/15)\eta L\varphi^5 H_c^{-4} + L \tan \beta H_c^{-1}$$
$$- \eta L \tan \beta H_c^{-4} - L\varphi H_c^{-1} - (1/3)L\varphi^3 H_c^{-1} - (2/15)L\varphi^5 H_c^{-1} \le 1. \tag{51}$$

The constraint on the geometry of the beam given by (16) is rearranged to

$$(\alpha/L)H + (1/L)R \sin \varphi \le 1; \tag{52}$$

and, by (38), it becomes

$$(\alpha/L)H + (1/L)R\varphi - (1/6L)R\varphi^3 + (1/120L)R\varphi^5 \le 1. \tag{53}$$

The last geometry constraint, $R \ge 330$, is not treated explicitly, since the numerical algorithm for the solution of the beam design problem requires upper and lower bounds on all the design variables; thus, the value 330 will be used as the lower bound on R.

The optimal beam design is obtained (after specifying the appropriate constants) by solving the signomial program of minimizing (25), subject to the constraints (30), (32), (40), (44), (46), (47), (51), and (53). The variables to be determined by the optimization are H, H_c, R, R_m, V, and φ. Note that an optimal solution to the signomial programming formulation of the design problem is acceptable only if the inequalities (30) and (46) hold as equations at optimum.

A few sample problems of optimal beam designs were solved by the computer code GGP, based on the generalized geometric (signomial) programming algorithm of Avriel, Dembo, and Passy (Ref. 5). These optimal design solutions are presented in the next section.

4. Sample Designs

Optimal beam configurations are sought for three different spans and loading conditions. The specified beam parameters are shown in Table 1. For the roof angle specified in Table 1, the corresponding constants are

$$A = 0.0367, \qquad B = 0.0794, \qquad C = 0.213.$$

In addition, the modulus of elasticity is assumed to be $E = 1.93 \times 10^6$ psi and a value of $\alpha = 1.5$ is taken in (16).

Optimal solutions were obtained by the computer code GGP in less than 10 seconds of CPU time on an IBM 370/168 computer. The optimal design variables are listed in Table 2.

It is interesting to observe the binding design constraints at optimum for the above cases (in addition to those which must be tight because they were originally equations). These are shown in Table 3. In Cases 1 and 2,

Table 1. Input parameters for beam optimization.

Case	Roof angle β (degrees)	Half-span L (inches)	Width b (inches)	Lamination thickness t (inches)	Load ω (lb/ft)
1	9.46	360	8.75	1.5	1200
2	9.46	240	6.75	1.5	1200
3	9.46	120	3.00	1.5	400

Table 2. Optimal design variables.

Case	Volume (ft^3)	φ (degrees)	H (inches)	H_c (inches)	R (inches)	R_m (inches)
1	184.32	3.84	70.7	68.3	1063	1099
2	71.05	5.01	50.3	47.0	860	886
3	5.93	6.16	19.5	17.6	330	340

constraint (44) is tight at optimum; and consequently (46), the defining relation for R_m, is also satisfied as an equation. In Case 3, however, (44) is no longer binding, and at first we obtained a solution in which both (44) and (46) were strict inequalities. Therefore, we reversed the sense of the inequality in (45) and (46) to

$$R_m \geq R + (1/2)H \tag{54}$$

and

$$RR_m^{-1} + (1/2)HR_m^{-1} \leq 1, \tag{55}$$

respectively, and (46) was replaced by (55) in the program. This change resulted in the above-listed optimal solution for Case 3 in which, of course, (55) held as an equation.

Table 3. Binding design constraints at optimum.

Case	Binding design constraints
1	Tension ⊥ grain stress (44); shear stress (47).
2	Tension ⊥ grain stress (44); shear stress (47).
3	Bending at tangent point (40); shear stress (47).

References

1. FOX, S. P., *Minimum-Depth of Double-Tapered Pitched Glued-Laminated Beams*, Information Report VP-X-73, Western Forest Products Laboratory, Vancouver, Canada.
2. BARRETT, J. D., FORSCHI, R. O., and FOX, S. P., *Perpendicular-to-Grain Strength of Douglas Fir.* Canadian Journal of Civil Engineering, Vol. 2, pp. 50–57, 1975.

3. LAMINATED TIMBER INSTITUTE OF CANADA, *Timber Design Manual,* Ottawa, Canada, 1973.
4. CANADIAN STANDARDS ASSOCIATION, *Code of Recommended Practice for Engineering Design in Timber,* Standard CSA-086, Rexdale, Ontario, Canada, 1970.
5. AVRIEL, M., and WILDE, D. J., *Optimal Condenser Design by Geometric Programming,* I & EC Process Design and Development, Vol. 6, pp. 256–262, 1967.
6. AVRIEL, M., DEMBO, R. S., and PASSY, U., *Solution of Generalized Geometric Programs,* International Journal of Numerical Methods in Engineering, Vol. 9, pp. 149–168, 1975.

21

Optimal Design of a Dry-Type Natural-Draft Cooling Tower by Geometric Programming[1]

J. G. ECKER[2] AND R. D. WIEBKING[3]

Abstract. In this paper, the optimal design of dry-type natural-draft cooling towers is investigated. Using physical laws and engineering design relations that govern the system, a rather detailed optimization model is developed. This model is then reformulated as a geometric programming problem. A primary consideration in this reformulation is how certain polynomial equations may be effectively replaced by inequalities. A numerical example follows.

Glossary of Symbols

A_{Fi}	total fin surface, sq ft
A_i	total inside tube surface, sq ft
A_o	total outside tube surface, sq ft
$\bar{A}_o$	total outside bare tube surface (finned tubes), sq ft
A_{To}	tot surface area of tower shell, sq ft
a_s	free flow area, sq ft
b	fin thickness, ft
$C_{F,Fi}$	fixed charges on air cooler fins, \$/yr

[1] This research was supported in part by the National Science Foundation, Grant No. MPS75-09443, and by the NATO Postdoctoral Fellowship Program in Science.

[2] Associate Professor, Center for Operations Research and Econometrics, Université Catholique de Louvain, Louvain, Belgium; and Department of Mathematical Sciences, Rensselaer Polytechnic Institute, Troy, New York.

[3] Engineering Consultant, Unternehmensberatung Schumann GmbH, Cologne, West Germany.

$C_{F,To}$	fixed charges on tower shell, \$/yr
$C_{F,Tu}$	fixed charges on air cooler tubes, \$/yr
$C_{O,P}$	operating cost of air cooler pump, \$/yr
c	distance between two adjacent fins, ft
$c_{p,a}$	specific heat of air, Btu/lb °F
D_e	equivalent diameter, ft
D_i	inside tube diameter, ft
D_o	outside tube diameter, ft
D_{To}	tower diameter, ft
D_v	equivalent volumetric diameter, ft
d	distance between centerlines of two adjacent banks of tubes in air cooler, ft
e	gap between fins of two adjacent banks of tubes, ft
F	cross-flow-temperature-difference correction factor
f	frictional surface, sq ft
G	greater terminal temperature difference, °F
H_{To}	tower height above air entrance, ft
h	fin height, ft
h_i	inside film coefficient, Btu/(sq ft) hr °F
h_o	outside film coefficient, Btu/(sq ft) hr °F
L	tube length, ft
l	tube wall thickness, ft
N	number of tubes in bank of air cooler
$\bar{N}$	number of vertical banks in air cooler
n	number of fins per ft. of tube length
q	rate of heat transfer, Btu/hr
S_l	center-to-center distance to the nearest tube in the next bank, ft
S_t	pitch in a bank, ft
t_{To}	tower shell thickness, ft
T	temperature, °R
U	lesser terminal temperature difference, °F
W	flow rate, lb/hr
ΔP_f	tower friction loss, psi
ΔP_l	tower leaving loss, psi
ΔP_o	pressure drop across air cooler coils, psi
ΔP_t	theoretical tower draft, psi
ΔT_a	change in air temperature, °F
ΔT_m	overall mean temperature drop between hot and cold fluid, °F
ΔT_{mi}	mean temperature drop through inside tube film, °F
ΔT_{mo}	mean temperature through outside tube film, °F
Ω	fin efficiency

Subscripts

a	air
w	water
c	cold
h	hot

1. Introduction

In this paper, we investigate an optimization model for the economic design of dry-type natural-draft cooling towers used in dissipating the waste heat of steam electrical plants. For the complete cycle from fuel to electricity, the efficiency of a fossil-fuel plant is about forty percent, while for nuclear plants the efficiency drops to nearly thirty percent. This means that a significant fraction of the energy in the fuel appears as waste heat. The usual method of carrying away this waste heat has been to use circulating water from a natural body of water. For ecological reasons, the resulting increase in the water temperature of the river, lake, or ocean bay may be unacceptable. Cooling towers are often used to transfer the heat of condensation of the turbine exhaust steam to the atmosphere. Most cooling towers currently being used are evaporative-type towers that result in a considerable loss of water to the atmosphere. Such towers are not practical in water-poor regions because of the large volume which they require. Also, even in regions where water is more plentiful the *wet plumes* associated with these towers often result in local fog and other undesirable conditions. Thus, dry-type cooling towers are now receiving considerable attention. Dry-type cooling towers transfer the heat of condensation of the turbine exhaust steam to the atmosphere with no evaporative loss of circulating water.

Unfortunately, relatively little information is available to assist in evaluating the performance and economics of dry cooling. In Ref. 1 and Ref. 2, various cooling methods are compared, and the broad economic problem of determining condensing system sizes with reference to dry-type cooling towers is considered. In Ref. 3 and Ref. 4, the authors describe the present state-of-the-art for dry cooling towers for thermal-electric generation. However, it appears that current analytical methods have not been used in the design investigations.

In this paper, we develop a rather detailed optimization model for the design of dry-type cooling towers. In Section 2, we consider the relevant physical and engineering principles. Most of these principles result in design relations that are stated as nonlinear equations. We have used standard engineering references to obtain many of the design relations, and our final model is certainly not the only one possible. Also, numerical values chosen for some of the physical constants in the model may need to be adjusted in a specific design application. In Section 3, we reformulate the optimization model as a geometric programming problem. A primary consideration in this reformulation is how certain polynomial equations may be effectively replaced by polynomial inequalities. Finally, in Section 4, we consider a specific application of the model and the resulting numerical solution.

Actually, our cooling tower model could be viewed as one component in an integrated condensing system for a steam electric plant consisting of the turbine exhaust annulus, a surface-type condenser, and a natural-draft dry-type cooling tower. Geometric programming has been used previously to investigate the first two components above (see Ref. 5, Ref. 6, and Ref. 7). Also, in Ref. 8, the integrated system is considered.

For a discussion of the basic concepts in geometric programming, the reader is referred to Ref. 9. We should also remark that, although we are considering the detailed design of dry-type natural-draft cooling towers, such an optimization approach is possible for many other similar problems. For examples of many such problems, see Ref. 10, especially the paper of M. Rijckaert which reviews many geometric programming applications.

2. Cooling Tower Design Relations

In this section, we develop the design relations for our optimization model using relevant physical laws and engineering principles. We consider an indirect system which utilizes water to reject the heat of condensation of turbine exhaust steam to the atmosphere. For simplicity, the shape of the cooling tower shell, which is usually hyperbolical, is assumed to be cylindrical. Figures 1, 2, and 3 describe the basic components of the tower. The warm circulating water enters the bottom of vertically mounted cooling coils, flows upward in the inner bank of coils to the top water boxes, and then is directed downward through the outer bank of coils. A detailed description of such towers is found in Ref. 3.

Since the cooling tower diameter is usually much larger than the tube diameter of the cooling coils, we may assume that the vertical banks of coils are parallel to each other with the same number of tubes per bank. We shall let the heat transfer surface be extended by radial fins with known thickness and spacing (see Fig. 2). Also, the tubes in adjacent banks will be arranged in triangular pitch as in Fig. 3.

We will consider a heat exchanger (air cooler) with fixed heat load, given water flow rate, and given inlet and outlet water temperatures. Some

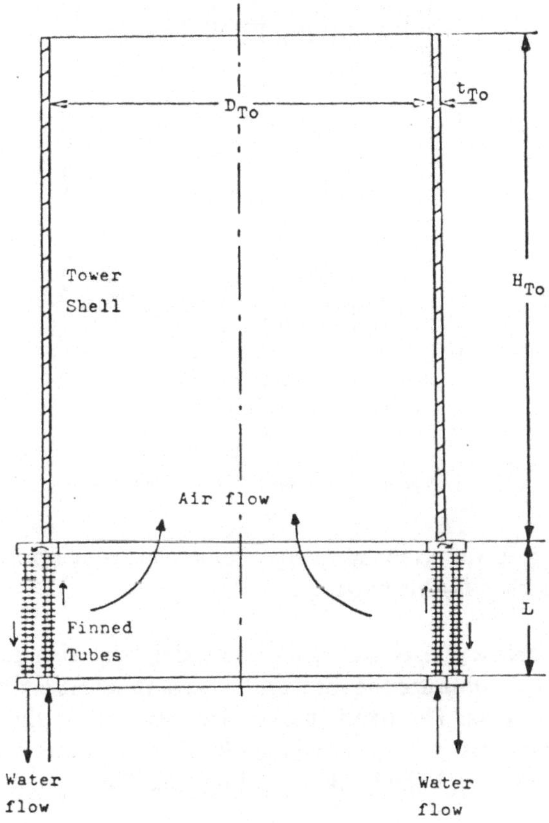

Fig. 1. Cooling tower model.

of the more important design variables in our model are listed below. Actually, these are the variables that we selected to be the independent variables in our model:

H_{To} = tower height above air entrance, ft
D_{To} = tower diameter, ft
N = number of tubes in bank of air coder,
L = tube length, ft
D_o = outside tube diameter, ft
D_i = inside tube diameter, ft,
h = fin height, ft,
W_a = air flow rate, lb/hr.

In the final model, certain auxiliary (dependent) variables will also appear.

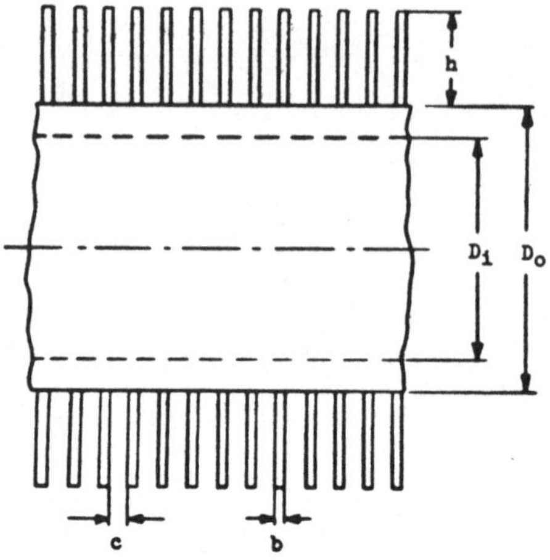

Fig. 2. Tube and fin nomenclature.

The reader is referred to the glossary of engineering symbols for definitions of all symbols used in this paper.

2.1. Objective Function. In our model, we wish to select design variables so as to minimize the total annual cost that consists of fixed charges plus operating costs. The fixed charges are incurred by the heat exchange tubes, the radial fins, and the tower shell. The operating costs result from pumping water through the bank of cooling coils. Thus, the total annual cost

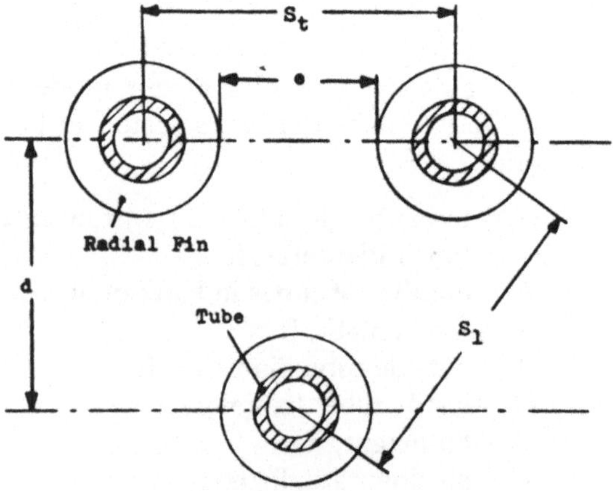

Fig. 3. Tube arrangement.

C is given by

$$C = C_{F,Tu} + C_{F,Fi} + C_{F,To} + C_{0,P}. \tag{1}$$

The fixed charges are assumed to be proportional to the surface area of the various components, while the operating cost is proportional to the pressure drop inside the heat exchange tubes. A simple calculation of surface areas A_o, A_{Fi} and the volume V_{To} yield, respectively,

$$C_{F,Tu} = c_1 D_o N L, \tag{2}$$

$$C_{F,Fi} = c_2 N L h(D_o + h), \tag{3}$$

$$C_{F,To} = c_3 D_{To} H_{To}, \tag{4}$$

for appropriate positive constants c_i depending in part on unit costs of material and depreciation rates. By considering the pressure drops through straight tubes, one can show as in Ref. 5 that

$$C_{O,P} = c_4 L / D_i^{4.8} N^{1.8}, \tag{5}$$

where $c_4 > 0$.

The design variable appearing in the objective function must satisfy certain engineering and physical principles. These principles will ultimately yield the constraints for our model.

2.2. Rate of Heat Transfer. We are considering a cooling tower with a fixed heat load, and we now examine its heat rejection capability. Heat transfer theory (Ref. 11 and Ref. 12) requires that the following three equations be satisfied:

$$q = c_{pa} W_a (T_{a,h} - T_{a,c}), \tag{6}$$

$$q = h_i A_i \Delta T_{mi}, \tag{7}$$

$$q = h_o(\Omega A_{Fi} + \bar{A}_o)\,\Delta T_{mo}. \tag{8}$$

In (7), the inside film coefficient h_i is calculated by the Dittus–Boelter correlation (Ref. 11) and has the form

$$h_i = (\text{const})(Re)^{0.8} / D_i, \tag{9}$$

where the Reynolds number Re is given by

$$Re = (\text{const}) / D_i N. \tag{10}$$

Using (9) and (10) and the fact that the inside tube area A_i is proportional to NLD_i, we see from (7) that

$$q \sim D_i^{-0.8} N^{0.2} L\, \Delta T_{mi}, \tag{11}$$

where here (and elsewhere in this paper) $u \sim v$ means that u is directly proportional to v. In order to eliminate many cumbersome physical and engineerrring constants, we will state several of the design relations as proportionalities. The constants of proportionality are readily determined and will not be given here. In Ref. 8, these constants arc precisely defined.

In (8), the outside film coefficient h_o is calculated using the Jameson correlation for transverse fins (Ref. 11), and

$$h_o \sim (D_e W_a / a_s)^{0.73} / D_e, \tag{12}$$

where the minimum cross-sectional flow area a_s is given by

$$a_s = L(\pi D_{To} - D_o N - 2bnNh) \tag{13}$$

and the equivalent diameter D_e is expressed as in Ref. 10 by

$$\begin{aligned} D_e &= (2/\pi)(\text{total outside surface})/(\text{projected perimeter}) \\ &= [2h(D_o + h) + D_o c]/(2h + c). \end{aligned} \tag{14}$$

Combining (8) and (12) yields

$$q \sim a_s^{-0.73} W_a^{0.73} D_e^{-0.27} (\Omega A_{Fi} + \bar{A}_o) \Delta T_{mo}. \tag{15}$$

The total fin surface area A_{Fi} and the total outside bare tube surface $\bar{A}_o$ are readily calculated and are given by

$$A_{Fi} = 2\pi n \bar{N} N L h (D_o + h), \tag{16}$$

$$\bar{A}_o = \pi n c \bar{N} N L D. \tag{17}$$

We will see in Section 3 that it will not be convenient to eliminate the variables a_s and D_e from (15) by using (13) and (14). Doing so would destroy the posynomial form (see Ref. 9, p. 2) of the right-hand side of (15).

2.3. Temperature Drop. The overall mean temperature drop ΔT_m between the hot and cold fluid in a heat exchanger is equal to the sum of the mean temperature drops through the inside and outside boundary layer of the tubes (Ref. 12); that is,

$$\Delta T_m = \Delta T_{mo} + \Delta T_{mi}, \tag{18}$$

where ΔT_m is given by

$$\Delta T_m = F(G - U)/\log(G/U), \tag{19}$$

with the crossflow temperature-difference correction factor F given in Ref. 11. In (19), the greater terminal temperature difference G is given by

$$G = T_{w,c} - T_{a,c}.$$

The temperatures $T_{w,c}$ and $T_{a,c}$ are fixed in our model, and therefore G is a constant. The lesser terminal temperature difference U is given by

$$U = T_{w,h} - T_{a,h}. \tag{20}$$

2.4. Tower Pressure Drop. In natural-draft cooling towers, the upward air motion is primarily brought about by a thermodynamic chimney effect. Aerodynamic effects of natural wind will not be considered in our model.

The theoretical tower draft ΔP_t must offset the pressure drops in the tower due to friction loss and leaving loss (ΔP_f and ΔP_l, respectively), as well as the pressure drop ΔP_o across the air cooler coils. Thus,

$$\Delta P_t = \Delta P_f + \Delta P_l + \Delta P_o. \tag{21}$$

The theoretical tower draft results from the difference between the specific weights of the surrounding atmosphere air and the heated air within the tower. It can be expressed (see Ref. 13) by

$$\Delta P_t = (0.0188)p_a H_{To}(T_{a,c}^{-1} - T_{a,h}^{-1}). \tag{22}$$

Using the relations in Ref. 13 for pressure drops due to friction, one can derive the proportionality law

$$\Delta P_f \sim W_a^{1.8} H_{To} T_{a,h}/D_{To}^{4.8}. \tag{23}$$

Also in Ref. 13, the pressure drop corresponding to the loss of kinetic energy of the air stream leaving the tower is expressed by

$$\Delta P_l \sim W_a^2 T_{a,h}/D_{To}^4. \tag{24}$$

The Gunter–Shaw correlation (Ref. 11) was selected for the calculation of the pressure drop across the coils and it yields

$$\Delta P_o \sim (W_a/a_s)^{\alpha} D_v^{\beta} (D_{To}/N)^{0.6} T_{a,m}, \tag{25-1}$$

where

$$T_{a,m} = (T_{a,h} + T_{a,c})/2 \tag{25-2}$$

and

$$\alpha = 1.8653 \quad \text{and} \quad \beta = -0.7347.$$

Here, the volumetric diameter is given by

$$D_v = 4V/f, \tag{26}$$

where V denotes the net free volume and is the volume between the center lines of two vertical banks of tubes less the volumes of the half-tubes and fins within the center lines (see Fig. 3). Thus,

$$V=(\pi/4)L\{4d(D_{To}-d)-N[D_0^2+4h(D_o+h)bn]\}, \tag{27}$$

where the distance d in Fig. 3 is given by

$$d=(\sqrt{3}/2)\pi D_{To}N^{-1}, \tag{28}$$

assuming

$$S_t=S_l.$$

The frictional surface f used in (26) is given by

$$f=(\bar{A}_o+A_{Fi})/\bar{N}.$$

From (16) and (17), we obtain

$$f=\pi nNL[cD_o+2h(D_o+h)]. \tag{29}$$

This completes our discussion of the basic design relations that we are going to use in our model.

Some variables in the above relations can be written as functions of the original design variables listed at the beginning of Section 2 and could be eliminated from the model. For example, a_s could be eliminated from (15) by using (13), but instead we will retain a_s as a variable in the problem. The following (dependent) variables will be used in the geometric programming formulation of out model in Section 3:

a_s = free flow area, sq ft,
D_e = equivalent diameter, ft,
D_v = equivalent volumetric diameter, ft,
U = lesser terminal temperature difference, °F,
$T_{a,h}$ = tempe. .ture of leaving hot air, °R,
$T_{a,m}$ = mean air temperature, °R,
ΔT_{mo} = mean temperature drop through outside tube film, °R,
ΔT_{mi} = mean temperature drop through inside tube film, °R,
ΔT_a = temperature increase of air, °R.

There are other design relations that could also be incorporated in our model. The effect of wind forces on the shell structure can be analyzed by membrane theory, and constraints on shell material strength would result. Furthermore, a minimum rise of the plume of warm air leaving the cooling tower can be required so that an elevated temperature inversion is penetrated. For simplicity, we will not develop such relations here but we remark that these relations do yield constraints that could be added to the geometric programming formulation of the model (see Ref. 8 for details).

In Section 3, we will introduce practical bounds on some of the design variables.

3. Geometric Programming Formulation

In geometric programming, all constraints must be formulated as inequalities of the form

$$\sum u_i(t) \leq 1 \qquad \text{or} \qquad \sum u_i(t) \geq 1,$$

where each $u_i(t)$ is a posynomial term in the positive vector variable t. The latter form is often called a *reversed* posynomial constraint. In Refs. 14–17, for example, the concept of condensing a reversed constraint is used to obtain solution methods for geometric programs with reversed constraints.

As we have seen, equality constraints arise quite naturally in practice. We could replace an equality constraint by two inequality constraints, but we will not use this approach here. Doing so introduces many reversed constraints that unnecessarily complicate the numerical solution of the model.

Some of the difficulties associated with replacing a posynomial equality constraint by two inequalities and then condensing the reversed constraint are illustrated in the following simple example. Consider a problem with the feasible region R defined by

$$R = \{t > 0 | t_1 + t_2 = 1, 10t_1 \leq 1\}.$$

Suppose that we replace

$$t_1 + t_2 = 1$$

with

$$t_1 + t_2 \leq 1 \qquad \text{and} \qquad t_1 + t_2 \geq 1$$

and condense the (latter) reversed constraint using arbitrarily selected weights, say

$$(\epsilon_1, \epsilon_2) = (1/2, 1/2)$$

for the arithmetic–geometric mean inequality

$$\sum u_i \geq \pi(u_i/\epsilon_i)^{\epsilon_i}.$$

The resulting condensed program has the feasible region

$$Q = \{t > 0 | t_1 + t_2 \leq 1, 2t_1^{1/2} t_2^{1/2} \geq 1, 10t_1 \leq 1\}$$

and is infeasible. However, if one selects weights ϵ_i according to

$$\epsilon_i = u_i(t^*)/\sum u_i(t^*),$$

where $t^* \in R$, then the resulting condensed program will be feasible. Finding such a feasible point t^* for an equality constrained problem is not always easy. More important, however, is the fact that the resulting condensed program may have only a single feasible point. In our example, the point

$$t^* = (1/10, 9/10)$$

is feasible, and the weights (1/10, 9/10) give a feasible condensed program with the *single* feasible point t^*. The point t^* is therefore optimal for the condensed program. Using this optimal solution t^* to define new weights according to the above rule will lead to the *same* condensed program. Thus, the condensation procedure will simply yield t^* at each subsequent iteration.

In this section, we will try to replace each design equation in our model by a single inequality. The decision as to which direction the inequality should take is easy for certain relations but is by no means obvious for all of them. For some relations, physical reasons and an engineering understanding of the problem guide us in selecting the appropriate inequality. In any case, we will try to replace the equality with an inequality whose direction is chosen so that the constraint is active at optimality.

We begin by considering Eq. (21) relating the theoretical draft to the pressure drops. Actually, the relation (21) can be written as

$$\Delta P_t \geq \Delta P_f + \Delta P_l + \Delta P_o, \tag{30}$$

because in (21) some additional pressure drops were neglected. In fact, numerical experience shows that ΔP_f is negligible relative to the other pressure drops and it could be deleted from the right-hand side of (30). If we introduce a new variable

$$\Delta T_a = T_{a,h} - T_{a,c} \tag{31}$$

and notice that

$$T_{a,c}^{-1} - T_{a,h}^{-1} = \Delta T_a / T_{a,h} T_{a,c},$$

then ΔP_t defined in (22) satisfies

$$\Delta P_t \sim H_{To}\, \Delta T_a\, T_{a,h}^{-1}. \tag{32}$$

Recall that the ambient air temperature $T_{a,c}$ is a constant. Using relations (30), (32), (23), (24), and (25) then yields

$$\begin{aligned} &c_5 T_{a,h}^2 W_a^{1.8}\, \Delta T_a^{-1}\, D_{To}^{-4.8} + c_6 T_{a,h}^2 W_a^2 D_{To}^{-4} H_{To}^{-1}\, \Delta T_a^{-1} \\ &+ c_7 T_{a,h} W_a^{\alpha} a_s^{-\alpha} D_v^{\beta} D_{To}^{0.6} N^{-0.6} T_{a,m} T_{a,h} H_{To}^{-1}\, \Delta T_a^{-1} \leq 1, \end{aligned} \tag{33}$$

for appropriate positive constants c_i. Recall that α and β were constants defined following (25).

The variable D_v in (33) must satisfy the relation defined by (26)–(29). To reformulate this relation as in inequality, we first remark that the pressure drop ΔP_o in (25) increases with decreasing volumetric equivalent diameter D_v. A higher theoretical draft must therefore be provided by increasing the tower height at an incremental capital cost. This argument can be formalized to show that (26) can be replaced by

$$D_v \leq 4V/f, \tag{34}$$

and equality will hold at optimality. Fortunately, V as defined in (27) contains only one term with a positive coefficient. Thus, using (34), (27), (28), and (29), we obtain the posynomial constraint

$$\begin{aligned} &c_8 D_0^2 N^2 D_{To}^{-2} + c_9 D_o N^2 D_v D_{To} + c_{10} N^2 h D_{To}^{-2}(D_o + h) \\ &\quad + c_{11} N^2 D_v h D_{To}^{-2}(D_o + h) + c_{12} N^{-1} \leq 1, \end{aligned} \tag{35}$$

with the positive constants c_i incorporating the various physical and engineering constants. In (35), notice that we have retained $D_o + h$ in two terms. We do this for a later convenience when we shall essentially replace $D_o + h$ by an auxiliary variable.

The variable a_s in (33) must satisfy Eq. (13). This relation is particularly interesting in that the proper direction of the inequality that will be driven to equality at optimality seems to depend on the numerical values of the coefficients c_i in the model. In fact, in some problems that we solve, the inequality

$$a_s \leq L(\pi D_{To} - D_o N - 2bnNh)$$

was required, while for other problems the opposite direction was chosen. The above inequality results in the posynomial constraint

$$c_{13} a_s L^{-1} D_{To}^{-1} + c_{14} D_o N D_{To}^{-1} + c_{15} h N D_{To}^{-1} \leq 1. \tag{36}$$

In our numerical example, we first use (36) in our solution of the model, but the data of the problem is such that it is inactive at optimality. The constraint is then reversed in a second run of the problem, and equality holds at optimality.

We now consider the heat-transfer relations discussed earlier in Section 2.2. The choice of the inequality in relations (7) and (8) is suggested by the fact that the heat-rejection capability of the tubes must be at least as great as the actual heat load q to be rejected. Thus, we rewrite (7) and (8) as

$$q \leq h_i A_i \, \Delta T_{mi}, \tag{37}$$

$$q \leq h_o(\Omega A_{Fi} + \bar{A}_o) \, \Delta T_{mo}, \tag{38}$$

respectively. The amount of heat that has to be rejected to the air stream through the tower is given by the right-hand side of relation (6). Actually, the tower shell itself has a capability to transfer heat, thus reducing the amount of heat that has to be rejected to the air stream. Thus, relation (6) can be reformulated as

$$q \geq c_{p,a} W_a \, \Delta T_a, \tag{39}$$

where the variable ΔT_a is defined in (31).

Relation (39) yields a constraint of the form

$$c_{16} W_a \, \Delta T_a \leq 1, \tag{40}$$

where

$$c_{16} = c_{p,a}/q.$$

As in the development of relation (11), Ineq. (37) results in the constraint

$$c_{17} D_i^{0.8} \, \Delta T_{mi}^{-1} L^{-1} N^{-0.2} \leq 1. \tag{41}$$

Using (15) and the definitions of A_{Fi} and $\bar{A}_o$ in (16) and (17), relation (38) yields a reversed constraint. Although it is not necessary, we will consider a simplification that avoids this reversed constraint. The heat transferred through the bare portion of the tubes is typically only a small fraction of the heat transferred through the fins (Ref. 11). Suppose that we estimate the ratio $\bar{A}_o/A_{Fi}$d *use an appropriate constant for the fin efficiency* Ω. The term $\Omega A_{Fi} + \bar{A}_o$ can be written as $A_{Fi}(\Omega + \bar{A}_o/A_{Fi})$, where $\Omega + \bar{A}_o/A_{Fi}$ is now a constant. With this simplification, (15), (16), and (38) yield the constraint

$$c_{18} a_s^{0.73} W_a^{-0.73} D_e^{0.27} (D_o + h)^{-1} N^{-1} L^{-1} h^{-1} \leq 1. \tag{42}$$

Because of the term $(D_o + h)^{-1}$, (42) is not a posynomial constraint, but this will be remedied below.

Using the definition of D_e in (14), we observe that (in practice) the distance c between adjacent fins is considerably less than the fin height h; see Fig. 2. We will therefore use the simplification

$$D_e = D_o + h. \tag{43}$$

We will see in our numerical example that this simplification has negligible effects in our model. Substituting D_e for $D_o + h$ in (42) yields a posynomial constraint. We also observe that $D_o + h$ appears in (35), and we also replace it by D_e. This is not necessary, but is probably desirable, since it reduces the degree of difficulty (see Ref. 9, p. 83).

To develop the constraints due to the temperature drop relations, we first need to approximate ΔT_m in (19) by a posynomial. We will use the method suggested in Ref. 9. Assuming that the range of variability of G/U is from 1 to 4.5 and that $F = 1$, one can use the approach in Ref. 9 to approximate ΔT_m by

$$\Delta T_m = 0.977 U^{0.4379} G^{0.5621}, \tag{44}$$

where G, defined in Section 2.3, is a constant in our model.

The basic relation for temperature drops (18) can actually be written as

$$\Delta T_m \geq \Delta T_{mo} + \Delta T_{mi} \tag{45}$$

since slight temperature drops due to dirt layers, for example, were neglected in (18). Thus, (44) and (45) yield the constraint

$$c_{19} \, \Delta T_{mo} \, U^{-0.4379} + c_{20} \, \Delta T_{mi} \, U^{-0.4379} \leq 1. \tag{46}$$

Examining the heat transfer relations, one can show that (46) will be driven to equality at optimality. In (46), the variable U must also satisfy relation (20). Notice that the only constraint thus far involving the variable U is given in (46) and U appears with a negative exponent in both terms. If we were to replace (20) by

$$U \geq T_{\omega,h} - T_{a,h},$$

this would result in a reversed constraint with a positive exponent for U. If this constraint is then condensed to form an approximating posynomial constraint, the resulting geometric program would not be canonical (Ref. 9, p. 169), since U would always appear with a negative exponent. The theory of degenerate programs in Ref. 9 would essentially allow us to delete the posynomial terms involving the variable U. However, we know that (46) is a crucial constraint in the model. Thus, we will choose the opposite direction for the inequality and obtain the constraint

$$c_{21} U + c_{22} T_{a,h} \leq 1. \tag{47}$$

Actually, if (47) were a strict inequality at optimality, then one could increase U and decrease each term in (46). This would allow an increase in ΔT_{mo} and ΔT_{mi} without violating (46). From (37) and (38), we observe that this permits a decrease in surface areas A_i and A_{Fi} with a corresponding decrease in cost. Thus, (47) will be active at optimality.

The two equations (31) and (25b) will be replaced by

$$T_{a,h} \geq T_{a,c} + \Delta T_a, \tag{48}$$

$$T_{a,m} \geq (T_{a,c} + T_{a,h})/2, \tag{49}$$

respectively. If either (48) or (49) used the opposite inequality, the resulting condensed program would not be canonical. The inequalities chosen in (48) and (49) are also suggested by physical reasons. Pressure drops in the tower increase with increasing $T_{a,h}$ and $T_{a,m}$ so that an increased tower height is necessary at increased cost.

We should also remark that the direction of the inequality chosen in (34) is necessary if the resulting condensed program is to be canonical.

The only remaining inequality relation is given by (43). We observe that $D_o + h$ appears linearly in the second term of the objective function, given by (3). Replacing $D_o + h$ by D_e and requiring that

$$D_e \geq D_o + h \tag{50}$$

will result in D_e being decreased in the minimization.

Finally, we shall impose constraints on the numerical values of some design variables. For example, D_o and D_i must certainly satisfy

$$D_o \geq D_i.$$

In fact, we will add the constraint

$$D_o - D_i \geq 2l, \tag{51}$$

where l is a minimum wall thickness. Also, two adjacent finned tubes must be spaced at least a certain minimum distance $e_{\min}$ apart. For this, we employ the constraint

$$\pi D_{To} N^{-1} - (D_o + 2h) \geq e_{\min}; \tag{52}$$

see Fig. 3. Both constraints (51) and (52) can be rewritten as posynomial constraints. In order to ensure that certain variables have realistic values, bounds may be placed on some design variables. In our numerical solution (Section 4), we use the constraints

$$D_o \leq 2 \text{ in.}, \qquad h \leq 1.5 \text{ in.}, \qquad D_{To} \leq 400 \text{ ft}, \qquad L \leq 100 \text{ ft}, \tag{53}$$

as well as a simplified structural constraint

$$H_{To}/D_{To} \leq 1.2. \tag{54}$$

Recalling that, in (3), (35), and (42), we replace $D_o + h$ by D_e, our final geometric programming model is given by (TGP) below.

(TGP): minimize (2)+(3)+(4)+(5)

subject to (33), (35), (36), pressure-drop constraints,
(40), 41), (42), heat-transfer constraints,
(46), (47), temperature-drop constraints,
(48), (49), (50), dependent-variable constraints,
(51)–(54), design-limit constraints,
and subject to all variables being positive.

In the next section, we discuss the result of applying this model to a specific example.

4. Numerical Example

Consider a natural-draft dry-type cooling tower with a constant heat load of $q = (0.2053)10^{10}$ Btu/hr for 7884 hr/yr. Suppose that water with a flow rate of

$$W_w = (1.055)10^8 \text{ lb/hr}$$

and an inlet temperature of $T_{w,h} = 580\,°\text{R}$ (120 °F) has to be cooled to $T_{w,c} = 560.5\,°\text{R}$ (10T.5 °F). Let the ambient air temperature be $T_{a,c} = 510\,°\text{R}$ (50 °F).

We assume that the fins are $b = 0.035$ in. thick and spaced $c = 0.090$ in. apart. We consider only $\bar{N} = 2$ banks of vertical tubes. In our formulation, we estimate the ratio $\bar{A}_o/A_{Fi}$ of bare tube surface area to fin surface area by

$$\bar{A}_o/A_{Fi} = 0.02.$$

We will need to check this estimate against the actual numerical value obtained upon solving the model. The minimum allowable tube wall thickness was given by $l = 0.083$ in., and the finned tubes must be at least $e_{\min} = 0.024$ in. apart.

The remaining numerical constants needed to determine all the coefficients c_i in the model are given in Ref. 8 and those details will not be discussed here.

The geometric programming model (TGP) has no reversed constraints and involves 17 variables in 37 terms. As discussed earlier, in the first run of the problem the posynomial constraint (36) was chosen for the defining relation for the variable a_s. In this case, the solution of the geometric programming model (TGP) yields an optimal cost of 671,925 \$/yr. However, constraint (36) associated with a_s was inactive at optimality; therefore, relation (13) does not hold. In fact, the left-hand side of (36) was equal to 0.9538 at the optimal values of the variables. All other constraints, excepting some of the design-limit constraints in (51)–(54), were active at optimality. In a second run of the problem, we therefore reversed the direction of the inequality in (36) and condensed the resulting constraint.

The results for this second run will now be summarized. The optimal values of the variables are given below with dimensions as defined in the glossary of terms:

$$\begin{array}{rrr}
H_{To} = 295.16, & h = 0.11684, & U = 39.063, \\
D_{To} = 400.00, & W_a = 0.27629 \times 10^9, & T_{ah} = 540.95, \\
N = 2519.2, & \Delta T_a = 30.962, & T_{am} = 525.47, \\
L = 85.421, & a_s = 58329, & \Delta T_{mo} = 38.935, \\
D_o = 0.16667, & D_e = 0.28351, & \Delta T_{mi} = 5.1750. \\
Di = 0.15285, & D_v = 0.03228, &
\end{array}$$

The optimal cost is 672,492 \$/yr.

For this solution of the model, the condensed constraint obtained from reversing the direction of the inequality in (36) was active. In fact, all constraints were active and satisfied within

$$\epsilon = 0.00005,$$

with the exception of some of the design-limit constraints [(51)–(54)]. If one uses the above values of the variables to calculate a_s according to its definition in (13), one obtains

$$a_s = 57{,}397.$$

Comparing this with the value of a_s obtained in our solution yields a difference of 1.6 percent. The variable N is of course restricted to be integer-valued and, for all practical purposes, can be rounded to the nearest integer.

Our estimate of

$$\bar{A}_o / A_{Fi} = 0.02$$

compares favorably with the actual numerical value of 0.019 obtained by using (16) and (17) to calculate this ratio for the optimal values of the variables. Also, in formulating our model, we used the simplification $D_e = D_o + h$. Checking this against the numerical values of the variables above shows that D_e given by (14) is only 1.3 percent less than $D_o + h$. The effect of these deviations on the optimal cost is negligible.

A slightly improved solution for the example can be obtained by using the current solution to make minor adjustments, but we will not discuss these details further.

5. Concluding Remarks

We should remark that the development of our final model and its numerical solution was an interactive process. The proper formulation of some constraints was not apparent, to us at least, until we considered some numerical results. Of course, a deeper understanding of the problem may have made some of our decisions easier but, as was the case involving the constraint on a_s, a certain amount of experimentation seems inevitable.

We do feel, however, that geometric programming is a natural framework in which to consider an optimization model for cooling-tower design. Very few assumptions were required to pose the problem as a geometric program. In this regard, not eliminating certain variables made the formulation easier.

Finally we remark that some variables in this application, such as the ambient air temperature, are fluctuating variables. Some of the stochastic aspects of the model are discussed in Ref. 8 and Ref. 18.

References

1. COSTARD, G., *Gunstigste Bemessung der Kondensationsanlage von Dampfturbinen*, Siemens Zeitschrift, Vol. 2, pp. 82–87, 1964.
2. OPLATKA, G., *Economics of Air-Cooled Condensers in a 750-MW Installation*, The Brown Boveri Review, Vol. 53, pp. 235–237, 1966.
3. ROSSIE, J. P., and CECIL, E. A., *Research on Dry-Type Cooling Towers for Thermal Electric Generation: Part I*, Project No. 16130 EES, Contract No. 14-12-823, US Government Printing Office, Washington, DC, 1970.
4. SMITH, E. C., and LARINOFF, M. N., *Power Plant Siting, Performance, and Economics with Dry Cooling Tower Systems*, American Power Conference, Chicago, Illinois, 1970.
5. AVRIEL, M., and WILDE, D. J., *Optimal Condenser Design by Geometric Programming*, I&EC Process Design and Development, Vol. 6, No. 1, 1967.
6. AVRIEL, M., and WILDE, D. J., *Engineering Design under Uncertainty*, I&EC Process Design and Development, Vol. 8, No. 1, 1969.
7. ECKER, J. G., and WIEBKING, R. D., *Optimal Selection of Steam Turbine Exhaust Annulus and Condenser Sizes by Geometric Programming*, Engineering Optimization, Vol. 2, No. 2, 1976.
8. WIEBKING, R. D., *Deterministic and Stochastic Geometric Programming Models for Optimal Engineering Design Problems in Electric Power Generation and Computer Solutions*, Rensselaer Polytechnic Institute, PhD Thesis, 1974.
9. DUFFIN, R. J., PETERSON, E. L., and ZENER, C., *Geometric Programming—Theory and Applications*, John Wiley and Sons, New York, New York, 1967.
10. AVRIEL, M., RIJCKAERT, M. J., and WILDE, D. J., Editors, *Optimization and Design*, Prentice Hall, Englewood Cliffs, New Jersey, 1973.
11. KERN, D. O., and KRAUS, A. D., *Extended Surface Heat Transfer*, McGraw-Hill Book Company, New York, New York, 1972.
12. MCADAMS, W. H., *Heat Transmission*, McGraw-Hill Book Company, New York, New York, 1954.
13. The Babcock and Wilcox Company, *Steam, Its Generation, and Use*, 37th Edition, 1963.
14. DUFFIN, R. J., *Linearizing Geometric Programs*, SIAM Review, Vol. 12, pp. 211–227, 1970.
15. AVRIEL, M., and WILLIAMS, A. C., *Complementary Geometric Programming*, SIAM Journal of Applied Mathematics, Vol. 19, pp. 125–141, 1970.
16. PASSY, U., *Condensing Generalized Polynomials*, Journal of Optimization Theory and Applications, Vol. 9, pp. 221–237, 1972.
17. DUFFIN, R. J., and PETERSON, E. L., *Geometric Programming with Signomials*, Journal of Optimization Theory and Applications, Vol. 11, pp. 3–35, 1973.
18. WIEBKING, R. D., *Optimal Engineering Design under Uncertainty by Geometric Programming*, Management Science, (to appear).

22

Bibliographical Note on Geometric Programming

M. J. RIJCKAERT[1] AND X. M. MARTENS[2]

Abstract. This note presents a list of published papers devoted to the theoretical development and practical implementation of geometric programming.

1. Introduction

In putting together this bibliographical note on geometric programming, we have followed the principles outlined below:

(i) As far as books are concerned, only those are mentioned which are completely or for a very substantial part devoted to geometric programming.

(ii) Only papers which have appeared in the literature and which deal specifically with the subject of geometric programming are included in this list. They were classified as either theoretical, computational, or application-oriented. If a paper deals with several aspects of geometric programming, the most dominant one is used as a basis for classification.

Finally, we realize that this list might not be completely exhaustive for the last type of papers, which might appear in a wide range of technical journals. The authors, therefore, will be grateful to receive information on missing papers.

2. Books

2.1. AVRIEL, M., RIJCKAERT, M. J., and WILDE, D. J., Editors, *Optimization and Design*, Prentice Hall, Englewood Cliffs, New Jersey, 1973.

[1] Professor, Instituut voor Chemie–Ingenieurstechniek, Katholieke Universiteit te Leuven, Leuven, Belgium.

[2] Doctoral Student, Instituut vor Chemie–Ingenieurstechniek, Katholieke Universiteit te Leuven, Leuven, Belgium.

2.2. BEIGHTLER, C. S., and PHILLIPS, D. T., *Applied Geometric Programming*, John Wiley and Sons, New York, New York, 1976.

2.3. DUFFIN, R. J., PETERSON, E. L., and ZENER, C. M., *Geometric Programming*, John Wiley and Sons, New York, New York, 1967.

2.4. WILDE, D. J., and BEIGHTLER, C. S., *Foundations of Optimization*, Prentice Hall, Englewood Cliffs, New Jersey, 1967.

2.5. ZENER, C. M., *Engineering Design by Geometric Programming*, John Wiley and Sons, New York, New York, 1971.

3. Papers on Theoretical Aspects of Geometric Programming

3.1. ABRAMS, R., and BUNTING, M., *Reducing Posynomial Programs*, SIAM Journal on Applied Mathematics, Vol. 27, pp. 629–640, 1974.

3.2. ABRAMS, R. A., *Consistency, Superconsistency, and Dual Degeneracy in Posynomial Programming*, Operations Research, Vol. 24, pp. 325–335, 1976.

3.3. AVRIEL, M., and WILLIAMS, A. C., *On the Primal and Dual Constraint Sets in Geometric Programming*, Journal of Mathematical Analysis and Applications, Vol. 32, pp. 684–688, 1970.

3.4. AVRIEL, M., and WILDE, D. J., *Stochastic Geometric Programming*, Proceedings of the International Mathematical Programming Symposium, Edited by H. W. Kuhn, Princeton University Press, Princeton, New Jersey, pp. 73–91, 1970.

3.5. AVRIEL, M., and WILLIAMS, A. C., *Complementary Geometric Programming*, SIAM Journal on Applied Mathematics, Vol. 19, pp. 125–141, 1970.

3.6. AVRIEL, M., *Fundamentals of Geometric Programming*, Applications of Mathematical Programming Techniques, Edited by M. Beale, English University Press, London, England, pp. 293–313, 1970.

3.7. BALLARD, D. H., JELINEK, C. O., and SCHINZINGER, R., *An Algorithm for the Solution of Constrained Polynomial Programming Problems*, Computing Journal, Vol. 17, pp. 261–266, 1974.

3.8. BEIGHTLER, C. S., CRISP, R. M., and MEIER, W. L., *Optimization by Geometric Programming*, Journal of Industrial Engineering, Vol. 19, pp. 117–120, 1968.

3.9. BEN-TAL, A., and BEN-ISRAEL, A., *Primal Geometric Programs Treated by Linear Programming*, SIAM Journal on Applied Mathematics, Vol. 30, pp. 538–556, 1976.

3.10. CHARNES, A., COOPER, W. W., and KORTANEK, K. O., *Semi-Infinite Programming, Differentiability, and Geometric Programming,* Aplikace Matematiky, Vol. 14, pp. 15–22, 1969.

3.11. DINKEL, J. J., and KOCHENBERGER, G., *A Note on Substitution Effects in Geometric Programming,* Management Science, Vol. 20, pp. 1141–1143, 1974.

3.12. DUFFIN, R. J., *Cost Minimization Problems Treated by Geometric Means,* Operations Research, Vol. 10, pp. 668–675, 1962.

3.13. DUFFIN, R. J., *Dual Programs and Minimum Cost,* SIAM Journal on Applied Mathematics, Vol. 10, pp. 119–123, 1962.

3.14. DUFFIN, R. J., and PETERSON, E. L., *Duality Theory for Geometric Programming,* SIAM Journal on Applied Mathematics, Vol. 14, pp. 1307–1349, 1966.

3.15. DUFFIN, R. J., and ZENER, C., *Geometric Programming, Chemical Equilibrium, and the Anti-Entropy Function,* Proceedings of the National Academy of Sciences, Vol. 63, pp. 629–638, 1969.

3.16. DUFFIN, R. J., *Linearizing Geometric Programs,* SIAM Review, Vol. 12, pp. 211–227, 1970.

3.17. DUFFIN, R. J., and PETERSON, E. L., *The Proximity of Algebraic Geometric Programming to Linear Programming,* Mathematical Programming, Vol. 3, pp. 250–254, 1972.

3.18. DUFFIN, R. J., and PETERSON, E. L., *Geometric Programs Treated with Slack Variables,* Applicable Analysis, Vol. 2, pp. 255–267, 1972.

3.19. DUFFIN, R. J., and PETERSON, E. L., *Reversed Geometric Programming Treated by Harmonic Means,* Indiana University Mathematics Journal, Vol. 22, pp. 531–550, 1972.

3.20. DUFFIN, R. J., and PETERSON, E. L., *Geometric Programming with Signomials,* Journal of Optimization Theory and Applications, Vol. 11, pp. 3–35, 1973.

3.21. EBEN, C. D., and FERRON, J. R., *A Conjugate Inequality for General Means with Applications to Extremum Problems,* AIChE Journal, Vol. 14, pp. 32–37, 1968.

3.22. EBEN, C. D., and FERRON, J. R., *Inequality Methods for Computation of Optimal Systems,* I&EC Fundamentals, Vol. 8, pp. 749–757, 1969.

3.23. ECKER, J. G., *Decomposition in Separable Geometric Programming,* Journal of Optimization Theory and Applications, Vol. 9, pp. 176–179, 1972.

3.24. GALE, D., *A Geometric Duality Theorem with Economic Interpretation*, Review of Economical Studies, Vol. 34, pp. 19–24, 1967.

3.25. GOCHET, W., and SMEERS, Y., *Constraint Sets of Geometric Programs Characterized by Auxiliary Problems*, SIAM Journal on Applied Mathematics, Vol. 29, pp. 708–718, 1975.

3.26. HEYMANN, M., and AVRIEL, M., *On a Decomposition for a Special Class of Geometric Programming Problems*, Journal of Optimization Theory and Applications, Vol. 3, pp. 392–409, 1969.

3.27. KLINGER, A., and MANGASARIAN, O., *Logarithmic Convexity and Geometric Programming*, Journal of Mathematical Analysis and Applications, Vol. 28, pp. 388–408, 1968.

3.28. KRAFT, O., *Geometric Programming as a Special Case of Dieter's Optimality Theory*, O. R. Verfahren, Vol. 6, pp. 121–128, 1969.

3.29. MINE, H., and OHNO, K., *Decomposition of Mathematical Programming Problems by Dynamic Programming and its Application to Block-Diagonal G. P.*, Journal of Mathematical Analysis and Applications, Vol. 32, pp. 370–385, 1970.

3.30. MORRIS, A. J., *Approximation and Complementary Geometric Programming*, SIAM Journal on Applied Mathematics, Vol. 22, pp. 527–531, 1972.

3.31. PASCUAL, L. D., and BEN-ISRAEL, A., *Constrained Maximization of Posynomials by Geometric Programming*, Journal of Optimization Theory and Applications, Vol. 5, pp. 73–80, 1970.

3.32. PASCUAL, L. D., and BEN-ISRAEL, A., *Vector-Valued Criteria in Geometric Programming*, Operations Research, Vol. 19, pp. 98–104, 1971.

3.33. PASSY, U., *Generalized Weighted Mean Programming*, SIAM Journal on Applied Mathematics, Vol. 20, pp. 763–778, 1971.

3.34. PASSY, U., *Signomial Geometric Programming Determining the Global Minimum*, Mathematical Programs for Activity Analysis, Edited by P. Van Moeseke, North Holland Publishing Company, Amsterdam, Holland, pp. 45–60, 1974.

3.35. PASSY, U., and WILDE, D. J., *Generalized Polynomial Optimization*, SIAM Journal on Applied Mathematics, Vol. 15, pp. 1344–1356, 1967.

3.36. PASSY, U., and WILDE, D. J., *Mass Action and Polynomial Optimization*, Journal of Engineering Mathematics, Vol. 3, pp. 325–335, 1969.

3.37. PETERSON, E. L., and ECKER, J. G., *Geometric Programming, a Unified Duality Theory for Quadratically Constrained Programs*, Bulletin of the American Mathematical Society, Vol. 74, pp. 316–321, 1968.

3.38. PETERSON, E. L., and ECKER, J. G., *Geometric Programming, Duality in Quadratic Programming, and LP-Approximation, II*, SIAM Journal on Applied Mathematics, Vol. 17, pp. 317–340, 1969.

3.39. PETERSON, E. L., *Symmetric Duality for Generalized Unconstrained Geometric Programming*, SIAM Journal on Applied Mathematics, Vol. 19, pp. 487–526, 1970.

3.40. PETERSON, E. L., and ECKER, J. G., *Geometric Programming, Duality in Quadratic Programming, and LP-Approximation, I*, Proceedings of the International Mathematical Programming Symposium, Edited by H. W. Kuhn, Princeton University Press, Princeton, New Jersey, pp. 445–480, 1970.

3.41. PETERSON, E. L., and ECKER, J. G., *Geometric Programming, Duality in Quadratic Programming, and LP-Approximation, III*, Journal of Mathematical Analysis and Applications, Vol. 29, pp. 367–383, 1970.

3.42. PETERSON, E. L., *The Decomposition of Large Generalized Geometric Programming Problems by Tearing*, Decomposition of Large-Scale Systems, Edited by D. M. Himmelblau, North Holland Publishing Company, Amsterdam, Holland, pp. 525–540, 1973.

3.43. PETERSON, E. L., *Geometric Programming and Some of its Extensions*, Optimization and Design, Edited by M. Avriel, M. J. Rijckaert, and D. J. Wilde, Prentice Hall, Englewood Cliffs, New Jersey, pp. 228–289, 1973.

3.44. PETERSON, E. L., *Geometric Programming—A Survey*, SIAM Review, Vol. 18, pp. 1–51, 1976.

3.45. PHILLIPS, D. T., and BEIGHTLER, C. S., *Functional Posynomials in Geometric Programming*, Proceedings of 23rd Annual AIIE National Conference, 1972.

3.46. PHILLIPS, D. T., *Geometric Programming with Slack Constraints and Degrees of Difficulty*, AIIE Transactions, Vol. 5, pp. 7–13, 1973.

3.47. PHILLIPS, D. T., and BEIGHTLER, C. S., *Geometric Programming, A Technical State-of-the-Art Survey*, AIIE Transactions, Vol. 5, pp. 97–112, 1973.

3.48. PHILLIPS, D. T., and REKLAITIS, G. V., *Constrained Derivatives and Equilibrium Conditions in Generalized Geometric Programming*, Journal of Engineering Mathematics, Vol. 8, pp. 311–315, 1974.

3.49. PHILLIPS, D. T., *Sensitivity Analysis in Generalized Polynomial Programming*, AIIE Transactions, Vol. 6, pp. 114–119, 1974.

3.50. RIJCKAERT, M. J., *Sensitivity Analysis in Geometric Programming*, Mathematical Programs for Activity Analysis, Edited by P. Van Moeseke, North Holland Publishing Company, Amsterdam, Holland, pp. 61–72, 1974.

3.51. RIJCKAERT, M. J., and HELLINCKX, L. J., *On the Application of Decomposition Techniques in Geometric Programming*, Decomposition of Large-Scale Systems, Edited by D. M. Himmelblau, North Holland Publishing Company, Amsterdam, Holland, pp. 517–524, 1973.

3.52. RIJCKAERT, M. J., *A Survey of Progress in Geometric Programming*, Cahiers du Centre d'Etude de Recherche Opérationelle, Vol. 16, pp. 369–382, 1974.

3.53. THEIL, H., *Substitution Effects in Geometric Programming*, Management Science, Vol. 19, pp. 25–30, 1972.

3.54. ZENER, C., *A Mathematical Aid in Optimizing Engineering Design*, Proceedings of the National Academy of Sciences, Vol. 47, pp. 537–539, 1961.

3.55. ZENER, C., *A Further Mathematical Aid in Optimizing Engineering Design*, Proceedings of the National Academy of Sciences, Vol. 48, pp. 518–522, 1962.

3.56. ZENER, C., and DUFFIN, R., *Optimization of Engineering Problems*, Westinghouse Engineer, Vol. 24, pp. 154–160, 1961.

4. Papers on Computational Aspects of Geometric Programming

4.1. AVRIEL, M., and WILLIAMS, A. C., *An Extension of Geometric Programming with Applications in Engineering Optimization*, Journal of Engineering Mathematics, Vol. 5, pp. 187–194, 1971.

4.2. AVRIEL, M., *Methods for Solving Signomials and Reverse Convex Programming Methods*, Optimization and Design, Edited by M. Avriel, M. J. Rijckaert, and D. J. Wilde, Prentice Hall, Englewood Cliffs, New Jersey, pp. 307–320, 1973.

4.3. AVRIEL, M., DEMBO, R., and PASSY, U., *Solution of Generalized Geometric Programs*, International Journal for Numerical Methods in Engineering, Vol. 9, pp. 149–169, 1975.

4.4. BECK, P. A., and ECKER, J. G., *A Modified Concave Simplex Algorithm for Geometric Programming,* Journal of Optimization Theory and Applications, Vol. 15, pp. 189–202, 1975.

4.5. BLAU, G. E., and WILDE, D. J., *A Lagrangian Algorithm for Equality Constrained Generalized Polynomial Optimization,* AIChE Journal, Vol. 17, pp. 235–240, 1971.

4.6. CHARNES, A., and COOPER, W. W., *A Convex Approximant Method for Nonconvex Extensions of Geometric Programming,* Proceedings of the National Academy of Sciences, Vol. 52, pp. 1361–1364, 1966.

4.7. DAWKINS, G. S., MCINNIS, B. C., and MOONAT, S. K., *Solution to Geometric Programming Problems by Transformation to Convex Programming Problems,* International Journal of Solids and Structures, Vol. 10, pp. 135–136, 1974.

4.8. DEMBO, R., *A Set of Geometric Programming Test Problems and Their Solutions,* Mathematical Programming, Vol. 10, pp. 192–214, 1976.

4.9. DINKEL, J. J., KOCHENBERGER, G. A., and MCCARL, B., *An Approach to the Numerical Solution of Geometric Programming,* Mathematical Programming, Vol. 7, pp. 181–190, 1974.

4.10. DINKEL, J. J., and KOCHENBERGER, G. A., *Discrete Solutions to Engineering Design Problems,* Journal of Engineering Mathematics, Vol. 9, pp. 29–38, 1975.

4.11. FRANK, C. J., *An Algorithm for Geometric Programming,* Recent Advances in Optimization Techniques, Edited by A. Lavi and T. Vogl, John Wiley and Sons, New York, New York, pp. 145–162, 1966.

4.12. GOCHET, W., LOUTE, E., and SOLOW, D., *Comparative Computer Results of Three Algorithms for Solving Prototype Geometric Programming Problems,* Cahiers du Centre d'Etude de Recherche Opérationelle, Vol. 16, pp. 461–486, 1974.

4.13. GOCHET, W., and SMEERS, Y., *On the Use of Linear Programs to Solve Prototype Geometric Programs,* Cahiers du Centre d'Etude de Recherche Opérationelle, Vol. 16, pp. 23–36, 1974.

4.14. KOCHENBERGER, G. A., WOOLSEY, R. E. D., and MCCARL, B. A., *On the Solution of Geometric Programming via Separable Programming,* Operations Research Quarterly, Vol. 24, pp. 285–296, 1973.

4.15. MCNAMARA, J. C., *A Solution Procedure for Geometric Programming,* Operations Research, Vol. 24, pp. 15–25, 1976.

4.16. PASCUAL, L. D., and BEN-ISRAEL, A., *On the Solution of Maximization Problems of Optimal Design by Geometric Programming*, Journal of Engineering Mathematics, Vol. 4, pp. 349–360, 1970.

4.17. PASSY, U., *Condensing Generalized Polynomials*, Journal of Optimization Theory and Applications, Vol. 9, pp. 221–237, 1972.

4.18. PASSY, U., and WILDE, D. J., *A Geometric Programming Algorithm for Solving Chemical Equilibrium Problems*, SIAM Journal on Applied Mathematics, Vol. 16, pp. 363–373, 1968.

4.19. REKLAITIS, G. V., and WILDE, D. J., *Differential Algorithm for Posynomial Programs*, Dechema Monographen, Band 67, pp. 503–542, 1971.

4.20. REKLAITIS, G. V., and WILDE, D. J., *Geometric Programming via a Primal Auxiliary Problem*, AIIE Transactions, Vol. 6, pp. 308–317, 1974.

4.21. RIJCKAERT, M. J., and MARTENS, X. M., *A Condensation Method for Generalized Geometric Programming*, Mathematical Programming, Vol. 11, pp. 89–93, 1976.

4.22. TEMPLEMAN, A. B., *On the Solution of Geometric Programs via Separable Programming*, Operations Research Quarterly, Vol. 25, pp. 184–185, 1974.

5. Papers on Applications of Geometric Programming

Chemical Engineering

5.1. AVRIEL, M., and WILDE, D. J., *Optimal Condenser Design by Geometric Programming*, I&EC Process Design and Development, Vol. 6, pp. 256–263, 1967.

5.2. AVRIEL, M., and WILDE, D. J., *Engineering Design under Uncertainty*, I&EC Process Design and Development, Vol. 8, pp. 124–131, 1969.

5.3. AVRIEL, M., and GUROVICH, V., *Optimal Location of Hydrogen Supply Centers to Minimize Distribution Costs*, Proceedings of the International Conference on Hydrogen Economy, Edited by T. N. Veziroglu, Miami, Florida, 1976.

5.4. BLAU, G., and WILDE, D. J., *Optimal System Design by Generalized Geometric Programming*, Canadian Journal of Chemical Engineering, Vol. 47, pp. 317–326, 1969.

5.5. HELLINCKX, L. J., and RIJCKAERT, M. J., *Minimization of Capital Investment for Batch Processes*, I&EC Process Design and Development, Vol. 10, pp. 422–423, 1971.

5.6. HELLINCKX, L. J., and RIJCKAERT, M. J., *Optimal Capacities of Production Facilities: An Application of Geometric Programming*, Canadian Journal of Chemical Engineering, Vol. 50, pp. 148–150, 1972.

5.7. KERMODE, R. I., *Geometric Programming: A Simple Efficient Optimization Technique*, Chemical Engineering, Vol. 18, pp. 97–100, 1967.

5.8. LAMONTE, R. R., and LEDERMAN, P. B., *The Uses and Limitations of Geometric Programming*, British Chemical Engineering and Process Technology, Vol. 17, pp. 34–37, 1972.

5.9. RIJCKAERT, M. J., and MARTENS, X. M., *Analysis and Optimization of the Williams–Otto Process by Geometric Programming*, AIChE Journal, Vol. 20, pp. 742–750, 1974.

Civil Engineering

5.10. DINKEL, J. J., and KOCHENBERGER, G. A., *On a Cofferdam Design Optimization*, Mathematical Programming, Vol. 6, pp. 114–117, 1974.

5.11. MORRIS, A. J., *Structural Optimization by Geometric Programming*, International Journal of Solids and Structures, Vol. 8, pp. 847–864, 1972.

5.12. MORRIS, A. J., *The Optimization of Statically Indeterminate Structures*, Proceedings of the AGARD Symposium on Structural Optimization, Neuilly-sur-Seine, France, Vol. 6, pp. 1–17, 1973.

5.13. MORRIS, A. J., *A Primal–Dual Method for Minimum Weight Design of Statically Determinate Structures with Several Systems of Load*, International Journal of Mechanical Science, Vol. 16, pp. 801–807, 1974.

5.14. MORRIS, A. J., *A Transformation for Geometric Programming Applied to the Minimum Weight Design of Statically Determinate Structure*, International Journal of Mechanical Science, Vol. 17, pp. 395–396, 1975.

5.15. NEGHABAT, F., and STARK, R. M., *A Cofferdam Design Optimization*, Mathematical Programming, Vol. 3, pp. 263–276, 1972.

5.16. TEMPLEMAN, A. B., *Structural Design for Minimum Cost Using the Method of Geometric Programming*, Proceedings of the Institute of Civil Engineering, Vol. 46, pp. 459–470, 1970.

5.17. TEMPLEMAN, A. B., *The Application of Geometric Programming to the Optimum Design of Bridge Structures*, Proceedings of the International Symposium on Optimization in Road Design, PRTC, London, England, pp. 125–132, 1971.

5.18. TEMPLEMAN, A. B., *Geometric Programming with Examples of the Optimum Design of Floor and Roof Systems*, Proceedings of the International Symposium on Computer Aided Structural Design, Peregrinus Press, London, England, pp. A1.1–A1.21, 1972.

5.19. TEMPLEMAN, A. B., and WINTERBOTTOM, S. K., *Structural Design Applications of Geometric Programming*, Proceedings of the AGARD Symposium on Structural Optimization, Neuilly-sur-Seine, France, Vol. 5, pp. 1–15, 1973.

5.20. TEMPLEMAN, A. B., *The Use of Geometric Programming Methods for Structural Optimization*, AGARD Symposium, Neuilly-sur-Seine, France, Vol. 3, pp. 1–17, 1974.

5.21. TEMPLEMAN, A. B., *Optimum Truss Design Using Approximating Functions*, Optimization in Structural Design, Edited by A. Sawczuk and Z. Mroz, Springer-Verlag, Berlin, Germany, pp. 327–349, 1975.

5.22. TEMPLEMAN, A. B., *Optimum Truss Design by Sequential Geometric Programming*, Journal on Structural Engineering, Vol. 3, pp. 155–163, 1976.

5.23. WILSON, A. J., TEMPLEMAN, A. B., and BRITCH, A. L., *Optimal Design of Drainage Systems*, Journal of Engineering Optimization, Vol. 1, pp. 111–125, 1974.

5.24. WILSON, A. J., and TEMPLEMAN, A. B., *An Approach to the Optimum Thermal Design of Office Buildings*, Building and Environment, Vol. 11, pp. 39–50, 1976.

Economics and Management

5.25. BALACHANDRAN, V., and GENSCH, D., *Solving the Marketing-Mix Problem Using Geometric Programming*, Management Science, Vol. 21, pp. 160–171, 1974.

5.26. KOCHENBERGER, G. A., *Inventory Model Optimization by Geometric Programming*, Decision Science, Vol. 2, pp. 193–205, 1971.

Electrical Engineering

5.27. SCHINZINGER, R., *Optimization in Electromagnetic System Design*, Recent Advances in Optimization Techniques, Edited by A. Lavi and T. Vogl, John Wiley and Sons, New York, New York, 1966.

5.28. ZENER, C., IEEE Transactions on Military Electronics, Vol. MLI–8, pp. 63–66, 1964.

Environmental Engineering

5.29. ECKER, J. G., and MCNAMARA, J. C., *Geometric Programming and the Preliminary Design of Industrial Waste Treatment Plants*, Water Resources Research, Vol. 7, pp. 18–22, 1971.

5.30. ECKER, J. G., *A Geometric Programming Model for Optimal Allocation of Stream-Dissolved Oxygen*, Management Science, Vol. 21, pp. 658–668, 1975.

Mechanical Engineering

5.31. BEIGHTLER, C. S., LO, T. C., and RYLANDER, H. G., *Optimal Design by Geometric Programming*, Transactions of the American Society of Mechanical Engineers, 1969.

5.32. DUFFIN, R. J., and ZENER, C., *Geometric Programming and the Darwin–Fowler Method in Statistical Mechanics*, Journal of Physical Chemistry, Vol. 74, pp. 2419–2423, 1970.

5.33. MANCINI, L. J., and PIZIALI, R. L., *Optimal Design of Helical Springs by Geometric Programming*, Journal of Engineering Optimization, Vol. 2, 1976.

5.34. PETROPOULOS, P., *Optimal Selection of Machining Rate Variables by Geometric Programming*, International Journal of Production Research, Vol. 11, pp. 305–314, 1973.

5.35. PHILLIPS, D. T., and BEIGHTLER, C. S., *Optimization in Tool Engineering Using Geometric Programming*, AIIE Transactions, Vol. 2, pp. 355–360, 1970.

5.36. RAGSDELL, K. M., and PHILLIPS, D. T., *Optimal Design of a Class of Welded Structures Using Geometric Programming*, Journal of Engineering for Industry, Vol. 98, pp. 1021–1025, 1976.

5.37. UNKLESBAY, K., STAATS, G. E., and CREGHTON, D. L., *Optimal Design of Journal Bearings*, International Journal of Engineering Science, Vol. 11, pp. 973–983, 1973.

Regional Science

5.38. DINKEL, J. J., KOCHENBERGER, G. A., and SEPPALA, Y., *On the Solution of Regional Planning Models via Geometric Programming*, Environmental Planning, Vol. 5, pp. 397–408, 1973.

5.39. KADAS, A. S., *An Application of Geometric Programming to Regional Economics*, Papers of the Regional Science Association, Vol. 34, pp. 95–106, 1975.

5.40. NIJKAMP, P., and PALINCK, J. H. P., *Interregional Model of Environmental Choice*, Papers of the Regional Science Association, Vol. 31, pp. 51–71, 1971.

5.41. NIJKAMP, P., and PALINCK, J. H. P., *A Dual Interpretation and Generalization of Entropy Maximization Models in Regional Science*, Papers of the Regional Science Association, Vol. 33, pp. 13–31, 1974.

Miscellaneous Topics

5.42. BOUCHEY, G. D., BEIGHTLER, C. S., and KOEN, B. V., *Optimization of Nuclear Systems by Geometric Programming*, Nuclear Science and Engineering, Vol. 44, pp. 267–272, 1971.

5.43. FEDEROWICZ, A. J., and MAZUMDAR, M., *Use of Geometric Programming to Maximize Reliability Achieved by Redundancy*, Operations Research, Vol. 16, pp. 948–954, 1968.

5.44. FOLKERS, J. S., *Preliminary Fleet Design by Geometric Programming*, International Shipbuilding Programming, Vol. 16, pp. 308–326, 1969.

5.45. FOLKERS, J. S., *Ship Optimization and Design*, Optimization and Design, Edited by M. Avriel, M. J. Rijckaert, and D. J. Wilde, Prentice Hall, Englewood Cliffs, New Jersey, pp. 221–226, 1973.

5.46. HALL, M. A., *Hydraulic Network Analysis Using Generalized Geometric Programming*, Networks, Vol. 6, pp. 105–131, 1976.

5.47. HALL, M., and PETERSON, E. L., *Traffic Equilibria Analyzed via Geometric Programming*, Proceedings of the International Symposium on Traffic Equilibrium Methods, Edited by M. Florian, Springer-Verlag, Berlin, Germany, 1976.

5.48. KELLY, D. W., *A Dual Formulation for Generating Information about Constrained Optima in Automated Design*, Computer Methods in Applied Mechanical Engineering, Vol. 5, pp. 339–353, 1975.

5.49. PASSY, U., *Modular Design, an Application of Structured Geometric Programming*, Operations Research, Vol. 18, pp. 441–453, 1970.

5.50. RIJCKAERT, M. J., *Engineering Applications of Geometric Programming*, Optimization and Design, Edited by M. Avriel, M. J. Rijckaert, and D. J. Wilde, Prentice Hall, Englewood Cliffs, New Jersey, pp. 196–220, 1973.

5.51. ROMANENKO, A. B., *Application of G.P. to the Problems of Minimization of Quadratic Integral Estimations* (in Russian), Bulletin of the Institute for Higher Education, Aviation Engineering, pp. 89–96, 1976.

5.52. ZENER, C., *Solar Powered Sea-Based Plants*, Optimization and Design, Edited by M. Avriel, M. J. Rijckaert, and D. J. Wilde, Prentice Hall, Englewood Cliffs, New Jersey, pp. 290–306, 1973.

Author Index

Subject Index